Innovative Materials for Next-Generation Defense Applications:

Cost, Performance, and Mass Production

Onur Güler
Karadeniz Technical University, Turkey

Müslim Çelebi
Karadeniz Technical University, Turkey

Abdullah Hasan Karabacak
Karadeniz Technical University, Turkey

IGI Global
Scientific Publishing
Publishing Tomorrow's Research Today

Vice President of Editorial	Melissa Wagner
Director of Acquisitions	Mikaela Felty
Director of Book Development	Jocelynn Hessler
Production Manager	Mike Brehm
Cover Design	Jose Rosado

Published in the United States of America by
IGI Global Scientific Publishing
701 East Chocolate Avenue
Hershey, PA, 17033, USA
Tel: 717-533-8845
Fax: 717-533-7115
Website: https://www.igi-global.com E-mail: cust@igi-global.com

Library of Congress Cataloging-in-Publication Data

Names: GuÌˆler, Onur, 1989- editor | CÌ§elebi, MuÌˆslim, 1990- editor | Karabacak, Abdullah Hasan, 1990- editor
Title: Innovative materials for next-generation defense applications : cost, performance, and mass production / edited by Onur GuÌˆler, MuÌˆslim CÌ§elebi, Abdullah Hasan Karabacak.
Description: Hershey, PA : IGI Global Scientific Publishing, 2025. | Includes bibliographical references and index. | Summary: "This book aims to provide more than a comprehensive review of materials for defense-it offers an in-depth look at how advanced materials are designed, tested, and optimized to meet the stringent requirements of defense applications"-- Provided by publisher.
Identifiers: LCCN 2025008279 (print) | LCCN 2025008280 (ebook) | ISBN 9798337309330 hardcover | ISBN 9798337309347 paperback | ISBN 9798337309354 ebook
Subjects: LCSH: Military engineering | Materials science
Classification: LCC UG147 .I55 2025 (print) | LCC UG147 (ebook) | DDC 623.028/4--dc23/eng/20250508
LC record available at https://lccn.loc.gov/2025008279
LC ebook record available at https://lccn.loc.gov/2025008280

British Cataloguing in Publication Data
A Cataloguing in Publication record for this book is available from the British Library.

Table of Contents

Preface xii

Chapter 1
Advanced Protective Coatings and Surface Treatments for Harsh Environments: Materials, Mechanisms, and Applications in Next-Generation Defense Systems 1
Niharika Das, Centurion University of Technology and Management, Odisha, India
Subhraraj Panda, Centurion University of Technology and Management, Odisha, India
Dipan Kumar Das, Centurion University of Technology and Management, Odisha, India
Santanu Kumar Nayak, Centurion University of Technology and Management, Odisha, India
Siddharth Parthasarathy, Centurion University of Technology and Management, Odisha, India

Chapter 2
Carbon-Based Metal-Free Nano-Materials: An Alternative Next-Generation Innovative Material for Future 35
Amit Kumar Dutta, Bangabasi Morning College, University of Calcutta, India

Chapter 3
High-Performance Aluminium Alloys for Aircraft Structures: Advances in Properties, Processing, and Manufacturing 61
Shalom Akhai, Maharishi Markandeshwar Engineering College, Maharishi Markandeshwar, Ambala, India
Syed Mohd Arif, Chandigarh Group of Colleges Jhanjeri, CEC-J, India
Rohan Dadhwal, Chandigarh Group of Colleges Jhanjeri, CSB-J, Mohali, India

Chapter 4
High-Performance Aluminium Alloys for Shipbuilding: Advances in Properties, Processing, and Manufacturing 107
Syed Mohd Arif, Chandigarh Group of Colleges Jhanjeri, CEC-J, India
Shalom S Akhai, Maharishi Markandeshwar Engineering College, Maharishi Markandeshwar, Ambala, India
Rohan Dadhwal, Chandigarh Group of Colleges Jhanjeri, CSB-J, Mohali, India

Chapter 5
Innovation of Bio-Inspired Materials for Next-Generation Defense Applications 145
G. Prasad, Chandigarh University, Punjab, India
Utsav Chakraborty, Chandigarh University, Punjab, India
Evance Saba Matete, Chandigarh University, Punjab, India
Kunwarpreet Singh Marwaha, Chandigarh University, Punjab, India
Rhoto Vero, Chandigarh University, Punjab, India

Chapter 6
Innovative Materials for Next-Generation Defence Applications: Implications for Civil Infrastructure in Cost, Performance, and Mass Production 181
Nidhya Rathinavel, PSG Institute of Technology and Applied Research, India
Arun Murugesan, PSG Institute of Technology and Applied Research, India
Karthick Jaisankar, PSG Institute of Technology and Applied Research, India
Mohammed Jafer Sathik Ali Abdul Hameed, PSG Institute of Technology and Applied Research, India

Chapter 7
Innovative Strategies in Lightweight Materials for High-Performance Defence Applications 225
G. Boopathy, Vel Tech Rangarajan Dr. Sagunthala R&D Institute of Science and Technology, Chennai, India
V. Srinivasan, Vel Tech Rangarajan Dr. Sagunthala R&D Institute of Science and Technology, India
Balaji Ganesan, Hindustan Institute of Technology and Science, India
Mohit Hemanth Kumar, Alliance University, India

Chapter 8
Nature-Inspired Design Strategies for Metal 3D-Printed Defense Components 273
Jagadeesh Baskaran, PSG Institute of Technology and Applied Research, India

Chapter 9
Phase Change Materials for Next-Generation Defense Applications 297
G. Prasad, Chandigarh University, Punjab, India

Chapter 10
Synergistic Effects of Sc and Ca on the Performance of Modified AZ61 Mg Alloy in Defense Application 315
Amit Tiwari, Suresh Gyan Vihar University, Jaipur, India
Anil Kumar, Delhi Skill and Entrepreneurship University, Delhi, India

Chapter 11
Ultra-Wide Bandgap ß-gallium Oxide (Ga2O3) Advancing Next-Generation Defense Technologies 353
Pramod Mandal, University of Pardubice, Czech Republic
Sudesna Roy, KIIT University, India

Compilation of References 399

About the Contributors 469

Index 475

Detailed Table of Contents

Preface xii

Chapter 1
Advanced Protective Coatings and Surface Treatments for Harsh Environments: Materials, Mechanisms, and Applications in Next-Generation Defense Systems 1
Niharika Das, Centurion University of Technology and Management, Odisha, India
Subhraraj Panda, Centurion University of Technology and Management, Odisha, India
Dipan Kumar Das, Centurion University of Technology and Management, Odisha, India
Santanu Kumar Nayak, Centurion University of Technology and Management, Odisha, India
Siddharth Parthasarathy, Centurion University of Technology and Management, Odisha, India

The creation of sophisticated protective coatings and surface treatments has been pushed by the need for high-performance materials in contemporary defense systems. These technologies are essential for increasing the functionality, dependability, and service life of structural elements that operate in harsh conditions such high temperatures, corrosive atmospheres, and mechanical stress. A thorough review of protective coating techniques, including as metallic, ceramic, polymeric, and nanocomposite systems, is given in this chapter. In addition to cutting-edge deposition methods including chemical vapor deposition (CVD), physical vapor deposition (PVD), sol-gel processing, and plasma spraying, emphasis is given on the fundamental mechanisms of protection, including barrier development, self-healing, anti-corrosion, and thermal resistance. To demonstrate real-world applications, case studies pertaining to aerospace, maritime, and military applications are examined.

Chapter 2
Carbon-Based Metal-Free Nano-Materials: An Alternative Next-Generation Innovative Material for Future 35
Amit Kumar Dutta, Bangabasi Morning College, University of Calcutta, India

Carbon-based metal-free nano-materials become ‘Revolutionary Materials’ today in material science domain, which have been incorporated into different industrial sectors and everyday consumer product to improve product efficiencies and performances across various domains. Various types of Carbon-nanomaterials such as fullerene, nano-diamond, different nano-sized carbon allotropes, more popular carbon nanotubes (CNT), graphene, graphene oxide (GO), reduced graphene oxide (rGO) nano-sheet, activated carbon granules, carbon quantum dots (CQDs), carbon dots (CDs), polymeric graphite carbon nitride (g-C3N4), carbonized polymer dots (CPDs) etc.have been successfully explored and exhibits outstanding performance towards some catalysis-chemical reactions. Most of them possesses largest surface-to-volume ratio, highly porous surface and hence highest catalyst-loading capacity. This chapter looks at the history of research on metal-free nano-materials and discuss how these 'Revolutionary Materials' will provide exciting future challenges and opportunity for the sake of society.

Chapter 3
High-Performance Aluminium Alloys for Aircraft Structures: Advances in Properties, Processing, and Manufacturing 61
Shalom Akhai, Maharishi Markandeshwar Engineering College, Maharishi Markandeshwar, Ambala, India
Syed Mohd Arif, Chandigarh Group of Colleges Jhanjeri, CEC-J, India
Rohan Dadhwal, Chandigarh Group of Colleges Jhanjeri, CSB-J, Mohali, India

This chapter explores aluminium alloys' vital role in next-gen defense systems, focusing on their evolution, processing, and aerospace/military applications. It traces their development from duralumin to advanced Al-Li and nanostructured variants, emphasizing strength-to-weight efficiency and corrosion resistance. Key processing innovations like thermomechanical treatments, additive manufacturing, and friction stir welding enable lightweight, high-performance components, such as AlSi10Mg brackets in the F-35 and GLARE® fuselage panels in the A380. Case studies highlight alloys like AA 7085 and Al-Li 2195 in military aircraft and spacecraft, enhancing fuel efficiency and durability. Challenges like galvanic corrosion, scandium scarcity, and energy-intensive production are analyzed, alongside solutions like AI-driven alloy design and closed-loop recycling. The chapter underscores sustainable practices, ensuring aluminium alloys remain essential for defense while advancing eco-friendly production.

Chapter 4
High-Performance Aluminium Alloys for Shipbuilding: Advances in Properties, Processing, and Manufacturing 107
Syed Mohd Arif, Chandigarh Group of Colleges Jhanjeri, CEC-J, India
Shalom S Akhai, Maharishi Markandeshwar Engineering College, Maharishi Markandeshwar, Ambala, India
Rohan Dadhwal, Chandigarh Group of Colleges Jhanjeri, CSB-J, Mohali, India

Aluminium alloys play a crucial role in modern shipbuilding, offering superior corrosion resistance, lightweight properties, and high structural integrity. This chapter explores key marine-grade aluminium alloys, including AA5083, AA5456, and AA6061, known for their excellent resistance to seawater corrosion and mechanical strength. High-strength alloys such as AA7020 and AA7075 are examined for their application in load-bearing structures. Surface protection technologies, including anodizing, cerium-based coatings, and microarc oxidation (MAO), enhance longevity and durability in harsh marine environments. Advanced processing techniques, such as friction stir welding (FSW) and additive manufacturing, enable improved fabrication efficiency and structural performance. Future research focuses on optimizing alloy composition, enhancing recyclability, and improving sustainability for next-generation marine vessels.

Chapter 5
Innovation of Bio-Inspired Materials for Next-Generation Defense Applications 145
G. Prasad, Chandigarh University, Punjab, India
Utsav Chakraborty, Chandigarh University, Punjab, India
Evance Saba Matete, Chandigarh University, Punjab, India
Kunwarpreet Singh Marwaha, Chandigarh University, Punjab, India
Rhoto Vero, Chandigarh University, Punjab, India

Bio-inspired materials are advancing next-generation military applications by emulating the distinctive features of natural systems to create lightweight, durable, and multifunctional solutions. This overview emphasizes recent progress in bio-inspired materials, encompassing its applications in armor systems, stealth technologies, adaptable structures, and energy-efficient gadgets. Significant developments encompass materials derived from the resilience of nacre, the self-repairing properties of biological tissues, and the concealment strategies of cephalopods. Challenges including scalability, cost-effectiveness, and material dependability under harsh conditions are also addressed. This study highlights the potential of bio-inspired materials to transform the defense sector and meet the changing requirements of contemporary conflict through the analysis of case studies and upcoming trends.

Chapter 6
Innovative Materials for Next-Generation Defence Applications: Implications for Civil Infrastructure in Cost, Performance, and Mass Production................ 181
Nidhya Rathinavel, PSG Institute of Technology and Applied Research, India
Arun Murugesan, PSG Institute of Technology and Applied Research, India
Karthick Jaisankar, PSG Institute of Technology and Applied Research, India
Mohammed Jafer Sathik Ali Abdul Hameed, PSG Institute of Technology and Applied Research, India

The chapter explores the integration of defence-grade materials into construction to enhance infrastructure resilience. Advanced materials like carbon fibre, Kevlar, graphene composites, and multifunctional geopolymers offer high strength, durability, and self-healing properties. Bio-inspired innovations, including mechanoluminescent, hydrophobic, and energy-autonomous materials, have successfully transitioned into civilian applications, improving seismic resilience, structural health monitoring, and environmental protection. Despite cost and scalability challenges, ongoing research in 4D printing, nanotechnology, and self-healing systems is driving broader implementation. Increased investment in advanced materials underscores their crucial role in developing sustainable, adaptive, and high-performance urban infrastructure.

Chapter 7
Innovative Strategies in Lightweight Materials for High-Performance Defence Applications 225
G. Boopathy, Vel Tech Rangarajan Dr. Sagunthala R&D Institute of Science and Technology, Chennai, India
V. Srinivasan, Vel Tech Rangarajan Dr. Sagunthala R&D Institute of Science and Technology, India
Balaji Ganesan, Hindustan Institute of Technology and Science, India
Mohit Hemanth Kumar, Alliance University, India

Defence needs lightweight composites because of their good strength to weight ratio, corrosion resistance, and thermal stability. Then this chapter studies the design and optimization of a number of matrix composites motivated by their mechanical and thermal performance in extreme environments. This highlights the importance of multiscale modelling, simulation and advanced manufacturing in improving composite reliability. This is applied to UVAs, armoured vehicles and naval systems for enhanced mobility and efficiency. Although there have been some impediments to production, the use of AI, nanomaterials, and sustainability will lead to scalable, high-performance solutions. These materials are then substantiated empirically for such critical missions. The chapter coordinates the relationship between the research based in the academia and the requirements by the defence industry, with strategies of new generation of materials having global sustainability goals and the operational needs.

Chapter 8
Nature-Inspired Design Strategies for Metal 3D-Printed Defense Components 273
Jagadeesh Baskaran, PSG Institute of Technology and Applied Research, India

Nature-inspired design offers a powerful paradigm for advancing the development of metal 3D-printed components in the defense sector. This chapter explores how biological principles such as hierarchical structuring, multifunctionality, and optimization can inform the design of robust, lightweight, and high-performance defense components. The chapter highlights recent research innovations, computational modeling approaches, and case studies, illustrating the transformative potential of biomimicry in defense applications.

Chapter 9
Phase Change Materials for Next-Generation Defense Applications 297
G. Prasad, Chandigarh University, Punjab, India

Phase Change Materials (PCMs) provide an innovative solution for heat regulation, storage of energy, and adaptive concealment in advanced defence applications. Phase Change Materials (PCMs) facilitate effective thermal control of military equipment, wearable systems, and infrastructure by taking in and releasing latent heat during phase transitions, hence improving operational reliability and troop comfort in harsh conditions. Furthermore, phase change materials enhance stealth technologies by regulating surface temperatures, diminishing infrared fingerprints, and facilitating adaptive thermal camouflage. This study examines recent breakthroughs in PCM technology, integration techniques for defence systems, and problems related to durability, scalability, and real-time responsiveness, emphasising their potential to substantially improve mission capability and survivability in contemporary conflict.

Chapter 10
Synergistic Effects of Sc and Ca on the Performance of Modified AZ61 Mg Alloy in Defense Application 315
Amit Tiwari, Suresh Gyan Vihar University, Jaipur, India
Anil Kumar, Delhi Skill and Entrepreneurship University, Delhi, India

This study investigates the synergistic effect of scandium and calcium on the microstructure and mechanical performance of modified AZ61 alloy. The additions of Sc and Ca enable a major refinement of the grain structure and improve the formation of thermally stable phases like Al3Sc and Mg2Ca and promote the distribution Mg17Al12 precipitate. The enhanced phase stability and precipitation behavior were also confirmed by microstructural and XRD analyses, which resulted in improved mechanical properties. Induced by a homogeneous precipitating phase and grain refinement, the modified AZ61 alloy displayed the maximum tensile strength for the aged at 200 °C and elongation found for 280 MPa and 15%, respectively. Sc and Ca retarded over-ageing effects at elevated temperatures (300°C and 400°C), resulting in much more stable tensile strength above 260 MPa and elongation 10% related to the stability of Al-Sc and Ca intermetallic phases. Fractography analysis showed a switch from brittle cleavage in the cast alloy to ductile dimple fracture the modified alloy at condition of high thermal.

Chapter 11
Ultra-Wide Bandgap ß-gallium Oxide (Ga2O3) Advancing Next-Generation Defense Technologies 353
Pramod Mandal, University of Pardubice, Czech Republic
Sudesna Roy, KIIT University, India

Ultra-wide bandgap β-Gallium oxide (Ga_2O_3) has gained great attention for next-generation defense applications due to its outstanding properties, including a wide energy bandgap (Eg) (~4.6-4.8 eV), high breakdown electric field (~8 MV/cm), high melting temperature (1795 ºC), and excellent chemical and thermal stability in harsh environments. These characteristics make it ideal for next-generation defense technologies such as solar-blind UV photodetectors (for missile and rocket plume detection), sensors (for surveillance), and power electronics (for energy storage and propulsion systems). This review explores the background, properties, and potential of β-Ga_2O_3 in next generation defense applications, also discussing various thin-film growth techniques, etc. Here, it also covers challenges related to growth, thermal management, and dopant material to enhance the optical and electrical properties. The review explored by outlining opportunities and future scope for β-Ga_2O^3-based next-generation devices for compact, and resilient defense systems.

Compilation of References 399

About the Contributors 469

Index 475

Preface

INTRODUCTION

The ever-evolving landscape of global security demands continuous advancement in defense technologies—and at the heart of this innovation lies material science. *Innovative Materials for Next-Generation Defense Applications: Cost, Performance, and Mass Production* presents a timely and comprehensive exploration of how advanced materials are revolutionizing defense systems. As editors, our goal with this volume is to provide a multifaceted understanding of materials that are engineered not only to meet the stringent technical specifications of modern defense applications but also to address cost-efficiency and scalability for mass production.

This book brings together contributions from experts across disciplines, offering in-depth analyses of high-strength alloys, advanced composites, ceramics, and surface coatings. Each chapter is crafted to highlight how key material properties—such as thermal stability, corrosion resistance, strength-to-weight ratio, and durability—play pivotal roles in enhancing the performance and reliability of defense technologies operating under extreme conditions.

What sets this volume apart is its dual focus: it is both a critical review of cutting-edge research and a guide to practical implementation. We emphasize not only material innovation but also the methods and strategies required for transitioning from laboratory-scale discoveries to real-world, deployable solutions. Whether it is through surface treatments, hybrid materials, or next-gen fabrication techniques, the insights offered here are meant to serve as a bridge between scientific advancement and industrial readiness.

Our intended audience is as diverse as the field itself—spanning academic researchers, defense engineers, materials scientists, R&D specialists, and graduate students alike. In particular, those working in sectors where structural integrity, environmental resistance, and lightweight construction are essential—such as aerospace, automotive, and heavy industries—will find this resource especially relevant.

As the defense sector continues to push the boundaries of what materials can do, it is our hope that this book will inspire further innovation and interdisciplinary collaboration. We believe that equipping readers with both the foundational knowledge and the forward-looking insights provided in this collection will contribute meaningfully to the development of safer, more efficient, and more resilient defense systems for the future.

ORGANIZATION OF THE BOOK

This edited volume begins with Chapter 1, which provides an in-depth analysis of advanced protective coatings and surface treatments designed for high-stress and corrosive environments—an essential frontier in next-generation defense systems. The chapter offers a comprehensive review of diverse coating technologies, from metallic to nanocomposite systems, while detailing the deposition methods such as CVD, PVD, sol-gel, and plasma spraying. Case studies in aerospace, maritime, and military sectors illustrate the real-world utility of these innovations in enhancing durability, thermal resistance, and corrosion protection.

Chapter 2 shifts the focus to revolutionary carbon-based metal-free nanomaterials, which are redefining materials science with their exceptional surface-to-volume ratios and catalytic potential. The discussion spans from fullerenes and graphene variants to quantum dots and carbon nitride structures, showcasing their immense promise in energy, electronics, and catalytic applications. The chapter also reflects on the historical evolution of these materials while anticipating the societal benefits of their future innovations.

In Chapter 3, the significance of high-performance aluminum alloys in aerospace applications is thoroughly explored, detailing their development from classical duralumin to advanced Al-Li alloys and nanostructured variants. The chapter addresses manufacturing advancements such as additive manufacturing and friction stir welding, while also confronting sustainability challenges through AI-assisted alloy design and recycling. Case studies including the F-35 and A380 demonstrate the material's critical role in optimizing fuel efficiency and structural integrity.

Chapter 4 continues the exploration of aluminum alloys, this time in the context of shipbuilding. It highlights marine-grade alloys like AA5083 and AA6061, focusing on their corrosion resistance, mechanical strength, and compatibility with cutting-edge processing techniques like FSW and MAO. The chapter underscores the ongoing efforts to enhance alloy recyclability and sustainability while maintaining the robust performance required in maritime defense applications.

The application of bio-inspired materials in defense systems is the subject of Chapter 5, which explores how nature-derived concepts such as self-repair, multi-functionality, and adaptive camouflage are revolutionizing armor systems, stealth technologies, and structural materials. Drawing from natural phenomena like nacre and cephalopod camouflage, the chapter also discusses technical challenges around scalability and material resilience in extreme environments.

Chapter 6 bridges defense and civil infrastructure, investigating how innovations like carbon fiber, Kevlar, and mechanoluminescent materials transition into urban settings to enhance resilience and sustainability. The integration of self-healing composites, energy-autonomous materials, and nanotechnology reflects the growing convergence between defense R&D and civil applications, particularly for smart, adaptive infrastructure systems.

Lightweight materials and their importance in enhancing mobility, energy efficiency, and protection in defense systems are central to Chapter 7. It examines matrix composites optimized for strength, thermal resistance, and multiscale modeling techniques. The chapter also highlights the role of AI, nanomaterials, and sustainable manufacturing in overcoming production challenges, with applications spanning UAVs, armored vehicles, and naval systems.

Chapter 8 delves into nature-inspired strategies for metal 3D printing, revealing how biological principles can lead to superior structural performance and material efficiency. With insights from hierarchical structuring and multifunctionality in nature, the chapter discusses computational modeling and real-world case studies that illustrate the transformative potential of biomimicry in additive manufacturing for defense.

Chapter 9 explores Phase Change Materials (PCMs) as a cutting-edge solution for thermal regulation and adaptive concealment in military environments. By leveraging latent heat transitions, PCMs enhance operational reliability and camouflage capabilities in extreme conditions. The chapter examines integration challenges and technological breakthroughs that are shaping the future of thermal management and stealth systems.

In Chapter 10, the synergistic influence of scandium and calcium on AZ61 magnesium alloys is examined, highlighting significant improvements in grain refinement, tensile strength, and high-temperature performance. Detailed analysis of phase formation and mechanical behavior reveals how these alloy modifications contribute to superior durability and thermal resistance, making them ideal candidates for demanding defense environments.

Finally, Chapter 11 introduces ultra-wide bandgap ß-gallium oxide (Ga_2O_3) as a game-changing material for future defense applications. The chapter explores its extraordinary electrical, thermal, and optical properties, which support high-performance devices such as UV photodetectors, power electronics, and thermal sensors. Despite challenges in growth and doping, Ga_2O_3 holds substantial promise for compact and resilient defense technologies.

CONCLUSION

As we bring this volume to a close, it becomes abundantly clear that the future of defense technology lies at the intersection of material science innovation, interdisciplinary collaboration, and sustainable engineering. The chapters assembled in this book reflect the breadth and depth of contemporary research driving the evolution of next-generation defense systems. From advanced protective coatings and bio-inspired designs to metal-free nanomaterials, phase change materials, and ultra-wide bandgap semiconductors, each contribution demonstrates how materials innovation is reshaping the defense landscape to meet the demands of performance, adaptability, and resilience in extreme environments.

The integration of civilian infrastructure applications and nature-derived design philosophies further underscores the transformative potential of these technologies beyond the battlefield. Whether it's enhancing the energy efficiency of military systems, extending the lifespan of aerospace and maritime components, or developing lightweight composites for high-mobility platforms, the insights provided here will serve both researchers and practitioners seeking forward-thinking strategies rooted in cutting-edge materials science.

We hope this volume serves not only as a reference but also as a catalyst for new inquiries, collaborations, and innovations that align with global sustainability imperatives while advancing defense capabilities. The challenges ahead are complex, but through the continued fusion of scientific rigor and creative inspiration, the path toward robust, intelligent, and future-ready defense systems is within reach.

Onur Güler
Karadeniz Technical University, Turkey

Müslim Çelebi
Karadeniz Technical University, Turkey

Abdullah Hasan Karabacak
Karadeniz Technical University, Turkey

Chapter 1
Advanced Protective Coatings and Surface Treatments for Harsh Environments:
Materials, Mechanisms, and Applications in Next-Generation Defense Systems

Niharika Das
https://orcid.org/0009-0008-4524-5198
Centurion University of Technology and Management, Odisha, India

Subhraraj Panda
https://orcid.org/0000-0003-4224-4663
Centurion University of Technology and Management, Odisha, India

Dipan Kumar Das
https://orcid.org/0000-0003-0224-3295
Centurion University of Technology and Management, Odisha, India

Santanu Kumar Nayak
https://orcid.org/0009-0004-2143-7655
Centurion University of Technology and Management, Odisha, India

Siddharth Parthasarathy
https://orcid.org/0000-0001-8319-3239
Centurion University of Technology and Management, Odisha, India

DOI: 10.4018/979-8-3373-0933-0.ch001

ABSTRACT

The creation of sophisticated protective coatings and surface treatments has been pushed by the need for high-performance materials in contemporary defense systems. These technologies are essential for increasing the functionality, dependability, and service life of structural elements that operate in harsh conditions such high temperatures, corrosive atmospheres, and mechanical stress. A thorough review of protective coating techniques, including as metallic, ceramic, polymeric, and nano-composite systems, is given in this chapter. In addition to cutting-edge deposition methods including chemical vapor deposition (CVD), physical vapor deposition (PVD), sol-gel processing, and plasma spraying, emphasis is given on the fundamental mechanisms of protection, including barrier development, self-healing, anti-corrosion, and thermal resistance. To demonstrate real-world applications, case studies pertaining to aerospace, maritime, and military applications are examined.

INTRODUCTION

Evolution of Coating Technologies

Innovations in materials science, industry, and the environment have propelled coating technology forward over the years. Although there were natural coatings at first, synthetic polymers improved their flexibility and resistance to corrosion. Nanotechnology's use of nanocomposite coatings transformed the sector. Recently, research has focused on low-VOC, ecologically friendly, and bio-based approaches to sustainability. Cost reductions, economies of scale, corrosion prevention, aesthetics, and environmental regulations are all aspects of coating technology advancement. New technologies such as powder coatings, waterborne technologies, and various epoxy systems are just some of the new technologies (Davis, 2003). Silicon-based technologies have shifted from specialty use to widespread application in coatings due to their custom chemistries and value-based performance (Witucki, 2003). With various techniques of deposition such as chemical vapor deposition and physical vapor deposition, thin-film coatings have played a central role in modern technology (Mbam et al., 2019). These technologies have particularly influenced the automotive industry, where improved paint chemistries and coating methods extended longevity and appearance (Akafuah et al., 2016). Coating technologies will also remain an area of research and development in the future because there is always a need and challenge in the production of thin-films (Mbam et al., 2019).

Importance of Protective Coatings in Defense Applications

Protective coatings play a significant role in defense operations because they promote longevity, performance, and endurance of defense vehicles, infrastructure, and equipment. They shield against hostile weather conditions, electromagnetic interference, chemical contamination, and ballistic effects. Nanocomposite and self-healing coatings strengthen structure, reduce maintenance, and optimize operating life. They increase combat efficiency and camouflage efficiency as well. Due to their superior heat barrier properties, anti-wear features, and resistance to corrosion, new materials like atomically thin 2D nanomaterials have applications as effective protective coatings in the defense and aviation industries (Anirudh S et al., 2023). The coatings can be applied to various components in the marine, military, and aerospace sectors. The coating markets are confronted with new issues due to increased usage of composites in aerospace and defense, which require particular methods that are distinct from those applied to metals (Armstrong, 2010). Advances in this field are driven by continued research on materials and coatings to be used for defense purposes, which is supported by institutions like the Defence Metallurgical Research Laboratory (Roy, 2023). For military defense, highly reflective protective silver coatings resistant to temperature, humidity, and salt fog are essential. With their corrosion protection, camouflage, and chemical agent resistance, protective coatings assist the military in maintaining force readiness and survival.

Key Challenges in Protective Coating Development

Protective coating development is hindered by issues like durability, environmental effect, and economy. Coatings need to be formulated with better adhesion, corrosion resistance, and wear properties to perform long-term in harsh environments. Environmental laws require low-VOC, non-toxic, and environmentally friendly coatings. Creating coatings with multiple uses and compatibility with substrates complicates design. Recent developments in protective coating development have solved key issues in various industries, such as cultural heritage preservation and graphene-based coatings. However, major issues in automotive, marine, and aerospace sectors include health risks, toxicity, and deformation caused by structural joint motions.

TYPES OF PROTECTIVE COATINGS

Polymer-Based Coatings

Figure 1. Process involving in polymer-based coatings (Nathanael & Oh, 2020)

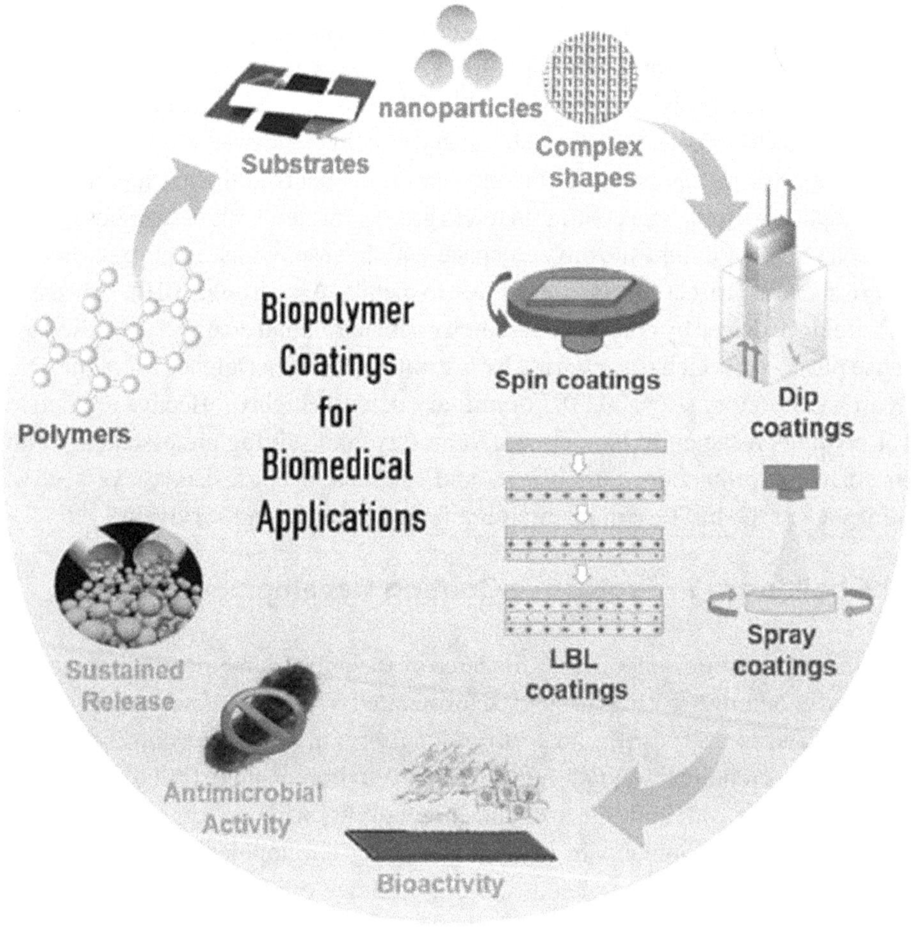

Polymer-based coatings are increasingly used in various sectors, including infrastructure, automotive, aerospace, and defense, due to their flexibility, adhesion, and resistance to environmental elements. Advances in polymer chemistry have led to high-performance coatings with improved durability, corrosion resistance, and self-healing properties. Nanotechnology has enhanced their functionality by adding mechanical strength, thermal stability, and barrier qualities. Eco-friendly and sustainable polymer coatings, such as water-based and bio-derived formulations, are gaining popularity. These coatings are composed of various polymeric compounds, with water-based formulations becoming more popular due to environmental concerns.

Ceramic and Oxide Coatings

While in the case of offering better protection from severe environmental conditions such as high temperatures, mechanical wear, and chemical corrosion, oxide and ceramic coatings play a crucial role. They possess better thermal stability, hardness, and oxidation resistance. They are typically composed of minerals such as titanium dioxide (TiO_2), zirconia (ZrO_2), and aluminum oxide (AlO_3). Ceramic coatings are being used in diverse industrial, military, and aircraft applications as thermal barrier coatings (TBCs) for turbine engines and jets to enhance component life and efficiency of energy. Oxide coatings like anodized aluminum or plasma-sprayed ceramic films for metals and alloys exhibit great corrosion resistance and electrical insulation. Despite their superior properties, limitations such as brittleness and complex manufacturing processes hinder their broader application. Oxide and ceramic coatings are imperative for future-generation high-performance protective coatings, however, since breakthroughs in nanostructured ceramics and hybrid coatings are helping break these limitationsFor metallic surfaces, ceramic oxide coatings have several benefits, including enhanced biocompatibility, wear resistance, and corrosion resistance (Nelson et al., 1981; Wang, 2023). Laser surface melting is one example of an advanced deposition process that keeps improving coating performance and quality (Chandra & Sarkar, 2019). Anodic spark deposition, which involves dielectric breakdown at the anode surface, can be used to deposit ceramic and oxide coatings on metal substrates.

Metallic and Alloy Coatings

Metallic and alloy coatings are used in various industries to improve surface mechanical properties, corrosion resistance, and longevity. These coatings, often made of metals like zinc, aluminum, chromium, and nickel, offer specialized qualities like increased strength, chemical resistance, and thermal stability. However, issues like cost, weight, and environmental deterioration persist. Emerging technologies like high-entropy alloys and nanostructured coatings are enabling next-generation metallic coatings with enhanced performance and sustainability. Various coating techniques have been developed, but problems like metal ion leakage, allergic reactions, and inadequate friction control persist.

Nanocomposite and Hybrid Coatings

Nanocomposite coatings and hybrid coatings are material technology developments, offering improved mechanical, barrier, and thermal properties. These coatings, made of nanomaterials and ceramic or polymeric materials, offer improved corrosion protection, durability, and wear resistance. Hybrid coatings are made up of organic and inorganic phases, offering light protection against chemical and aggressive weather. They also offer antibacterial properties, self-healing, and improved adhesion properties. They have applications in a broad variety of industries such as medicinal implants, automotive, and aerospace (Hamid, 2014; Fernando, 2009). Nanocomposite coatings can provide improved scratch resistance, UV protection, and corrosion protection (Fernando, 2009; Dennis et al., 2015). Multifunctional coatings are made possible due to the development of a "interphase" zone with modified properties that is created due to the addition of nanoparticles into polymeric matrices (Dennis et al., 2015). 30-60 GPa hardness has been achieved using nanocomposite hard coatings made of nanocrystalline grains in an amorphous matrix for wear protection (Patscheider, 2003). Reasons for the creation of these coatings are sustainability issues, chromium replacement needs, and performance improvement under harsh environments (Dennis et al., 2015).

Self-Healing and Smart Coatings

Through auto-repairing of defects and responding to environmental stimulus, smart coatings and self-healing are emerging high-technology materials that can enhance durability and performance of protective surfaces. In repair of structure integrity and prevention of secondary failure, self-healing coatings consist of vascular networks, dynamic polymer chains, or microcapsules that discharge the curing chemicals when subjected to mechanical abuse. Their use in armor structures, protective aircraft paint, and ships offers improved performance and reliability under severe operating conditions. Potential for use of smart anticorrosive and self-healing coatings to enhance material performance and service life has drawn monumental attention. Damage-healing and functionality-recovery features are imparted by these coatings through the use of micro/nanocapsules containing corrosion inhibitors, healing agents, and catalysts (Kartsonakis et al., 2024). Active corrosion protection is made possible by stimuli-responsive properties activated by temperature, pH, and humidity (Udoh et al., 2022). Multifunctional coatings can offer lubrication, corrosion protection, and antibacterial properties based on natural phenomena (Zvonkina & Hilt, 2014). Recent developments include eco-friendly smart coatings with dual-responsive superhydrophobic and anti-biofouling properties using biomimetic stimuli-responsive mesoporous polydopamine microspheres (Ni et al., 2021). These films show the ability to self-heal under pH or near-infrared stimuli, with potential in biomedical and anti-fouling purposes. There are ongoing efforts aimed at designing enhanced self-healing systems and at making low-cost solutions for industrial applications across different fields (Kartsonakis et al., 2024).

ADVANCED SURFACE TREATMENTS FOR DEFENSE MATERIALS

Chemical Treatments (Anodizing, Phosphating, etc.)

Figure 2. Chemical treatments of defense material by anodizing.

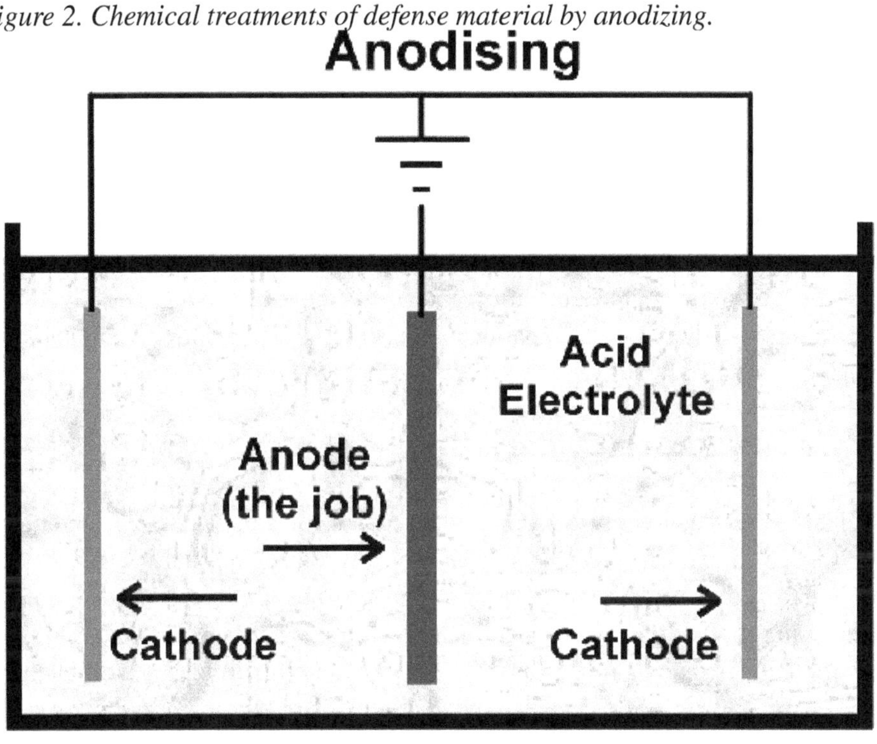

Chemical treatments like anodizing and phosphating are widely used to improve the surface properties of metals, including wear, adhesion, and corrosion resistance. Anodizing is the most prevalent treatment used on aluminum, an electrochemical process that forms a porous, hard oxide coating that may be sealed or dyed for functional or aesthetic reasons. Phosphating, on the other hand, is the chemical process of chemical reaction in turning a metal surface into a coating of phosphate using phosphoric acid solutions. The process offers an excellent foundation for corrosion protection as well as for paint adhesion. Other processes involving chemicals and providing similar protection advantages include chromating and passivation, which are frequently developed to meet unique mechanical and environmental needs. These surface alterations play a crucial role in increasing the lifespan of parts used in defense, automotive, and aerospace industries. In improvement of metal surface properties, paint adhesion, and anticorrosive characteristics, chemical surface treatments such as phosphating and anodizing are commonly utilized (Petschel & Hart, 2000). The most popular preparation for ferrous and non-ferrous metals, phosphorating, provides superior lubrication, adhesion, and resistance to corrosion and wear (Narayanan, 2005). Anodizing produces ornamental and protective oxide coatings that have a variety of uses in industry and architecture, particularly for aluminum and its alloys (Sheasby et al., 1987). Anodizing processes enhance the biological response, corrosion resistance, and film stability of titanium implants (Disegi, 1997). Numerous changes have been made to these treatments to accommodate evolving industrial demands, especially in the process, appliance, and automotive industries (Narayanan, 2005). Because of their efficiency and adaptability, phosphate conversion coatings and anodizing continue to be essential in a variety of sectors, even in the face of environmental concerns and the development of substitute techniques (Petschel & Hart, 2000; Narayanan, 2005). Anodizing is a surface treatment that works well for making magnesium alloys more resistant to corrosion. The adherence and biocompatibility of hydroxyapatite coatings on biomedical metal implants can be enhanced by surface activation and pretreatment methods.

Thermal and Plasma Coatings

Advanced surface treatment technologies such as thermal and plasma coatings are employed to enhance a material's resistance to corrosion, hardness, and heat stability in severe environments. Thermal spray coatings produce a protective film by depositing molten or semi-molten materials on a substrate. The precise deposition of ceramics, metals, or composites onto a range of surfaces is facilitated by plasma coatings, a type of thermal spraying that involves a high-energy plasma arc to melt and propel coating materials. They significantly extend the service life of components by protecting them against high temperatures, oxidation, and mechanical degradation. The defense, automotive, and aerospace industries all widely utilize them. More advanced surface treatment techniques referred to as thermal spray coatings are employed in order to enhance material properties and protect components from severe environments (Sampath & Mccune, 2000). Various methods like plasma spray, high-velocity oxy-fuel (HVOF), and detonation gun deposition can be employed to deposit these coatings (Sprague & Smidt, 2007). A protective coating with thicknesses from 25 μm up to several millimeters is formed by melting and fast depositing metallic, ceramic, or polymer materials on a substrate (Fauchais & Vardelle, 2012). Thermal spray coatings are employed to find solutions for issues such as corrosion protection, wear resistance, and thermal insulation in a range of industries, including biomedical, automotive, and aerospace (Gaur & Kamari, 2024). The functional life of materials exposed to heat, oxidation, and mechanical degradation is significantly extended by these coatings (Sampath & Mccune, 2000). Thermal spray method is a convenient method for enhancing material properties and adding new materials to high-performance uses due to its versatility and low cost (Fauchais & Vardelle, 2012). In extreme environments, thermal spray coatings can provide protection from hot corrosion, wear, and abrasion.

Laser and Ion Beam Surface Modifications

Laser and ion beam surface modifications are advanced techniques that improve the mechanical, chemical, and tribological properties of materials by changing the material surface structure at the micro and nanoscale. High-energy laser beams are used in laser surface modification to create local melting, quick solidification, or ablation, which increases hardness, wear resistance, and corrosion protection. In contrast, ion beam modification makes use of accelerated ions to pierce the material's surface and create defects or changes in composition that enhance strength, adhesion, and resistance to oxidation. These techniques are widely used in biomedical, aerospace, and defense to increase the longevity and performance of vital parts in harsh conditions. Material characteristics can be effectively shaped through surface changes using lasers and ion beams. These techniques may increase the hardness, roughness, and wettability of polymers (Velardi et al., 2010), as well as the corrosion and wear resistance of metals (Picraux & Pope, 1984; McCafferty et al., 1987). The creation of unique surface characteristics and metastable phases made possible by ion implantation, ion beam mixing, and laser treatment can enhance both surface and bulk properties (Appleton et al., 1984). Material characteristics can be effectively shaped through surface changes using lasers and ion beams. These techniques may increase the hardness, roughness, and wettability of polymers (Velardi et al., 2010), as well as the corrosion and wear resistance of metals (Picraux & Pope, 1984; McCafferty et al., 1987). The creation of unique surface characteristics and metastable phases made possible by ion implantation, ion beam mixing, and laser treatment can enhance both surface and bulk properties (Appleton et al., 1984). Through the provision of precise control of composition and microstructure, such techniques facilitate easier production of new materials with properties that are quite different from thermodynamic equilibrium (Picraux & Pope, 1984). Uses encompass improving the performance of metal components in a range of technical fields and modifying the electrical properties of semiconductors (Appleton et al., 1984; McCafferty et al., 1987). The high level of control of these methods also allows for systematic investigation of atomic-scale microstructure-property relationships (Picraux & Pope, 1984).

Electrochemical and Electroless Deposition

The two primary methods for depositing thin metal coatings in various industrial applications are electrochemical and electroless deposition. An external electrical current is used in electrochemical deposition, which is also termed as electroplating, to plate a metal over a conductive substrate. This method is widely used in fields such as electronics, aerospace, and automotive for corrosion resistance, wear protection, and enhanced conductivity. It offers precise control over homogeneity and thickness. Electroless deposition, conversely, is an autocatalytic process of depositing metal on a substrate through a reduction reaction by means of a chemical reaction without relying on an outside source of electrical energy. Such a process plays an important part in applications like PCBs as well as electromagnetic shielding because such a process comes particularly handy while covering non-conductive materials like polymers as well as ceramics. Both deposition processes are vital to advanced manufacturing since they enable the development of high-performance coatings with properties that are uniquely tailored to meet specific functional requirements. In metal coating production, electrochemical and electroless deposition are vital processes. Electroless deposition employs chemical reducing agents, whereas electrodeposition employs electric current to reduce metal ions (Paunovic & Schlesinger, 1998). Electroplating gives high deposition rates and bath stability, but it is not able to coat complex shapes uniformly. While having less deposition rate, electroless plating has a more uniform coating (Narayanan & Seshadri, 2008). Electroless nickel plating, discovered in 1946, has since then increased its significance, particularly in industries such as protective coatings and printed circuits (Schlesinger, 2011). Several factors, such as substrate material, pretreatment, thickness of the deposit, and coating properties, influence the degree to which electroless coatings resist corrosion (Narayanan & Seshadri, 2008). Although electroless deposition is increasingly gaining acceptance, further research is needed to understand reaction mechanisms, improve coating properties, and develop environmentally friendly procedures (Djokić, 2002). In the 20th century, electroless deposition—technology used since centuries—underwent tremendous advancements in science as well as in industry.

MATERIAL SELECTION AND DESIGN CONSIDERATIONS

Performance Requirements for Military Applications

Figure 3. Requirements for military applications

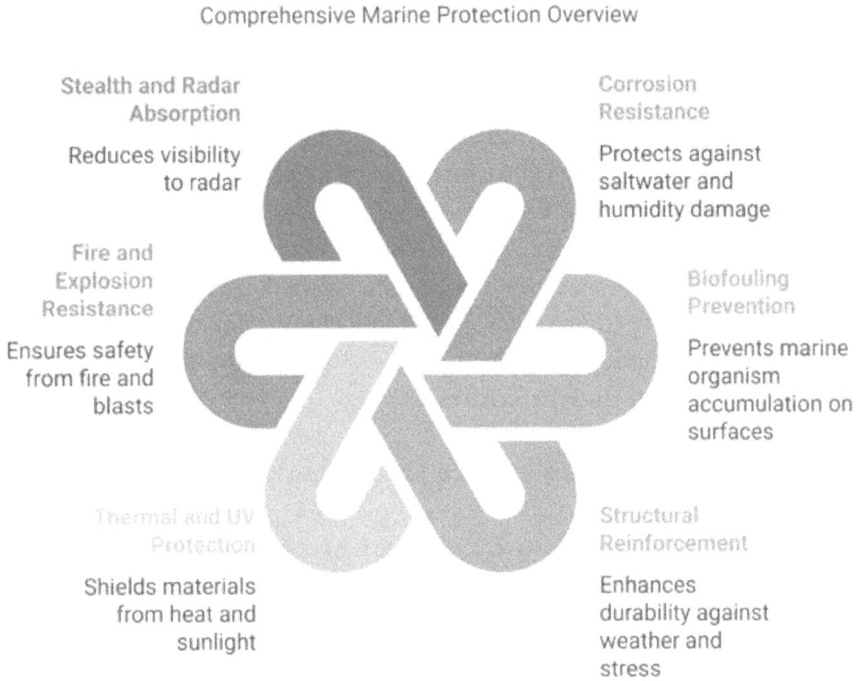

Materials and coatings that endure rugged operating and environmental conditions and retain reliability, durability, and efficacy are needed to achieve performance specifications for military use. Military-grade materials, due to exposure to abrasive settings, seawater, and extreme climates, need high resistance to wear and corrosion. Due to exposure to fuels, chemicals, and extreme temperature fluctuations, thermal and chemical stability is also necessary. In order to withstand mechanical stress, high-speed impacts, and ballistic attacks, the material must have mechanical strength as well as impact resistance. For enhancing stealth properties, materials should also have low acoustic, infrared, or radar signatures. Coatings must provide electromagnetic shielding in electronic and communication systems to prevent interference and maintain signal integrity. Lastly, cost-effectiveness and scalability of manufacturing are critical factors that ensure materials can be made efficiently enough to meet rigorous military specifications. Military uses require seismic sensors, lasers, microelectronics, and high-performance screens. Active matrix liquid crystal displays (AMLCDs) with sunlight-readable, high-resolution color screens are replacing previous technologies used in military equipment (Hopper & Desjardins, 1995). With specific requirements for output power, wavelength, and beam quality, CO2 lasers are under investigation for various military uses, including missile guidance and rangefinders (Harney, 1989). To meet the demands of advanced radar, communications, and electronic warfare systems, heterostructure microelectronics like HBTs and HEMTs are under development (Greiling et al., 1995). High sensitivity to arbitrary vibrations, broad dynamic range, and configurable sensitivity are required for seismic sensors in security and military applications. Rather than overall sensitivity being improved, designers are focusing on reducing threshold sensitivity levels and designing customized sensors for specific applications (Pakhomov et al., 2005). High-speed microelectronics with the ability to resist ionizing radiation and operate in extreme environments are required for forthcoming military systems. There are real military applications for wearable physiological monitoring to enhance soldier health and performance.

Corrosion and Wear Resistance Optimization

Optimization of corrosion and wear resistance is important to improve the durability and long life of a material, predominantly in harsh conditions such as military, marine, and aerospace applications. To negate degradation due to chemical, electrochemical, and mechanical interactions, advanced coatings as well as surface treatments are needed. For the purpose of improving resistance to oxidation, moisture, and abrasive wear, methods like thermal spraying, plasma coatings, electrochemical and electroless deposition, and nanocomposite reinforcements are generally applied. Besides, protective functionality of coatings is also boosted through the addition of corrosion inhibitors, self-healing polymers, and ceramic particles. Material selection, surface modification, and eco-friendly protective coatings like chromium-free coatings are also major contributors to performance enhancement. High-performance materials can be created by industries that satisfy high durability standards while minimizing maintenance costs and maximizing service life through the incorporation of these new methods. Improvement of corrosion and wear resistance in different materials has been the focus of recent research. The optimal coating parameters for improved corrosion and wear properties for electroless Ni-P-Cu coatings were determined by weighted principal component analysis combined with grey relational analysis (Supriyo Roy & P. Sahoo, 2014). In order to enhance corrosion and wear resistance in titanium alloys for orthopedic implants without compromising biocompatibility, a computational model using cohesive energy, oxide formation energy, and other descriptors was developed (Siony et al., 2022). Corrosion behavior was explored in a wide range of conditions, whereas artificial neural networks and genetic algorithms were used to optimize wear performance in AISI 1040 steel (Agrawal & Mukhopadhyay, 2022). Amorphous and amorphous/nanocrystalline microstructures demonstrate the optimum combined properties in amorphous Fe-Cr-Nb-B alloys, for which the amorphous to crystalline phase balance was found to be necessary to optimize corrosion and wear resistance (Koga et al., 2020).

Environmental and Chemical Stability

For materials employed in challenging applications such as the aerospace, military, and industrial industries, chemical and environmental stability is a basic requirement. Severe temperatures, high humidity, ultraviolet radiation, and severe chemical environments—such as corrosive gases, acids, and solvents—need to be endured by materials. High-tech surface coating and treatments, such as fluoropolymer film, ceramic-based paint, and chemically inert nanocomposites, are applied to enhance resistance to degradation. In addition, through responding to environmental changes, self-healing materials and adaptive coatings with self-healing abilities could also enhance lifetime further. To minimize the effect on the environment while maintaining high performance requirements, the development of sustainable and green protective coatings such as water-based coatings and chromium-free corrosion inhibitors is also increasingly significant. Materials may maintain their structural integrity, functionality, and reliability under harsh operating conditions for many years by ensuring superior environmental and chemical stability. Environmental stability is a significant consideration when selecting materials for military use. Materials have the tendency to degrade in a variety of ways upon exposure to external conditions, including corrosion, erosion, and thermal instability (Clark, 1993). Economic factors, manufacturing processes, environmental resistance, and performance requirements all need to be reconciled in the selection process (Farag, 1997). Military textiles require durability and protection against fire, ballistic, chemical, biological, and nuclear attacks (Adanur & Tewari, 1997). Greedy algorithm method is one of the recent advancements in material screening techniques that can quickly assess thermodynamic stability under different conditions (Twyman et al., 2022). The Pilling-Bedworth ratio, particularly significant in conditions with abundant oxygen, can be employed to assess whether materials are capable of self-passivation. Material design that can possibly meet performance requirements when exposed to harsh conditions encountered in military application depends on environmental stability.

Adhesion and Durability Considerations

Efficient corrosion protection and long-term functionality of protective films rely significantly on adhesion and durability. By forming a lasting barrier that repels moisture, oxygen, and corrosive species from reaching the substrate, strong adhesion by the coating results in a stable film. Blistering, delamination, or cracking is caused by lack of adherence and can weaken a coating's resistance to corrosion. On the other hand, durability refers to the ability of the coating to withstand environmental stresses over time, such as UV radiation, temperature fluctuations, mechanical wear, and chemical exposure. To enhance adhesion and durability and deliver long-term protection under harsh service conditions, high-performance coatings are typically formulated with tailored chemistries and surface treatments. To maximize adhesion and the overall life of the coating, attention to surface preparation, judicious coating material selection, and precise application procedures are essential. Adhesion and durability are two critical corrosion prevention coatings. Protective capability has to be tested by sustaining adhesion when subjected to the environment (Dickie, 1994). Surface roughness affects adhesion and corrosion spread, yet coating-metal interactions might not be suitably described by classical pull-off tests (Croll, 2020). The strength of adhesion is significantly affected by coating processes; electrodeposition is superior to thermal spraying for aluminum alloy 7175 (lî u et al., 2023). Adhesion failure is considered the first stage of corrosion of coated metal, and wet adhesion plays a significant role in the prevention of corrosion (Funke, 1985). Effective surface preparation, hydrophobic paints, and barrier pigments are the characteristics that enhance adhesion and resistance to corrosion (Funke, 1985; Croll, 2020). At the initial stages of corrosion-driven loss of adhesion, the paint resin composition and interface chemistry play significant roles (Dickie, 1994). For optimal adhesion and resistance in corrosive conditions, the selection of the optimal coating process and materials is crucial (lî u et al., 2023).

MANUFACTURING AND SCALABILITY

Mass Production Challenges in Protective Coatings

Figure 4. Figure representing mass production challenges in protective coatings

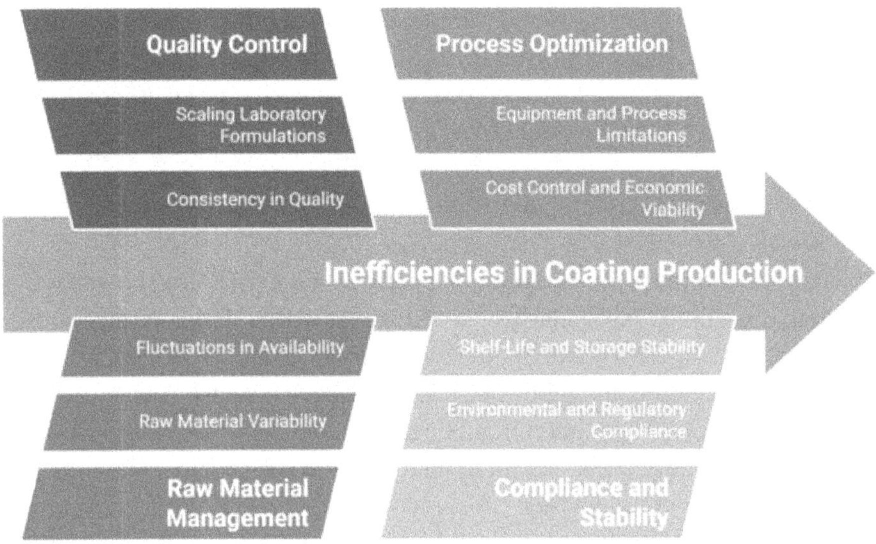

Scalability of manufacturing methods, cost, and consistency of quality are among the issues with mass production of protective coatings. Strict control over raw material choice, formulation, and application process is required to achieve uniform composition and function with large-scale production. Flaws that compromise protection can arise from differences in coating thickness, adhesion properties, and curing conditions. In addition, environmental regulations restrict the use of harmful chemicals by promoting the design of environmentally appropriate replacements without sacrifice in performance. High-throughput manufacturing requires effective manufacturing processes like roll-to-roll coating, spray-on coating in a controlled automatic way, or advanced curing technology such as UV or plasma curing. During large-scale production of protective coatings, sustaining large-scale repeatability as well as the balance between performance, cost, and regulatory requirements remains a central issue. There are various hurdles to overcome in the production of protective coatings in large quantities for different industries. Protective coatings for marine, automotive, and aerospace applications must withstand corrosion and operate under harsh environmental conditions (Fayomi et al., 2020). Threats to human health and the environment are posed by the application of toxic substances like chromium in automobile paints and tributyltin in marine antifouling paints (Fayomi et al., 2020). One of the biggest issues in aerospace is that structural joint motions can stretch and warp coatings (Fayomi et al., 2020). Enameling poses special difficulties for the ceramics industry, particularly for complex surfaces (Georgilas & Tourassis, 2007). Scientists are exploring new materials such as graphene-based coatings that possess superior mechanical strength, impermeability, and chemical resistance to overcome these challenges (Nine et al., 2015). Another critical focus for improving protective coatings in industrial processes is the design of smart coatings and optimizing coating processes (Makhlouf, 2014).

Cost-Effective Synthesis and Deposition Techniques

Economical synthesis and deposition methods are critical to the general use of protective coatings, particularly in applications where large surface areas or high-volume parts need to be treated. Conventional methods like spray coating, dip coating, and electroplating are commonly practiced because they are relatively simple and can be scaled up, but minimizing waste and maximizing material usage is necessary to minimize costs. Recent developments in sol-gel processing, chemical vapor deposition (CVD), and thermal spraying have made it possible to produce high-performance coatings with reduced material and energy usage. Moreover, the use of nanomaterials and hybrid formulations provides improved protective properties at thinner thicknesses, further reducing material costs. Processes such as roll-to-roll processing and robotic automated application systems enhance throughput and uniformity, making them viable for industrial-scale production. Finally, the most suitable and cost-effective synthesis and deposition technique relies on the substrate material, needed protective performance, and working environment, with a view to maximum protection at minimal expense. This overview discusses cost-effective methods for synthesizing and depositing protective coatings. Dip-coating, spin-coating, and spray-coating are some of the methods discussed, which provide satisfactory film uniformity and low surface roughness (Butt, 2022). Plasma-based processes are featured as cost-effective for high-temperature protective coatings (Monteiro & Davies, 1998). For metallic bone implants, plasma spray methods offer cost-effective coatings, whereas electrodeposition and aerosol deposition enable the development of controlled microstructures (Moore et al., 2017). The novel electrostatic spray assisted vapor deposition (ESAVD) technique is introduced as a cost-effective, versatile process for making films and coatings with special microstructures, high deposition efficiency, and good adhesion (Choy, 2000). The process can be carried out in open atmosphere without advanced reactors or vacuum facilities, which renders it a very promising method for many applications. In general, these articles stress the need to balance cost-effectiveness with coating quality and performance.

Industrial Scale-Up Strategies

Under strict operating and regulation conditions, industrial scaling up approaches to protective coating and treatment in military materials require an inflexible strategy that balances production efficiency, performance, and reliability. Defensive coatings with superior resistance to corrosion, abrasion, chemical exposure, and harsh climatic conditions must be implemented for military use. To upscale these advanced formulations, lab-scale research—like self-healing systems, multifunctional films, or nanocomposite coatings—has to be adapted for high-throughput manufacturing without compromising its protective properties. Employment of flexible and modular production systems can be achieved with automatics for precise application as well as integration of non-destructive in-line quality control methods is crucial strategies to ensure consistency and surface free from defects. Huge pieces such as military trucks, aircraft, and ships are ideally suited to surface preparation methods such as grit blasting or plasma treatment. Use of environmentally friendly alternatives to antiquated practices such as chromate-based paints is also necessitated by the need for strict compliance with sustainability regulations and military specifications (e.g., MIL-SPEC). To ensure reliable protection at scale, material scientists, process engineers, and defense contractors need to collaborate closely to scale up defense coatings successfully. Developments in protective coatings for aerospace and defense are exemplified by recent studies. Makhlouf (2014) suggests that smart coatings are a promising technology for the automotive and aerospace sectors. By means of 3D CFD models, scale-up strategies for coating furnaces in particle production have been defined, proposing new benchmarks for greater production capacity (Liu et al., 2020). Critical examination has been performed on techniques for generating materials and coatings for extreme environments such as high temperatures, abrasion, and corrosion (Urbina et al., 2018). Due to their outstanding corrosion resistance, anti-wear properties, and thermal barrier characteristics, atomically thin 2D nanomaterials such as MXenes, graphene, and transition metal dichalcogenides are promising as protective coatings. The materials function properly between temperatures ranging from -10°C to 1100°C and are apt for applications in various aerospace, naval, and military technology parts (Anirudh et al., 2023).

Quality Control and Standardization

As far as offering consistent performance, reliability, and strict conformity to harsh military conditions are concerned, standardization and quality control of protective coating and treatment of military defense materials are essential. The coatings must have stringent mechanical strength, chemical stability, corrosion protection, and environmental durability conditions because defense use is that vital. At different stages, e.g., inspection of raw materials, assessment of surface preparation, tracking of application process, inspection after treatment using methods such as adhesion test, thickness gauging, salt spray test, thermal cycling, etc., systematic inspection and testing fall under the control of quality. Treatments and coatings will need to have standard standards for different defense platforms and suppliers according to standardization, which is facilitated by guidelines such as MIL-SPEC, ISO, and ASTM. Quality control in large production is also enhanced by the implementation of traceability protocols, electronic documentation, and in-line measurement systems. The defense industry can guarantee the high performance required for mission-essential equipment by applying strict quality control and standardization processes, ensuring operational readiness, life cycle, and safety across a wide range of austere environments. Corrosion protection is most important in defense and industrial uses, particularly in hostile environments such as NASA launch facilities (Calle et al., 2010). Effective corrosion control requires standardization, utilization of the latest knowledge, and proper information transfer (Schwenk, 1986). The boundaries of conventional protective coatings due to environmental laws are being met by the emergence of smart coating technologies that have the ability to detect corrosion damage autonomously, block, and self-heal (Calle et al., 2010). A holistic approach considering factors such as coating material selection, surface preparation, application standards, quality control processes, and appropriate requirements is needed to achieve economic protection of steel structures (Chandler & Bayliss, 1985). Focusing on the development of creative solutions that are resilient to harsh conditions and compliant with evolving regulatory demands, continued research and development on corrosion prevention seeks to enhance the sustainability and longevity of materials in corrosive environments (Calle et al., 2010; Schwenk, 1986).

APPLICATIONS IN DEFENSE AND MILITARY SYSTEM

Aerospace and Aviation Coatings

In military and defense systems, protective coatings are essential, especially in aerospace and aviation applications where materials are subjected to harsh operating environments. High-altitude UV exposure, abrupt temperature changes, aerodynamic friction, and corrosive conditions like air containing salt and chemical pollutants are all things that these coatings are designed to resist. Coatings on military aircraft improve mechanical stress resistance, corrosion protection, and radar cross-section reduction by including stealth or radar-absorbing materials. Resistance to gasoline, hydraulic fluids, and de-ice chemicals and good adhesion and flexibility are provided by specialized formulations such as polyurethane and epoxy-based coatings. In addition, in order to ensure maximum service life, jet engine components are coated with thermal barrier materials to prevent oxidation as well as heat damages. To reduce visibility in many spectrums, camouflage and infrared signature reduction coatings are also used. With all of this in mind, protective coatings are essential to structural integrity, lowering maintenance costs, and improving mission readiness and survivability in hostile environments in aerospace and aviation defense systems. In aerospace and military systems, protective coatings are essential for dealing with corrosion prevention, durability, and environmental issues. Advances in technology have resulted in enhanced coating performance, such as increased fluid resistance, flexibility, and adhesion (Smith et al., 2000). Future applications of smart coatings, like corrosion-sensing functional nanocoatings and self-healing nanostructured coatings, appear promising (Ferreira et al., 2011). Competition to create new materials, surface pre-treatments, and top coats is what drives the aerospace industry (Makhlouf, 2014). Nonetheless, the search for alternatives to traditional chemical treatments and processes that could result in pollution has been prompted by environmental issues (Smith, 1995). New treatments and preventive measures are being developed to maintain structural integrity while meeting future environmental regulations as the industry adapts to stricter environmental regulations (Smith, 1995; Makhlouf, 2014).

Naval and Marine Protection Systems

Figure 5. High performance military coating

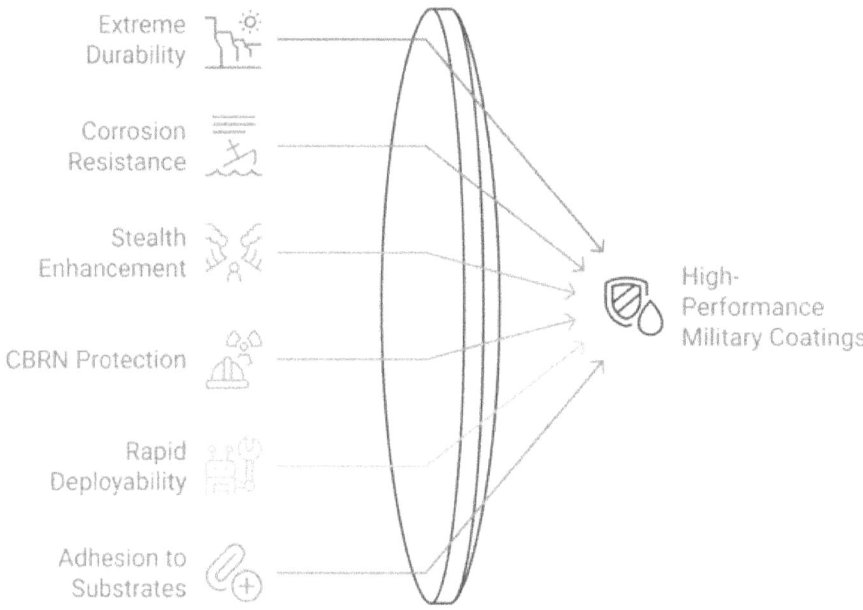

Marine and naval protective coatings are necessary in defense and military systems to protect infrastructure and ships from highly corrosive marine environments. Seawater exposure, high humidity, biofouling, and mechanical stress on warships, submarines, aircraft carriers, and offshore defense installations can significantly compromise their structural integrity and efficiency To prevent corrosion, reduce drag, and increase fuel efficiency, advanced marine paints are used to coat hulls and underwater equipment. They consist of epoxy or polyurethane-based barrier coats, anti-corrosive primer, and antifouling paint. On top of this, specialized coatings are used to maintain hydrodynamic performance and resist the growth of marine life. On top of this, to reduce detectability, navy warships are painted with stealth material that absorbs radar or sonar waves. Tough naval specifications have to be met by these protection systems, which also have to withstand prolonged time exposures with extremely low maintenance. All these considerations, marine and naval defense system coatings deliver lifetime, battle-readiness, and ruggedness in some of the most extreme operating conditions. Defense and military systems, particularly marine and naval protection systems, are being revolutionized through the use of nanotechnology. It enhances fuel efficiency and reduces emissions through the production of lighter but stronger armor materials for tanks, automobile parts, and airplanes (Mayyadah S. Abed & Z. Jawad, 2022; E. Sadiku et al., 2019). Armor vehicles, uniforms, and unmanned aerial vehicles are just a few of the planes and defense systems that make use of nanocomposites (E. Sadiku et al., 2019). Nanotechnology-enhanced infrared sensor technology is essential to military and civilian defense systems because it provides passive tracking and detection for missile and space defense (Latika). Becker, S. R. (2006). According to F. Calabrese et al. (2012), the Damage Control System (DCS) is a crucial information-retrieval and equipment-control system that is integrated into ship management systems in naval applications. It enables personnel to identify, assess, and manage dangerous situations, improving shipboard safety and security.

Land-Based Military Equipment and Vehicles

Marine and naval protective coatings play a vital role in defense and military systems to preserve infrastructure and ships subject to highly corrosive marine environments. Ongoing exposure to seawater, humidity, biofouling, and mechanical abrasion on warships, submarines, aircraft carriers, and offshore defense platforms can severely undermine their structural integrity and operational performance. To avoid corrosion, minimize drag, and maximize fuel efficiency, advanced marine coatings are applied on hulls and underwater parts. Such coating includes epoxy or polyurethane-based barrier coats, anti-corrosive primer, and antifouling paint. Furthermore, there are special coatings utilized to ensure hydrodynamic performance as well as resisting marine organism growth. In addition, to limit detectability, naval warships are coated with stealth materials which absorb radar or sonar. Durable naval requirements have to be satisfied by these defense mechanisms, which must also survive for extended durations of time with minimum maintenance. All things considered, marine and naval defense system coatings provide lifespan, combat capability, and durability in some of the harshest operating conditions. Defense and military systems, especially those involving marine and naval security, are being revolutionized by nanotechnology. It enables the manufacture of lighter yet tougher materials for armor, automobiles, and aircraft, making fuel more efficient and lowering emissions. (E. Sadiku et al., 2019; Mayyadah S. Abed & Z. Jawad, 2022). Suits, armored tanks, and unmanned vehicles are just some of the armed forces and aviation equipment using nanocomposites (E. Sadiku et al., 2019). Nanotechnology-enhanced infrared sensor technology is essential to military and civilian defense systems because it provides passive tracking and detection for missile and space defense (Latika). Becker, S. R. (2006). According to F. Calabrese et al. (2012), the Damage Control System (DCS) is a crucial information-retrieval and equipment-control system that is integrated into ship management systems in naval applications. It enables personnel to identify, assess, and manage dangerous situations, improving shipboard safety and security.

Protective Gear and Personal Armor

Marine and naval protective coatings play a vital role in defense and military systems to preserve infrastructure and ships subject to highly corrosive marine environments. Ongoing exposure to seawater, humidity, biofouling, and mechanical abrasion on warships, submarines, aircraft carriers, and offshore defense platforms can severely undermine their structural integrity and operational performance. To avoid corrosion, minimize drag, and maximize fuel efficiency, advanced marine coatings are applied on hulls and underwater parts. Such coating includes epoxy or polyurethane-based barrier coats, anti-corrosive primer, and antifouling paint. Furthermore, there are special coatings utilized to ensure hydrodynamic performance as well as resisting marine organism growth. In addition, to limit detectability, naval warships are coated with stealth materials which absorb radar or sonar. Durable naval requirements have to be satisfied by these defense mechanisms, which must also survive for extended durations of time with minimum maintenance. All things considered, marine and naval defense system coatings provide lifespan, combat capability, and durability in some of the harshest operating conditions. Defense and military systems, especially those involving marine and naval security, are being revolutionized by nanotechnology. It enables the manufacture of lighter yet tougher materials for armor, automobiles, and aircraft, making fuel more efficient and lowering emissions. (E. Sadiku et al., 2019; Mayyadah S. Abed & Z. Jawad, 2022). Suits, armored tanks, and unmanned vehicles are just some of the armed forces and aviation equipment using nanocomposites (E. Sadiku et al., 2019).

FUTURE TRENDS AND INNOVATIONS

Next-Generation Self-Healing Coatings

Thousands of lives have been saved due to the vast change in military protective equipment, including the integration of lightweight ballistic materials in lieu of metal armor in vests and helmets (Bhatnagar & Lang, 2006). Anthropometric data, psychological factors, and gender-related needs are considered in designing contemporary body armor (Sparks, 2012). With enhanced mechanical, chemical, and physical properties along with solutions to discomfort and bulk, nanotechnology offers promising advances in protective clothing (Thilagavathi et al., 2008). Existing military protective equipment needs to balance comfort and mobility while satisfying a variety of competing requirements, including protection against biological and chemical agents (Sparks, 2012). Chances of multidisciplinary research and teaching in areas like textile engineering, clothing design, and physiology are offered by the creation of protective apparel systems (Shanley et al., 1993). Development of innovative materials and design methods to optimize protection, comfort, and functionality is also being researched (Sparks, 2012; Thilagavathi et al., 2008).

Integration of AI and Nanotechnology in Coating Design

The marriage of nanotechnology and artificial intelligence (AI) is driving monumental advancements across many fields, particularly materials science and medicine. Deep learning and machine learning AI techniques are optimizing the efficacy and accuracy of nanotechnology applications (Naik & Jagtap, 2024). This convergence has accelerated material discovery, nanoscale characterization, and the development of personalized medical applications (Izanker et al., 2023In the coatings sector, machine learning and artificial intelligence are enhancing overall design capability, shortening development timelines, and optimizing features like tribological, mechanical, and corrosion protection (Verma & Khanna, 2022). The integration of nanotechnology and artificial intelligence (AI) is improving the development of therapeutic and diagnostic nanomaterials, improving patient data capture, solving tumor heterogeneity challenges, and streamlining material properties for increased potency in cancer therapy (Adir et al., 2019). The union bears tremendous accomplishments in numerous areas despite the challenges.

CONCLUSION AND OUTLOOK

Future Research Directions

Future research directions in protective coating and treatment are more focused on creating clever, multifunctional, and eco-friendly solutions to satisfy the demands of contemporary defense and industrial applications. Important research fields include the development of environmentally friendly coatings that limit their harmful effects on the environment and human health by using waterborne, solvent-free, or bio-based chemicals. Ultra-thin coatings with improved mechanical strength, self-cleaning qualities, and high barrier capabilities are becoming possible because to developments in nanotechnology. Researchers are also creating damage-sensing and self-healing coatings that can detect structural fatigue and automatically patch microcracks to prolong the life of vital components. Adaptive coatings, which react to environmental cues like temperature, humidity, or chemical exposure, are another intriguing field. Additionally, the use of AI and machine learning in coating design and performance prediction is opening up new avenues for quick development and improvement. Enabling next-generation coatings that are affordable, adaptable to industrial applications, and provide exceptional protection, durability, and utility is the goal here. Many sectors rely on protective coatings, and research is being conducted to develop new materials and methods. Graphene-based coatings have emerged as a potentially multipurpose material due to its exceptional mechanical strength, impermeability, and chemical resistance (Nine et al., 2015). Competitiveness in the automotive and aerospace industries drives rapid innovation in materials, surface pre-treatments, and top coatings (Makhlouf, 2014). Smart coatings are gaining popularity despite ongoing challenges for industrial applications (Makhlouf, 2014). Laws that forbid the use of particular chemicals and processes in metal finishing because of environmental concerns have prompted the aerospace sector to search for new protection plans that satisfy future environmental standards. Future research should concentrate on developing eco-friendly alternatives, improving coating performance, and exploring novel materials like graphene for a variety of applications, such as anti-fouling, corrosion resistance, and flame retardancy (Nine et al., 2015; Smith, 1995).

REFERENCES

Abdel Hamid, Z. (2014). Review article: Composite and nanocomposite coatings. *J Metall Eng*, *3*, 29–42.

Abed, M. S., & Jawad, Z. A. (2022). Nanotechnology for defence applications. In *Nanotechnology for electronic applications* (pp. 187–205). Springer Nature Singapore. DOI: 10.1007/978-981-16-6022-1_10

Abed, M. S., & Jawad, Z. A. (2022). Nanotechnology for defence applications. In *Nanotechnology for electronic applications* (pp. 187–205). Springer Nature Singapore. DOI: 10.1007/978-981-16-6022-1_10

Adir, O., Poley, M., Chen, G., Froim, S., Krinsky, N., Shklover, J., Shainsky-Roitman, J., Lammers, T., & Schroeder, A. (2020). Integrating artificial intelligence and nanotechnology for precision cancer medicine. *Advanced Materials*, *32*(13), 1901989. DOI: 10.1002/adma.201901989 PMID: 31286573

Agrawal, R., & Mukhopadhyay, A. (2023). The use of machine learning and metaheuristic algorithm for wear performance optimization of AISI 1040 steel and investigation of corrosion resistance. *Proceedings of the Institution of Mechanical Engineers. Part J, Journal of Engineering Tribology*, *237*(4), 701–717. DOI: 10.1177/13506501221094416

Akafuah, N. K., Poozesh, S., Salaimeh, A., Patrick, G., Lawler, K., & Saito, K. (2016). Evolution of the automotive body coating process—A review. *Coatings*, *6*(2), 24. DOI: 10.3390/coatings6020024

Aljibori, H. S., Alamiery, A., & Kadhum, A. A. H. (2023). Advances in corrosion protection coatings: A comprehensive review. *Int. J. Corros. Scale Inhib*, *12*(4), 1476–1520.

Anirudh, S., Krishnamurthy, S., Kandasubramanian, B., & Kumar, P. (2023). Probing into atomically thin layered nano-materials protective coating for aerospace and strategic defence application–A review. *Journal of Alloys and Compounds*, *968*, 172203. DOI: 10.1016/j.jallcom.2023.172203

Appleton, B. R., Sartwell, B., Peercy, P. S., Schaefer, R., & Osgood, R. (1985). Ion-beam-and laser-induced surface modifications. *Materials Science and Engineering*, *70*, 23–51. DOI: 10.1016/0025-5416(85)90267-8

Arango, S., Peláez-Vargas, A., & García, C. (2012). Coating and surface treatments on orthodontic metallic materials. *Coatings*, *3*(1), 1–15. DOI: 10.3390/coatings3010001

Armstrong, G. (2010). Engineered coatings for composites and polymers used in defence and aerospace: now and the future. *Transactions of the IMF, 88*(4), 179-184. DOI: 10.1179/002029610X12734927543102

Artesani, A., Di Turo, F., Zucchelli, M., & Traviglia, A. (2020). Recent advances in protective coatings for cultural heritage–an overview. *Coatings*, *10*(3), 217. DOI: 10.3390/coatings10030217

Ashby, M., Cope, E., & Cebon, D. (2013). Materials selection for engineering design. In *Informatics for materials science and engineering* (pp. 219–244). Butterworth-Heinemann. DOI: 10.1016/B978-0-12-394399-6.00010-2

Ashby, M., Cope, E., & Cebon, D. (2013). Materials selection for engineering design. In *Informatics for materials science and engineering* (pp. 219–244). Butterworth-Heinemann. DOI: 10.1016/B978-0-12-394399-6.00010-2

Becker, L. (2006, February). Influence of IR sensor technology on the military and civil defense. In *Quantum Sensing and Nanophotonic Devices III* (Vol. 6127, pp. 180–194). SPIE. DOI: 10.1117/12.640529

Becker, L. (2006, February). Influence of IR sensor technology on the military and civil defense. In *Quantum Sensing and Nanophotonic Devices III* (Vol. 6127, pp. 180–194). SPIE. DOI: 10.1117/12.640529

Bhatnagar, A., & Lang, D. (2006). Military and law enforcement applications of lightweight ballistic materials. In *Lightweight Ballistic Composites* (pp. 364–397). Woodhead Publishing. DOI: 10.1533/9781845691554.2.364

Bhatnagar, A., & Lang, D. (2006). Military and law enforcement applications of lightweight ballistic materials. In *Lightweight Ballistic Composites* (pp. 364–397). Woodhead Publishing. DOI: 10.1533/9781845691554.2.364

Butt, M. A. (2022). Thin-film coating methods: A successful marriage of high-quality and cost-effectiveness—a brief exploration. *Coatings*, *12*(8), 1115. DOI: 10.3390/coatings12081115

Calabrese, F., Cataldo, M., Corallo, A., De Pascalis, A., Mancarella, L., Ostuni, L., & Zizzari, A. A. (2012). Damage Control System: an application for ship safety and security. *IFAC Proceedings Volumes, 45*(27), 103-108.

Calabrese, F., Cataldo, M., Corallo, A., De Pascalis, A., Mancarella, L., Ostuni, L., & Zizzari, A. A. (2012). Damage Control System: an application for ship safety and security. *IFAC Proceedings Volumes, 45*(27), 103-108.

Calle, L., Hintze, P., Li, W., & Buhrow, J. (2010). Smart coatings for autonomous corrosion detection and control. In *AIAA SPACE 2010 Conference & Exposition* (p. 8877). DOI: 10.2514/6.2010-8877

Chandler, K. A., & Bayliss, D. A. (1985). Corrosion protection of steel structures.

Choy, K. L. (2000). Innovative and cost effective deposition of coatings using ES-AVD method. *Surface Engineering*, *16*(6), 465–468.

Clark, D. E. (1993). Environmental stability of materials—Issues and opportunities. *MRS Bulletin*, *18*(9), 25–31. DOI: 10.1557/S0883769400037994

Clark, D. E. (1993). Environmental stability of materials—Issues and opportunities. *MRS Bulletin*, *18*(9), 25–31. DOI: 10.1557/S0883769400037994

Croll, S. G. (2020). Surface roughness profile and its effect on coating adhesion and corrosion protection: A review. *Progress in Organic Coatings*, *148*, 105847. DOI: 10.1016/j.porgcoat.2020.105847

Davis, J. (2003). Coating technology: The future: New and emerging technologies. *Materials Performance*, *42*(11), 24–27.

Dennis, R. V., Patil, V., Andrews, J. L., Aldinger, J. P., Yadav, G. D., & Banerjee, S. (2015). Hybrid nanostructured coatings for corrosion protection of base metals: A sustainability perspective. *Materials Research Express*, *2*(3), 032001. DOI: 10.1088/2053-1591/2/3/032001

Dickie, R. A. (1994). Paint adhesion, corrosion protection, and interfacial chemistry. *Progress in Organic Coatings*, *25*(1), 3–22. DOI: 10.1016/0300-9440(94)00500-1

Disegi, J. A. (1997, April). Anodizing treatments for titanium implants. In *Proceedings of the 1997 16 Southern Biomedical Engineering Conference* (pp. 129-132). IEEE. DOI: 10.1109/SBEC.1997.583231

Djokić, S. S. (2002). Electroless deposition of metals and alloys. In *Modern aspects of electrochemistry* (pp. 51–133). Springer US. DOI: 10.1007/0-306-47604-5_2

Djokić, S. S., & Cavallotti, P. L. (2010). Electroless deposition: Theory and applications. *Electrodeposition. Theory into Practice*, •••, 251–289.

Fauchais, P., & Vardelle, A. (2012). Thermal sprayed coatings used against corrosion and corrosive wear. *Advanced plasma spray applications, 10*, 34448.

Fayomi, O. S. I., Agboola, O., Akande, I. G., & Emmanuel, A. O. (2020, December). Challenges of coatings in aerospace, automobile and marine industries. In *AIP Conference Proceedings* (Vol. 2307, No. 1). AIP Publishing. DOI: 10.1063/5.0033579

Fayomi, O. S. I., Agboola, O., Akande, I. G., & Emmanuel, A. O. (2020, December). Challenges of coatings in aerospace, automobile and marine industries. In *AIP Conference Proceedings* (Vol. 2307, No. 1). AIP Publishing. DOI: 10.1063/5.0033579

Fernando, R. H. (2009). Nanocomposite and nanostructured coatings: Recent advancements.

Ferreira, M. G. S., Zheludkevich, M. L., & Tedim, J. (2011). Advanced protective coatings for aeronautical applications. In *Nanocoatings and ultra-thin films* (pp. 235–279). Woodhead Publishing. DOI: 10.1533/9780857094902.2.235

Friedl, K. E. (2018). Military applications of soldier physiological monitoring. *Journal of Science and Medicine in Sport*, *21*(11), 1147–1153. DOI: 10.1016/j.jsams.2018.06.004 PMID: 29960798

Funke, W. (1985). The role of adhesion in corrosion protection by organic coatings. *Journal of the oil and colour chemists' association, 68*(9), 229-232.

Gaur, U. P., & Kamari, E. (2024). Applications of thermal spray coatings: A review. *J. Therm. Spray Eng*, *4*, 106–114. DOI: 10.52687/2582-1474/405

Georgilas, I. P., & Tourassis, V. D. (2007, December). Quality issues in enameling of ceramic industry products. In *2007 IEEE International Conference on Industrial Engineering and Engineering Management* (pp. 1226-1230). IEEE. DOI: 10.1109/IEEM.2007.4419387

Greiling, P., Kirkpatrick, C., & Valentine, G. (1995). Military applications for heterostructure microelectronics technology. *Solid-State Electronics*, *38*(9), 1559–1567. DOI: 10.1016/0038-1101(95)00069-6

Harrison, W. N., Moore, D. G., & Richmond, J. C. (1947). *Review of an investigation of ceramic coatings for metallic turbine parts and other high-temperature applications* (No. NACA-TN-1186).

Hopper, D. G., & Desjardins, D. D. (1995, June). Requirements for AMLCDs in US military applications. In *Cockpit Displays II* (Vol. 2462, pp. 130–141). SPIE. DOI: 10.1117/12.210949

Huynh, V., Ngo, N. K., & Golden, T. D. (2019). Surface activation and pretreatments for biocompatible metals and alloys used in biomedical applications. *International Journal of Biomaterials*, *2019*(1), 3806504. DOI: 10.1155/2019/3806504 PMID: 31275394

Izanker, S. V., Dhole, A., & Kumar, P. (2023, November). Navigating the Nexus: Exploring the Fusion of AI and Nanotechnology for Cutting-Edge Advances. In *2023 1st DMIHER International Conference on Artificial Intelligence in Education and Industry 4.0 (IDICAIEI)* (Vol. 1, pp. 1-5). IEEE.

Kartsonakis, I. A., Kontiza, A., & Kanellopoulou, I. A. (2024). Advanced Micro/Nanocapsules for Self-Healing Coatings. *Applied Sciences (Basel, Switzerland)*, *14*(18), 8396. DOI: 10.3390/app14188396

Koga, G. Y., Ferreira, T., Guo, Y., Coimbrão, D. D., Jorge, A. M.Jr, Kiminami, C. S., Bolfarini, C., & Botta, W. J. (2021). Challenges in optimizing the resistance to corrosion and wear of amorphous Fe-Cr-Nb-B alloy containing crystalline phases. *Journal of Non-Crystalline Solids*, *555*, 120537. DOI: 10.1016/j.jnoncrysol.2020.120537

Liu, M., Chen, Z., Chen, M., Shao, Y., Liu, B., & Tang, Y. (2020). Scale-up strategy study of coating furnace for TRISO particle fabrication based on numerical simulations. *Nuclear Engineering and Design*, *357*, 110413. DOI: 10.1016/j.nucengdes.2019.110413

Makhlouf, A. S. H. (2014). Protective coatings for automotive, aerospace and military applications: Current prospects and future trends. In *Handbook of Smart Coatings for Materials Protection* (pp. 121–131). Woodhead Publishing. DOI: 10.1533/9780857096883.1.121

Makhlouf, A. S. H. (2014). Protective coatings for automotive, aerospace and military applications: Current prospects and future trends. In *Handbook of Smart Coatings for Materials Protection* (pp. 121–131). Woodhead Publishing. DOI: 10.1533/9780857096883.1.121

Mbam, S. O., Nwonu, S. E., Orelaja, O. A., Nwigwe, U. S., & Gou, X. F. (2019). Thin-film coating; historical evolution, conventional deposition technologies, stress-state micro/nano-level measurement/models and prospects projection: A critical review. *Materials Research Express*, *6*(12), 122001. DOI: 10.1088/2053-1591/ab52cd

McCafferty, E. E., Hubler, G. K., Natishan, P. M., Moore, P. G., Kant, R. A., & Sartwell, B. D. (1987). Naval research laboratory surface modification program: Ion beam and laser processing of metal surfaces for improved corrosion resistance. *Materials Science and Engineering*, *86*, 1–17. DOI: 10.1016/0025-5416(87)90438-1

Monteiro, O., & Davies, B. (1998). Cost Effective Plasma-Based Process for Depositing a Wide Variety of High Temperature Protective Coatings. *Materials Technology*, *13*(1), 3–4. DOI: 10.1080/10667857.1998.11752756

Moore, B., Asadi, E., & Lewis, G. (2017). Deposition methods for microstructured and nanostructured coatings on metallic bone implants: A review. *Advances in Materials Science and Engineering*, *2017*(1), 5812907. DOI: 10.1155/2017/5812907

Moore, R., Torres, H., DeLuca, R., Gentile, M., Zadrovicz, K. C., Albrecht, M., & Kohli, S. (2023, June). Salt spray resistant silver coatings for aerospace and defense applications. In *Advanced Optics for Imaging Applications: UV through LWIR VIII* (Vol. 12530, pp. 46–55). SPIE. DOI: 10.1117/12.2663465

Naik, G. G., & Jagtap, V. A. (2024). Two heads are better than one: Unravelling the potential Impact of Artificial Intelligence in nanotechnology. *Nano TransMed*, 100041.

Nartita, R., Ionita, D., & Demetrescu, I. (2021). Sustainable coatings on metallic alloys as a nowadays challenge. *Sustainability (Basel)*, *13*(18), 10217. DOI: 10.3390/su131810217

Nathanael, A. J., & Oh, T. H. (2020). Biopolymer coatings for biomedical applications. *Polymers*, *12*(12), 3061. DOI: 10.3390/polym12123061 PMID: 33371349

Nelson, R. L., Ramsay, J. D. F., Woodhead, J. L., Cairns, J. A., & Crossley, J. A. A. (1981). The coating of metals with ceramic oxides via colloidal intermediates. *Thin Solid Films*, *81*(4), 329–337. DOI: 10.1016/0040-6090(81)90517-4

Ni, X., Gao, Y., Zhang, X., Lei, Y., Sun, G., & You, B. (2021). An eco-friendly smart self-healing coating with NIR and pH dual-responsive superhydrophobic properties based on biomimetic stimuli-responsive mesoporous polydopamine microspheres. *Chemical Engineering Journal*, *406*, 126725. DOI: 10.1016/j.cej.2020.126725

Nine, M. J., Cole, M. A., Tran, D. N., & Losic, D. (2015). Graphene: A multipurpose material for protective coatings. *Journal of Materials Chemistry. A, Materials for Energy and Sustainability*, *3*(24), 12580–12602. DOI: 10.1039/C5TA01010A

Nine, M. J., Cole, M. A., Tran, D. N., & Losic, D. (2015). Graphene: A multipurpose material for protective coatings. *Journal of Materials Chemistry. A, Materials for Energy and Sustainability*, *3*(24), 12580–12602. DOI: 10.1039/C5TA01010A

Pakhomov, A., Pisano, D., Sicignano, A., & Goldburt, T. (2005, May). High performance seismic sensor requirements for military and security applications. In *Unattended Ground Sensor Technologies and Applications VII* (Vol. 5796, pp. 117–124). SPIE. DOI: 10.1117/12.604358

Patscheider, J. (2003). Nanocomposite hard coatings for wear protection. *MRS Bulletin*, *28*(3), 180–183. DOI: 10.1557/mrs2003.59

Peckerar, M. C., & Neidert, R. E. (2005). High-speed microelectronics for military applications. *Proceedings of the IEEE, 71*(5), 657–666. DOI: 10.1109/PROC.1983.12647

Picraux, S. T., & Pope, L. E. (1984). Tailored surface modification by ion implantation and laser treatment. *Science, 226*(4675), 615–622. DOI: 10.1126/science.226.4675.615 PMID: 17774924

Roy, M. (2023). Materials and Coatings for Defence Application. *Defence Science Journal, 73*(2), 119–120. DOI: 10.14429/dsj.73.18849

Roy, S., & Sahoo, P. (2014). Parametric optimization of corrosion and wear of electroless Ni-P-Cu coating using grey relational coefficient coupled with weighted principal component analysis. *International Journal of Mechanical and Materials Engineering, 9*(1), 1–15. DOI: 10.1186/s40712-014-0010-y

Sadiku, E. R., Agboola, O., Mochane, M. J., Fasiku, V. O., Owonubi, S. J., Ibrahim, I. D., & Ndamase, A. S. (2019). The use of polymer nanocomposites in the aerospace and the military/defence industries. In *Polymer nanocomposites for advanced engineering and military applications* (pp. 316–349). IGI Global. DOI: 10.4018/978-1-5225-7838-3.ch011

Salman, S. A., & Okido, M. (2013). Anodization of magnesium (Mg) alloys to improve corrosion resistance. *Corrosion prevention of magnesium alloys*, 197-231.

Sampath, S., & McCune, R. (2000). Thermal-spray processing of materials. *MRS Bulletin, 25*(7), 12–13. DOI: 10.1557/mrs2000.117

Sankara Narayanan, T. S. N. (2005). Surface pretretament by phosphate conversion coatings-A review. *Reviews on Advanced Materials Science, 9*, 130–177.

Sankara Narayanan, T. S. N., & Seshadri, S. K. (2008). Electro-and electroless plated coatings for corrosion protection.

Shanley, L. A., Slaten, B. L., & Shanley, P. S. (1993). Military protective clothing: Implications for clothing and textiles curriculum and research. *Clothing & Textiles Research Journal, 11*(3), 55–59. DOI: 10.1177/0887302X9301100308

Sheasby, P. G., Wernick, S., & Pinner, R. (1987). Surface treatment and finishing of aluminum and its alloys. Volumes 1 and 2.

Singh, H., Chatha, S. S., Sidhu, H. S., & Sharma, K. (2011). Hot-corrosion resistance of alloy and composite coatings: A review. In *National Conference on Advancements and Futuristic Trends in Mechanical and Materials Engineering(*Oct. 7-8, 2011*).*

Siony, N., Vuong, L., Lundaajamts, O., & Kadkhodaei, S. (2022). Computational design of corrosion-resistant and wear-resistant titanium alloys for orthopedic implants. *Materials Today. Communications*, *33*, 104465. DOI: 10.1016/j.mtcomm.2022.104465

Smith, C. J. (1995). Advances in protective coatings and processes for aerospace applications. *Aircraft Engineering*, *67*(5), 13–16. DOI: 10.1108/eb037598

Smith, C. J. E., Higgs, M. S., & Baldwin, K. R. (2000). Advances in protective coatings and their application to ageing aircraft. In *Applied Vehicle Technology Panel (AVT) Workshop.Proceedings of RTO meeting* (pp. 15-1).

Sparks, E. (Ed.). (2012). *Advances in military textiles and personal equipment.* Elsevier. DOI: 10.1533/9780857095572

Thilagavathi, G., Raja, A. S. M., & Kannaian, T. (2008). Nanotechnology and protective clothing for defence personnel. *Defence Science Journal*, *58*(4), 451–459. DOI: 10.14429/dsj.58.1667

Titu, A. M., Ravai-Nagy, S., & Pop, A. B. (2023). Research on the influence of Coating technologies on Adhesion Anti-corrosion Layers in the case of Al7175 Aluminum Alloy. *Coatings*, *13*(6), 1054. DOI: 10.3390/coatings13061054

Twyman, N. M., Walsh, A., & Buonassisi, T. (2022). Environmental stability of crystals: A greedy screening. *Chemistry of Materials*, *34*(6), 2545–2552. DOI: 10.1021/acs.chemmater.1c02644 PMID: 35431438

Udoh, I. I., Shi, H., Daniel, E. F., Li, J., Gu, S., Liu, F., & Han, E. H. (2022). Active anticorrosion and self-healing coatings: A review with focus on multi-action smart coating strategies. *Journal of Materials Science and Technology*, *116*, 224–237. DOI: 10.1016/j.jmst.2021.11.042

Ulaeto, S. B., Ravi, R. P., Udoh, I. I., Mathew, G. M., & Rajan, T. P. D. (2023). Polymer-based coating for steel protection, highlighting metal–organic framework as functional actives: A review. *Corrosion and Materials Degradation*, *4*(2), 284–316. DOI: 10.3390/cmd4020015

Urbina, M., Rinaldi, A., Cuesta-Lopez, S., Sobetkii, A., Slobozeanu, A. E., Szakalos, P., Qin, Y., Prakasam, M., Piticescu, R.-R., Ducros, C., & Largeteau, A. (2018). The methodologies and strategies for the development of novel material systems and coatings for applications in extreme environments-a critical review. *Manufacturing Review*, *5*, 5. DOI: 10.1051/mfreview/2018006

Velardi, L., Lorusso, A., Paladini, F., Siciliano, M. V., Di Giulio, M., Raino, A., & Nassisi, V. (2010). Modification of polymer characteristics by laser and ion beam. *Radiation Effects and Defects in Solids*, *165*(6-10), 637–642. DOI: 10.1080/10420151003729516

Verma, J., & Khanna, A. S. (2023). Digital advancements in smart materials design and multifunctional coating manufacturing. *Physics Open*, *14*, 100133. DOI: 10.1016/j.physo.2022.100133

Wang, F., Sun, J., & Xu, T. (2023). OXIDE CERAMIC COATINGS ON ORTHOPAEDIC IMPLANTS: A REVIEW OF TIO2 AND AL2O3. *Ceramics-Silikáty*, *67*(4), 525–542. DOI: 10.13168/cs.2023.0052

Webster, D. C., Chisholm, B. J., & Stafslien, S. J. (2007). Mini-review: Combinatorial approaches for the design of novel coating systems. *Biofouling*, *23*(3), 179–192. DOI: 10.1080/08927010701250948 PMID: 17653929

Wirtz, G. P., Brown, S. D., & Kriven, W. M. (1991). Ceramic coatings by anodic spark deposition. *Materials and Manufacturing Processes*, *6*(1), 87–115. DOI: 10.1080/10426919108934737

Witucki, G. L., & Pajk, T. (2004). The evolution of silicon-based technology in coatings. *PITTURE E VERNICI-EUROPE-*, *80*, 7-17.

Yang, N. (2022). Special Issue "Functionalities of Polymer-Based Nanocomposite Films and Coatings". *Coatings*, *12*(9), 1245. DOI: 10.3390/coatings12091245

Zvonkina, I. J., & Hilt, M. (2014). Strategies for developing multi-functional, self-healing coatings for corrosion prevention and other functions. In *Handbook of Smart Coatings for Materials Protection* (pp. 105–120). Woodhead Publishing. DOI: 10.1533/9780857096883.1.105

Chapter 2
Carbon-Based Metal-Free Nano-Materials:
An Alternative Next-Generation Innovative Material for Future

Amit Kumar Dutta
https://orcid.org/0000-0002-1235-766X
Bangabasi Morning College, University of Calcutta, India

ABSTRACT

Carbon-based metal-free nano-materials become 'Revolutionary Materials' today in material science domain, which have been incorporated into different industrial sectors and everyday consumer product to improve product efficiencies and performances across various domains. Various types of Carbon-nanomaterials such as fullerene, nano-diamond, different nano-sized carbon allotropes, more popular carbon nanotubes (CNT), graphene, graphene oxide (GO), reduced graphene oxide (rGO) nano-sheet, activated carbon granules, carbon quantum dots (CQDs), carbon dots (CDs), polymeric graphite carbon nitride (g-C3N4), carbonized polymer dots (CPDs) etc.have been successfully explored and exhibits outstanding performance towards some catalysis-chemical reactions. Most of them possesses largest surface-to-volume ratio, highly porous surface and hence highest catalyst-loading capacity. This chapter looks at the history of research on metal-free nano-materials and discuss how these 'Revolutionary Materials' will provide exciting future challenges and opportunity for the sake of society.

DOI: 10.4018/979-8-3373-0933-0.ch002

INTRODUCTION

Carbon is one of the most abundant elements on the Earth and also most useful in the Periodic Table. Carbon brushes, Graphite, a more stable form of Carbon, are well known essential component of electric motors and have been extensively utilised as lubricant (C.E. Morstein, 2021). Again, activated charcoal or activated carbon granules are very popular and have been widely used in cosmetics, personal care products and water filtration process over the past three decades (W. A. Helbig, 1946). Now-a-days, so many research works have been concentrated to develop alternative cost-effective and next-generation multi-functional materials with the help of these virtually unlimited and free carbon-based sources, though so many metal-based multi-functional nanomaterials provide incredible benefits to the society and have already been used in everyday consumer products. Metal-based nanomaterials, while offering unique properties, present drawbacks like potential toxicity, environmental concerns, and limitations in sustainable synthesis and recycling. Moreover, carbon-based materials usually possess unique structural dimension, outstanding electrical conductivity, excellent mechanical, thermal, optical properties, enabled them attracted significant interest in diverse area. Carbon also exists in a wide variety of allotropic forms with 0D to 3D structures. Again, in the field of nano-technology, when the dimension of a material be so small, in the nano-scaled range, new interesting properties have been generated such as interesting chemical and physical properties, large specific surface area, large specific surface to volume ratio, greater surface tension, greater surface activity, high catalyst-loading capacity and better light absorption capacity. As a whole, carbon-based nano-materials become 'Revolutionary Materials' which have been developed for enormous attainable applications and can be utilised in daily life and everyday consumer products (Figure 1) (R. Verma, 2024). Various types of Carbon-nanomaterials such as fullerene, nano-diamond, different nano-sized carbon allotropes, more popular carbon nanotubes (CNT), graphene, graphene oxide (GO), reduced graphene oxide (rGO) nano-sheet, carbon nano-fibers, carbon quantum dots (CQDs), carbon dots (CDs), polymeric graphite carbon nitride (g-C_3N_4), carbonized polymer dots (CPDs) etc. have been thoroughly investigated and started to be commercialized for the sake of society.

Now-a-days, with tremendous development of science and technology, so many catalysis-research-works have been devoted on innovation and development of interesting technologies to be utilised in daily life and everyday consumer products and also to solve two major problems facing world, day by day demand of energy source and increasing pollution with advancement of human activities/civilization (A. K. Dutta, 2012, A. K. Dutta, 2014). Very recent, photo-catalysis have been widely applied for degradation of various organic pollutants under solar and visible light illumination and has been successfully utilised in waste-water treatment, agricultural pesticide-residue treatment, air purification. Researchers have also paid considerable attention to increase hydrogen-fuel production, an alternative non-carbon energy source, in large scale way using efficient catalyst material even under natural solar-light illumination (A. K. Dutta, 2023, R. Banerjee, 2021, P. Mondal, 2021). Integration of such mechanism of nano-photocatalysis technology into different consumer products such as agriculture, food-safety, medicine, pharmacy, cosmetics, personal care product etc. has new era of innovation and enhancing product efficiencies and performances. As a catalyst material, researcher have now paid considerable attention on virtually unlimited and free carbon-based sources (also non-toxic bio-compatible) which have become suitable candidates for full-spectrum-solar-light driven (320-800 nm) photocatalysis. Metal free carbon-based quantum dots (CQDs), carbon dots (CDs), Graphene, GO, CNT can do light-harvesting over very broad spectral range from UV to NIR (X. Wang, 2017, Y. E. Serge-Correales, 2022). Again, polymeric graphite carbon nitride (g-C_3N_4), as an organic semiconductor, can strongly prevent the recombination of photo-generated electron–hole pairs, enhance the electron-transport behaviour and improve their photo-electrocatalysis performances (S. Cao, 2014). Some commercial painting materials, when mixed with photo-catalytic materials, such as carbon-based 2D graphene, graphene oxide (GO), rGO nano-sheet, CNT, Carbon QDs, g-C_3N_4 supported semiconductor composites, have been widely utilised for improving the air quality and eliminate bad-odors through degradation of carbon monoxide (CO), nitrogen and sulphur containing organic compounds (NO_x), even volatile organic and inorganic pollutants (VOCs) and converted to harmless substances under solar or indoor light irradiation that come into contact with surface of them (J. Zhao, 2003, F. Nawaz, 2024). Photo-catalytic oxidation (PCO) technology has been used in air purification technology that utilise light energy to breakdown pollutants, bacteria and to remove VOCs (VOCs mineralization) and improve air quality. Nano-catalyst thin-films coating roof-tiles, glazed ceramic tiles, concrete, glass windows, anti-fogging and anti-reflecting glasses, mirrors can prevent the accumulation of deposited dust, dirt or contaminant particles on its surface, keeping them clean. Nano-photocatalytic self-cleaning surface coatings have widely been used in exterior building construction materials, tent-materials, transport cars, trains, planes etc. in which sunlight or indoor light have been used to break down pollutants that can be washed away with rain/water (L. Yuan, 2011). Using these self-cleaning surface coatings phenomenon, photocatalytic disinfection has broadly been applied to protect healthcare-associated infections in hospitals, public areas like bathrooms and train stations, and cleaning areas in the food industry (S. Kang, 2007). single-walled carbon nanotubes (SWCNTs) have been demonstrated as strong anti-microbial agent, can entire the cell wall and cytoplasmic membrane of micro-organisms and viruses, that survive on surfaces for a long time in most cases and can able to induce the oxidative interference of proteins, metabolic activities, or damage in bacterial DNA and enzyme structures (S. E. Abo-Neima, 2020). Now-a-days, red/ NIR emissive CDs have been widely used in cancer treatment which can enhance photo-catalytic response towards reactive oxygen species (ROS) generation and destroy cancerous cell's structure. Simultaneously, the deep penetration power and photo-thermal effect of NIR-light raises the temperature of tumors and encourage photo-thermal therapy (PTT), photo-dynamic therapy (PDT) where light-to-heat conversion efficiency kill cancer cells (C. Han, 2024). Within the past few years, activated charcoal nanomaterials have been actively used in cosmetic products. Due to its adsorbing properties, ability to adsorb impurities, oil, and toxins from the skin, it is being used in all sorts of beauty products from face masks to skin cleansers and even soaps. In some cosmetic products such as creams, lotions, gel, sprays, sticks etc., Carbon Dots (CDs) have been included to create UV-filters that block ultraviolet (UV) radiation and exhibit high sun protection factor (SPF). Carbon nanotubes, graphene/graphene oxide, and fullerenes, as an emerging class of novel materials, can exhibit a considerable antimicrobial activity, have been incorporated in cosmetic products formulation and also serve as photo-sensitizer in photo-catalysis process (M. Azizi-Lalabadi, 2020).

Figure 1. Illustration of different application of carbon nano-materials

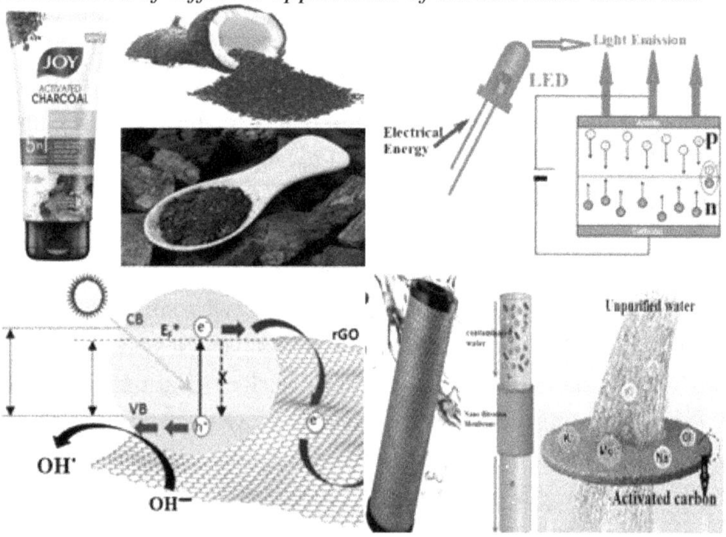

Now-a-days carbon-nanomaterials have been widely used in light-emitting diodes (LEDs) to improve LED technology and device performance. Carbon Dots are highly stable, possess excellent photoluminescence behaviour, show full-spectrum photoluminescence properties (Full-colour-light emitting) and have been successfully fabricated in electroluminescent light-emitting diodes (CD-LED) films with improved brightness. In 2011, Wang. et. al first constructed white LEDs using CDs which possess multiple emission state (Red, blue, green) and have been successfully used to fabricate full-colour and white light emitting films (F. Wang, 2011). LEDs emit light when a positive charge (hole) and a negative charge (electron) do re-combination. The efficient CD-LEDs possess suitable environment in which adjacent hole-rich (p-type) and electron-rich (n-type) regions are created in the nano-materials resulting in p-n junction. When the appropriate voltage is applied to the nano-materials, electrons and holes are driven towards each other to re-combine, release energy in the form of photon and emit light (for solid-state lighting). Very similar to reverse osmosis (RO) purification, nano-fiber membrane-based filtration process has been successfully employed in commercial water purification system to provide safe drinking-water. Analogous to 'Safe Drinking Water Foundation, Canada', Department of Science and Technology, Govt of India has implemented 'Nano Mission' to convey collaborative research and development projects for nanotechnology-based water purification process. Research groups have paid great attention to develop activated carbon-based nano-fiber membranes because activated carbon granules have been widely used in water filtration process over the past three decades. Carbon-based nanomaterials are highly porous material and due to present of different functional groups on single sheet surface, this nano-materials acts as versatile absorbent used for the removal of micro-pollutants, finer-contaminates, heavy metals such as lead, fluoride, chlorine and inorganic contaminants from water-bodies. Simultaneously its good reducing properties can do bacterial reduction, chlorine-toxic chemicals reduction, maintain healthy total dissolved solids (TDS) as per WHO and improve taste, odor of drinking water, provide point-of-use (POU) water-purification for domestic use.

Keeping in mind the multiple application to the society, in this chapter, it has been discussed on details about the fundamental principle of photo-catalytic technologies based on different semiconducting nano-materials. The possible mechanism of the photo-catalytic processes has also been explored through the detection of reactive oxygen species (ROS). How the proposed nano-materials have been commercialized for the sake of society based on their efficiencies and availability in different field such as in photo-active devices for decomposition and remediation of organic pollutants in waste-water treatment, hydrogen-fuel generation as alternative energy source from water splitting, self-cleaning of deposited dust or contaminant particles from surfaces, air purification have been deliberated. This chapter also presented detailed information on a photo-catalysis based number of commercial products such as solar panel, self-cleaning glasses, tiles, concrete, paint materials etc.

MATERIALS

Carbon Nanotube (CNT)

After diamond, graphite and fullerene, the most popular nano-sized carbon nanotube (CNT) has been discovered as 4th allotrope of Carbon. This is graphitic allotropes of carbon because the basic structure of CNT follows Graphite-like honeycomb lattice where the C-atoms are arranged in a regular hexagonal pattern (Figure 2a). When that single graphite sheet has inherent tendency to rolling up (curled-up) into a hollow cylindrical tube (Figure 2b) with diameter in the range of 2-25 nm (naturally occurring), they are called carbon nanotube. This rolled-up layer structure of sp^2 carbon atoms were first discovered by Radushkevich et. al. (1952) during hydrogenation reaction of CO over Fe catalyst. This type of unusual carbon structure formation is commonly known as multi-walled carbon nanotubes (MWCNTs), consists of several concentric tubes having outer diameter in the range of 2-25 nm, inner diameter, *ca.* 1-3 nm and interlayer distance about 0.35 nm. More useful single-walled carbon nanotubes (SWCNTs), consisting of only one cylindrical tube (diameter, *ca.* 0.4-5.0 nm) were first synthesized in 1993 by Iijima et. al. Similar to graphite sheet, CNT show extraordinary mechanical (tensile) strength, surprisingly high thermal and electrical conductivity which may arise from strongest in-plane carbon-carbon bonds arranged in a hexagonal lattice structure and delocalised Π-electrons in the sp^2 hybridised carbon atoms. However, CNTs are chemically more reactive than the flat graphite sheet and the chemical reactivity arises from the π-orbital mismatch caused by the surface curvature and in fact, the narrower CNTs are chemically more reactive. Again, depending on the cylindrical-tube-diameter, the metallic to semiconducting behaviour has been generated ie electrical conductivity is highly diameter dependent (Eg increases as the diameter decreases), an effect of quantum confinement effect. CNTs have been widely utilized in nano-electronics, integrated circuits of different electronic devices.

Figure 2. Schematic diagram a) basic structure and b) hollow-cylindrical structure of carbon nanotube (CNT)

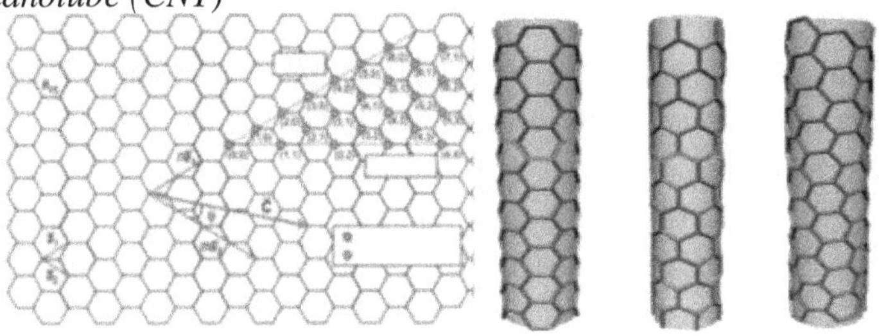

Graphene, Graphene Oxide (GO) and Reduced Graphene Oxide (rGO)

Figure 3. a) Photography of synthesized graphene nanomaterials, b) schematic diagram of Graphene structure

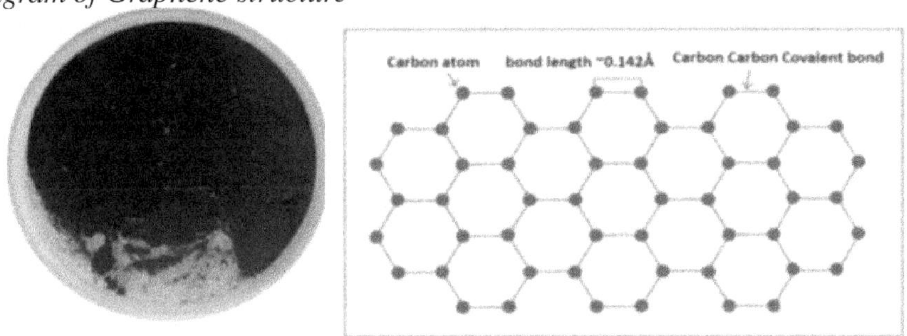

Another widespread carbon nanomaterials are graphene along with its derivatives graphene oxide (GO) and reduced graphene oxide (rGO). Graphene nano-materials is commonly known as wondering material owing to possessing of strong electrical, thermal conductivity and high mechanical strength. Graphene is harder than diamonds and lightweight but 200 times stronger than steel (Figure 3a). Graphene is also graphitic allotropes, follow graphite-like honeycomb lattice structure where the C-atoms are arranged with strongest in-plane carbon-carbon bonds in a regular hexagonal pattern (Figure 3b) to form a single layer structure so that it can be considered as single- layered Graphite material and can be easily synthesized through cleavage of Graphite. The hot news of first 'Graphene-sheet isolation with thickness of 18-108 nm' was first appeared in *Science* (Oct, 2004). A. Geim and K. Novoselor at the University of Manchester won Nobel Prize in Physics in 2010 for the ground breaking experiments on the 2D-graphene material. Graphene has the largest surface area i.e. surface-to-volume ratio due to presence of unique 2D flat-layered structure with either single-layer nano-sheet of sp^2 hybridised carbon atoms densely packed to form 2D hexagonal honeycomb-like structure or several coupled-layers of this hexagonal structure. As a result, large number of carbon atoms are exposed to its surface with large number of active sites which can promote highest catalyst-loading capacity for chemical functionalization to boost catalytic reactions. Moreover, Graphene is the most conductive material on Earth and has the potential to improve flexible touch screens, fast data transmission, fast charging in cell phone, electronic devices. Generally, carbon-based materials usually possess outstanding electrical conductivity due to presence of Π-conjugated electrons in the sp^2 hybridised domains which can acts as excellent electron acceptor, provides ultrafast electron transfer in nano-composite and help to enhance catalytic and light-induced catalytic performance of nano-composites. As a whole, the Π-conjugated structure can effectively accelerate the electron transfer, scavenge photo-generated electron from semiconductor surface, enhance photo-generated charge-carrier separation and electron mobility or migration efficiency. This delocalised Π-electron cloud also responsible for Π-plasmon resonances and extensively used in plasmonic (LSPR)-based sensing application in health or biomedical, environmental and food testing systems over the years (B.D. Gupta, 2019, S. Akhai, 2024).

Figure 4. Proposed mechanism of graphene-sheet preparation from Graphite through exfoliation

Graphene can be easily synthesized through cleavage of Graphite (Figure 4). Several mechanical cleavage processes i.e. Top-down break-down process of bulk graphite where pi-stacked layers (bound by Vander Waals forced) are broken to form thin layers through thermal exfoliation of bulk graphite. Ultra-sonication-assisted liquid exfoliation, where dispersing graphite in a proper liquid medium can produce graphene by sonication in a process known as Liquid Phase Exfoliation. Also, some bottom-up methods are available such as chemical vapor deposition, silicon evaporation, epitaxial growth, unzipping multiwalled carbon nanotubes, using an electric arc etc. to produce graphene but production of cheap high-quality graphene at industrial scale and large-areas need expensive equipment, the synthesis is performed at high temperatures, and it is rather limited.

Graphene oxide (GO) is one derivative of Graphene which has similar 2D hexagonal structure but the single-layer of carbon atoms is covalently functionalized with different oxygen containing functional groups. Instead, GO can be produced in desirable quantities rather rapidly and at low cost. The oxygen containing functional groups allow GO to be processed easily in dispersion and confer it both high colloidal stability in water and a unique set of mechanical, colloidal or optical properties. These GO properties can be further tuned using chemical engineering, most of these methods use strong oxidants, such as potassium permanganate, GO presents significant amounts of defects in its crystalline network. Defect engineering, specifically vacancy defects, Stone-Wales defects, and edge-plane terminations, plays a crucial role in tailoring the physicochemical properties of the carbon-based nanomaterials These defects significantly influence mechanical, electrical, and chemical characteristics, making them essential for diverse applications. Therefore, GO conductive properties are far below those of graphene, although its optical and mechanical properties suffer a lesser impact.

Figure 5. Overall illustration of graphene, graphene oxide and reduced-graphene oxide preparation

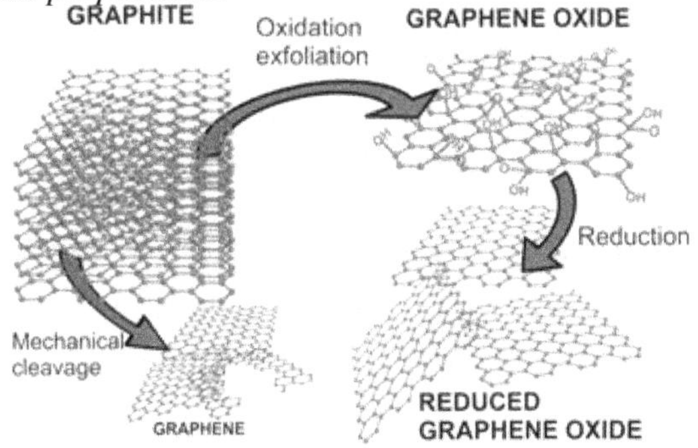

GO have been synthesized by chemical oxidation of natural Graphite or CNT by strong oxidants (Figure 5). Hydroxyl, epoxide, carbonyl groups have been introduced on the carbon basal plane and carboxyl group at edges through chemical exfoliation of bulk graphite flakes. As a result, this oxygen containing functional groups have been inserted in the sp^2 structure of Graphite layer disrupting sp^2 bonding network and create structural defect, generate distinctive properties, changes in mechanical strength, stability, increase hydrophilic behavior or dispersibility and as a whole reactivity of GO. The properties of GO can be further tuned using chemical engineering and have been applied in different field including solar cells, sensors, supercapacitors, generation of neurons, cellular migration, drug delivery, membranes, multifunctional gels, water purification and many more. In comparison to Graphene, GO show low electrical conductivity depends on the degree of oxidation due to disruption of conductive-carbon-network rather, GO exhibits semi-conducting behavior and widely utilized in light-induced catalytic process, solar-sell device. Moreover, O-groups in GO make them more bio-compatible material, more oxygen-riched structure, can easily bind with different bio-molecules, polymers, fluorescence molecules and have been widely utilized in different bio-medical applications such as DNA binding, nano-carrier drug delivery, fluorescence-probe-cell-imaging etc. Reduced graphene oxide (rGO) has easily been synthesized from GO through reduction by thermal, photo-thermal or chemical way under inert or reducing atmosphere. Several oxygen-functional groups, present in GO, can facilitate the reduction process and desirable quantities rGO have been obtained rather rapidly and at low-cost. High level of reduction is more preferable to researchers because it creates more structural defects, causes drastic changes in the structural properties, mechanical strength, stability, as a whole reactivity and reaching to very similar properties of pristine graphene with high carbon to oxygen ratio. During reduction process, their occurs elimination of oxygen containing compounds in GO structure and corresponding C/O ratio increases, and when C/O will be very high, almost zero oxygen contain, rGO properties closer to Pristine Graphene and can be regarded as chemically derived Graphene-like. As the large-scale Graphene production is limited in the sense of thermodynamic parameter, instead rGO is more widely utilized as excellent electrical-conductor, supercapacitor. Additionally, different O-functional groups present in RGO can increases hydrophilic behaviour and make them some-what soluble in water. In this way, RGO regain graphene-like properties with some aqueous dispersion, becoming a very good compromise between graphene and GO and have been extensively incorporated in innovative applications within a wide range of science and technology branches. Moreover, during reduction process, specific surface area also increases in such a way it possesses high energy storage capacity and has been utilised in anode and cathode material of rechargeable lithium-ion batteries. Unlike, metal-based materials, these metal-free counterparts also display excellent HER (Hydrogen Evolution Reaction), OER (Oxygen Evolution Reaction), mechanism in batteries due to their superior catalytic activity and stability. However, metal-free materials, particularly those based on carbon, are gaining prominence for their environmental friendliness and potential for lower cost. Sometimes, defect-engineering and surface modification have been easily incorporated to induce semiconducting, catalysis properties through changing the electronic structure of semiconductors (band-structure) and generate defective energy levels near the conduction-band (CB) or near the valence-band (VB) for better light absorption and electron-hole pair separation which can boost photo-chemical reactions (Figure 6). Using this photo-chemical mechanism, detection and sensing technology (working under UV Photo-activation) of various gases/molecules have been developed by Ellis *et al.* where combined RGO with polymeric graphitic carbon nitride composite have been utilised that can capable of sensing of oxygen in the range of 300-100000 ppm following photo-redox mechanism that appeared during oxygen reduction on RGO surface under UV irradiation that increases the conductance of the carbon nitride/RGO sensor.

Figure 6. Band potential position of reduced graphene-oxide (rGO) semiconductors against normal hydrogen electrode (NHE) as a reference

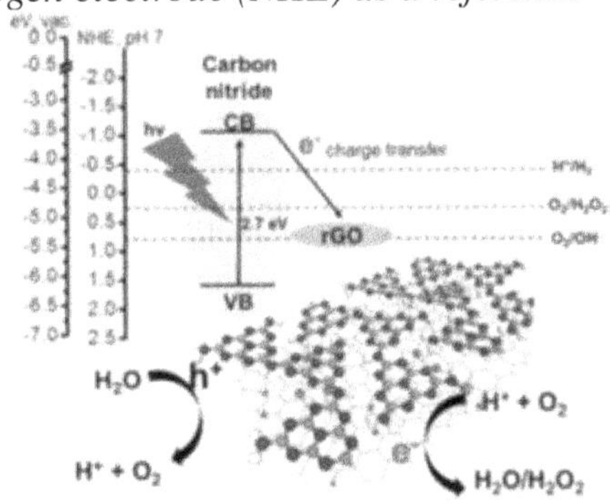

Carbon Quantum Dots (CQDs), Carbon Dots (CDs), Polymeric Graphite Carbon Nitride (g-C_3N_4) and Carbonized Polymer Dots (CPDs)

Carbon-based nanomaterials also possess unique optical properties. Carbon quantum dots (CQDs), carbon dots (CDs), polymeric graphite carbon nitride (g-C_3N_4), carbonized polymer dots (CPDs) with size below 10 nm show excellent photoluminescence behaviour (quantum confinement effect) and have been widely applied in luminous bodies. Within very small dimension, abundant sp^2/sp^3 Carbon sources promote integrity of pi-conjugation/ conjugated system and improve photoluminescence capabilities. These nanomaterials also possess large numbers of surface functional groups such as hydroxyl (-OH), carboxyl (-COOH), epoxy (-O-), amino ($-NH_2$) which are generated during synthesis process from nitrogen and oxygen containing raw materials and incorporated into the typical pi-electron network. These surface functional groups are mainly responsible for unique optical properties. Again, doping with heteroatom N, O, P, S, or Boron has been proven to be a practical approach to tuning the optical properties of the Carbon-nanomaterials. Yang's group had prepared red/ NIR emissive carbon quantum dots (CQDs), deep red emissive carbonized polymer dots (CPDs) where efficient conjugated pi-system have been generated with N-heterocycle and aromatic rings of polymer chain through delocalisation of electrons. As a result, they exhibit exceptional sunlight-harvesting capacities, and improve broad visible light absorption.

Carbon dots (CDs) also are promising materials and a new type of carbon-based nanomaterials which also hold wide application in daily life and everyday consumer products. CDs are primarily a quasi-spherical 0D carbon nanomaterial (below 10 nm), mainly consist of carbon, hydrogen and oxygen but majority of sp^2 or sp^3 bonded carbon clusters are present in CDs. they show excellent full-spectrum photoluminescence properties (Full-colour-light emitting) and have potential application in light-emitting diodes (LEDs) for solid-state lighting, have been used to fabricate a white-light emitting diode (white LEDs). CDs have been accidentally discovered by Xu et.al during arc-synthesis and electrophoretic purification of single walled carbon nanotubes, after that Sun et. al. successfully synthesized CDs through top-down breakdown of larger carbon structure graphite flakes and CNT (Figure 7).

Figure 7. General mechanism of carbon dots (CDs) preparation

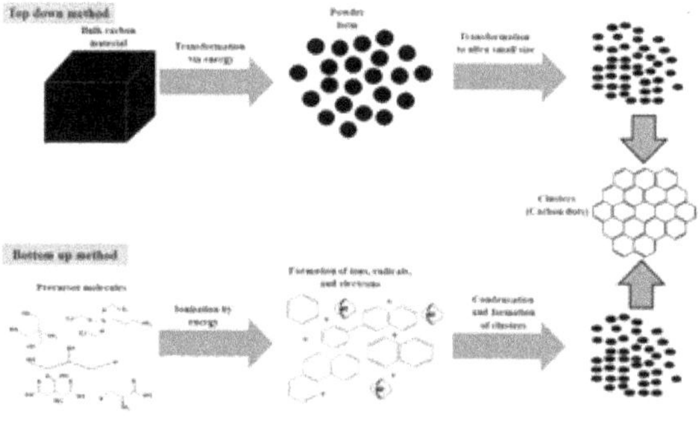

Owing to the small particle size (below 10 nm) of CDs, the photoluminescence properties mostly appear from quantum confinement effect but their chemical composition, variety of functional group that are joined to the carbon core structure, can enhance the photoluminescence capabilities. The colour of fluorescence is also affected by electronic band-gap transition of conjugated pi-domain of carbon core, surface defect states, local fluorophores attachment to CDs surface and element doping. Following surface engineering, Nguyen et. had reported full-colour-light-emitting CDs where colour of the fluorescence had been modulated using different surface functional group attachment to carbon core which possess multiple emission states (red, green and blue emission) (Figure 8a). The surface functional groups act as fluorophores, which are dominant factor of photoluminescence behaviour where they can change the PL emission state through increasing the internal HOMO energy level and as a whole decreasing the energy gap between HOMO and LUMO energy levels. In this way, CDs can emit light in the whole visible spectrum and can open the door for wide range of practical applications (bio-sensing, bio-imaging and bio-medicine) (A. Saravanan, 2020). When CDs have been synthesized from different O- and N-containing raw materials which typically contain different functional groups including hydroxyl (-OH), carboxyl (-COOH), carbonyl (-CO), amino (-NH_2) etc. even different amino groups, obtained surface-functionalised CDs with multiple emission states. Nitrogen of pyridinic N, pyrrolic N, graphitic N, amino N can affect the energy level position through incorporating various trapping states (Figure 8b) within the band structure of carbon core (caused by different chemical bonds in both carbon core or functional group attached to the CDs surface) and change the emission wavelength of the CDs. During synthesis process, increased C=O/ C=N-related surface states are main responsible for the photoluminescence shift of CDs from blue to red. This type of N-containing functional group incorporation is most easy and attracted great attention to researchers owing to their similar atomic size to the carbon atom and similar electronic structure, strong electronegativity and more availability of electrons. that can entire into and disrupt the pi-electron network of carbon cluster of CDs altering the electrical structure enhance the fluorescence performance improve these material's quantum yield (QY).

Figure 8. a) Full-spectrum photoluminescence properties and b) band-structure of carbon dots (CDs)

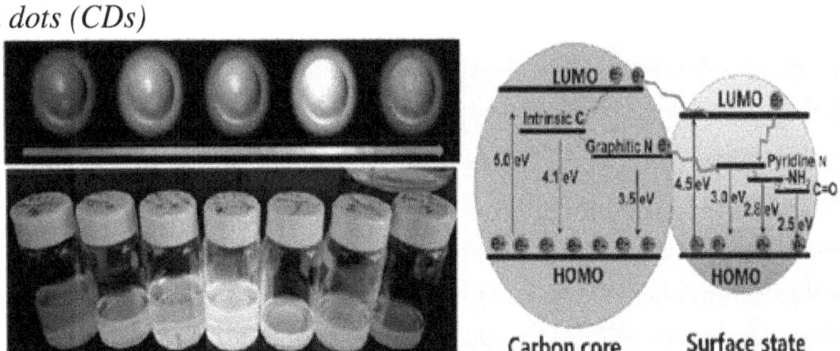

Recently, large-scale CDs have been very easily synthesized through eco-friendly way from almost any plant-based organic precursor compounds such as leaves, seeds, tree bark, vegetables, fruits, wheat flour, any bio-degradable waste or any chemical compounds that have C, H and O bonds present in them such as glucose, urea, ascorbic acids, citric acids etc. (Figure 9). In this way, variety of functional group have been incorporated and can be manipulated to change the properties specially photoluminescence capabilities.

Figure 9. Schematic diagram for synthesis process of carbon dots (CDs)

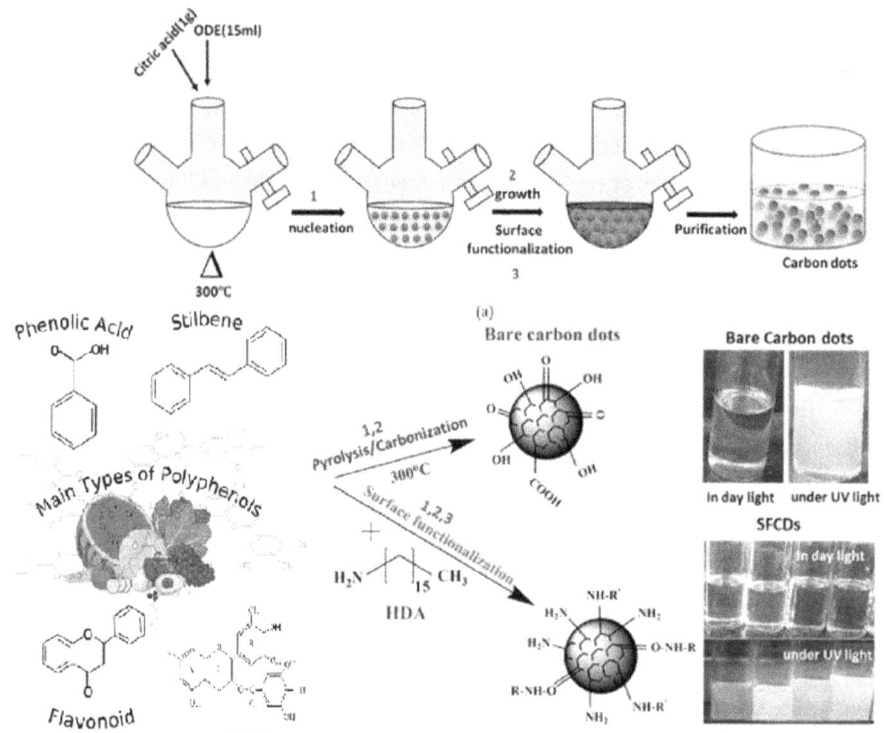

More O-containing functional group on the CDs can increase solubility in polar solvent and make them more water dispersibility. when the polarity of the solvent increases, the emission wavelength of CDs has shifted towards red region make them red-emitting carbon dots (RCDs). Again, the molecular emission states of CDs can also be varied through increasing surface defect (Defect engineering) and degree of oxidation. Higher the degree of surface oxidation, their occurs more surface defects, oxygen vacancy and decreasing the photon transition band gap, resulting in the red-shift of emission wavelength, triggering one-photon and NIR induced (800-1200 nm) two-photon high-colour-purity red fluorescence emission. Sometimes, CD's high degree of crystallinity promotes Π-Π* and n-Π* transition and results exceptional solar-light harvesting capacities, tuneable photo-luminescence, photo-dynamic effect and effective photo-exited electron transfer properties. N-doped CDs (heteroatom such as P or S or N or Boron are doped with CDs) show outstanding photon up-conversion (UC properties) ability to convert long wavelength (550-850 nm) NIR-light to short wavelength (440-500 nm) UV-Vis light for enhancing full spectral-range (320-800 nm) solar energy harvesting and full-spectrum-solar-light driven application (B.S. Richards, 2021). Now-a-days, red/NIR emissive CDs have been widely used in cancer treatment which can enhance photo-catalytic response towards reactive oxygen species (ROS) generation and destroy cancerous cell's structure. Simultaneously, the deep penetration power and photo-thermal effect of NIR-light raises the temperature of tumors and encourage photo-thermal therapy (PTT), photo-dynamic therapy (PDT) where light-to-heat conversion efficiency kill cancer cells.

polymeric graphite carbon nitride (g-C_3N_4) is another commonly available carbon-based metal-free nano-materials made of only carbon and nitrogen. Through simple solvent free thermal poly-condensation reaction, the innovative organic conjugated-polymer structures have been created in such a way that strong aromatic C-N heterocyclic ring and sp^2 hybridisation gives rise to a π-conjugated unique electronic energy band structure with a low band gap energy of 2.7-2.8 eV corresponding to visible light-harvesting. In this way, g-C_3N_4 becomes well-known organic semiconductor (n-type), metal-free Π-conjugated semiconductor and exhibits very similar efficiency for conducting visible and solar-light-driven photo-catalysis and photo-chemical reactions compared to metal-based semiconducting nano-materials. Wang et. al first reported that this unique electronic structure shows excellent response in the visible-light in the range up to 460 nm and exhibits outstanding photo-catalytic performance which can be regarded as next-generation innovative photo-catalyst materials. Graphitic phase carbon nitrides have two major sub-structure, Heptazine-based polymer layer (Heptazine-based g-C_3N_4 HCN) and Triazine-based polymer layer (Triazine-based g-C_3N_4 TCN). HCN and TCN form a heterojunction, where charge transfers from HCN to TCN can improve photocatalytic performance. Similar to graphene, the parallel planes (The graphitic planes containing carbon, nitrogen) are constructed from basic tectonic units, tris-S-triazine rings connected with planner amino groups (Figure 10). Sometimes, planar tri-angular core groups are created from three fused triazine rings with three hydrogens at the corners. After that, thermal poly-condensation can generate polymeric graphite carbon nitride (g-C_3N_4) connected via this tris-triazine-based patterns. The crystal sheet of each g-C_3N_4 layer is stacked on top of each other and bound by Vander Waals forced forming various architectures and adjustable electronic structures. Thermal exfoliation or thermal etching synthesis process can do finer adjustment of that electronic and optical properties through gradual decreasing the thickness of the material and band gap also. In this way ultrathin g-C_3N_4 nanosheet has been synthesized as next-generation photocatalyst with a suitable full spectral-range (320-800 nm) light-response.

Figure 10. Chemical and band-structure of polymeric graphite carbon nitride (g-C_3N_4)

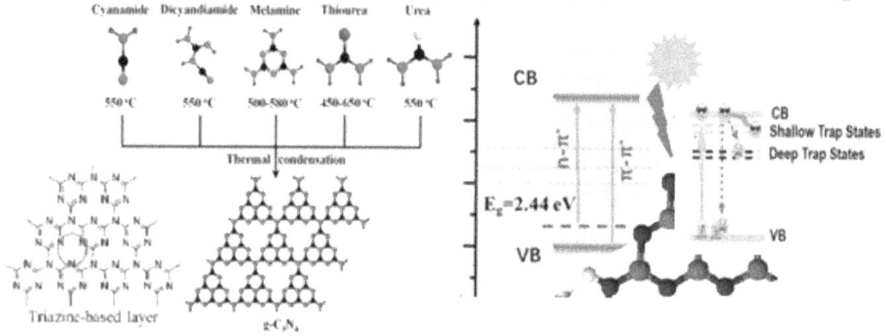

Occasionally, polymeric carbon nitride (PCN) has been synthesized from other nitrogen-riched precursor compounds, Urea, cyanamide, dicyanodiamide, Melamine (an abundant Amino (-NH_2) functionalised), through polycondensation at elevated temperature 350-575°C. (M. Weers, 2024). Melamine, Melem, melem-tetramer (Melon) becomes very popular to recent research and exhibits outstanding photocatalytic performances owing to possessing abundant amino (-NH_2) surface functional group. Again, Cyanamide is a liquid precursor, allow highly ordered way of polymerisation, less defect containing and most important, it can lead to the formation of variety of nano-structures, fine tuning of properties. Very recently discovered extra-ordinary 'Dark Photo-catalysis' (V.W. Lau, 2017) has been attributed by Cyanamide-functionalized Carbon-nitride where so beautiful structural network can form ultra-long-lived radicals (photo-induced states) when exposed to solar-light irradiation in the presence of an electron donor. The radicals can store their trapped electrons for long-time and release them on demand even in the dark, when light goes out, to participate in photo-chemical reactions. In this way, photocatalytic activity has been increased, where activity is restricted to only daytime or be affected due to fluctuations in solar radiation due to changes in weather conditions. According to Lau et. al., a fantastic melon derived C_3N_4 can produce long-lived 'trapped' electrons as reducing equivalent which possess lifetime from tens of minutes to hours after light-irradiation, store the electrons even up to hours and release them on demand when coupled to an electrocatalyst to complete time-delayed reactions.

Again, amino (-NH_2) group riched-C_3N_4 becomes more useful to the society because excellent electron withdrawing effect of -NH_2 group can improve the electronic structure and photocatalytic performances. Lone pair electrons of Nitrogen can participate in electronic band structure and create additional lone-pair electronic states in between pi-bonding electronic HOMO and LUMO (Figure 10) which play important role in narrowing the band gap energy and extending the light absorption range from 460 nm to 660 nm (Y. Chen, 2024). Again, n-π* electron transition can effectively reduce charge re-combination by increasing the repulsion among lone-pair electrons owing to their reduced inter-distance. In this way, in such type of engineered C_3N_4 nanosheets, the energetically more favourable n-π* electron transition can increase life-time of excited state through suppressing charge recombination, prolong shallow electron trapping lifetime (lifetimes: 565.8ps) and resulting in significantly enhance photo-activity. Recently, researchers have synthesized nitrogen-rich polymeric C3N4 (through thermolysis of N-rich molecular precursor) in which, abundant lone pair electrons and large number of n-π* electronic transition have been utilized to increase photocatalytic performance through overcoming weak visible light harvesting and high charge re-combination.

Application of Carbon-Based Metal-Free Nano-Materials in Daily Life and Everyday Consumer Products

In the field of agriculture, food-safety, water-purification, pharmacy, cosmetics, personal care etc., nano-sized man-made engineered materials have been extensively utilized which exhibits unique properties that differ from natural bulk materials. Several metal-free carbon-based Graphene, GO, CNT, CQDs NPs etc. and their composite have been successfully explored and exhibits outstanding performance towards some catalysis-chemical reactions for multi-functional photo-catalysis application. In the field of environmental pollution, the nano-photocatalysis technology provides excellent support through the degradation of various organic pollutants under solar and visible light illumination and has been successfully utilized in waste-water treatment, agricultural pesticide-residue treatment, air purification (A. K. Dutta, 2023). Photocatalytic nano-fiber membrane-based filtration process has been successfully employed in commercial water purification system to provide safe drinking-water. Integration of such nano-photocatalysis mechanism into different consumer products such as UV-protective textiles, personal care products; self-cleaning building construction materials, roof-tiles, glazed ceramic tiles, concrete, paint, transports; anti-bacterial medical textiles, clothing, food-packaging; anti-fogging, anti-reflective glasses, mirrors, solar panel etc. have been successfully attributed for the sake of society. Of late, to utilize efficiently solar-light, researcher have now paid great attention on full-spectrum-solar-light driven photocatalytic system ie. full spectral-range (320-800 nm) light-driven photocatalysis and have been achieved with the help of metal free carbon-based quantum dots (CQDs), carbon dots (CDs), Graphene, GO, CNT which can do light-harvesting over very broad spectral range from UV to NIR (X. Wang, 2017). Again, Graphene has the largest surface area i.e. surface-to-volume ratio due to presence of unique 2D flat-layered structure, large number of carbon atoms are exposed to its surface with large number of active sites which can promote highest catalyst-loading capacity for chemical functionalization to boost photocatalytic reactions. Moreover, carbon-based materials usually possess outstanding electrical conductivity due to presence of Π-conjugated electrons in the sp2 hybridized domains which can acts as excellent electron acceptor and transporter in nano-composite and help to enhance photo-catalytic performance of nano-composites. CNT, Carbon quantum dots (CQDs) acts as photo-generated electron trap to suppress charge carrier re-combination, longer their lifetime and thus facilitating the effective utilization of that photo-induced charge-carriers in nano-photocatalysis process. Unlike traditional macroscopic carbon material, carbon quantum dots (CQDs), polymeric graphite carbon nitride (g-C3N4), carbonized polymer dots (CPDs) etc. with a size below 10 nm, itself can directly absorb light in the UV-Vis region and have been attributed as interesting metal-free organic semiconductor photo-catalyst.

Now a days, several air-purifier devices have been developed to be utilized in daily life based on nano-photocatalysis technology, where air pollutants are converted to harmless substances in the presence of nano-catalyst under solar or indoor light irradiation. In these devices, photo-catalytic oxidation (PCO) has been incorporated to purify air where nano-catalyst material can oxidize the gaseous compounds such as carbon monoxide, nitrogen oxides, volatile organic compounds (VOCs), other inorganic gases etc. that come into contact with surface of them. Very recent, numerous carbon-based material such as 2D graphene, graphene oxide (GO), rGO nano-sheet, CNT, Carbon QDs, polymeric graphite carbon nitride (g-C3N4) supported semiconductor composites have been fabricated to enhance PCO technology and used in air purification technology that utilize light energy to breakdown pollutants, bacteria and to remove VOCs (VOCs mineralization) and improve air quality. Again, some wall paints mixed with photo-catalytic materials have been widely applied for improving the air quality through degradation of volatile organic and inorganic pollutants present in air or deposited on paint surface.

Now-a-days, with tremendous development of science and technology, so many research works have been devoted on innovation and development of textile material with more features like more comfort, durable, hygienic, anti-odor, anti-microbial, anti-static, breathability, improved softness, better tear strength, water/spill repellence and wrinkle-resistance but without altering the inherent properties of the textiles including flexibility and washability. Nano-carbon embedded textile clothing, when exposed to ultraviolet or visible light, the photo-catalytic characteristic of the semi-conducting nanoparticle allows it to act as self-cleaning agent by photo-catalytic degradation of various color stains. Even, photo-catalytic self-cleaning ability have been retained after several cycles of home-washing. Silver-based nano-materials, which have well-known anti-microbial and photo-sensitizer property, have been immobilized to develop sun-light driven cleaning and anti-bacterial characteristic and fabrics of socks, inner-wears etc. embedded with Nano-carbon and silver-hetero-structure can protect against odor and infection. Another nano-textile research-works develop nano-fibre with diameters in the few nm range, which has been directly use to wove a cloth. Nano-fibers have been generated from different nano-sized polymeric materials. Textile-garments using nano-fiber yarn can enhance comfort, breathability, moisture-management, anti-bacterial properties. Sporting goods, swimming suit, even PPE kits, face masks, surgical goods etc. using nano-fiber yarn have been benefited with photocatalytic self-cleaning.

Same mechanism has been applied on nano-thin-films-based advanced coating which are now increasingly used as protective coating in various applications such as in exterior building construction materials, tent-materials, roof-tiles, glazed ceramic tiles, concrete, transport vehicles etc. and can prevent the accumulation of deposited dust, dirt or contaminant particles on its surface, keeping them clean. Nano-photocatalytic self-cleaning Nano-carbon-based surface coatings also utilise sunlight or indoor light to break-down deposited greasy and viscous sediment that can be washed away with rain/water. Moreover, Surface coating on glasses, mirrors, camera lenses, touchscreens etc. assistance with mano-catalysts, also exhibits anti-reflecting, anti-fogging properties ensuing a perfect view. Based on solar photo-catalysis, self-cleaning surfaces have widely been used in photovoltaic or solar panels with improved transmittance of natural sunlight and reflectance properties, since these materials are exposed to sunlight and natural rainfall.

Using these self-cleaning surface coatings phenomenon, photocatalytic disinfection has broadly been applied to protect healthcare-associated infections in hospitals, public areas like washrooms, transport stations, cleaning areas in the food industry etc. Similar to well-known anti-microbial Silver-based medicines, carbon-based nano-materials can also degrade the cell-wall and cytoplasmic membrane of micro-organisms and viruses, that survive on surfaces for a long time, through the production of reactive oxygen species (ROS) such as hydroxyl radicals and hydrogen peroxide. Recently, single-walled carbon nanotubes (SWCNTs) have been demonstrated as strong anti-microbial agent, can entire bacterial cell wall and exhibits significant impact on bacterial cell-membrane.

During development of Cancer treatment, heterogeneous nano-photocatalysis have been employed to provide advantage of catalysis chemical reaction on growing tumor cells and to accelerate depletion of metabolism of cancer cells. Owing to extremely smaller in size, nano-materials can be easily absorbed by the cells in biological system and easily travel into blood vessels, organs within the body. Following nano-photocatalytic mechanism, nano-materials can encourage photodynamic therapy (PDT) where, they act as photo-sensitizer and can generate reactive oxygen species (ROS) upon exposure to light. That ROS cause apoptotic and necrotic cell death and trigger oxidative damage of the amino acids of protein, fatty acids in lipids and the nucleo-bases in DNA of tumor cells. Now-a-days, red-light emitting carbon nano-particles with high fluorescence quantum yield, NIR-light harvesting N-doped carbon dots have been used in cancer treatment. In this case, bio-nanomedicine and photo-catalysis have been merged to provide advantage of both bio-imaging marker and photo-dynamic agents.

Now-a-days, photocatalytic nano-fiber based membranes often used in the removal of micro-pollutants, finer-contaminates from water-bodies. Analogous to 'Safe Drinking Water Foundation, Canada', Department of Science and Technology, Govt of India has implemented 'Nano Mission' to convey collaborative research and development projects for nanotechnology-based water purification process. When nano-photocatalysis technology is incorporated into membrane-based filtration process, ie. filtration-photocatalysis integrated method both separation and decomposition of unwanted pollutants have been carried out simultaneously and provide high quality drinking water to millions of people in developing countries. Very similar to reverse osmosis (RO) purification, nano-fiber membrane-based filtration process has been successfully employed in commercial water purification system and has been extensively used for softening of water, removal of both physical, chemical and biological contaminants leading to the formation of safe drinking-water. Incorporation of photocatalytic activated carbon granules or Graphene into this nano-fiber membranes can enhance photo-catalytic disinfection through effective killing of wide range of micro-organisms such as virus, bacteria, parasites, algae, pathogens and provide improved water filtration. Some research groups have paid great attention to develop activated carbon-based nano-fiber membranes because activated carbon granules have been widely used in water filtration process over the past three decades. Carbon-based nanomaterials are highly porous material and due to present of different functional groups on single sheet surface, this nano-materials acts as versatile absorbent used for the removal of heavy metals, lead, fluoride, chlorine and inorganic contaminants. Simultaneously its good reducing properties can do bacterial reduction, chlorine-toxic chemicals reduction and improve taste, odor of drinking water, provide point-of-use (POU) water-purification for domestic use.

CONCLUSION

From the present study of the status of carbon-researches around the Globe, it can be concluded that carbon-based metal-free nano-materials got enormous importance in our modern life. Such innovative materials have successfully been incorporated into different industrial sectors and everyday consumer products such as agriculture, food-safety, medicine, pharmacy, cosmetics, personal care product etc. to revolutionize and improve product efficiencies and performances across various domains. The proposed nano-materials have also successfully been commercialized for the sake of society based on their efficiencies and availability in different field such as in photo-active devices for decomposition and remediation of organic pollutants in waste-water treatment, hydrogen-fuel generation as alternative energy source from water splitting, self-cleaning of deposited dust or contaminant particles from surfaces, air purification and bio-medical applications.

ACKNOWLEDGMENT

The Author is grateful to Prof. Bibhutosh Adhikary, Department of Chemistry, Indian Institute of Engineering Science and Technology (IIEST), Shibpur, for helpful discussions. The Author is indebted to UGC, India for financial support through the MRP [PSW–066/15–16 (ERO)]. The author also acknowledges the MHRD (India) for providing instrumental facilities to the Department of Chemistry, IIEST Shibpur and RUSA Scheme of the Department of Higher Education, Government of West Bengal, for providing instrumental facilities to the Department of Chemistry, Bangabasi Morning College, Kolkata, India.

REFERENCES

Abo-Neima, S. E., Motaweh, H. A., & Elsehly, E. M. (2020). Antimicrobial activity of functionalised carbon nanotubes against pathogenic microorganisms. *IET Nanobiotechnology / IET*, *14*(6), 457–464. DOI: 10.1049/iet-nbt.2019.0342 PMID: 32755954

Akhai, S., Walia, J., Bhullar, B. S., & Wadhwa, A. S. (2024). Advances in Bandgap and Conductivity Engineering of Graphene-Based Semiconductors. In Carbon-Based Materials and Environmental Remediation: Graphene, Biochar, and More (pp. 211-228). IGI Global Scientific Publishing. DOI: 10.4018/979-8-3693-8257-8.ch007

Azizi-Lalabadi, M., Hashemi, H., Feng, J., & Jafari, S. M. (2020). Carbon nanomaterials against pathogens; the antimicrobial activity of carbon nanotubes, graphene/graphene oxide, fullerenes, and their nanocomposites. *Advances in Colloid and Interface Science*, *284*, 102250–102267. DOI: 10.1016/j.cis.2020.102250 PMID: 32966964

Banerjee, R., Pal, A., Ghosh, D., Ghosh, A. B., Nandi, M., & Biswas, P. (2021). Improved photocurrent response, photostability and photocatalytic hydrogen generation ability of CdS nanoparticles in presence of mesoporous carbon. *Materials Research Bulletin*, *134*, 111085–111098. DOI: 10.1016/j.materresbull.2020.111085

Cao, S., & Yu, J. (2014). g-C_3N_4-Based Photocatalysts for Hydrogen Generation. *The Journal of Physical Chemistry Letters*, *5*(12), 2101–2107. DOI: 10.1021/jz500546b PMID: 26270499

Chen, Y., Cheng, Y., Liu, Y., Wang, Y., Qu, Y., Jiang, D., Qin, Z., & Yuan, Y. (2024). Achieving long-lived shallow trapping states in carbon nitride through the n-π*electronic transition for enhanced photocatalytic hydrogen generation. *Applied Catalysis B: Environmental, 342*, 123453–123467. DOI: 10.1016/j.apcatb.2023.123453

Dutta, A. K. (2023). Metal Chalcogenides-Based Multifunctional Nanomaterials for Solar-Light-Driven Photo-Catalysis and Electro-Chemical Sensing. In Kumar, A. (Ed.), *Innovative Multifunctional Nanomaterial for Photocatalysis, Sensing, and Imaging* (pp. 1–16). IGI Global Scientific Publishing., DOI: 10.4018/978-1-6684-8743-3.ch001

Dutta, A. K., Maji, S. K., & Adhikary, B. (2014). γ–Fe2O3 nanoparticles: An easily recoverable effective photo-catalyst for the degradation of rose bengal and methylene blue dyes in the waste-water treatment plant. *Materials Research Bulletin, 49*, 28–34. DOI: 10.1016/j.materresbull.2013.08.024

Dutta, A. K., Maji, S. K., Srivastava, D. N., Mondal, A., Biswas, P., Paul, P., & Adhikary, B. (2012). Synthesis of FeS and FeSe Nanoparticles from a Single Source Precursor: A Study of Their Photocatalytic Activity, Peroxidase-Like Behavior, and Electrochemical Sensing of H2O2. *ACS Applied Materials & Interfaces, 4*(4), 1919–1927. DOI: 10.1021/am300408r PMID: 22448782

Gupta, B. D., Pathak, A., & Semwal, V. (2019). Carbon-Based Nanomaterials for Plasmonic Sensors: A Review. *Sensors (Basel), 19*(16), 3536. DOI: 10.3390/s19163536 PMID: 31412590

Han, C., Kundu, B. K., Liang, Y., & Sun, Y. (2024). Near-Infrared Light-Driven Photocatalysis with an Emphasis on Two-Photon Excitation: Concepts, Materials, and Applications. *Advanced Materials, 36*(5), 2307759–2307772. DOI: 10.1002/adma.202307759 PMID: 37703435

Helbig, W. A. (1946). Activated carbon. *Journal of Chemical Education, 23*(2), 98–104. DOI: 10.1021/ed023p98

Iijima, S., & Ichihashi, T. (1993). Single-shell carbon nanotubes of 1-nm diameter. *Nature, 363*(6430), 603–605. DOI: 10.1038/363603a0

Kang, S., Pinault, M., Pfefferle, L. D., & Elimelech, M. (2007). Single-walled carbon nanotubes exhibit strong antimicrobial activity. *Langmuir, 23*(17), 8670–8673. DOI: 10.1021/la701067r PMID: 17658863

Lau, V. W., Klose, D., Kasap, H., Podjaski, F., Pignié, M. C., Reisner, E., Jeschke, G., & Lotsch, B. V. (2017). Dark Photocatalysis: Storage of Solar Energy in Carbon Nitride for Time-Delayed Hydrogen Generation. *Angewandte Chemie International Edition in English*, *56*(2), 510–514. DOI: 10.1002/anie.201608553 PMID: 27930846

Mondal, P., Satra, J., Srivastava, D. N., Bhadu, G. R., & Adhikary, B. (2021). $Pd^{\delta+}$-Mediated Surface Engineering of AgMnO4 Nanorods as Advanced Bifunctional Electrocatalysts for Highly Efficient Water Electrolysis. *ACS Catalysis*, *11*(6), 3687–3703. DOI: 10.1021/acscatal.0c05638

Morstein, C. E., & Dienwiebel, M. (2021). Graphite lubrication mechanisms under high mechanical load. *Wear*, *477*, 203794. DOI: 10.1016/j.wear.2021.203794

Nawaz, F., Ali, M., Ahmad, S., Yong, Y., Rahman, S., Naseem, M., Hussain, S., Razzaq, A., Khan, A., Ali, F., Balushi, R., Al-Hinaai, M., & Ali, N. (2024). Carbon based nanocomposites, surface functionalization as a promising material for VOCs (volatile organic compounds) treatment. *Chemosphere*, *364*, 143014–143025. DOI: 10.1016/j.chemosphere.2024.143014 PMID: 39121955

Radushkevich, L. V. (1952). On the Structure of Carbon Formed by Thermal Decomposition of Carbon Oxide at an Iron Contact. *J. Phys. Chem. Russ.*, *26*, 88–95.

Richards, B. S., Hudry, D., Busko, D., Turshatov, A., & Howard, I. A. (2021). Photon Upconversion for Photovoltaics and Photocatalysis: A Critical Review. *Chemical Reviews*, *121*(15), 9165–9195. DOI: 10.1021/acs.chemrev.1c00034 PMID: 34327987

Saravanan, A., Maruthapandi, M., Das, P., Ganguly, S., Margel, S., Luong, J. H. T., & Gedanken, A. (2020). Applications of N-Doped Carbon Dots as Antimicrobial Agents, Antibiotic Carriers, and Selective Fluorescent Probes for Nitro Explosives. *ACS Applied Bio Materials*, *3*(11), 8023–8031. DOI: 10.1021/acsabm.0c01104 PMID: 35019541

Serge-Correales, Y. E., Ullah, S., Ferreira-Neto, E. P., Rojas-Mantilla, H. D., Hazra, C., & Ribeiro, S. J. L. A. (2022). UV-visible-NIR active smart photocatalytic system based on NaYbF4:Tm3+ upconverting particles and Ag_3PO_4/H_2O_2 for photocatalytic processes under light on/light off conditions. *Materials Advances*, *3*(6), 2706–2715. DOI: 10.1039/D1MA01028J

Verma, R., Akhai, S., & Wadhwa, A. S. (2024). Use of smart materials in physiotherapy. In *Revolutionizing healthcare treatment with sensor technology* (pp. 300–319). IGI Global Scientific Publishing. DOI: 10.4018/979-8-3693-2762-3.ch019

Wang, F., Chen, Y. H., Liuand, C. Y., & Ma, D. (2011). White light-emitting devices based on carbon dots' electroluminescence. *Chemical Communications (Cambridge)*, *47*(12), 3502–3504. DOI: 10.1039/c0cc05391k PMID: 21311781

Wang, X., Wang, F., Sang, Y., & Liu, H. (2017). Full-Spectrum Solar-Light-Activated Photocatalysts for Light–Chemical Energy Conversion. *Advanced Energy Materials*, *7*(23), 1700473–1700489. DOI: 10.1002/aenm.201700473

Weers, M., Seggern, A. R., Vocke, H., Taffa, D. H., & Wark, M. (2024). Two Ways to more NH_2-Groups: Formation of Polymeric Carbon Nitride via Melem Tetramer Nano Sheets or Supramolecular Assembly of Melamine and Cyanuric Acid for Applications as Photocatalyst. *ACS Applied Nano Materials*, *7*(1), 1402–1415. DOI: 10.1021/acsanm.3c05516

Yuan, L., Dai, J., Fan, X., Song, T., Tao, Y. T., Wang, K., Xu, Z., Zhang, J., Bai, X., Lu, P., Chen, J., Zhou, J., & Wang, Z. L. (2011). Self-Cleaning Flexible Infrared Nanosensor Based on Carbon Nanoparticles. *ACS Nano*, *5*(5), 4007–4013. DOI: 10.1021/nn200571q PMID: 21466218

Zhao, J., & Yang, X. (2003). Photocatalytic oxidation for indoor air purification: A literature review. *Building and Environment*, *38*(5), 645–654. DOI: 10.1016/S0360-1323(02)00212-3

Chapter 3 High-Performance Aluminium Alloys for Aircraft Structures: Advances in Properties, Processing, and Manufacturing

Shalom Akhai
https://orcid.org/0000-0002-7533-457X
Maharishi Markandeshwar Engineering College, Maharishi Markandeshwar, Ambala, India

Syed Mohd Arif
https://orcid.org/0000-0002-0821-317X
Chandigarh Group of Colleges Jhanjeri, CEC-J, India

Rohan Dadhwal
https://orcid.org/0009-0001-0832-534X
Chandigarh Group of Colleges Jhanjeri, CSB-J, Mohali, India

ABSTRACT

This chapter explores aluminium alloys' vital role in next-gen defense systems, focusing on their evolution, processing, and aerospace/military applications. It traces their development from duralumin to advanced Al-Li and nanostructured variants, emphasizing strength-to-weight efficiency and corrosion resistance. Key processing innovations like thermomechanical treatments, additive manufacturing, and friction stir welding enable lightweight, high-performance components, such as AlSi10Mg brackets in the F-35 and GLARE® fuselage panels in the A380. Case studies highlight alloys like AA 7085 and Al-Li 2195 in military aircraft and space-

DOI: 10.4018/979-8-3373-0933-0.ch003

craft, enhancing fuel efficiency and durability. Challenges like galvanic corrosion, scandium scarcity, and energy-intensive production are analyzed, alongside solutions like AI-driven alloy design and closed-loop recycling. The chapter underscores sustainable practices, ensuring aluminium alloys remain essential for defense while advancing eco-friendly production.

INTRODUCTION

Chapter Highlights

- **Comprehensive Evolution of Aluminium Alloys in Aerospace -** The chapter traces the historical advancement of aluminium alloys from early aviation to next-generation aerospace platforms, demonstrating how evolving compositions like Al-Li and Al-Mg-Sc alloys have enhanced strength-to-weight ratios and fatigue resistance.
- **Detailed Characterization of Alloy Systems -** In-depth analysis of 2xxx, 7xxx, and Al-Li alloy systems is presented, with focus on their mechanical properties, corrosion behavior, and microstructural transformations under various thermomechanical and heat treatment conditions.
- **Integration of Nanostructured and Scandium-Based Alloys -** Nanostructuring via severe plastic deformation and scandium addition are shown to significantly improve grain refinement, weldability, and fatigue life—making them suitable for critical load-bearing structures.
- **Advanced Processing Techniques for Performance Optimization -** The paper discusses innovative processes such as friction stir welding (FSW), additive manufacturing (AM), and retrogression and re-aging (RRA) that enhance joint integrity and component performance in aerospace-grade aluminium alloys.
- **Manufacturing Innovations with Emphasis on Digitalization -** Emphasis is placed on precision CNC machining, hybrid aluminium-composite structures, and digital manufacturing (AI-driven alloy design, digital twins) that enable lightweight, fuel-efficient aircraft with improved production efficiency.
- **Application-Oriented Case Studies in Modern Aircraft -** Real-world case studies (Airbus A380, Boeing 777X, F-35 Lightning II) illustrate the strategic implementation of advanced aluminium alloys in fuselage panels, wing ribs, and structural brackets to achieve weight savings and durability.
- **Property Enhancement through Alloy Design and Coatings -** Technical evaluation of alloying strategies and surface engineering (e.g., cerium-based

coatings, MAO) highlights improvements in corrosion resistance, thermal stability, and long-term performance under harsh operational environments.

- **Sustainability and Recycling Challenges Addressed -** The chapter explores closed-loop recycling, low-carbon smelting, and biodegradable coating technologies to address environmental concerns associated with traditional aluminium alloy production.
- **Vision for Future Aluminium Alloys in Aerospace -** Insights into machine-learning-assisted alloy discovery, multifunctional material systems, and embedded sensor-enabled smart components present a roadmap for next-gen sustainable and intelligent aircraft

Figure 1. Graphical abstract

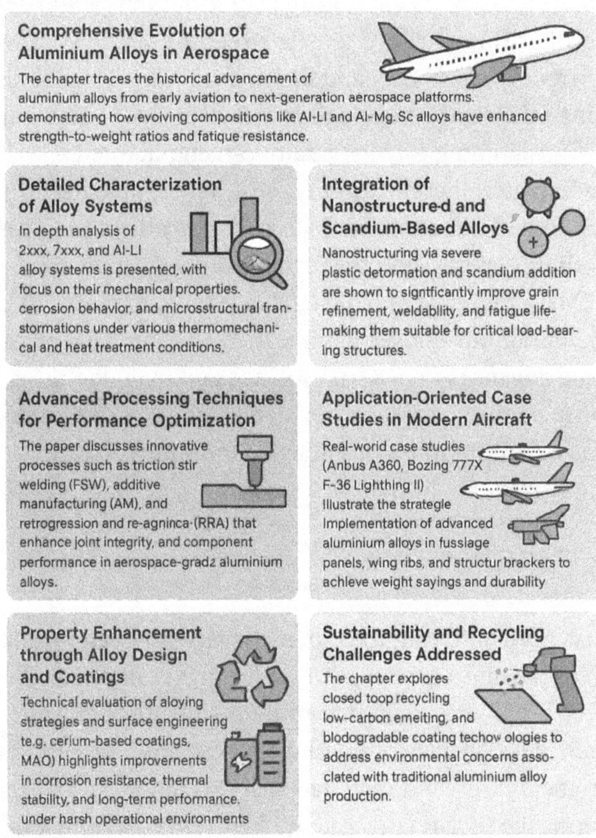

Aluminium alloys are crucial in aerospace engineering due to their strength-to-weight ratio, corrosion resistance, and adaptability. Recent advancements in alloy development, processing techniques, and manufacturing innovations have significantly impacted aircraft performance. Emerging trends include lithium-enhanced alloys, additive manufacturing, and hybrid structures (Kumar et al., 2024a; Kumar et al., 2024b; Khang & Akhai, 2025). The evolution of aluminium alloys from WWII-era 7075-T6 to modern Al-Li alloys highlights their enduring relevance in modern aviation. Traditional alloys dominate airframes for their balance of strength and workability, while advanced alloys like Lithium reduce density by 3-10% and enhance weldability and strength. Nanostructured alloys refine grain through severe plastic deformation (SPD) for improved mechanical properties and fatigue life. Advancements in processing techniques include thermomechanical processing, heat treatment, additive manufacturing, and friction stir welding. Manufacturing innovations include precision machining, hybrid structures, automation, and digital manufacturing. Al-Li alloys achieve tensile strengths >600 MPa with 5–7% density reduction, corrosion resistance, and thermal stability. Case studies include the Airbus A380 using GLARE for fuselage panels, Boeing 777X using AA 7085 for wing ribs, and F-35 Lightning II using additive-manufactured AlSi10Mg brackets. Challenges include material compatibility, cost, environmental impact, and recycling innovations. Future directions include next-gen alloys, smart manufacturing, closed-loop recycling systems, and low-carbon processing methods (Gialanella & Malandruccolo, 2020; Florea et al., 2024; Thavasilingam et al., 2025).

In conclusion, aluminium alloys continue to evolve, meeting aerospace demands through synergistic advances in material science and manufacturing. As the industry prioritizes sustainability, ongoing innovation in alloy design and eco-friendly processes will ensure aluminium's prominence in future aircraft.

HISTORICAL PERSPECTIVE

Aluminium has been a driving force in aerospace innovation since WWII, with post-war research focusing on strength and fatigue resistance. The evolution of alloys, such as 7055-T77, has been driven by a shift towards fuel efficiency and emission reduction. The superior strength-to-weight ratio of aluminium has led to advancements in metallurgy and manufacturing techniques, transforming aircraft design and performance (Thareja & Akhai, 2017; Sharma & Akhai, 2019; Akhai et al., 2021).

Early Use of Aluminium in Aviation

The Wright brothers' historic flight in 1903 marked the beginning of powered aviation, but early aircraft primarily used wood and fabric. However, the demand for stronger and lighter materials quickly led to aluminium's adoption. By the 1920s, aluminium alloys began replacing wooden components in aircraft structures due to their durability, ease of fabrication, and corrosion resistance.

One of the first major uses of aluminium in aviation was in the Junkers J1, an all-metal aircraft built in 1915. This aircraft demonstrated the feasibility of aluminium structures, but it wasn't until the 1930s that aluminium alloys became standard in the aerospace industry. The introduction of duralumin (an aluminium-copper alloy) revolutionized aircraft construction, offering improved strength and reduced weight (Melhem et al., 2024; Duclos et al., 2024).

World War II and Aluminium Advancements

During World War II, the need for high-performance aircraft drove rapid advancements in aluminium alloy development. The demand for stronger, lighter, and more durable materials led to the widespread use of the 2xxx (Al-Cu) and 7xxx (Al-Zn-Mg) series alloys. The 7075-T6 alloy, developed in the 1940s, became a game-changer due to its exceptional strength, which was comparable to that of steel but at a significantly lower weight.

The Boeing B-29 Superfortress and the North American P-51 Mustang were two iconic aircraft that heavily relied on aluminium alloys for their airframes. These alloys provided the necessary mechanical properties for high-speed and high-altitude operations while ensuring structural integrity (Baker, 2024; McIntyre, 2024).

Post-War Developments and the Jet Age

Following World War II, aluminium alloy development continued at an accelerated pace. The rise of jet-powered aircraft required materials that could withstand higher operational stresses and temperature fluctuations. The aerospace industry responded by refining heat treatment processes and alloy compositions to enhance fatigue resistance and corrosion protection. The 1950s and 1960s saw the emergence of high-strength aluminium alloys such as 2024-T3 and 7050-T74, which were used extensively in commercial and military aircraft. These alloys featured improved damage tolerance and fracture toughness, making them ideal for critical structural components such as fuselages and wings (Winstone, 2000; Mouritz, 2012).

One notable example was the Boeing 707, the first successful commercial jetliner, which utilized aluminium alloys extensively in its airframe. This aircraft demonstrated the importance of lightweight materials in improving fuel efficiency and performance (Simons, 2021; Willingham, 2019).

The Space Race and Aluminium's Role

The 1960s and 1970s marked the beginning of space exploration, further pushing the boundaries of aluminium alloy applications. NASA's Apollo program relied heavily on aluminium-lithium (Al-Li) alloys due to their reduced density and high strength. The Saturn V rocket, which carried astronauts to the Moon, incorporated aluminium alloys in its structural framework to achieve the necessary weight savings for space travel. The use of aluminium alloys extended beyond space exploration and into supersonic aviation. The Concorde, a joint British-French supersonic passenger aircraft, utilized aluminium alloys capable of withstanding extreme aerodynamic heating at speeds exceeding Mach 2 (Dougherty, 2010; Sobel, 2023; Green et al., 2024).

Advancements in Aluminium Alloys: 1980s–2000s

As aerospace engineering progressed, the demand for higher performance materials led to continuous innovations in aluminium alloy compositions. The 1980s and 1990s saw the introduction of new alloying elements such as lithium, scandium, and zirconium, which enhanced mechanical properties and corrosion resistance.

Aluminium-lithium alloys, such as AA 2090 and AA 2195, became particularly attractive due to their ability to reduce weight by 3–10% while maintaining excellent strength and fatigue resistance. These alloys found applications in military aircraft and space vehicles, including NASA's Space Shuttle External Tank, which utilized Al-Li 2195 to achieve significant weight savings (Rao & Padal, 2022; Ablakatov et al., 2022). Additionally, advances in thermomechanical processing and heat treatment techniques further improved the performance of traditional 2xxx and 7xxx series alloys. The development of tailored alloys like 7055-T77 provided enhanced corrosion resistance, which was crucial for extending aircraft service life (Ablakatov et al., 2022; Zhou et al., 2021).

21st Century: The Rise of Hybrid Materials and Digital Manufacturing

In the 21st century, aluminium alloys have continued to evolve to meet the growing demands of the aerospace industry. While composite materials such as carbon fiber-reinforced polymers (CFRP) have gained popularity, aluminium alloys remain indispensable for structural components where metallic properties are essential.

Aircraft manufacturers such as Airbus and Boeing have adopted hybrid structures that combine aluminium alloys with composites to optimize performance. For example, the Airbus A350 incorporates GLARE (glass-fiber-reinforced aluminium) in its fuselage panels, leveraging the benefits of both materials. Similarly, Boeing's 777X utilizes high-strength aluminium alloys such as AA 7085 in wing structures to achieve weight reductions without compromising structural integrity (Majumdar et al., 2024). Another major advancement in recent years is the integration of additive manufacturing (AM) into aluminium alloy processing. Techniques such as selective laser melting (SLM) enable the production of complex geometries with minimal material waste, paving the way for lightweight and optimized aerospace components (Akhai, 2023; Khang & Akhai, 2024; Akhai & Khang, 2024). The F-35 Lightning II fighter jet exemplifies this trend, incorporating additively manufactured aluminium brackets to reduce overall weight and assembly complexity (Landry & Bihlman, 2023).

Aluminium's Role in Sustainable Aviation

With growing concerns over carbon emissions and environmental sustainability, the aerospace industry is actively exploring greener manufacturing practices and recycling strategies for aluminium alloys. The development of closed-loop recycling systems and low-carbon smelting techniques is helping to reduce the environmental footprint of aluminium production. Additionally, research into high-performance aluminium alloys with improved thermal stability and corrosion resistance is contributing to more durable and long-lasting aircraft components. Innovations such as cerium-based coatings and microarc oxidation (MAO) treatments are extending the lifespan of aluminium structures, reducing maintenance costs and material waste. The historical evolution of aluminium alloys in aerospace engineering highlights their continued importance in aircraft design and performance. From early duralumin to modern Al-Li and hybrid structures, aluminium alloys have consistently adapted to the changing demands of the industry. **Table 1** provides literature review table summarizing the historical evolution of high-performance aluminium alloys for aircraft structures (Raj et al., 2021; Zhou et al., 2021; Aglawe et al., 2023; Li et al., 2023).

Table 1. Literature review on the evolution of high-performance aluminium alloys for aircraft structures over time

Era	Key Developments	Alloy Series/ Examples	Aircraft/Applications	Technological Impact
Early 1900s	First use of aluminium in aviation	Early aluminium alloys (Duralumin)	Junkers J1 (1915)	Demonstrated feasibility of aluminium in aircraft structures, replacing wood and fabric
1920s–1930s	Adoption of aluminium for structural applications	2xxx series (Al-Cu alloys)	Boeing 247, Douglas DC-3	Improved strength-to-weight ratio and corrosion resistance, enabling metal airframes
World War II (1940s)	High-strength aluminium alloys developed for military aircraft	7075-T6 (Al-Zn-Mg-Cu)	B-29 Superfortress, P-51 Mustang	Provided superior strength and durability for high-speed and high-altitude performance
1950s–1960s (Jet Age)	Advanced aluminium alloys with enhanced fatigue resistance	2024-T3, 7050-T74	Boeing 707, Douglas DC-8	Enabled commercial jetliners with improved longevity and fuel efficiency
1960s–1970s (Space Race & Supersonic Era)	Aluminium alloys used in space and high-speed applications	Al-Li alloys (AA 2090, AA 2195), High-temperature Al alloys	Saturn V, Apollo Lunar Module, Concorde	Reduced weight and improved performance in extreme conditions
1980s–1990s	Development of high-performance Al-Li and scandium alloys	AA 2099, AA 2195, Al-Mg-Sc alloys	Space Shuttle External Tank, Military aircraft	3–10% weight reduction with enhanced fatigue resistance
2000s–2010s	Hybrid materials, improved corrosion resistance, and digital manufacturing	GLARE®, AA 7055-T77, AA 7085	Airbus A350, Boeing 777X	Integration of composites and aluminium for lightweight, durable structures
2020s–Present	Additive manufacturing, sustainable production, and AI-driven alloy design	AlSi10Mg (AM), Next-gen Al-Li alloys	F-35 Lightning II (AM parts), AI-optimized alloys	Complex geometries with reduced waste, eco-friendly processing methods

The **evolution of aluminium alloys** has played a **pivotal role** in the advancement of **aircraft structures**, significantly influencing **aerodynamic efficiency, structural integrity, and overall performance**. Over the past century, **metallurgical innovations** have led to the **development of high-strength, lightweight, and corrosion-resistant aluminium alloys** tailored for specific aerospace applications.

1. **Early 1900s – The Dawn of Aluminium in Aviation:**
    - The **first use of aluminium alloys** in aircraft construction began with **Duralumin**, an **Al-Cu alloy** that replaced traditional wood and fabric structures.
    - The **Junkers J1 (1915)** was the **first all-metal aircraft**, showcasing **aluminium's potential** in aviation.
2. **1920s–1930s – Aluminium as the Structural Backbone:**
    - The introduction of the **2xxx series (Al-Cu alloys)** enabled **stronger, corrosion-resistant airframes**, marking the transition to **full metal aircraft**.
    - Aircraft like the **Boeing 247 and Douglas DC-3** benefited from **enhanced strength-to-weight ratios**, making aviation more reliable and efficient.
3. **World War II (1940s)– High-Strength Aluminium for Military Aircraft:**
    - The need for **stronger, more durable aircraft** led to the development of **7075-T6 (Al-Zn-Mg-Cu alloys)**.
    - Aircraft such as the **B-29 Superfortress and P-51 Mustang** achieved **superior performance, high-speed maneuverability, and durability**.
4. **1950s–1960s – The Jet Age and Commercial Aviation Growth:**
    - The introduction of **fatigue-resistant aluminium alloys** like **2024-T3 and 7050-T74** supported the expansion of **commercial aviation**.
    - Aircraft such as the **Boeing 707 and Douglas DC-8** relied on these materials for **improved fuel efficiency and extended service life**.
5. **1960s–1970s – The Space Race and Supersonic Innovations:**
    - With the development of **Al-Li alloys (AA 2090, AA 2195) and high-temperature aluminium alloys**, aerospace engineering pushed new frontiers.
    - These materials were used in **Saturn V, Apollo Lunar Module, and Concorde**, offering **weight savings and extreme condition durability**.
6. **1980s–1990s – Aluminium-Lithium and Scandium-Based Alloys:**
    - **AA 2099, AA 2195, and Al-Mg-Sc alloys** provided a **3–10% weight reduction** while enhancing **fatigue resistance and toughness**.
    - These alloys were integral in **military aircraft and the Space Shuttle's External Tank**.
7. **2000s–2010s – Hybrid Materials and Digital Manufacturing:**
    - The introduction of **GLARE® (glass-fiber-reinforced aluminium) and advanced aluminium alloys (AA 7055-T77, AA 7085)** revolutionized **aircraft structures**.

- Used in aircraft like the **Airbus A350 and Boeing 777X**, these materials combined **lightweight properties with improved corrosion resistance**.

8. **2020s–Present – Additive Manufacturing and AI-Driven Design:**
   - **Next-gen Al-Li alloys and AlSi10Mg (AM)** have facilitated **complex geometries, reduced material waste, and enhanced eco-friendly processing methods**.
   - These alloys are used in cutting-edge aircraft such as the **F-35 Lightning II** and **AI-optimized aerospace materials** for next-generation aviation.

The **continuous evolution** of aluminium alloys has **transformed the aerospace industry**, enabling the development of **lighter, stronger, and more durable aircraft**. The **integration of AI, additive manufacturing, and sustainable production techniques** is shaping the **future of aerospace materials**, ensuring that **high-performance aluminium alloys remain at the forefront of aviation innovation**.

ALLOY DEVELOPMENT AND MICROSTRUCTURAL DESIGN

Traditional alloys like 2xxx and 7xxx are dominant in airframes due to their balance of strength and workability. Advanced alloys like AA 2099 and AA 2195 use lithium to reduce density and scandium for enhanced weldability and strength.

- **Traditional Alloys:** The 2xxx (Al-Cu) and 7xxx (Al-Zn-Mg) series aluminium alloys have been the cornerstone of airframe construction for decades, offering an optimal balance of strength, toughness, and workability. These alloys have undergone significant development and optimization, resulting in a range of tempers and compositions tailored to specific aerospace applications (Aglawe et al., 2023; Shen et al., 2023).
- **Advanced Alloys:** Recent advances in alloy development have focused on reducing density while maintaining or enhancing mechanical properties. Lithium additions have been particularly effective in this regard, with alloys such as AA 2099 and AA 2195 exhibiting density reductions of 3-10% compared to traditional aluminium alloys. This reduction in density, combined with improved strength-to-weight ratios, has made lithium-containing alloys attractive for use in next-generation aircraft (Hajjioui et al., 2023).
- Scandium-containing alloys (Al-Mg-Sc) have also shown significant promise, offering enhanced weldability, strength, and corrosion resistance. These alloys have been developed for use in structural components, such as fuse-

lage and wing skins, where their improved properties can be fully exploited (Rhamdani et al., 2025; Syrvatka et al., 2022; Langan & Dorin, 2023).

- **Nanostructured Alloys:** The application of severe plastic deformation (SPD) techniques has enabled the development of nanostructured aluminium alloys with significantly improved mechanical properties. Grain refinement via SPD processes, such as equal-channel angular pressing (ECAP) and high-pressure torsion (HPT), can result in grain sizes in the nanoscale range (<100 nm). This reduction in grain size leads to enhanced strength, ductility, and fatigue life, making nanostructured alloys attractive for use in demanding aerospace applications (Lowe et al., 2021).

ADVANCES IN PROCESSING TECHNIQUES

Significant advancements have been made in processing techniques to enhance the properties and performance of aluminium alloys. **Table 2** discusses advanced manufacturing and processing technologies for aluminium alloys, highlighting key advancements in aerospace applications. It also discusses cutting-edge processing methods for high-performance alloys and modern techniques for enhancing aluminium alloy performance (Ojo & Taban, 2023; Singh et al., 2025; Wadhwa et al., 2025; Akhai et al., 2025; Singh & Akhai, 2024).

Table 2. Advanced manufacturing and processing technologies for aluminium alloys and their applications

Processing Technique	Overview	Key Benefits	Applications
Thermomechanical Processing (TMP)	A combination of controlled rolling, forging, and heat treatment processes that refine grain structures, enhancing mechanical properties like strength, toughness, and ductility.	- Improves mechanical properties (higher strength, toughness, and fatigue resistance). - Enhances microstructural stability for better performance. - Reduces residual stresses, improving overall durability.	- Commonly used in 2xxx (Al-Cu) and 7xxx (Al-Zn-Mg) series alloys. - Aerospace components like fuselage panels, wing skins, and landing gear structures.
Heat Treatment (Retrogression and Re-Aging - RRA)	A specialized heat treatment technique involving controlled heating and cooling cycles to improve corrosion resistance and strength in high-strength aluminium alloys.	- Enhances stress corrosion cracking resistance in 7xxx-series alloys. - Optimizes mechanical properties while maintaining structural integrity. - Extends service life of aerospace components.	- 7xxx-series alloys used in aircraft wings, fuselage frames, and structural elements. - Applications in military and commercial aircraft requiring high corrosion resistance.
Additive Manufacturing (AM)	Layer-by-layer fabrication of aluminium components using techniques like Selective Laser Melting (SLM) and Electron Beam Melting (EBM).	- Enables production of complex geometries with minimal material waste. - Reduces production time and cost compared to traditional machining. - Allows for lightweight, high-performance component design.	- Aerospace structures such as brackets, engine components, and support structures. - Customized and optimized designs for UAVs and advanced aircraft.
Friction Stir Welding (FSW)	A solid-state joining process that utilizes a non-consumable rotating tool to generate frictional heat, allowing aluminium alloys to be welded without melting.	- Produces high-strength, defect-free joints with minimal distortion. - Preserves mechanical properties by avoiding issues like porosity and cracking seen in fusion welding. - Ideal for joining lightweight aluminium-lithium (Al-Li) alloys.	- Structural joining in aerospace applications (fuel tanks, fuselage panels, and wing sections). - Used in high-performance aircraft where traditional welding weakens materials.

The aerospace and industrial sectors **rely heavily on advanced manufacturing and processing techniques** to enhance the **mechanical properties, durability, and performance** of aluminium alloys. These **cutting-edge methods** allow for the **production of lightweight, high-strength components** essential for aircraft structures, engines, and support systems. Key innovations include **thermomechanical**

processing, heat treatment, additive manufacturing, and friction stir welding, each playing a crucial role in optimizing aluminium alloy performance.

1. **Thermomechanical Processing (TMP):**
   - **TMP is a combination of controlled rolling, forging, and heat treatment** that refines **grain structures** to improve **strength, toughness, and fatigue resistance**.
   - By **reducing residual stresses and enhancing microstructural stability**, TMP **increases the durability and reliability** of aluminium components.
   - This method is widely applied in **high-strength 2xxx (Al-Cu) and 7xxx (Al-Zn-Mg) series alloys**, making it ideal for **fuselage panels, wing skins, and landing gear structures** in aerospace applications.
2. **Heat Treatment (Retrogression and Re-Aging - RRA):**
   - **RRA is a specialized heat treatment process** that enhances **corrosion resistance and mechanical properties**, particularly in **7xxx-series alloys**.
   - This method **optimizes stress corrosion cracking resistance**, ensuring **longer service life and structural integrity** for aerospace components.
   - It is commonly used in **aircraft wings, fuselage frames, and structural elements** for both **military and commercial aviation**, where **high strength and corrosion resistance** are essential.
3. **Additive Manufacturing (AM):**
   - **AM, including Selective Laser Melting (SLM) and Electron Beam Melting (EBM), enables layer-by-layer fabrication** of aluminium components with **complex geometries**.
   - By **minimizing material waste, reducing production time, and lowering costs**, AM facilitates the creation of **lightweight, high-performance aerospace components**.
   - This technique is particularly valuable for **aerospace brackets, engine components, and UAV support structures**, where **customization and weight reduction** are critical.
4. **Friction Stir Welding (FSW):**
   - **FSW is a solid-state joining process** that generates frictional heat using a rotating tool, allowing aluminium alloys to be welded **without melting**.
   - This technique **produces high-strength, defect-free joints**, avoiding common issues like **porosity and cracking** seen in traditional welding methods.

- FSW is ideal for **joining aluminium-lithium (Al-Li) alloys**, widely used in **fuel tanks, fuselage panels, and wing sections** of modern aircraft.

These advanced manufacturing and processing technologies are redefining the capabilities of aluminium alloys, ensuring that aerospace and industrial applications can benefit from stronger, lighter, and more durable materials. By continuously refining these processes, the industry is paving the way for next-generation aircraft and high-performance aluminium-based structures (Ojo & Taban, 2023; Singh et al., 2025; Wadhwa et al., 2025; Akhai et al., 2025; Singh & Akhai, 2024).

MANUFACTURING INNOVATIONS

The aerospace industry has witnessed significant advancements in manufacturing technologies, enabling the production of complex, high-performance aluminium structures. **Table 3** provides a clear and detailed comparison of **manufacturing innovations** in aluminium alloy production, focusing on their benefits and real-world applications in aerospace.

Table 3. High-precision manufacturing and automation in aluminium alloy aircraft production

Manufacturing Innovation	Overview	Key Benefits	Applications
Precision Machining (CNC Technology)	Computer Numerical Control (CNC) machining allows for highly accurate and repeatable production of aluminium components with tight tolerances. Advanced multi-axis machining enables complex geometries with superior surface finish and minimal defects.	- Enhances dimensional accuracy and surface quality. - Reduces material waste and improves component reliability. - Enables high-speed production with minimal manual intervention.	- Critical aerospace components such as engine mounts, brackets, and fittings. - High-performance aircraft structures requiring precise tolerances.
Hybrid Structures (Aluminium-Composites Integration)	Co-cured aluminium-composite panels combine aluminium's strength and corrosion resistance with composites' lightweight and stiffness, leading to more fuel-efficient aircraft.	- Reduces overall aircraft weight, improving fuel efficiency. - Enhances fatigue resistance and structural integrity. - Optimizes strength-to-weight ratio for improved performance.	- Airbus A350 (fuselage and wing structures). - Military and commercial aircraft seeking lightweight, high-strength solutions.
Automation (Robotic Assembly Lines)	Robotics and automated systems enhance efficiency and consistency in aerospace manufacturing, particularly in large-scale component assembly such as wing skins and fuselage sections.	- Minimizes human error and enhances production accuracy. - Increases manufacturing speed and reduces labor costs. - Improves workplace safety by reducing manual handling.	- Aerospace assembly lines for aircraft fuselages, wings, and structural components. - Mass production of aluminium alloy-based aircraft parts.
Digital Manufacturing (AI & Machine Learning Integration)	AI-driven simulations and digital twin technology optimize manufacturing parameters, improving alloy design and production efficiency. Real-time monitoring enables defect detection and adaptive process control.	- Optimizes material properties for enhanced performance. - Reduces production costs by minimizing defects and rework. - Enables predictive maintenance, improving operational efficiency.	- Smart manufacturing of next-generation aluminium alloys. - Digital twin technology in aerospace production and quality assurance.

The manufacturing and automation of aluminium alloys in aircraft production have undergone transformative advancements, allowing for higher precision, efficiency, and reliability in aerospace components. This progress is largely driven by CNC machining, hybrid structures, automation, and AI-integrated digital manufacturing. These innovations reduce material waste, improve structural integrity, and enhance production speed, making modern aluminium alloy aircraft more fuel-efficient, lightweight, and structurally robust.

1. **Precision Machining (CNC Technology):**
   - Computer Numerical Control (CNC) machining is a key innovation that enables the manufacture of highly accurate and repeatable aluminium components with extremely tight tolerances.
   - Multi-axis CNC machining facilitates the creation of complex geometries with a superior surface finish and minimal defects, crucial for critical aerospace components such as engine mounts, brackets, and fittings.
   - This high-precision approach ensures dimensional accuracy, reduces material waste, and improves component reliability, allowing for the mass production of lightweight yet durable aluminium aircraft structures.
2. **Hybrid Structures (Aluminium-Composites Integration):**
   - The integration of aluminium alloys with composite materials has led to the development of co-cured aluminium-composite panels, which combine aluminium's strength and corrosion resistance with the lightweight and high stiffness of composites.
   - This innovation significantly reduces aircraft weight, improves fuel efficiency, and enhances fatigue resistance, leading to longer operational life and lower fuel consumption.
   - The Airbus A350 is a prime example of an aircraft utilizing hybrid aluminium-composite structures in its fuselage and wing sections, setting a benchmark for future fuel-efficient commercial and military aircraft.
3. **Automation (Robotic Assembly Lines):**
   - Robotic and automated assembly lines have transformed aerospace manufacturing, particularly in the production of large aluminium alloy structures, such as wing skins and fuselage sections.
   - Automation reduces human error, enhances production accuracy, and improves consistency, ensuring higher-quality aerospace components.
   - Additionally, it increases manufacturing speed, lowers labor costs, and enhances workplace safety, minimizing the risks associated with manual handling of large aircraft components.

4. **Digital Manufacturing (AI & Machine Learning Integration):**
   - Artificial Intelligence (AI) and machine learning are revolutionizing aircraft manufacturing by optimizing material properties, predicting defects, and improving production efficiency.
   - Digital twin technology, which creates virtual models of aircraft components, enables real-time monitoring, predictive maintenance, and adaptive process control, reducing production costs and minimizing defects.
   - These innovations contribute to the development of next-generation aluminium alloys, ensuring stronger, more durable, and environmentally friendly aircraft structures.

By integrating precision machining, hybrid structures, automation, and digital manufacturing, the aerospace industry has significantly improved the performance, efficiency, and cost-effectiveness of aluminium alloy aircraft production. These advancements align with the future vision of sustainable, high-performance aerospace solutions, ensuring continued innovation in aircraft design and manufacturing.

PROPERTY ENHANCEMENTS

The pursuit of improved performance, reliability, and lifespan has driven significant advancements in enhancing the properties of aluminium alloys. These enhancements have been achieved through innovative alloy design, processing techniques, and surface engineering. **Table 4** discusses the enhancement of high-performance aluminium alloys for aerospace applications, focusing on mechanical, corrosion, and thermal properties. It also discusses the impact of alloy composition on these properties and their applications in aircraft structures (Wadhwa & Akhai, 2014; Arora & Akhai, 2015; Sharma et al., 2024; Akhai et al., 2021).

Table 4. Key aluminium alloy property enhancements and their aerospace applications

Property Enhancement	Overview	Key Benefits	Specific Aluminium Alloys	Applications
Mechanical Properties (High-Strength & Lightweight Alloys)	Aluminium-lithium (Al-Li) alloys offer superior mechanical properties with tensile strengths exceeding 600 MPa and density reductions of 5–7%. Strengthening mechanisms involve precipitates like δ' (Al_3Li) and T1 (Al_2CuLi), refining grain structure and enhancing fatigue resistance.	- Increases strength while reducing overall weight. - Enhances fatigue resistance and toughness. - Optimized for lightweight, high-performance structures.	**AA 2099, AA 2195, AA 2050**	- Aircraft fuselage, wing structures, and landing gear components. - Spacecraft applications requiring high strength-to-weight ratios.
Corrosion Resistance (Surface Protection Technologies)	Cerium-based coatings and microarc oxidation (MAO) provide superior corrosion resistance. Cerium-based coatings act as protective barriers, while MAO creates a durable oxide layer that prevents corrosion ingress.	- Extends component lifespan by preventing corrosion-related failures. - Enhances environmental resistance for long-term performance. - Reduces maintenance costs and downtime.	**AA 7075-T73, AA 7050-T76, AA 2024-T351**	- Aircraft fuselage, structural joints, and external components exposed to harsh environments. - Marine and aerospace applications requiring high corrosion resistance.
Thermal Stability (High-Temperature Performance)	The addition of zirconium (Zr) and manganese (Mn) forms dispersoids that stabilize the microstructure at elevated temperatures. These dispersoids pin grain boundaries, preventing excessive grain growth and degradation.	- Maintains mechanical properties at high temperatures. - Enhances resistance to oxidation and thermal fatigue. - Improves material longevity under extreme conditions.	**AA 2618, AA 2219, AA 5083**	- Aerospace engine components, heat exchangers, and high-temperature structural parts. - Hypersonic aircraft and spacecraft requiring superior thermal stability.

The advancement of high-performance aluminium alloys has been driven by the need for lighter, stronger, and more durable materials in aerospace applications. These advancements focus on three key areas: mechanical properties, corrosion resistance, and thermal stability. The table categorizes these property enhancements, explaining their significance, associated benefits, and specific aluminium alloys utilized in various aerospace components.

1. **Mechanical Properties:**
   - The addition of lithium (Li) in aluminium-lithium (Al-Li) alloys has resulted in materials with high tensile strengths (>600 MPa) while achieving a 5–7% density reduction.
   - These alloys offer higher fatigue resistance and improved toughness, making them ideal for structural aircraft components where weight reduction is critical.
   - Specific Al-Li alloys like AA 2099, AA 2195, and AA 2050 have been developed to optimize these mechanical properties and are widely used in modern aircraft and spacecraft structures.
2. **Corrosion Resistance:**
   - Aluminium alloys used in aerospace applications are subjected to harsh environmental conditions, requiring surface protection to prevent corrosion-induced failures.
   - Cerium-based coatings and Microarc Oxidation (MAO) are advanced surface treatments that create protective layers on aluminium, increasing its lifespan and durability.
   - Alloys such as AA 7075-T73, AA 7050-T76, and AA 2024-T351 have been specifically engineered for superior corrosion resistance, making them ideal for aircraft fuselage and external structural components.
3. **Thermal Stability:**
   - Aircraft components operating at high temperatures, such as engine parts and heat exchangers, require alloys that maintain their structural integrity under extreme heat.
   - The addition of zirconium (Zr) and manganese (Mn) to aluminium alloys results in dispersoids that stabilize the microstructure, preventing grain growth and thermal degradation.
   - AA 2618, AA 2219, and AA 5083 are commonly used for high-temperature aerospace applications due to their enhanced thermal resistance and oxidation stability.

By optimizing mechanical strength, corrosion resistance, and thermal stability, modern aluminium alloys continue to play a vital role in aerospace engineering. The integration of these alloys in aircraft fuselage, wing structures, landing gear, and engine components demonstrates their versatility and critical contribution to aviation advancements.

CASE STUDIES

The application of advanced aluminium alloys and manufacturing technologies has been instrumental in the development of modern aircraft. These materials and processes have enabled significant improvements in performance, weight reduction, and cost efficiency, making them indispensable in the aerospace industry. The following case studies highlight the successful integration of these technologies in various aerospace programs, showcasing their impact on aircraft design, manufacturing, and performance. Advanced aluminium materials are being integrated into leading aircraft programs, such as the Airbus A380, Boeing 777X, F-35 Lightning II, Embraer E-Jets, Lockheed Martin C-130J, Bombardier Global 7500, Gulfstream G650, Airbus A350 XWB, Boeing 787 Dreamliner, and Dassault Falcon 8X. These materials offer fatigue resistance, weight reduction, corrosion resistance, and damage tolerance, as well as improved wing structures for fuel efficiency and range. The use of additive-manufactured brackets in aerospace manufacturing has shown potential for weight reduction, part count reduction, design flexibility, and cost efficiency. The use of high strength-to-weight ratio, corrosion resistance, formability, and weldability in fuselage skins has enhanced efficiency in regional aircraft design. These materials are set to revolutionize the industry and set new benchmarks for wing spars in wide-body aircraft (Raj et al., 2021; Sivamaran et al., 2021; Rao & Padal, 2022; Parveez et al., 2022; Li et al., 2023). **Table 5** discusses the application of advanced aluminium alloys in modern aircraft, their case studies, innovative applications in design and manufacturing, their impact on aerospace performance and efficiency, and their integration in leading aircraft programs.

Table 5. Integration of advanced aluminium materials in leading aircraft programs

Aircraft Model	Material Used	Application	Key Advantages	Industry Impact
Airbus A380	GLARE	Fuselage Panels	Fatigue resistance, weight reduction, corrosion resistance, damage tolerance	Pioneered hybrid materials in commercial aviation
Boeing 777X	AA 7085	Wing Ribs	High compressive strength, lightweight, corrosion resistance, machinability	Improved wing structures for fuel efficiency and range
F-35 Lightning II	AlSi10Mg (AM)	Additive-Manufactured Brackets	Weight reduction, part count reduction, design flexibility, cost efficiency	Showcased potential of AM in aerospace manufacturing
Embraer E-Jets	AA 6013-T6	Fuselage Skins	High strength-to-weight ratio, corrosion resistance, formability, weldability	Enhanced efficiency in regional aircraft design

continued on following page

Table 5. Continued

Aircraft Model	Material Used	Application	Key Advantages	Industry Impact
Lockheed Martin C-130J	AA 7055	Bulkheads	High strength, stress corrosion cracking resistance, lightweight, machinability	Set new standards for bulkhead durability
Bombardier Global 7500	AA 7449	Wing Spars	High strength, fatigue resistance, lightweight, machinability	Advanced wing spar technology in business jets
Gulfstream G650	AA 7085	Floor Beams	High strength, corrosion resistance, lightweight, machinability	Improved floor beam performance in long-range jets
Airbus A350 XWB	AA 2139	Wing Ribs	High strength, fatigue resistance, lightweight, machinability	Revolutionized wing rib design for commercial aircraft
Boeing 787 Dreamliner	AA 7050	Wing Spars	High strength, stress corrosion resistance, lightweight, machinability	Set new benchmarks for wing spars in wide-body aircraft
Dassault Falcon 8X	AA 2024	Fuselage Skins	High strength, fatigue resistance, lightweight, formability	Advanced fuselage materials in business aviation

Airbus A380: GLARE Fuselage Panels

The Airbus A380, the world's largest commercial passenger aircraft, represents a significant milestone in the application of advanced materials in aviation. One of the most notable innovations in the A380 is the use of GLARE (glass-fiber-reinforced aluminium) for its fuselage panels. GLARE is a hybrid material that combines the benefits of aluminium alloys with those of composite materials, offering a unique combination of properties that are ideal for aerospace applications.

Material Composition and Properties

GLARE consists of alternating layers of thin aluminium alloy sheets (typically AA 2024 or AA 7475) and glass-fiber-reinforced epoxy layers. The aluminium layers provide strength and ductility, while the glass-fiber layers enhance fatigue resistance and damage tolerance. The adhesive bonding between the layers ensures a robust and durable structure.

Advantages of GLARE

- **Fatigue Resistance:** GLARE exhibits superior fatigue resistance compared to traditional aluminium alloys, making it ideal for fuselage panels that are subjected to cyclic loading during flight.
- **Weight Reduction:** The use of GLARE in the A380's fuselage panels has resulted in significant weight savings, contributing to improved fuel efficiency and reduced emissions.
- **Corrosion Resistance:** The glass-fiber layers provide a barrier against corrosion, enhancing the longevity of the fuselage structure.
- **Damage Tolerance:** GLARE's hybrid structure allows it to withstand impact damage better than conventional aluminium alloys, improving the safety and reliability of the aircraft.

Application in the A380

The upper fuselage panels of the A380 are made from GLARE, covering a significant portion of the aircraft's structure. The material's lightweight and high-performance characteristics have enabled Airbus to achieve a balance between structural integrity and weight reduction, which is critical for the A380's operational efficiency.

Impact on Aerospace Industry

The successful implementation of GLARE in the A380 has demonstrated the potential of hybrid materials in aerospace applications. It has paved the way for further research and development of advanced materials that combine the best properties of metals and composites.

Boeing 777X: AA 7085 Wing Ribs

The Boeing 777X, the latest iteration of Boeing's successful 777 series, incorporates advanced aluminium alloys in its wing structure to achieve superior performance and efficiency. One of the key materials used in the 777X is AA 7085, a high-strength aluminium alloy selected for its exceptional compressive strength and lightweight properties.

Material Composition and Properties

AA 7085 is a high-strength, precipitation-hardened aluminium alloy that contains zinc, magnesium, and copper as its primary alloying elements. The alloy's high compressive strength is attributed to the formation of fine precipitates, such as MgZn2, during the aging process. These precipitates provide strengthening mechanisms that enhance the alloy's mechanical properties.

Advantages of AA 7085

- **High Compressive Strength:** AA 7085's exceptional compressive strength makes it ideal for wing ribs, which must withstand significant loads during flight.
- **Lightweight:** The alloy's low density contributes to weight reduction, which is critical for improving fuel efficiency and reducing operating costs.
- **Corrosion Resistance:** AA 7085 exhibits good corrosion resistance, ensuring the longevity of the wing structure.
- **Machinability:** The alloy's machinability allows for the production of complex wing rib geometries with high precision.

Application in the 777X

The wing ribs of the Boeing 777X are made from AA 7085, providing the necessary strength and stiffness to support the aircraft's large, folding wingtips. The use of AA 7085 has enabled Boeing to design a lightweight, high-performance wing structure that enhances the aircraft's aerodynamic efficiency and range.

Impact on Aerospace Industry

The adoption of AA 7085 in the 777X highlights the importance of advanced aluminium alloys in modern aircraft design. The alloy's unique combination of properties has set a new standard for wing structures, encouraging further innovation in materials science and engineering.

F-35 Lightning II: Additive-Manufactured Brackets

The F-35 Lightning II, a fifth-generation multirole combat aircraft, incorporates additive manufacturing (AM) technologies to produce lightweight, high-performance components. One of the most notable applications of AM in the F-35 is the production of aluminium brackets using selective laser melting (SLM).

Material Composition and Properties

The brackets are made from AlSi10Mg, a high-strength aluminium alloy that contains silicon and magnesium. The alloy's composition provides excellent mechanical properties, including high strength, good thermal conductivity, and low density. The SLM process involves the layer-by-layer deposition of aluminium powder, which is fused together using a high-powered laser to create complex geometries.

Advantages of Additive Manufacturing

- **Weight Reduction:** AM enables the production of lightweight components with optimized geometries, reducing the overall weight of the aircraft.
- **Part Count Reduction:** The ability to produce complex, integrated structures reduces the number of individual parts, simplifying assembly and reducing potential failure points.
- **Design Flexibility:** AM allows for the creation of intricate internal structures and features that cannot be achieved using traditional manufacturing methods.
- **Cost Efficiency:** By reducing material waste and simplifying production processes, AM can lower manufacturing costs.

Application in the F-35

The F-35's AM brackets are used in various parts of the aircraft, including the engine and airframe structures. The use of AM has enabled the production of components that are not only lighter but also more durable and reliable, contributing to the aircraft's overall performance and mission readiness.

Impact on Aerospace Industry

The successful implementation of AM in the F-35 has demonstrated the potential of this technology to transform aerospace manufacturing. It has opened up new possibilities for the design and production of high-performance components, paving the way for further advancements in additive manufacturing.

Embraer E-Jets: AA 6013-T6 Fuselage Skins

Embraer's E-Jets series, which includes the E170, E175, E190, and E195, utilizes advanced aluminium alloys in their fuselage skins to achieve a balance between strength, weight, and corrosion resistance. One of the key materials used in these aircraft is AA 6013-T6, a high-strength aluminium alloy that offers excellent mechanical properties and formability.

Material Composition and Properties

AA 6013-T6 is a heat-treatable aluminium alloy that contains magnesium, silicon, and copper as its primary alloying elements. The alloy's high strength is achieved through a combination of solid solution strengthening and precipitation hardening. The T6 temper further enhances the alloy's mechanical properties, making it suitable for aerospace applications.

Advantages of AA 6013-T6

- **High Strength-to-Weight Ratio:** AA 6013-T6 offers an excellent strength-to-weight ratio, making it ideal for fuselage skins that must withstand aerodynamic loads.
- **Corrosion Resistance:** The alloy's composition provides good resistance to corrosion, ensuring the longevity of the fuselage structure.
- **Formability:** AA 6013-T6 can be easily formed into complex shapes, allowing for the production of streamlined fuselage skins.
- **Weldability:** The alloy's weldability simplifies the assembly process, reducing production time and costs.

Application in the E-Jets

The fuselage skins of Embraer's E-Jets are made from AA 6013-T6, providing the necessary strength and durability to support the aircraft's structure. The use of this alloy has enabled Embraer to design a lightweight, high-performance fuselage that enhances the aircraft's efficiency and range.

Impact on Aerospace Industry

The adoption of AA 6013-T6 in the E-Jets highlights the importance of advanced aluminium alloys in regional aircraft design. The alloy's unique combination of properties has set a new standard for fuselage skins, encouraging further innovation in materials science and engineering.

Lockheed Martin C-130J Super Hercules: AA 7055 Bulkheads

The Lockheed Martin C-130J Super Hercules, a military transport aircraft, incorporates advanced aluminium alloys in its bulkheads to achieve superior strength and durability. One of the key materials used in the C-130J is AA 7055, a high-strength aluminium alloy that offers exceptional mechanical properties and resistance to stress corrosion cracking.

Material Composition and Properties

AA 7055 is a high-strength aluminium alloy that contains zinc, magnesium, and copper as its primary alloying elements. The alloy's high strength is achieved through a combination of solid solution strengthening and precipitation hardening. The alloy's resistance to stress corrosion cracking is attributed to its optimized composition and heat treatment.

Advantages of AA 7055

- **High Strength:** AA 7055 offers exceptional strength, making it ideal for bulkheads that must withstand significant loads.
- **Stress Corrosion Cracking Resistance:** The alloy's resistance to stress corrosion cracking ensures the longevity of the bulkhead structure.
- **Lightweight:** The alloy's low density contributes to weight reduction, which is critical for improving fuel efficiency and reducing operating costs.
- **Machinability:** AA 7055's machinability allows for the production of complex bulkhead geometries with high precision.

Application in the C-130J

The bulkheads of the C-130J are made from AA 7055, providing the necessary strength and durability to support the aircraft's structure. The use of this alloy has enabled Lockheed Martin to design a lightweight, high-performance bulkhead that enhances the aircraft's efficiency and mission readiness.

Impact on Aerospace Industry

The adoption of AA 7055 in the C-130J highlights the importance of advanced aluminium alloys in military aircraft design. The alloy's unique combination of properties has set a new standard for bulkheads, encouraging further innovation in materials science and engineering.

Bombardier Global 7500: AA 7449 Wing Spars

The Bombardier Global 7500, a long-range business jet, incorporates advanced aluminium alloys in its wing spars to achieve superior strength and fatigue resistance. One of the key materials used in the Global 7500 is AA 7449, a high-strength aluminium alloy that offers exceptional mechanical properties and resistance to fatigue.

Material Composition and Properties

AA 7449 is a high-strength aluminium alloy that contains zinc, magnesium, and copper as its primary alloying elements. The alloy's high strength is achieved through a combination of solid solution strengthening and precipitation hardening. The alloy's resistance to fatigue is attributed to its optimized composition and heat treatment.

Advantages of AA 7449

- **High Strength:** AA 7449 offers exceptional strength, making it ideal for wing spars that must withstand significant loads.
- **Fatigue Resistance:** The alloy's resistance to fatigue ensures the longevity of the wing spar structure.
- **Lightweight:** The alloy's low density contributes to weight reduction, which is critical for improving fuel efficiency and reducing operating costs.
- **Machinability:** AA 7449's machinability allows for the production of complex wing spar geometries with high precision.

Application in the Global 7500

The wing spars of the Global 7500 are made from AA 7449, providing the necessary strength and durability to support the aircraft's structure. The use of this alloy has enabled Bombardier to design a lightweight, high-performance wing spar that enhances the aircraft's efficiency and range.

Impact on Aerospace Industry

The adoption of AA 7449 in the Global 7500 highlights the importance of advanced aluminium alloys in business jet design. The alloy's unique combination of properties has set a new standard for wing spars, encouraging further innovation in materials science and engineering.

Gulfstream G650: AA 7085 Floor Beams

The Gulfstream G650, a long-range business jet, incorporates advanced aluminium alloys in its floor beams to achieve superior strength and durability. One of the key materials used in the G650 is AA 7085, a high-strength aluminium alloy that offers exceptional mechanical properties and resistance to corrosion.

Material Composition and Properties

AA 7085 is a high-strength aluminium alloy that contains zinc, magnesium, and copper as its primary alloying elements. The alloy's high strength is achieved through a combination of solid solution strengthening and precipitation hardening. The alloy's resistance to corrosion is attributed to its optimized composition and heat treatment.

Advantages of AA 7085

- **High Strength:** AA 7085 offers exceptional strength, making it ideal for floor beams that must withstand significant loads.
- **Corrosion Resistance:** The alloy's resistance to corrosion ensures the longevity of the floor beam structure.
- **Lightweight:** The alloy's low density contributes to weight reduction, which is critical for improving fuel efficiency and reducing operating costs.
- **Machinability:** AA 7085's machinability allows for the production of complex floor beam geometries with high precision.

Application in the G650

The floor beams of the G650 are made from AA 7085, providing the necessary strength and durability to support the aircraft's structure. The use of this alloy has enabled Gulfstream to design a lightweight, high-performance floor beam that enhances the aircraft's efficiency and range.

Impact on Aerospace Industry

The adoption of AA 7085 in the G650 highlights the importance of advanced aluminium alloys in business jet design. The alloy's unique combination of properties has set a new standard for floor beams, encouraging further innovation in materials science and engineering.

Airbus A350 XWB: AA 2139 Wing Ribs

The Airbus A350 XWB, a long-range, wide-body commercial aircraft, incorporates advanced aluminium alloys in its wing ribs to achieve superior strength and fatigue resistance. One of the key materials used in the A350 XWB is AA 2139, a high-strength aluminium alloy that offers exceptional mechanical properties and resistance to fatigue.

Material Composition and Properties

AA 2139 is a high-strength aluminium alloy that contains copper, magnesium, and silver as its primary alloying elements. The alloy's high strength is achieved through a combination of solid solution strengthening and precipitation hardening. The alloy's resistance to fatigue is attributed to its optimized composition and heat treatment.

Advantages of AA 2139

- **High Strength:** AA 2139 offers exceptional strength, making it ideal for wing ribs that must withstand significant loads.
- **Fatigue Resistance:** The alloy's resistance to fatigue ensures the longevity of the wing rib structure.
- **Lightweight:** The alloy's low density contributes to weight reduction, which is critical for improving fuel efficiency and reducing operating costs.
- **Machinability:** AA 2139's machinability allows for the production of complex wing rib geometries with high precision.

Application in the A350 XWB

The wing ribs of the A350 XWB are made from AA 2139, providing the necessary strength and durability to support the aircraft's structure. The use of this alloy has enabled Airbus to design a lightweight, high-performance wing rib that enhances the aircraft's efficiency and range.

Impact on Aerospace Industry

The adoption of AA 2139 in the A350 XWB highlights the importance of advanced aluminium alloys in commercial aircraft design. The alloy's unique combination of properties has set a new standard for wing ribs, encouraging further innovation in materials science and engineering.

Boeing 787 Dreamliner: AA 7050 Wing Spars

The Boeing 787 Dreamliner, a long-range, wide-body commercial aircraft, incorporates advanced aluminium alloys in its wing spars to achieve superior strength and durability. One of the key materials used in the 787 Dreamliner is AA 7050, a high-strength aluminium alloy that offers exceptional mechanical properties and resistance to stress corrosion cracking.

Material Composition and Properties

AA 7050 is a high-strength aluminium alloy that contains zinc, magnesium, and copper as its primary alloying elements. The alloy's high strength is achieved through a combination of solid solution strengthening and precipitation hardening. The alloy's resistance to stress corrosion cracking is attributed to its optimized composition and heat treatment.

Advantages of AA 7050

- **High Strength:** AA 7050 offers exceptional strength, making it ideal for wing spars that must withstand significant loads.
- **Stress Corrosion Cracking Resistance:** The alloy's resistance to stress corrosion cracking ensures the longevity of the wing spar structure.
- **Lightweight:** The alloy's low density contributes to weight reduction, which is critical for improving fuel efficiency and reducing operating costs.
- **Machinability:** AA 7050's machinability allows for the production of complex wing spar geometries with high precision.

Application in the 787 Dreamliner

The wing spars of the 787 Dreamliner are made from AA 7050, providing the necessary strength and durability to support the aircraft's structure. The use of this alloy has enabled Boeing to design a lightweight, high-performance wing spar that enhances the aircraft's efficiency and range.

Impact on Aerospace Industry

The adoption of AA 7050 in the 787 Dreamliner highlights the importance of advanced aluminium alloys in commercial aircraft design. The alloy's unique combination of properties has set a new standard for wing spars, encouraging further innovation in materials science and engineering.

Dassault Falcon 8X: AA 2024 Fuselage Skins

The Dassault Falcon 8X, a long-range business jet, incorporates advanced aluminium alloys in its fuselage skins to achieve superior strength and durability. One of the key materials used in the Falcon 8X is AA 2024, a high-strength aluminium alloy that offers exceptional mechanical properties and resistance to fatigue.

Material Composition and Properties

AA 2024 is a high-strength aluminium alloy that contains copper, magnesium, and manganese as its primary alloying elements. The alloy's high strength is achieved through a combination of solid solution strengthening and precipitation hardening. The alloy's resistance to fatigue is attributed to its optimized composition and heat treatment.

Advantages of AA 2024

- **High Strength:** AA 2024 offers exceptional strength, making it ideal for fuselage skins that must withstand significant loads.
- **Fatigue Resistance:** The alloy's resistance to fatigue ensures the longevity of the fuselage skin structure.
- **Lightweight:** The alloy's low density contributes to weight reduction, which is critical for improving fuel efficiency and reducing operating costs.
- **Formability:** AA 2024 can be easily formed into complex shapes, allowing for the production of streamlined fuselage skins.

Application in the Falcon 8X

The fuselage skins of the Falcon 8X are made from AA 2024, providing the necessary strength and durability to support the aircraft's structure. The use of this alloy has enabled Dassault to design a lightweight, high-performance fuselage that enhances the aircraft's efficiency and range.

Impact on Aerospace Industry

The adoption of AA 2024 in the Falcon 8X highlights the importance of advanced aluminium alloys in business jet design. The alloy's unique combination of properties has set a new standard for fuselage skins, encouraging further innovation in materials science and engineering.

CHALLENGES AND LIMITATIONS

The challenges of aluminium-composite joints include material compatibility issues like galvanic corrosion, high cost of scandium, and the need for alternative strengthening elements. Additionally, aluminium smelting is energy-intensive, necessitating innovations in recycling and energy-efficient processing.

Material Compatibility: Addressing Galvanic Corrosion in Aluminium-Composite Joints

One of the critical challenges in aluminium alloy applications, particularly in aerospace and automotive industries, is material compatibility when bonded with composites. Aluminium alloys, including Al-Mg-Sc, often interact with carbon-fiber-reinforced polymers (CFRPs) or other advanced composites (Montemor, 2016; Kant et al., 2019; Paz Martínez-Viademonte et al., 2020; Zhu et al., 2023). While such hybrid structures provide enhanced strength-to-weight ratios, they also introduce concerns related to galvanic corrosion. Galvanic corrosion occurs when two dissimilar materials—such as aluminium and carbon fiber—come into electrical contact in the presence of an electrolyte, such as moisture in the atmosphere. Since aluminium is less noble than carbon fiber, it undergoes accelerated corrosion, leading to structural degradation. The problem is exacerbated in marine and aerospace environments, where humidity, salt exposure, and temperature variations intensify the corrosion process. To mitigate these effects, researchers are developing advanced protective coatings, barrier layers, and anodization techniques. Protective coatings such as ceramic nanocoatings, polymeric barriers, and hybrid composite surface treatments are being investigated to enhance longevity. Additionally, advancements in adhesive bonding techniques and interface engineering are improving material compatibility, reducing direct metal-composite contact, and extending the lifespan of such hybrid structures (Song et al., 2021; Jiang et al., 2020).

Cost Constraints: The Scarcity and Expense of Scandium

Scandium is one of the most valuable elements in aluminium alloy development, significantly enhancing the mechanical properties of Al-Mg alloys. However, its limited global availability poses a significant economic challenge. The production of scandium oxide (Sc_2O_3) is costly due to its low concentration in primary ores, which makes extraction difficult and expensive. Most of the world's scandium supply is a byproduct of titanium, uranium, and rare earth element mining, further limiting its availability. Due to these constraints, scandium-reinforced aluminium alloys remain cost-prohibitive for widespread use in consumer applications. The aerospace and defense industries, which demand high-performance materials, can justify the expense, but mass adoption in automotive, construction, and general engineering sectors remains limited (Ghosh et al., 2022; Junior et al., 2021; Chi et al., 2023; Wang & Cheng, 2011). To address this, researchers are investigating alternative alloying elements, such as zirconium (Zr), yttrium (Y), and cerium (Ce), which can replicate some of scandium's strengthening effects at a lower cost. Additionally, advancements in electrochemical extraction and secondary recovery from industrial waste streams are being explored to make scandium sourcing more sustainable and economical (Kumar et al., 2023; Czerwinski, 2020).

Environmental Impact: Energy-Intensive Processing and Recycling Challenges

Aluminium production is an energy-intensive process, primarily due to the Hall-Héroult smelting method, which consumes vast amounts of electricity. Producing one ton of primary aluminium requires approximately 14,000 kWh of electricity, contributing to significant carbon dioxide (CO_2) emissions. This environmental impact is a growing concern, especially in light of global sustainability goals and carbon neutrality initiatives. While aluminium is highly recyclable, the recycling rate of high-performance aluminium alloys remains suboptimal due to material degradation, contamination, and the presence of microalloying elements like scandium. Standard recycling processes, such as remelting, often lead to loss of alloying elements, requiring additional refining steps to maintain performance characteristics (Kumar et al., 2023; Czerwinski, 2020). Innovations in closed-loop recycling systems, molten salt electrolysis, and low-carbon smelting technologies are being explored to reduce energy consumption and improve material circularity. Emerging methods such as direct solid-state recycling and electrolytic scandium recovery hold promise for reducing reliance on primary ore extraction and minimizing the environmental footprint of aluminium alloy production (Saevarsdottir et al., 2020).

FUTURE DIRECTIONS

Machine learning is revolutionizing alloy discovery, enabling the development of multi-functional materials with superior performance. Smart manufacturing integrates embedded sensors in aluminium structures for real-time structural health monitoring and predictive maintenance. Advancements in closed-loop recycling systems and low-carbon processing methods make aluminium manufacturing more sustainable.

Next-Generation Aluminium Alloys: Accelerating Discovery with Machine Learning

The future of aluminium alloy development is being shaped by machine learning (ML) and artificial intelligence (AI)-driven materials discovery. Traditional alloy design has relied on experimental trial-and-error methods, which are time-consuming and expensive. However, computational materials science is revolutionizing this process by rapidly screening and predicting optimal alloy compositions.

By leveraging big data and predictive modeling, researchers can identify new high-performance aluminium alloys that exhibit superior strength, corrosion resistance, and thermal stability. ML algorithms analyze vast datasets of material properties, phase diagrams, and processing conditions, accelerating the development of novel compositions with enhanced mechanical and functional properties (Abbas et al., 2025; Motia et al., 2024; Abbass et al., 2025; Akhai, 2024; Akhai & Kumar, 2024). For example, computational thermodynamics (CALPHAD) and genetic algorithms are being employed to design multi-component aluminium alloys with tailored microstructures. These approaches enable the discovery of new materials with enhanced fatigue resistance, lightweight characteristics, and improved manufacturability.

Smart Manufacturing: Real-Time Structural Health Monitoring with Embedded Sensors

The integration of smart manufacturing techniques is set to revolutionize how aluminium structures are monitored and maintained. Embedded sensors within aluminium components can provide real-time data on structural integrity, detecting stress, fatigue, and corrosion before failure occurs. These sensors, including fiber-optic strain gauges, piezoelectric transducers, and nanomaterial-based coatings, enable predictive maintenance strategies. Instead of relying on periodic inspections, industries such as aerospace, automotive, and civil engineering can use real-time monitoring to extend service life and enhance safety. For instance, in aircraft fuselage panels made from aluminium alloys, embedded self-sensing coatings can detect microcracks and corrosion initiation, triggering maintenance alerts before significant damage occurs. Similarly, in automotive chassis components, smart sensors can assess thermal expansion and mechanical stress, optimizing performance under varying driving conditions.

Sustainability: Closed-Loop Recycling and Low-Carbon Processing Innovations

Sustainability remains a critical focus in the aluminium industry, with efforts directed at developing closed-loop recycling systems and reducing carbon emissions. Traditional aluminium recycling faces challenges such as oxidation losses, alloy contamination, and energy inefficiencies, limiting the recyclability of high-performance alloys. To address these issues, new electrometallurgical processes, such as plasma-assisted recycling and direct alloy-to-alloy remelting, are being developed. These approaches enable the recovery of microalloying elements like scandium and zirconium, ensuring that recycled aluminium maintains its original properties. Additionally, advancements in low-carbon smelting technologies, such as inert anode electrolysis, offer a path toward environmentally friendly aluminium production. Companies like Rio Tinto and Alcoa are pioneering the development of inert anode cells, which eliminate carbon emissions from the aluminium smelting process. These innovations are expected to significantly reduce the carbon footprint of primary aluminium production. Another exciting frontier in sustainability is the exploration of biodegradable coatings and nanocomposite materials to replace traditional anodization and chemical treatments. These eco-friendly surface modifications can enhance corrosion resistance while minimizing environmental impact (Verma et al., 2024; Akhai, 2024; Akhai & Wadhwa, 2024; Akhai & Khang, 2024; Mala et al., 2025; Akhai & Abbass, 2025; Khang & Akhai 2025).

CONCLUSION

- Aluminium alloys have significantly influenced the development of aircraft structures over the past century. From the early 1900s with Duralumin, to the jet age and commercial aviation growth, aluminium alloys have been used for various applications. The evolution of aluminium alloys has led to lighter, stronger, and more durable aircraft, with the integration of AI, additive manufacturing, and sustainable production techniques shaping the future of aerospace materials.
- Advanced manufacturing and processing techniques are crucial in the aerospace and industrial sectors for enhancing the mechanical properties, durability, and performance of aluminium alloys. Key innovations include thermomechanical processing, heat treatment, additive manufacturing, and friction stir welding. These techniques improve the durability, reliability, and performance of aluminium alloys, paving the way for next-generation aircraft and high-performance aluminium-based structures.

- The manufacturing and automation of aluminium alloys in aircraft production have improved precision, efficiency, and reliability. This is driven by CNC machining, hybrid structures, automation, and AI-integrated digital manufacturing. These innovations reduce material waste, improve structural integrity, and enhance production speed, making modern aircraft more fuel-efficient, lightweight, and structurally robust. The integration of these technologies aligns with the future vision of sustainable, high-performance aerospace solutions.
- Hybrid materials like GLARE, AA 7085, and AlSi10Mg brackets have been successfully applied in various aircraft programs, demonstrating their impact on weight reduction, fatigue resistance, corrosion resistance, and cost efficiency. The Airbus A380, Boeing 777X, F-35 Lightning II, and Embraer E-Jets showcase the benefits of advanced aluminium alloys in aircraft design. These alloys have enhanced aircraft safety, durability, fuel efficiency, and environmental sustainability. Military aviation programs, such as the Lockheed Martin C-130J and Bombardier Global 7500, have also leveraged high-performance aluminium alloys for superior strength, stress corrosion resistance, and fatigue resistance.
- The future of aluminium alloy development and application is driven by a convergence of materials science, computational modeling, smart manufacturing, and sustainability initiatives. While challenges such as galvanic corrosion, scandium scarcity, and energy-intensive processing pose obstacles, ongoing innovations in alternative alloying elements, real-time health monitoring, and closed-loop recycling are paving the way for a more efficient and eco-friendly aluminium industry.
- By leveraging AI-driven alloy discovery, embedded sensor technology, and sustainable processing methods, the next generation of aluminium materials will not only enhance mechanical performance and structural reliability but also contribute to a greener and more resource-efficient future.

ABBREVIATIONS

AA Aluminium Alloy
A380 Airbus A380 Aircraft
AISI American Iron and Steel Institute
Al Aluminium
Al-Cu Aluminium-Copper
Al-Li Aluminium-Lithium
Al-Mg-Sc Aluminium-Magnesium-Scandium

Al-Si Aluminium-Silicon
AM Additive Manufacturing
AI Artificial Intelligence
B-29 Boeing B-29 Superfortress
CFRP Carbon Fiber Reinforced Polymer
CNC Computer Numerical Control
DC-3 Douglas DC-3 Aircraft
ECAP Equal-Channel Angular Pressing
EBM Electron Beam Melting
FSW Friction Stir Welding
GLARE Glass Laminate Aluminium Reinforced Epoxy
HPT High-Pressure Torsion
MAO Microarc Oxidation
ML Machine Learning
NASA National Aeronautics and Space Administration
P-51 North American P-51 Mustang
RRA Retrogression and Re-Aging
Sc Scandium
SLM Selective Laser Melting
SPD Severe Plastic Deformation
TMP Thermomechanical Processing
UAV Unmanned Aerial Vehicle
WWII World War II
XWB Extra Wide Body (Airbus A350 XWB)

REFERENCES

Abbas, M., Akhai, S., Abbas, U., Jafri, R., & Arif, S. M. (2025). AI-enabled sustainable urban planning and management. In *Real-World Applications of AI Innovation* (pp. 233–260). IGI Global Scientific Publishing.

Abbass, M., Abbas, U., Jafri, R., Arif, S. M., & Akhai, S. (2025). AI and Machine Learning Applications in Sustainable Smart Cities. In *Sustainable Smart Cities and the Future of Urban Development* (pp. 1–32). IGI Global Scientific Publishing.

Ablakatov, I. K., Baiserikov, B. M., Ismailov, M. B., & Nurgozhin, M. R. (2022). Aluminum-lithium alloys: types, properties, application, and production technologies. Overview. *Kompleksnoe Ispolzovanie Mineralnogo Syra= Complex use of mineral resources, 323*(4), 5-14.

Aglawe, K., Giri, S., Dhande, M., & Shelare, S. (2023, September). Application of aluminum alloys in aviation industry: A review. In *AIP Conference Proceedings* (Vol. 2800, No. 1). AIP Publishing. DOI: 10.1063/5.0163002

Akhai, S. (2023). Navigating the Potential Applications and Challenges of Intelligent and Sustainable Manufacturing for a Greener Future. *Evergreen, 10*(4), 2237–2243. DOI: 10.5109/7160899

Akhai, S. (2024). A Review on Optimizations in μ-EDM Machining of the Biomedical Material Ti6Al4V Using the Taguchi Method: Recent Advances Since 2020. *Latest Trends in Engineering and Technology*, 395-402.

Akhai, S. (2024). Innovating Healthcare Delivery: Harnessing Drones for Medical Services. In *Modelling of Virtual Worlds Using the Internet of Things* (Vol. 1, pp. 232-250). CRC Press.

Akhai, S. (2024). Smart Home Environment for Society: An Overview of IoT-Enabled Air Conditioners. In *Modelling of Virtual Worlds Using the Internet of Things* (Vol. 1, pp. 74-87). CRC Press.

Akhai, S., & Abbass, M. (2025). Toward Resilient Futures: The Role of AI-Driven Strategies in Climate Adaptation. In *Nexus of AI, Climatology, and Urbanism for Smart Cities* (pp. 325-340). IGI Global Scientific Publishing.

Akhai, S., Abbass, M., Kaur, P., & Kaur, T. (2025). Digital Transformation Across Generations: Robotics and AI in Action. In *Impacts of Digital Technologies Across Generations* (pp. 23-40). IGI Global Scientific Publishing.

Akhai, S., & Khang, A. (2024). Energy Efficiency and Human Comfort: AI and IoT Integration in Hospital HVAC Systems. *Medical Robotics and AI-Assisted Diagnostics for a High-Tech Healthcare Industry*, 93-108.

Akhai, S., & Khang, A. (2024). Innovations in Medical Manufacturing: A Review of 3D Printing, Robotics, and Internet of Things (IoT). In *The Quantum Evolution* (Vol. 1, pp. 226-241). CRC Press.

Akhai, S., & Kumar, V. (2024). Quantum resilience and distributed trust: The promise of blockchain and quantum computing in defense. In *Sustainable Security Practices Using Blockchain, Quantum and Post-Quantum Technologies for Real Time Applications* (pp. 125-153). Springer Nature Singapore.

Akhai, S., Singh, S., Kaur, S., & Kaur, K. (2024). Advancing Aerospace Inspection With Terahertz Technology. In *New Innovations in AI, Aviation, and Air Traffic Technology* (pp. 195-211). IGI Global. DOI: 10.4018/979-8-3693-1954-3.ch010

Akhai, S., Srivastava, P., Sharma, V., & Bhatia, A. (2021). Investigating weld strength of AA8011-6062 alloys joined via friction-stir welding using the RSM approach. [). IOP Publishing.]. *Journal of Physics: Conference Series*, *1950*(1), 012016. DOI: 10.1088/1742-6596/1950/1/012016

Akhai, S., & Wadhwa, A. S. (2024). Wireless networking solutions for climate control and comfort in modern air conditioning systems. In *Intelligent Networks* (pp. 165–181). CRC Press. DOI: 10.1201/9781003541363-11

Alamdari, H. (2017). Aluminium production process: Challenges and opportunities. *Metals*, *7*(4), 133. DOI: 10.3390/met7040133

Arora, N., & Akhai, S. (2015). Reclaiming EN-14b steel grade implements by hardfacing. *International Journal of Scientific Research*, *4*(10), 14–16.

Baker, D. (2024). *US Air Power 1945-1990: Volume 1-US Fighters and Fighter-Bombers, 1945-1949* (Vol. 2). Helion and Company.

Chi, J., Wang, C., Zhou, G., Fu, X., Chen, X., Yin, X., Zhang, Z., & Wang, Y. (2023). A critical review on separation and extraction of scandium from industrial wastes: Methods, difficulties, and mechanism. *Journal of Environmental Chemical Engineering*, *11*(5), 111068. DOI: 10.1016/j.jece.2023.111068

Czerwinski, F. (2020). Cerium in aluminum alloys. *Journal of Materials Science*, *55*(1), 24–72. DOI: 10.1007/s10853-019-03892-z PMID: 33066314

Dougherty, N. (2010). Simulated Altitude Liquid Rocket Testing at the USAF AEDC: Atlas/Mercury, Titan/Gemini, Saturn/Apollo, Space Shuttle, ISS. In *46th AIAA/ASME/SAE/ASEE Joint Propulsion Conference & Exhibit* (p. 6982). DOI: 10.2514/6.2010-6982

Duclos, A., Bernardi, E., Robbiola, L., Deshayes, C., de Souza Machado, L., Martini, C., Chiavari, C., Balbo, A., Escobar, C., Guilminot, E., Malard, B., & Brunet, M. (2024). Duralumin Alloys in World War II Heritage Aircraft: Correlations Between Manufacturing Parameters and Alloys' Properties. *Heritage*, *7*(12), 6772–6790. DOI: 10.3390/heritage7120313

Florea, M., Constantin, V. S., Bucur, A. C., Faur, R., Predu, D., & Cazacu, A. (2024). Materials and Structures Used in Aeronautics: Present and Future Perspectives. *INCAS Bulletin, 16*(4).

Ghosh, A., Dhiman, S., Gupta, A., & Jain, R. (2022). Process evaluation of scandium production and its environmental impact. *Environments (Basel, Switzerland)*, *10*(1), 8. DOI: 10.3390/environments10010008

Gialanella, S., Malandruccolo, A., Gialanella, S., & Malandruccolo, A. (2020). Alloys for aircraft structures. *Aerospace alloys*, 41-127.

Green, J. L., Cooke, D., Beckman, A. B. W., Ramos, K., & Donahue, B. B. (2024). Maximizing the Potential of NASA's Super Heavy Launch Vehicles: Lessons From Saturn V. In *AIAA AVIATION FORUM AND ASCEND 2024* (p. 4879).

Hajjioui, E. A., Bouchaâla, K., Faqir, M., & Essadiqi, E. (2023). A review of manufacturing processes, mechanical properties and precipitations for aluminum lithium alloys used in aeronautic applications. *Heliyon*, *9*(3), e12565. DOI: 10.1016/j.heliyon.2022.e12565 PMID: 36895401

Hasanov, R. I. (2023). Towards a sustainable future in aluminium production: Environmental and economic benefits of revolutionary inert anode technology. *Journal of Sustainability Science and Management*, *18*(10), 176–186. DOI: 10.46754/jssm.2023.10.012

Jiang, B., Chen, Q., & Yang, J. (2020). Advances in joining technology of carbon fiber-reinforced thermoplastic composite materials and aluminum alloys. *International Journal of Advanced Manufacturing Technology*, *110*(9), 2631–2649. DOI: 10.1007/s00170-020-06021-2

Junior, A. B., Espinosa, D. C. R., Vaughan, J., & Tenório, J. A. S. (2021). Recovery of scandium from various sources: A critical review of the state of the art and future prospects. *Minerals Engineering*, *172*, 107148. DOI: 10.1016/j.mineng.2021.107148

Kant, R., Chauhan, P. S., Bhatt, G., & Bhattacharya, S. (2019). Corrosion monitoring and control in aircraft: A review. *Sensors for Automotive and Aerospace Applications*, 39-53.

Khang, A., & Akhai, S. (2024). Green intelligent and sustainable manufacturing: key advancements, benefits, challenges, and applications for transforming industry. *Machine Vision and Industrial Robotics in Manufacturing*, 405-417.

Khang, A., & Akhai, S. (2025). E-Waste and Lithium-ion Battery Recycling Insights for Sustainable Transportation. In *Driving Green Transportation System Through Artificial Intelligence and Automation: Approaches, Technologies and Applications* (pp. 203-230). Springer Nature Switzerland. DOI: 10.1007/978-3-031-72617-0_11

Khang, A., & Akhai, S. (2025). E-Waste and Lithium-ion Battery Recycling Insights for Sustainable Transportation. In *Driving Green Transportation System Through Artificial Intelligence and Automation: Approaches, Technologies and Applications* (pp. 203-230). Springer Nature Switzerland. DOI: 10.1007/978-3-031-72617-0_11

Kumar, H., Wadhwa, A. S., Akhai, S., & Kaushik, A. (2024). Parametric analysis, modeling and optimization of the process parameters in electric discharge machining of aluminium metal matrix composite. *Engineering Research Express*, *6*(2), 025542. DOI: 10.1088/2631-8695/ad4ba9

Kumar, H., Wadhwa, A. S., Akhai, S., & Kaushik, A. (2024). Parametric optimization of the machining performance of Al-SiCp composite using combination of response surface methodology and desirability function. *Engineering Research Express*, *6*(2), 025505. DOI: 10.1088/2631-8695/ad38ff

Kumar, V., Angra, S., & Singh, S. (2023). Influence of rare earth elements on aluminium metal matrix composites: A review. *Mater. Phys. Mech*, *51*(2), 1–20.

Landry, S. J., & Bihlman, W. (2023). Aerospace systems automation. In *Springer Handbook of Automation* (pp. 1115–1137). Springer International Publishing. DOI: 10.1007/978-3-030-96729-1_51

Langan, T. J., & Dorin, T. (2023). Scandium in Commercial Wrought Aluminum Alloys. *Rare Earth Metals and Minerals Industries: Status and Prospects*, 359-389.

Li, S. S., Yue, X., Li, Q. Y., Peng, H. L., Dong, B. X., Liu, T. S., Yang, H.-Y., Fan, J., Shu, S.-L., Qiu, F., & Jiang, Q. C. (2023). Development and applications of aluminum alloys for aerospace industry. *Journal of Materials Research and Technology*, *27*, 944–983. DOI: 10.1016/j.jmrt.2023.09.274

Li, S. S., Yue, X., Li, Q. Y., Peng, H. L., Dong, B. X., Liu, T. S., Yang, H.-Y., Fan, J., Shu, S.-L., Qiu, F., & Jiang, Q. C. (2023). Development and applications of aluminum alloys for aerospace industry. *Journal of Materials Research and Technology*, *27*, 944–983. DOI: 10.1016/j.jmrt.2023.09.274

Lowe, T. C., Valiev, R. Z., Li, X., & Ewing, B. R. (2021). Commercialization of bulk nanostructured metals and alloys. *MRS Bulletin*, *46*(3), 265–272. DOI: 10.1557/s43577-021-00060-0

Majumdar, S., Sinha, A., Das, A., Datta, P., & Nag, D. (2024). An Insight View of Evolution of Advanced Aluminum Alloy for Aerospace and Automotive Industry: Current Status and Future Prospects. *Journal of The Institution of Engineers (India): Series D*, 1-18.

Mala, S., Deep, U., Mala, E., & Akhai, S. (2025). Digital Health Technologies for Preventive Cardiology. In *Impact of Digital Solutions for Improved Healthcare Delivery* (pp. 379–400). IGI Global.

McIntyre, T. B. (2024). *The Heroes We Needed: The B-29ers Who Ended World War II and My Fight to Save the Forgotten Stories of the Greatest Generation*. Helion and Company.

Melhem, G. N., Munroe, P. R., & Vithal, A. (2024). *Supersonic Effects on Rivets*. CRC Press. DOI: 10.1201/9781003516903

Menzie, W. D., Barry, J. J., Bleiwas, D. I., Bray, E. L., Goonan, T. G., & Matos, G. (2010). *The global flow of aluminum from 2006 through 2025*. US Department of the Interior, US Geological Survey. DOI: 10.3133/ofr20101256

Montemor, M. F. (2016). Corrosion issues in joining lightweight materials: A review of the latest achievements. *Physical Sciences Reviews*, *1*(2), 20150011. DOI: 10.1515/psr-2015-0011

Motia, K., Kumar, R., & Akhai, S. (2024). AI and Smart Manufacturing: Building Industry 4.0. In *Modern Management Science Practices in the Age of AI* (pp. 1-28). IGI Global.

Mouritz, A. P. (2012). *Introduction to aerospace materials*. Elsevier.

Ojo, O. O., & Taban, E. (2023). Post-processing treatments–microstructure–performance interrelationship of metal additive manufactured aerospace alloys: A review. *Materials Science and Technology*, *39*(1), 1–41. DOI: 10.1080/02670836.2022.2130530

Parveez, B., Kittur, M. I., Badruddin, I. A., Kamangar, S., Hussien, M., & Umarfarooq, M. A. (2022). Scientific advancements in composite materials for aircraft applications: A review. *Polymers*, *14*(22), 5007. DOI: 10.3390/polym14225007 PMID: 36433134

Paz Martínez-Viademonte, M., Abrahami, S. T., Hack, T., Burchardt, M., & Terryn, H. (2020). A review on anodizing of aerospace aluminum alloys for corrosion protection. *Coatings*, *10*(11), 1106. DOI: 10.3390/coatings10111106

Raj, R. J., Selvam, P., & Pughalendi, M. (2021). A review of aluminum alloys in aircraft and aerospace industry. *J. Huazhong Univ. Sci. Technol*, *1671*, 4512.

Rao, M. S., & Padal, K. B. (2022, November). New developments of aluminium alloys for future generation applications-A review. In. AIP Conference Proceedings: Vol. 2648. *No. 1* (p. 030011). AIP Publishing LLC. DOI: 10.1063/5.0114421

Rhamdani, A. R., Rhamdhani, M. A., Brooks, G., Pownceby, M. I., Nugraha Thaha, Y., Abbott, T. B., Grandfield, J., & Hartley, C. (2025). The production of rare earth based magnesium and aluminium alloys–a review. *Mineral Processing and Extractive Metallurgy Review*, *46*(1), 44–67. DOI: 10.1080/08827508.2023.2243009

Saevarsdottir, G., Kvande, H., & Welch, B. J. (2020). Reducing the carbon footprint: aluminium smelting with changing energy systems and the risk of carbon leakage. In *Light metals 2020* (pp. 726–734). Springer International Publishing. DOI: 10.1007/978-3-030-36408-3_98

Sharma, V., & Akhai, S. (2019). Mechanical Behaviour of Fly Ash Reinforced Aluminum Composite Prepared by Casting. *Journal of Advanced Research in Mechanical Engineering and Technology*, *6*(1&2), 23–26. DOI: 10.24321/2454.8650.201904

Sharma, V. K., Kumar, P., Akhai, S., Kumar, V., & Joshi, R. S. (2024). Corrosion inhibition analysis on cerium induced hydrophobic surface of Al-6061/SiC/Al2O3 hybrid composites. *Proceedings of the Institution of Mechanical Engineers. Part B, Journal of Engineering Manufacture*, *238*(14), 2126–2138. DOI: 10.1177/09544054231222048

Sharma, V. K., Kumar, P., Kumar, D., Chalisgaonkar, R., & Akhai, S. (2024). Characterization of Al 6061/Al2O3/SiC composites with cerium oxide: corrosion analysis and microstructural insights. *Composites Theory and Practice, 24.*

Shen, G., Chen, X., Yan, J., Fan, L., Yang, Z., Zhang, J., & Guan, R. (2023). A review of progress in the study of Al-Mg-Zn (-Cu) wrought alloys. *Metals*, *13*(2), 345. DOI: 10.3390/met13020345

Simons, G. M. (2021). *Boeing 737: The World's Most Controversial Commercial Jetliner*. Air World.

Singh, A., Sharma, S. K., & Batish, A. (2025). Dynamic recrystallization during solid state friction stir welding/processing/additive manufacturing: Mechanisms, microstructure evolution, characterization, modeling techniques and challenges. *Critical Reviews in Solid State and Material Sciences*, *50*(1), 77–135. DOI: 10.1080/10408436.2024.2391333

Singh, S., & Akhai, S. (2024, July). Electronics in precision control for mechanical engineering: A brief overview. In *AIP Conference Proceedings* (Vol. 3121, No. 1). AIP Publishing. DOI: 10.1063/5.0221464

Sivamaran, V., Azaruddin, S., Sivaprasad, K., Ravikumar, K., Kumar, V. M., & undefined, . (2021). A Short Review on Applications of Aluminium Composites: Automotive, Aerospace and aircraft, Rail transport, and Marine transport industry. [JPIE]. *Journal of Production & Industrial Engineering*, *2*(2). Advance online publication. DOI: 10.26706/jpie.2.2.20211202

Sobel, S. G. (2023). The Hidden Chemistry behind the Apollo 11 Mission. *After Apollo: Cultural Legacies of the Race to the Moon*, 29.

Song, G. L., Zhang, C., Chen, X., & Zheng, D. (2021). Galvanic activity of carbon fiber reinforced polymers and electrochemical behavior of carbon fiber. *Corrosion Communications*, *1*, 26–39. DOI: 10.1016/j.corcom.2021.05.003

Syrvatka, V., Rabets, A., Gromyko, O., Luzhetskyy, A., & Fedorenko, V. (2022). Scandium–microorganism interactions in new biotechnologies. *Trends in Biotechnology*, *40*(9), 1088–1101. DOI: 10.1016/j.tibtech.2022.02.006 PMID: 35346528

Thareja, P., & Akhai, S. (2017). Processing parameters of powder aluminium-fly ash P/M composites. *Journal of advanced research in manufacturing, material science & metallurgical engineering, 4*(3&4), 24-35.

Thavasilingam, K., Sakthimurugan, D., & Giridharan, K. (2025). Aluminum alloys for aircraft structures. In *Aerospace Materials* (pp. 385–404). Elsevier.

Verma, R., Akhai, S., & Wadhwa, A. S. (2024). Use of smart materials in physiotherapy. In *Revolutionizing healthcare treatment with sensor technology* (pp. 300–319). IGI Global. DOI: 10.4018/979-8-3693-2762-3.ch019

Wadhwa, A. S., & Akhai, S. (2014). Comparison of Surface Hardening Techniques for En 353 Steel Grade. *International Journal of Emerging Technology and Advanced Engineering*, *4*(10), 194–203.

Wadhwa, A. S., Akhai, S., Abbass, M., Chouksey, A., Tiwari, S., & Taneja, T. (2025). Computational Intelligence-Driven Design and Optimization of Polyurethane Belt-Type Oil Skimmer for Sustainable Manufacturing Using Solidworks 3D CAD. In *Using Computational Intelligence for Sustainable Manufacturing of Advanced Materials* (pp. 445-464). IGI Global Scientific Publishing.

Wang, W., & Cheng, C. Y. (2011). Separation and purification of scandium by solvent extraction and related technologies: A review. *Journal of Chemical Technology and Biotechnology*, *86*(10), 1237–1246. DOI: 10.1002/jctb.2655

Willingham, F. (2019). Boeing 707 group: A history. *Air Power History*, *66*(1), 56–57.

Winstone, M. (2000). Technical Evaluation Report. *New Metallic Materials for the Structure of Aging Aircraft.*

Zhou, B., Liu, B., & Zhang, S. (2021). The advancement of 7xxx series aluminum alloys for aircraft structures: A review. *Metals*, *11*(5), 718. DOI: 10.3390/met11050718

Zhu, H., Leng, M., Jin, G., & Miao, H. (2023). A Review of Research on Galvanic Corrosion of Aluminum Alloys. *Fluid Dynamics & Materials Processing, 19*(7).

Chapter 4

High-Performance Aluminium Alloys for Shipbuilding: Advances in Properties, Processing, and Manufacturing

Syed Mohd Arif
https://orcid.org/0000-0002-0821-317X
Chandigarh Group of Colleges Jhanjeri, CEC-J, India

Shalom S Akhai
https://orcid.org/0000-0002-7533-457X
Maharishi Markandeshwar Engineering College, Maharishi Markandeshwar, Ambala, India

Rohan Dadhwal
https://orcid.org/0009-0001-0832-534X
Chandigarh Group of Colleges Jhanjeri, CSB-J, Mohali, India

ABSTRACT

Aluminium alloys play a crucial role in modern shipbuilding, offering superior corrosion resistance, lightweight properties, and high structural integrity. This chapter explores key marine-grade aluminium alloys, including AA5083, AA5456, and AA6061, known for their excellent resistance to seawater corrosion and mechanical strength. High-strength alloys such as AA7020 and AA7075 are examined for their application in load-bearing structures. Surface protection technologies, including anodizing, cerium-based coatings, and microarc oxidation (MAO), enhance longevity and durability in harsh marine environments. Advanced processing techniques,

DOI: 10.4018/979-8-3373-0933-0.ch004

such as friction stir welding (FSW) and additive manufacturing, enable improved fabrication efficiency and structural performance. Future research focuses on optimizing alloy composition, enhancing recyclability, and improving sustainability for next-generation marine vessels.

INTRODUCTION

Abbreviations

FSW Friction Stir Welding
GMAW Gas Metal Arc Welding
TIG Tungsten Inert Gas (welding)
SLM Selective Laser Melting (additive manufacturing)
EBM Electron Beam Melting (additive manufacturing)
MAO Microarc Oxidation
HPDC High-Pressure Die Casting
LCA Life Cycle Assessment
IMO International Maritime Organization
Al Aluminium
Mg Magnesium
Si Silicon
Zn Zinc
Cu Copper
AA Aluminium Alloy (used as prefix, e.g., AA5083)
ASTM American Society for Testing and Materials
PEO Plasma Electrolytic Oxidation

Common Alloy Series and Example Alloys

5xxx Aluminium-Magnesium Alloys
6xxx Aluminium-Magnesium-Silicon Alloys
7xxx Aluminium-Zinc-Magnesium-Copper Alloys
AA5083 Aluminium Alloy 5083 (Al-Mg)
AA5456 Aluminium Alloy 5456 (Al-Mg)
AA6061 Aluminium Alloy 6061 (Al-Mg-Si)
AA6063 Aluminium Alloy 6063 (Al-Mg-Si)
AA7020 Aluminium Alloy 7020 (Al-Zn-Mg)
AA7075 Aluminium Alloy 7075 (Al-Zn-Mg-Cu)

Other Related Abbreviations

LCA Life Cycle Assessment
LCC Life Cycle Cost
PEO Plasma Electrolytic Oxidation
HPDC High-Pressure Die Casting
LPDC Low-Pressure Die Casting

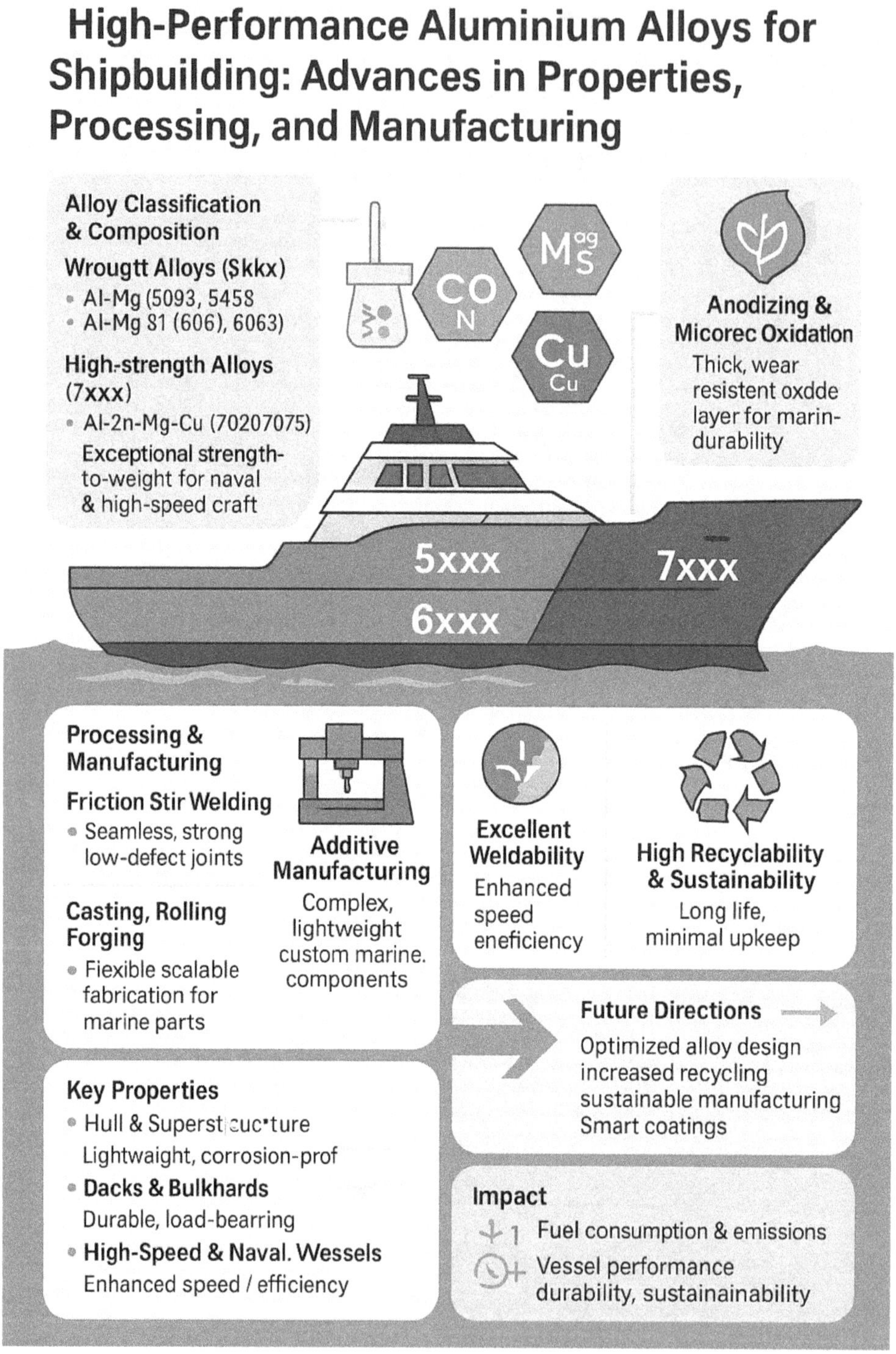

Figure 1. High-performance aluminium alloys for shipbuilding

The shipbuilding industry has undergone significant transformations over the past century, driven in part by advances in materials science. Among the materials that have revolutionized shipbuilding, aluminium alloys stand out for their unique combination of properties, including lightweight characteristics (Jongejan et al., 2015), high strength (Emmanuel et al., 2021), and excellent corrosion resistance (Maritimer, 1967). These attributes make aluminium alloys particularly suitable for marine environments, where exposure to seawater, humidity, and mechanical stresses can degrade traditional materials like steel.

Aluminium alloys offer several advantages over conventional shipbuilding materials (Thareja & Akhai, 2016; Kumar et al., 2024a; Kumar et al., 2024b). Their lightweight nature (Jongejan et al., 2015) reduces fuel consumption and increases speed, making them ideal for high-speed marine vessels and naval ships (Srivatsav et al., 2021). Additionally, their inherent resistance to corrosion minimizes maintenance costs and extends the lifespan of marine structures (Jongejan et al., 2015). The key properties of aluminium alloys, including their lightweight nature, high strength, and corrosion resistance (Sharma & Akhai, 2019; Thareja & Akhai, 2017). This makes Aluminium an attractive choice for shipbuilders.

Aluminium alloys can be broadly classified into two categories: wrought alloys, which are processed by mechanical means such as rolling or forging, and cast alloys, produced by pouring molten aluminium into molds. The selection of alloy depends on the specific requirements of the vessel, including weight constraints, exposure to corrosive environments, and structural demands. Processing techniques for aluminium alloys include welding, surface treatment, and heat treatment. Welding aluminium requires specific techniques, such as Gas Metal Arc Welding and Tungsten Inert Gas welding, due to its thermal conductivity and reactivity (Akhai et al., 2021; Arora & Akhai, 2015; Singh & Akhai, 2015; Kumar et al., 2023). Surface treatments, such as anodizing or applying protective coatings, enhance the durability of aluminium structures exposed to seawater. Heat treatment can also be used to enhance mechanical properties by heating the alloy to a specific temperature followed by rapid cooling. Aluminium alloys find extensive applications across various types of vessels, including naval ships (Srivatsav et al., 2021), due to their lightweight nature and corrosion resistance. Their use has become increasingly prevalent in the shipbuilding industry, driving innovation and efficiency in marine vessel design and construction.

This chapter explores the key properties, classification, processing techniques, and applications of marine-grade aluminium alloys, while also addressing challenges and future trends in the field.

KEY PROPERTIES OF MARINE-GRADE ALUMINIUM ALLOYS

Marine-grade aluminum alloys are specifically designed for demanding marine environments, offering unique properties for shipbuilding and various applications. Their exceptional strength-to-weight ratio provides excellent mechanical strength while being significantly lighter than steel (Jongejan et al., 2015; Maritimer, 1967), enhancing stability and reducing power consumption (Maritimer, 1967). The corrosion resistance of alloys like those in the 5xxx and 6xxx series is a critical advantage, thanks to the natural formation of a protective oxide layer (Jongejan et al., 2015). Additional protection can be achieved through surface treatments and coatings. Furthermore, the weldability of alloys like AA6061 facilitates the creation of complex structures with minimal defects (A Guide to Marine Grade Aluminum, 2022), enhanced by advanced techniques like friction stir welding (Wahid et al., 2019). Overall, marine-grade aluminum alloys combine strength, corrosion resistance, and weldability, making them an attractive choice for marine applications (Tjahjanti et al., 2013). **Table 1** provides a comprehensive overview of the key properties and benefits of marine-grade aluminium alloys, including their strength, durability, and performance, and their applications in shipbuilding.

Table 1. Essential characteristics of marine-grade aluminium alloys

Property	Description	Benefits in Marine Applications	Notable Alloys
Strength-to-Weight Ratio	High strength while being significantly lighter than steel.	- Reduces overall vessel weight, improving speed and fuel efficiency. - Enhances maneuverability and structural integrity.	5083, 6061, 6082
Corrosion Resistance	Forms a natural oxide layer that protects against saltwater and marine environments.	- Extends vessel lifespan. - Reduces maintenance costs. - Prevents pitting and stress corrosion cracking.	5083, 5456, 5052
Weldability	Excellent weldability, allowing efficient construction and seamless joints.	- Enables the creation of complex structures. - Reduces defects and enhances durability.	6061, 5083, 5383
Surface Treatment Compatibility	Can be anodized or coated to enhance corrosion resistance and durability.	- Improves longevity and aesthetic appeal. - Provides additional protection in extreme conditions.	All marine-grade aluminium alloys

continued on following page

Table 1. Continued

Property	Description	Benefits in Marine Applications	Notable Alloys
Fabrication Versatility	Easily formed into various shapes via rolling, extrusion, and machining.	- Enables diverse shipbuilding applications. - Reduces manufacturing complexity and cost.	5052, 6061, 5086
Impact on Performance	Lightweight nature reduces fuel consumption and emissions.	- Increases efficiency and sustainability. - Meets environmental regulations.	5083, 6061, 5456

Strength-to-Weight Ratio

One of the most significant advantages of aluminum alloys is their high strength-to-weight ratio (Sharma & Akhai 2019; Continental Steel & Tube Company, 2020). This property is critical in shipbuilding, where reducing weight without compromising structural integrity can lead to improved fuel efficiency and higher speeds (Maritimer, 1967).

- **Comparison with Steel:** Aluminum alloys are significantly lighter than steel—up to 60% less in weight (Jongejan et al., 2015)—while still providing comparable or even superior strength (Kasten, 2025). This weight reduction is particularly advantageous for vessels that require high speed and agility, such as naval ships and fast ferries (Emmanuel et al., 2021).
- **Marine-Grade Alloys:** Alloys in the 5xxx and 6xxx series are particularly noted for their strength-to-weight characteristics (Zimmerman, 2018). For instance, the 5083 alloy is widely used in shipbuilding due to its excellent mechanical strength combined with lightweight properties, making it suitable for hulls and superstructures (Properties and Applications of 5052 Aluminum, 2025).
- **Impact on Performance:** The lightweight nature of aluminum alloys contributes to lower fuel consumption (Maritimer, 1967). This efficiency is increasingly important as environmental regulations tighten globally, pushing shipbuilders to seek materials that enhance fuel economy while maintaining performance standards (Tański & Snopiński, 2017).

Corrosion Resistance in Marine Environments

Corrosion resistance is a critical property of marine-grade aluminum alloys, essential for withstanding the harsh conditions they face (Maritimer, 1967).

- **Natural Oxide Layer:** Aluminum alloys naturally form a protective oxide layer when exposed to air, enhancing their corrosion resistance (Jongejan et al., 2015). This layer acts as a barrier against environmental factors (Maritimer, 1967).
- **Performance in Saltwater:** In marine environments, alloys like AA5083 and AA5456 exhibit exceptional performance due to their magnesium and manganese content, which enhance resistance to pitting and stress corrosion cracking (Srivatsav et al., 2021).
- **Surface Treatments:** The corrosion resistance of aluminum can be further enhanced through surface treatments like anodizing, which increases the thickness of the natural oxide layer (Jongejan et al., 2015). Anodic-cathodic protection can also be employed (Sinyavskii & Kalinin, 2005).

Weldability and Fabrication

The weldability of aluminum alloys is crucial in shipbuilding, enabling the creation of complex structures (A Guide to Marine Grade Aluminum, 2022).

- **Excellent Weldability:** Alloys such as AA6061 are known for their excellent weldability, allowing for efficient construction processes (A Guide to Marine Grade Aluminum, 2022).
- **Advanced Welding Techniques:** Advanced welding techniques like friction stir welding have further improved the quality and efficiency of aluminum alloy fabrication (Wahid et al., 2019). FSW produces high-strength joints with minimal distortion (Wahid et al., 2019).
- **Fabrication Versatility:** Marine-grade aluminum alloys can be easily formed into various shapes through rolling, extrusion, and machining, allowing for tailored components in marine applications (Glaser & Shockley, 1970).

CLASSIFICATION OF MARINE ALUMINUM ALLOYS

Marine aluminum alloys are classified based on their composition and properties. The classification includes wrought alloys, specifically the 5xxx and 6xxx series (Srivatsav et al., 2021), and high-strength alloys from the 7xxx series. The 5xxx series alloys, like AA5083 and AA5456, are used in shipbuilding due to their high magnesium content (Srivatsav et al., 2021). The 6xxx series, including AA6061, offers a balance of strength, weldability, and machinability (A Guide to Marine Grade Aluminum, 2022). The 7xxx series alloys, like AA7020 and AA7075, are known for their high strength and are used in load-bearing structures (Srivatsav et al., 2021). **Table 2** provides a comprehensive classification and properties of marine aluminium alloys, including their composition, properties, and applications, as well as their types, uses, and comparisons of marine-grade alloys by series.

Table 2. Comparison of marine-grade aluminium alloys by series and their uses

Alloy Series	Key Alloys	Composition	Key Properties	Common Applications
5xxx	AA5083	Al-Mg	High corrosion resistance; good weldability	Hulls, decks, superstructures
	AA5456	Al-Mg	High strength; excellent weldability	Structural components
6xxx	AA6061	Al-Mg-Si	Good mechanical properties; heat-treatable	Frames, supports
	AA6063	Al-Mg-Si	Excellent extrudability; aesthetic finish	Architectural elements
7xxx	AA7020	Al-Zn	High strength; lightweight	Naval ships
	AA7075	Al-Zn	Exceptional strength-to-weight ratio	High-speed vessels

Wrought Aluminium Alloys (5xxx & 6xxx Series)

5xxx Series Alloys

The 5xxx series alloys are primarily composed of aluminium and magnesium, which significantly enhances their strength and corrosion resistance. Notable examples include:

- **AA5083**: This alloy is highly regarded for its excellent corrosion resistance and is commonly used in shipbuilding for hulls, decks, and superstructures. Its ability to withstand harsh marine environments makes it a preferred choice for various marine applications.
- **AA5456**: Similar to AA5083, this alloy also exhibits high strength and good weldability. It is often used in applications requiring enhanced performance under stress, making it suitable for critical structural components in ships.

The high magnesium content in these alloys contributes to their outstanding resistance to seawater corrosion, making them ideal for vessels operating in challenging marine conditions.

6xxx Series Alloys

The 6xxx series alloys, which include AA6061 and AA6063, are characterized by their balanced properties of strength, weldability, and machinability. Key features include:

- **AA6061**: Known for its versatility, this alloy is commonly used in various marine applications due to its good mechanical properties and excellent weldability. It is often utilized for structural components such as frames and supports
- **AA6063**: This alloy is particularly valued for its aesthetic finish and is often used in architectural applications on ships. Its good extrudability allows for the creation of complex shapes needed in modern ship designs.

The 6xxx series alloys are heat-treatable, which further enhances their mechanical properties, making them suitable for a wide range of marine applications.

High-Strength Aluminium Alloys (7xxx Series)

The 7xxx series aluminium alloys are known for their exceptional strength and are primarily used in load-bearing structures. Key examples include:

- **AA7020**: This alloy offers high strength and is often employed in naval ships where structural integrity is crucial. Its lightweight nature aids in reducing overall vessel weight while maintaining performance.
- **AA7075**: Renowned for its high strength-to-weight ratio, AA7075 is frequently used in high-speed vessels and military applications. Its ability to withstand significant loads makes it ideal for critical components that require durability without adding excess weight.

These high-strength alloys are particularly valuable in applications where weight reduction is essential without compromising structural integrity. Their use enhances the performance of naval ships and other high-speed marine vessels.

Wrought Aluminum Alloys (5xxx & 6xxx Series)

The 5xxx series alloys are primarily composed of aluminum and magnesium, which significantly enhances their strength and corrosion resistance (Endre & Popoviç, 2006). Notable examples include:

- **AA5083**: This alloy is highly regarded for its excellent corrosion resistance and is commonly used in shipbuilding for hulls, decks, and superstructures (A Guide to Marine Grade Aluminum, 2022). Its ability to withstand harsh marine environments makes it a preferred choice for various marine applications (Aluminium in Marine Applications – Aluminium Alloys Used in Boat Building, 2020).
- **AA5456**: Similar to AA5083, this alloy also exhibits high strength and good weldability. It is often used in applications requiring enhanced performance under stress, making it suitable for critical structural components in ships.
- The high magnesium content in these alloys contributes to their outstanding resistance to seawater corrosion (Endre & Popoviç, 2006), making them ideal for vessels operating in challenging marine conditions.

The 6xxx series alloys, which include AA6061 and AA6063, are characterized by their balanced properties of strength, weldability, and machinability (Srivatsav et al., 2021). Key features include:

- **AA6061**: Known for its versatility, this alloy is commonly used in various marine applications due to its good mechanical properties and excellent weldability (Wahid et al., 2019). It is often utilized for structural components such as frames and supports.
- **AA6063**: This alloy is particularly valued for its aesthetic finish and is often used in architectural applications on ships. Its good extrudability allows for the creation of complex shapes needed in modern ship designs (Guttmann & Dubost, 2006).

The 6xxx series alloys are heat-treatable, which further enhances their mechanical properties, making them suitable for a wide range of marine applications.

High-Strength Aluminum Alloys (7xxx Series)

- The 7xxx series aluminum alloys are known for their exceptional strength and are primarily used in load-bearing structures (Emmanuel et al., 2021). Key examples include:

- **AA7020**: This alloy offers high strength and is often employed in naval ships where structural integrity is crucial. Its lightweight nature aids in reducing overall vessel weight while maintaining performance.
- **AA7075**: Renowned for its high strength-to-weight ratio, AA7075 is frequently used in high-speed vessels and military applications. Its ability to withstand significant loads makes it ideal for critical components that require durability without adding excess weight (Sokoluk et al., 2019).

These high-strength alloys are particularly valuable in applications where weight reduction is essential without compromising structural integrity. Their use enhances the performance of naval ships and other high-speed marine vessels (Tański & Snopiński, 2017).

PROCESSING AND MANUFACTURING TECHNOLOGIES

The processing and manufacturing technologies employed in the fabrication of marine-grade aluminum alloys play a crucial role in determining the performance and longevity of shipbuilding components. Traditional casting and forming techniques are crucial for fabricating aluminum alloy components, producing complex shapes and structures with precision (Kumar et al., 2024; Sharma et al., 2024a; Sharma et al., 2024b; Sharma et al., 2024c). Friction Stir Welding is a game-changing technology in shipbuilding, producing high-quality joints with minimal distortion and defects (Ogata et al., 2014). FSW is particularly beneficial for joining thick plates and complex geometries. Additive manufacturing, or 3D printing, is gaining traction in the shipbuilding industry, enabling lightweight, customized components with reduced material waste (Kim et al., 2023; Akhai & Khang, 2024; Khang & Akhai, 2024; Wadhwa et al., 2025; Akhai & Wadhwa, 2024; Akhai 2023). Aluminum alloys are well-suited for additive manufacturing due to their excellent mechanical properties and compatibility with advanced printing techniques (Dhinakaran et al., 2020). **Table 3** discusses manufacturing technologies for marine aluminium alloys, their production methods, key techniques, advanced methods for shipbuilding aluminium alloys, and comparison of processing technologies for marine-grade aluminium.

Table 3. Manufacturing technologies for marine aluminium alloys

Technology	Method	Key Features	Advantages	Common Marine Applications
Casting	Die Casting	High-pressure injection of molten aluminium into molds	- High dimensional accuracy - Smooth surface finish	High-volume production of complex parts
	Low-Pressure Die Casting	Lower pressure filling for reduced porosity	- Improved mechanical properties - Controlled solidification	Structural components requiring durability
	Sand Casting	Molten aluminium poured into sand molds	- Flexible part sizes - Cost-effective for low-volume production	Large marine components
Forming	Extrusion	Heated aluminium forced through a die to form long shapes	- Uniform cross-sections - Strong structural integrity	Beams, frames, and support structures
	Rolling	Aluminium ingots pressed into sheets or plates	- Precision thickness control - Versatile applications	Ship hulls, decks, and panels
	Forging	Shaping aluminium through compressive forces	- High-strength components - Enhanced grain structure	Load-bearing marine components
Welding	Friction Stir Welding (FSW)	Solid-state welding using friction-generated heat	- Strong joints with minimal defects - Reduced heat-affected zone	Thick plate welding, hull construction
Additive Manufacturing	3D Printing (Selective Laser Melting - SLM)	Layer-by-layer fabrication of aluminium components	- Customization and complex geometries - Reduced material waste	Spare parts, lightweight structures, intricate components

Casting and Forming Techniques

Casting and forming techniques are fundamental processes in the production of aluminum alloy components, allowing for the creation of complex shapes and structures with high precision (Kaufman & Rooy, 2004).

Traditional Casting Methods

- **Die Casting**: This method involves injecting molten aluminum into a mold under high pressure (Nunes et al., 2023). The process is characterized by several key steps:
  - **Die Preparation**: The steel die is preheated and coated with a release agent to facilitate easy ejection of the casting (Setyarini et al., 2019).
  - **Injection**: Molten aluminum is injected into the die cavity at pressures ranging from 10 to 175 megapascals, ensuring that intricate details are filled effectively (Nunes et al., 2023).
  - **Cooling and Ejection**: Once the cavity is filled, pressure is maintained until solidification occurs. The die is then opened, and the casting is ejected.

Die casting produces components with excellent surface finish and dimensional accuracy, making it suitable for high-volume production of complex parts used in marine applications.

- **Low-Pressure Die Casting**: This technique employs lower pressure levels (0.3 to 1.5 bar) to fill the mold, which reduces porosity formation during casting (Nunes et al., 2023). The process includes phases such as riser tube filling, cavity filling, and pressure holding, allowing for controlled solidification and improved mechanical properties.
- **High-Pressure Die Casting**: HPDC enhances production efficiency by minimizing air entrapment through rapid filling of the die cavity (Nunes et al., 2023). This method can be further refined using vacuum-assisted techniques to reduce porosity and improve component strength.
- **Sand Casting**: Although less common for marine applications, sand casting allows for the production of large components by pouring molten aluminum into sand molds. This method is more flexible in terms of part size but typically results in lower precision compared to die casting.

Forming Techniques

Forming processes such as extrusion, rolling, and forging are also essential in shaping aluminum alloys for marine applications:

- **Extrusion**: This process involves forcing heated aluminum through a die to create long shapes with uniform cross-sections. It is commonly used for producing structural components like beams and frames [0#x].

- **Rolling**: Rolling techniques are employed to create sheets or plates from aluminum ingots, which can then be cut or shaped for specific applications [0#x].
- **Forging**: Forging involves shaping metal using compressive forces, resulting in components with enhanced mechanical properties due to refined grain structures [0#x].

Friction Stir Welding for Shipbuilding

Friction stir welding has emerged as a revolutionary technology in the shipbuilding industry, particularly for joining aluminum alloys (Thomas & Nicholas, 1997). This solid-state welding technique offers several advantages (Raval & Judal, 2020):

- **High-Quality Joints**: FSW produces strong joints with minimal distortion and defects compared to traditional welding methods (Kotari et al., 2020). The process involves rotating a non-consumable tool that generates heat through friction while moving along the joint line, effectively softening the material without melting it (Khalaf et al., 2022).
- **Ideal for Thick Plates**: FSW is particularly advantageous for joining thick plates and complex geometries commonly found in shipbuilding applications (Ogata et al., 2014). This capability allows for greater design flexibility while maintaining structural integrity.
- **Reduced Heat-Affected Zone**: Since FSW does not involve melting the base material, it minimizes the heat-affected zone, reducing the risk of thermal distortion and maintaining desirable mechanical properties (Ramesha et al., 2021).
- **Environmental Benefits**: FSW is also considered more environmentally friendly than traditional welding methods since it does not require filler materials or shielding gases.

Additive Manufacturing in Marine Applications

Additive manufacturing, commonly known as 3D printing, is gaining traction in the shipbuilding industry due to its ability to produce lightweight, customized components with reduced material waste.

- **Customization**: Additive manufacturing allows for rapid prototyping and customization of parts tailored to specific requirements. This flexibility can lead to innovative designs that optimize performance while minimizing weight.

- **Material Efficiency**: Unlike traditional subtractive manufacturing processes that often result in significant material waste, additive manufacturing builds components layer by layer, significantly reducing waste (Sharma et al., 2022).
- **Compatibility with Aluminum Alloys**: Aluminum alloys are well-suited for additive manufacturing due to their excellent mechanical properties and compatibility with advanced printing techniques. Processes such as selective laser melting enable precise control over component geometry and material properties.
- **Applications**: In marine contexts, additive manufacturing can be used for producing spare parts, complex geometries that would be difficult or impossible to fabricate using traditional methods, and even entire structural components that contribute to overall vessel performance.

SURFACE PROTECTION AND CORROSION RESISTANCE

The durability and longevity of marine-grade aluminum alloys are significantly enhanced through various surface protection and corrosion resistance techniques. The selection of appropriate surface treatment is crucial due to the aggressive nature of marine environments, where aluminum alloys are prone to corrosion (Panda, 2021). Following is an exploration of critical methods:

- **Anodizing and Cerium-based Coatings:** Anodizing is an electrochemical process that creates a protective oxide layer on the aluminum surface, enhancing corrosion resistance and durability (Benmohamed et al., 2022; Sun et al., 2020). Cerium-based coatings can provide additional protection (Benmohamed et al., 2022).
- **Microarc Oxidation:** MAO creates a hard, wear-resistant oxide layer on aluminum alloys, particularly beneficial in harsh marine environments (Sun et al., 2020).
- **Anti-fouling Treatments:** These treatments prevent marine organism growth on ship hulls, ensuring long-term performance of aluminum alloy structures (Fayomi et al., 2020). They are often combined with protective coatings (Cerchier et al., 2020). The growth of organisms on wetted surfaces can increase weight and consequently fuel consumption in boats (Cerchier et al., 2020).
- **Superhydrophobic Coatings:** Coatings can be created to provide good adherence to the substrate and abrasion resistance, acting as a barrier to marine atmospheric corrosion (Lu et al., 2014).

- **Organic Coatings:** Organic coatings for marine applications must have great corrosion protection and antifouling performance (Song & Feng, 2020). The presence of defects in an organic coating is the root cause of corrosion damage (Song & Feng, 2020).
- **PEO Coatings:** Plasma electrolytic oxidation is a surface treatment that increases corrosion and wear resistance of light alloys and allows the introduction of particles into the coating (Cerchier et al., 2020).

Table 4 discusses surface protection and corrosion resistance methods for marine aluminium alloys, compares corrosion resistance techniques, provides protective treatments for marine-grade alloys, discusses corrosion prevention and surface protection in marine applications, and discusses advanced coatings and treatments for marine aluminium durability.

Table 4. Protective treatments for marine-grade aluminium alloys

Method	Process Description	Key Benefits	Common Marine Applications
Anodizing	Electrochemical process forming a thick oxide layer on aluminium	Enhances corrosion resistance, improves adhesion for coatings, allows for coloring	Ship superstructures, marine fittings
Cerium-Based Coatings	Environmentally friendly alternative to chromate coatings	Provides corrosion protection, forms a strong barrier, non-toxic	Hulls, structural components
Microarc Oxidation (MAO)	High-voltage electrochemical oxidation process creating a wear-resistant layer	Increases hardness, improves wear resistance, excellent performance in harsh conditions	Propellers, marine engine parts
Anti-Fouling Coatings	Protective coatings with biocides to prevent marine organism growth	Reduces biofouling, maintains hydrodynamic efficiency, lowers fuel consumption	Ship hulls, underwater structures
Protective Coatings	Barrier coatings applied to aluminium surfaces	Prevents corrosion and physical damage, extends material lifespan	Decks, exposed marine components

Anodizing

As you mentioned, anodizing is indeed a widely used electrochemical process to improve corrosion resistance (Benmohamed et al., 2022; Sun et al., 2020) and durability.

- **Formation of Oxide Layer:** It creates a thick, protective oxide layer on the aluminum surface. This layer is highly resistant to corrosion and wear, which is beneficial in marine environments (Егоркин et al., 2020). The corrosion process is controlled by ion diffusion throughout the coatings (Sun et al., 2020).

- **Enhanced Adhesion:** The anodized surface provides excellent adhesion for paints and other coatings, further enhancing protection against environmental factors (Егоркин et al., 2020).
- **Coloring Options:** Anodizing can add color for aesthetic purposes (Benmohamed et al., 2022).

Cerium-Based Coatings

Cerium-based coatings are gaining traction as a replacement for chromate coatings (Kanani et al., 2014).

- **Eco-Friendly Solution:** They are environmentally friendly, lacking harmful substances like chromium.
- **Barrier Protection:** These coatings create a barrier preventing corrosive elements from reaching the metal (Kanani et al., 2014). Research suggests cerium salts can inhibit corrosion by covering cathodic intermetallic sites on aluminum alloys.
- **Application Methods:** Application includes dip coating or spray application, and the coating's thickness and uniformity are optimized through controlled parameters.

Microarc Oxidation for Durability

MAO is an advanced technique for a hard, wear-resistant oxide layer (Zhang et al., 2014).

- **Hardness and Wear Resistance:** MAO generates a thicker oxide layer than anodizing, enhancing wear resistance for abrasive conditions (Zhang et al., 2020).
- **Effective in Harsh Conditions:** MAO-treated surfaces show excellent corrosion resistance in harsh marine environments, being exposed to both abrasive particles and corrosive seawater (Zhang et al., 2014). The surface of the coating can become more compact and smooth with the increase of the cathode voltage (Shen et al., 2017).
- **Customization:** The process allows customization of the oxide layer's properties by adjusting parameters like voltage and electrolyte composition (Shen et al., 2017).

APPLICATIONS OF ALUMINIUM ALLOYS IN SHIPBUILDING

Aluminum alloys are indeed gaining popularity in shipbuilding thanks to their properties (Glaser & Shockley, 1970). Following are few key applications:

- **Hulls and Superstructures:** Aluminum alloys are commonly used in passenger ships, ferries, and naval vessels (Tjahjanti et al., 2013). The weight savings can be significant, around 35% to 45% in hulls and 55% to 65% in superstructures (Jongejan et al., 2015). This reduction improves speed, maneuverability, and payload capacity (Muribwathoho et al., 2022).
- **Decks, Bulkheads, and Load-Bearing Components:** Aluminum alloys contribute to the overall strength and durability of vessels (Hosseinabadi & Khedmati, 2021). High-strength alloys like AA7075 are often used in critical load-bearing applications (Hosseinabadi & Khedmati, 2021).
- **High-Speed Marine Vessels and Naval Ships:** Aluminum alloys are well-suited for these vessels, reducing fuel consumption (Hosseinabadi & Khedmati, 2021).
- **Offshore Structures**: Aluminum alloys corrode much slower than steel in seawater environments, reducing maintenance costs (Jongejan et al., 2015).

In summary, aluminum alloys offer a compelling combination of lightweight construction and corrosion resistance for diverse marine applications (Maritimer, 1967). Their use can lead to more efficient, durable, and sustainable vessels (Aluminium for Modern Marine Applications, 2025). **Table 5** summarizes the applications of aluminium alloys in shipbuilding, highlighting their advantages, specific use cases, and the impact on vessel performance, efficiency, and sustainability. **Table 6** provides a comprehensive overview of the applications of aluminium alloys in shipbuilding, highlighting their advantages, specific use cases, and the impact on vessel performance, efficiency, and sustainability.

Hull Structures and Superstructures

As you mentioned, the hull and superstructure are critical for buoyancy, stability, and structural integrity, and aluminum alloys are increasingly favored for these applications (Maritimer, 1967).

Lightweight Characteristics

- Aluminum's density (~2.7 g/cm^3) compared to steel (~7.85 g/cm^3) leads to significant weight savings: 35-45% in hulls and 55-65% in superstructures (Muribwathoho et al., 2022).
- Benefits of weight reduction:
  - Improved speed and maneuverability, making aluminum ideal for high-speed ferries, patrol boats, and naval ships (Tański & Snopiński, 2017).
  - Increased payload capacity without compromising structural integrity.
  - Enhanced fuel efficiency, important for reducing costs and meeting environmental regulations.

Corrosion Resistance

- Aluminum alloys naturally form a protective oxide layer, preventing further corrosion (Muribwathoho et al., 2022).
- The addition of magnesium and manganese enhances this property in marine environments.
- Aluminum corrodes at a much lower rate (around 1 micrometer per year) compared to steel (up to 120 micrometers per year) (Maritimer, 1967).
- Superior corrosion resistance extends the lifespan of marine structures and lowers maintenance costs (Muribwathoho et al., 2022).

Commonly Used Alloys

- AA5083: High magnesium content, excellent corrosion resistance, and weldability, making it suitable for hulls (Zhang et al., 2019).
- AA5456: Similar to AA5083, used where high strength and corrosion resistance are needed.
- AA6061: Good balance of strength, weldability, and machinability, suitable for superstructures.
- AA6063: Often used in extrusions, ideal for complex superstructure designs.

Applications in Vessel Types

- Passenger Ships and Ferries: Lightweight aluminum allows for larger passenger capacities and improved fuel efficiency (Radić et al., 2020).
- Naval Vessels: Corrosion resistance and strength are ideal for military applications where durability and performance are critical.

- Offshore Support Vessels: Benefit from reduced weight and enhanced corrosion resistance (Muribwathoho et al., 2022).

Decks, Bulkheads, and Load-Bearing Components

Aluminum alloys are not just for hulls and superstructures; they're also used for decks, bulkheads, and load-bearing components (Muribwathoho et al., 2022).

Enhanced Strength and Durability

- Aluminum alloys offer excellent mechanical properties for load-bearing applications (Muribwathoho et al., 2022).
- High-strength alloys like AA7075 are used in critical structural components (Endre & Popoviç, 2006).
  - AA7075: Known for its high strength-to-weight ratio.
  - AA5086: High magnesium content provides excellent strength and corrosion resistance, ideal for decks and bulkheads.

Design Flexibility

- Lightweight nature allows for innovative designs that optimize space without sacrificing stability or strength (Glaser & Shockley, 1970).
- Modular Construction: Ease of fabrication enables modular techniques, reducing assembly time and costs.
- Complex Geometries: Malleability allows for creating complex shapes and structures, enhancing functionality and aesthetics.

Reduced Maintenance Costs

- The inherent corrosion resistance of aluminum reduces maintenance needs compared to steel (Maritimer, 1967).
- Anti-Fouling Treatments: Can be applied to aluminum decks and bulkheads to prevent marine organism growth, further reducing maintenance.
- Longevity: Durability ensures decks and bulkheads remain functional and aesthetically pleasing for extended periods.

High-Speed Marine Vessels and Naval Ships

Aluminum alloys are particularly well-suited for high-speed marine vessels and naval ships (Hosseinabadi & Khedmati, 2021).

Speed and Agility

- Lightweight nature enhances acceleration and maneuverability, ideal for high-speed ferries, patrol boats, and naval vessels (Hosseinabadi & Khedmati, 2021).
- High-Speed Ferries: Aluminum hulls and superstructures enable speeds up to 40 knots or more.
- Patrol Boats: Agility of aluminum vessels allows for quick response times.

Fuel Efficiency

- Reducing vessel weight leads to lower fuel consumption (Aluminium for Modern Marine Applications, 2025).
- Environmental Compliance: Aluminum vessels can comply with IMO's High-Speed Code requirements, which focus on reducing emissions and improving fuel efficiency.
- Operational Cost Savings: Lower fuel consumption translates into significant cost savings.

Compliance with Standards

Aluminum ships are designed to meet stringent safety and performance standards. The material's ability to withstand torsional, flexural, compression, and impact loads further solidifies its suitability for high-speed applications (Hosseinabadi & Khedmati, 2021).

- **Fire Safety**: Aluminum alloys have a high melting point and do not emit toxic fumes when exposed to fire, making them a safe choice for high-speed vessels.
- **Structural Integrity**: The material's strength and durability ensure that vessels can withstand the rigors of high-speed travel and harsh marine environments (Muribwathoho et al., 2022).

Applications in Naval Ships

Naval ships benefit from the use of aluminum alloys in several ways:

- **Reduced Radar Signature**: Aluminum's low density and reflective properties help reduce the radar signature of naval vessels, enhancing stealth capabilities.

- **Enhanced Survivability**: The material's resistance to corrosion and fatigue ensures that naval ships remain operational in challenging conditions (Maritimer, 1967).
- **Improved Performance**: The lightweight nature of aluminum allows for faster speeds and greater maneuverability, which are critical in military operations (Hosseinabadi & Khedmati, 2021).

Emerging Applications and Future Trends

Aluminum alloys are popular in several fields like aircraft structures (Akhai, 2024; Akhai et al., 2024; Khang & Akhai, 2025). As the shipbuilding industry evolves, new applications for aluminum alloys are emerging (Glaser & Shockley, 1970).

- **Hybrid Vessels**: Hybrid vessels benefit from the lightweight nature of aluminum, allowing for greater energy efficiency and extended range.
- **Autonomous Ships**: Aluminum alloys are well-suited for autonomous ships due to their strength, corrosion resistance, and ease of fabrication (Muribwathoho et al., 2022).
- **Sustainable Shipbuilding**: Aluminum alloys are highly recyclable, making them an environmentally friendly choice for future marine vessels (Muribwathoho et al., 2022).

Table 5. Applications of aluminium alloys in shipbuilding, highlighting their advantages, specific use cases, and the impact on vessel performance, efficiency, and sustainability

Application	Key Advantages	Common Alloys Used	Specific Use Cases	Impact on Vessel Performance
Hull Structures	- Lightweight (35-45% lighter than steel) - Superior corrosion resistance	AA5083, AA5456, AA6061	- Passenger ships - Ferries - Naval vessels - Offshore support vessels	- Improved speed and maneuverability - Increased payload capacity - Enhanced fuel efficiency
Superstructures	- Lightweight (55-65% lighter than steel) - Excellent weldability	AA5083, AA6061, AA6063	- Cruise ships - Naval ships - High-speed ferries	- Reduced weight for better stability - Enhanced design flexibility - Lower maintenance costs
Decks	- High strength-to-weight ratio - Corrosion resistance	AA5086, AA6061, AA7075	- Passenger ships - Cargo ships - Naval vessels	- Improved durability - Reduced maintenance - Enhanced load-bearing capacity
Bulkheads	- Lightweight - Design flexibility - Corrosion resistance	AA5083, AA5086, AA6061	- Commercial ships - Naval ships - Offshore platforms	- Optimized space utilization - Lower lifecycle costs - Enhanced structural integrity
Load-Bearing Components	- High strength - Durability - Corrosion resistance	AA7075, AA5086, AA6061	- Naval ships - High-speed vessels - Offshore structures	- Improved structural integrity - Reduced weight for better performance - Lower maintenance
High-Speed Marine Vessels	- Lightweight - High strength - Fuel efficiency	AA5083, AA5456, AA7075	- Patrol boats - High-speed ferries - Naval ships	- Enhanced speed and agility - Lower fuel consumption - Compliance with IMO standards
Naval Ships	- Reduced radar signature - Enhanced survivability - Lightweight	AA5083, AA5456, AA7075	- Destroyers - Frigates - Aircraft carriers	- Improved stealth capabilities - Greater maneuverability - Extended operational life
Hybrid Vessels	- Lightweight - Energy efficiency - Recyclability	AA5083, AA6061, AA7075	- Hybrid-electric ships - Autonomous vessels	- Reduced emissions - Extended range - Lower operational costs

continued on following page

Table 5. Continued

Application	Key Advantages	Common Alloys Used	Specific Use Cases	Impact on Vessel Performance
Autonomous Ships	- Lightweight - Durability - Ease of fabrication	AA5083, AA6061, AA7075	- Unmanned cargo ships - Autonomous patrol boats	- Enhanced efficiency - Reduced operational costs - Improved safety
Sustainable Shipbuilding	- Recyclability - Corrosion resistance - Lightweight	AA5083, AA6061, AA7075	- Eco-friendly passenger ships - Green cargo ships	- Reduced environmental impact - Lower lifecycle costs - Compliance with regulations

Table 6. Detailed breakdown of the applications of aluminium alloys in shipbuilding

Hull Structures	
Aspect	**Details**
Lightweight	35-45% lighter than steel, improving speed, maneuverability, and payload capacity.
Corrosion Resistance	Natural oxide layer and alloying elements (e.g., magnesium) enhance resistance to seawater.
Common Alloys	AA5083, AA5456, AA6061
Use Cases	Passenger ships, ferries, naval vessels, offshore support vessels.
Performance Impact	Improved fuel efficiency, increased speed, and extended lifespan.
Superstructures	
Aspect	Details
Lightweight	55-65% lighter than steel, reducing overall vessel weight.
Weldability	Excellent weldability allows for complex designs and modular construction.
Common Alloys	AA5083, AA6061, AA6063
Use Cases	Cruise ships, naval ships, high-speed ferries.
Performance Impact	Enhanced stability, reduced maintenance, and improved design flexibility.
Decks	
Aspect	Details
Strength-to-Weight	High strength-to-weight ratio ensures durability without adding excessive weight.
Corrosion Resistance	Resists corrosion from seawater and humidity, reducing maintenance needs.
Common Alloys	AA5086, AA6061, AA7075
Use Cases	Passenger ships, cargo ships, naval vessels.
Performance Impact	Improved load-bearing capacity and reduced lifecycle costs.
Bulkheads	

continued on following page

Table 6. Continued

Hull Structures	
Aspect	**Details**
Aspect	Details
Design Flexibility	Lightweight nature allows for innovative designs and efficient space utilization.
Corrosion Resistance	Resists corrosion, ensuring long-term durability.
Common Alloys	AA5083, AA5086, AA6061
Use Cases	Commercial ships, naval ships, offshore platforms.
Performance Impact	Optimized space utilization and enhanced structural integrity.
Load-Bearing Components	
Aspect	Details
High Strength	High-strength alloys like AA7075 provide exceptional load-bearing capacity.
Durability	Resists fatigue and corrosion, ensuring long-term performance.
Common Alloys	AA7075, AA5086, AA6061
Use Cases	Naval ships, high-speed vessels, offshore structures.
Performance Impact	Improved structural integrity and reduced maintenance costs.
High-Speed Marine Vessels	
Aspect	Details
Lightweight	Reduces weight, enhancing speed and agility.
Fuel Efficiency	Lower fuel consumption due to reduced weight.
Common Alloys	AA5083, AA5456, AA7075
Use Cases	Patrol boats, high-speed ferries, naval ships.
Performance Impact	Faster speeds, lower operational costs, and compliance with IMO standards.
Naval Ships	
Aspect	Details
Reduced Radar Signature	Low density and reflective properties enhance stealth capabilities.
Survivability	Resists corrosion and fatigue, ensuring operational readiness.
Common Alloys	AA5083, AA5456, AA7075
Use Cases	Destroyers, frigates, aircraft carriers.
Performance Impact	Improved stealth, maneuverability, and extended operational life.
Hybrid Vessels	
Aspect	Details
Energy Efficiency	Lightweight nature improves energy efficiency and extends range.
Recyclability	Highly recyclable, supporting sustainable shipbuilding practices.
Common Alloys	AA5083, AA6061, AA7075

continued on following page

Table 6. Continued

Hull Structures	
Aspect	**Details**
Use Cases	Hybrid-electric ships, autonomous vessels.
Performance Impact	Reduced emissions, lower operational costs, and compliance with environmental regulations.
Autonomous Ships	
Aspect	Details
Ease of Fabrication	Lightweight and easy to fabricate, supporting advanced designs.
Durability	Resists corrosion and fatigue, ensuring long-term performance.
Common Alloys	AA5083, AA6061, AA7075
Use Cases	Unmanned cargo ships, autonomous patrol boats.
Performance Impact	Enhanced efficiency, reduced operational costs, and improved safety.
Sustainable Shipbuilding	
Aspect	Details
Recyclability	Nearly 75% of all aluminium ever produced is still in use, supporting sustainability.
Corrosion Resistance	Reduces maintenance and extends lifespan, lowering environmental impact.
Common Alloys	AA5083, AA6061, AA7075
Use Cases	Eco-friendly passenger ships, green cargo ships.
Performance Impact	Reduced environmental impact, lower lifecycle costs, and compliance with regulations.

CHALLENGES AND FUTURE TRENDS

The shipbuilding industry is enhancing the recyclability of aluminum alloys to reduce environmental impact (Muribwathoho et al., 2022). Future alloy design aims to optimize composition for improved mechanical properties, corrosion resistance (Maritimer, 1967), and weldability. Advanced computational tools are being used to predict and refine performance. Emerging technologies like hybrid manufacturing and smart coatings are expected to revolutionize aluminum alloy fabrication in shipbuilding. The industry faces challenges while exploring innovative trends to enhance sustainability, performance, and manufacturing processes (Scipioni et al., 2023).

Enhancing Sustainability and Recyclability

Sustainability has become a significant focus in the shipbuilding industry, with an increasing emphasis on the recyclability of materials used in construction (Kappenthuler & Seeger, 2020).

- **Recycled Aluminum Alloys**: The development of high-strength recycled aluminum alloys, such as Speira's VIA Maris Njørdal (VIA Maris Njørdal, 2024), represents a significant advancement in sustainable shipbuilding. This alloy not only offers a 15% higher strength compared to traditional options like VIA Maris 5083 but also allows for a reduction in material usage by up to 15%. Such innovations contribute to lower environmental impacts while maintaining performance standards (Duigou et al., 2016).
- **Life Cycle Assessment**: The use of life cycle assessments is becoming integral in evaluating the environmental impact of materials throughout their lifecycle (Önal et al., 2020). Companies like Speira are utilizing LCA methodologies to create Environmental Product Declarations for their products, ensuring transparency and accountability in sustainability efforts.
- **Circular Economy**: The shift towards a circular economy is crucial for enhancing sustainability in the aluminum sector (Scipioni et al., 2023). By recovering aluminum at the end of its life cycle and reintroducing it into production processes, the industry can significantly reduce emissions associated with primary aluminum production (Raabe et al., 2022). Recycling aluminum requires only 5% of the energy needed to produce new aluminum from raw materials.
- **Industry Collaboration**: There is a growing recognition of the need for collaboration across the value chain to optimize recycling rates and develop circular business models that benefit all stakeholders involved (Hadi et al., 2021). This approach can help address the projected increase in demand for aluminum due to population growth and the transition to a greener economy.

Alloy Design for Improved Performance

Optimizing alloy composition is indeed crucial for enhancing mechanical properties, corrosion resistance (Endre & Popoviç, 2006), and weldability.

- **Advanced Computational Tools:** These tools are becoming essential for predicting how compositional changes affect performance, enabling targeted development of new alloys (Hosseinabadi & Khedmati, 2021).

- **Focus on Mechanical Properties:** Improving strength-to-weight ratios while enhancing corrosion and fatigue resistance is a key goal.
- **Weldability Improvements:** Enhancing weldability through alloy design is essential for ensuring structural integrity (Wahid et al., 2019).

Emerging Technologies in Marine Aluminum Fabrication

The emergence of new technologies promises to revolutionize how aluminum components are fabricated.

- **Hybrid Manufacturing:** Combining traditional methods with additive manufacturing allows for greater design flexibility and material efficiency. This approach enables the production of complex geometries previously unattainable (Kumar et al., 2020).
- **Smart Coatings:** The development of smart coatings that respond dynamically to environmental conditions is an exciting trend. These coatings can provide real-time feedback on corrosion levels or wear, enabling proactive maintenance strategies (Hasan et al., 2024).
- **Additive Manufacturing:** Additive manufacturing (3D printing) is gaining traction in shipbuilding, enabling rapid prototyping and customization while reducing material waste (Muribwathoho et al., 2022).

The 5xxx and 6xxx series alloys are popular choices for marine applications due to their strength and corrosion resistance (Kumar et al., 2020). The use of aluminum in marine transport is increasing due to its lightness and corrosion resistance (Glaser & Shockley, 1970).

CONCLUSIONS

Aluminium alloys have revolutionized the shipbuilding industry by offering an exceptional balance of lightweight properties, high strength, and corrosion resistance. These materials have enabled the construction of more fuel-efficient, durable, and high-performance vessels. Advances in processing techniques, such as friction stir welding (FSW) and additive manufacturing (3D printing), have further enhanced the efficiency, structural integrity, and adaptability of aluminium components. Additionally, cutting-edge surface protection technologies, including anodizing and microarc oxidation (MAO), ensure the longevity of aluminium alloys in harsh marine environments.

Looking forward, the future of marine-grade aluminium alloys will be shaped by optimized alloy compositions, enhanced recyclability, and sustainable manufacturing practices. As the industry moves towards greener alternatives, emerging fabrication technologies and innovative coatings will continue to drive advancements in shipbuilding, leading to next-generation vessels that are lighter, stronger, and more environmentally friendly. Key conclusions include:

- Marine-grade aluminium alloys are indispensable in shipbuilding due to their high strength-to-weight ratio, corrosion resistance, and excellent weldability. Their continued advancement will improve vessel performance while reducing environmental impact.
- Classification into 5xxx, 6xxx, and 7xxx series alloys highlights the diversity of aluminium materials available for shipbuilding. Each alloy series offers unique advantages, ensuring optimal performance in various marine applications.
- Processing and manufacturing technologies, such as die casting, extrusion, FSW, and additive manufacturing, are critical in determining the performance characteristics of aluminium alloys. These advancements enable precision engineering, material efficiency, and design flexibility in marine applications.
- Surface protection and corrosion resistance techniques, including anodizing, cerium-based coatings, MAO, and anti-fouling treatments, play a crucial role in safeguarding aluminium alloys from harsh marine environments. Ongoing research will further enhance durability and sustainability.
- Applications of aluminium alloys in shipbuilding continue to expand with technological innovations. Their use in hulls, superstructures, decks, bulkheads, and load-bearing components has led to lighter, faster, and more efficient vessels, including naval ships, high-speed crafts, and next-generation hybrid and autonomous vessels.
- Sustainability and recyclability are becoming central to the future of aluminium alloys in shipbuilding. Adopting circular economy principles, eco-friendly coatings, and energy-efficient manufacturing will ensure that aluminium remains a preferred material for marine applications.

As materials science and marine engineering continue to evolve, aluminium alloys will play a pivotal role in developing more advanced, resilient, and sustainable ships. By harnessing the unique properties of these alloys and integrating emerging technologies, the shipbuilding industry is set to enter a new era of efficiency, durability, and environmental responsibility.

REFERENCES

Егоркин, В. С., Медведев, И. М., Sinebryukhov, S. L., Vyaliy, I. E., Gnedenkov, A. S., Nadaraia, K. V., Izotov, N. V., Mashtalyar, D. V., & Гнеденков, С. В. (2020). Atmospheric and Marine Corrosion of PEO and Composite Coatings Obtained on Al-Cu-Mg Aluminum Alloy. *Materials (Basel)*, *13*(12), 2739. DOI: 10.3390/ma13122739 PMID: 32560318

Akhai, S. (2023). Navigating the Potential Applications and Challenges of Intelligent and Sustainable Manufacturing for a Greener Future. *Evergreen*, *10*(4), 2237–2243. DOI: 10.5109/7160899

Akhai, S. (2024). Innovating Healthcare Delivery: Harnessing Drones for Medical Services. In *Modelling of Virtual Worlds Using the Internet of Things* (Vol. 1, pp. 232-250). CRC Press.

Akhai, S., & Khang, A. (2024). Innovations in Medical Manufacturing: A Review of 3D Printing, Robotics, and Internet of Things (IoT). In *The Quantum Evolution* (Vol. 1, pp. 226-241). CRC Press.

Akhai, S., Singh, S., Kaur, S., & Kaur, K. (2024). Advancing Aerospace Inspection With Terahertz Technology. In *New Innovations in AI, Aviation, and Air Traffic Technology* (pp. 195-211). IGI Global. DOI: 10.4018/979-8-3693-1954-3.ch010

Akhai, S., Srivastava, P., Sharma, V., & Bhatia, A. (2021). Investigating weld strength of AA8011-6062 alloys joined via friction-stir welding using the RSM approach. [). IOP Publishing.]. *Journal of Physics: Conference Series*, *1950*(1), 012016. DOI: 10.1088/1742-6596/1950/1/012016

Akhai, S., & Wadhwa, A. S. (2024). Recent Advances in Bio-Tribology From Joint Lubrication to Medical Implants: A Review. *Journal of Materials*, *2*(2), 125–135.

Aluminium for modern marine applications. (2025). *Hydro*. https://www.hydro.com/en-SE/aluminium/industries/marine-offshore

Arora, N., & Akhai, S. (2015). Reclaiming EN-14b steel grade implements by hardfacing. *International Journal of Scientific Research*, *4*(10), 14–16.

Benmohamed, M., Benmounah, A., Haddad, A., & Yahi, S. (2022). The effect of inhibiting molybdate used in anodizing-conversion treatment to improve corrosion protection of AA2030 aluminum alloy in different steps. *Journal of Engineering and Applied Sciences (Asian Research Publishing Network)*, *69*(1), 40. Advance online publication. DOI: 10.1186/s44147-022-00090-8

Cerchier, P., Pezzato, L., Gennari, C., Moschin, E., Moro, I., & Dabalà, M. (2020). PEO coating containing copper: A promising anticorrosive and antifouling coating for seawater application of AA 7075. *Surface and Coatings Technology, 393*, 125774. DOI: 10.1016/j.surfcoat.2020.125774

Continental Steel & Tube Company. (2020, March 26). *A guide to understanding marine grade aluminum*. https://continentalsteel.com/blog/marine-grade-aluminum/

Dhinakaran, V., Stalin, B., Ravichandran, M., Balasubramanian, M., Chairman, C. A., & Pritima, D. (2020). Wire Arc Additive Manufacturing Perspectives and Recent Developments. *IOP Conference Series. Materials Science and Engineering, 988*(1), 12102. DOI: 10.1088/1757-899X/988/1/012102

Duigou, J. L., Gulbrandsen-Dahl, S., Vallet, F., Söderberg, R., Eynard, B., & Perry, N. (2016). Optimization and lifecycle engineering for design and manufacture of recycled aluminium parts. *CIRP Annals, 65*(1), 149–152. DOI: 10.1016/j.cirp.2016.04.111

Emmanuel, A. O., Fayomi, O. S. I., & Akande, I. G. (2021). Aluminium Alloys as Advanced Materials: A short communication. *IOP Conference Series. Materials Science and Engineering, 1107*(1), 12024. DOI: 10.1088/1757-899X/1107/1/012024

Endre, R., & Popoviç, M. (2006). Problems and prospect of Al-Mg alloys application in marine constructions. *Metalurgija, 12*(4), 297. https://scindeks.ceon.rs/article.aspx?artid=0354-63060604297R

Fayomi, O. S. I., Agboola, O., Akande, I. G., & Emmanuel, A. O. (2020). Challenges of coatings in aerospace, automobile and marine industries. *AIP Conference Proceedings*, 020038. Advance online publication. DOI: 10.1063/5.0033579

Glaser, K. F., & Shockley, G. R. (1970, April 21). Aluminum In Marine Construction. *Offshore Technology Conference*. https://doi.org/DOI: 10.4043/1197-MS

Hadi, S. P., Prabawani, B., & Qomariah, A. (2021). Circular Initiatives for Industrial Sustainability. *E3S Web of Conferences, 317*, 1078. https://doi.org/DOI: 10.1051/e3sconf/202131701078

Hasan, H. R., Musamih, A., Salah, K., Jayaraman, R., Omar, M., Arshad, J., & Boscovic, D. (2024). Smart agriculture assurance: IoT and blockchain for trusted sustainable produce. *Computers and Electronics in Agriculture, 224*, 109184. DOI: 10.1016/j.compag.2024.109184

Hosseinabadi, O. F., & Khedmati, M. R. (2021). A review on ultimate strength of aluminium structural elements and systems for marine applications. *Ocean Engineering, 232*, 109153. DOI: 10.1016/j.oceaneng.2021.109153

Jongejan, M., Mackintosh, A., & de Kluijver, D. (2015). Cost and Weight Reduction by Choosing Aluminium for Offshore Structures. *Abu Dhabi International Petroleum Exhibition and Conference*. https://doi.org/DOI: 10.2118/177449-MS

Kanani, M., Danaee, I., & Maddahy, M. H. (2014). Microstructural characteristics and corrosion behavior of cerium oxide conversion coatings on AA6063. *Materials and Corrosion*, *65*(11), 1073–1079. DOI: 10.1002/maco.201307539

Kappenthuler, S., & Seeger, S. (2020). Holistic evaluation of the suitability of metal alloys for sustainable marine construction from a technical, economic and availability perspective. *Ocean Engineering*, *219*, 108378. DOI: 10.1016/j.oceaneng.2020.108378

Kasten, M. (2025). Aluminum Strength vs Steel Strength. http://www.kastenmarine.com/alumVSsteel.htm

Kaufman, J. G., & Rooy, E. L. (2004). *Aluminum Alloy Castings: Properties, Processes And Applications*. http://ci.nii.ac.jp/ncid/BA7486842X

Khalaf, H. I., Al-Sabur, R., Abdullah, M. E., Kubit, A., & Derazkola, H. A. (2022). Effects of Underwater Friction Stir Welding Heat Generation on Residual Stress of AA6068-T6 Aluminum Alloy. *Materials (Basel)*, *15*(6), 2223. DOI: 10.3390/ma15062223 PMID: 35329680

Khang, A., & Akhai, S. (2024). Green intelligent and sustainable manufacturing: key advancements, benefits, challenges, and applications for transforming industry. *Machine Vision and Industrial Robotics in Manufacturing*, 405-417.

Khang, A., & Akhai, S. (2025). E-Waste and Lithium-ion Battery Recycling Insights for Sustainable Transportation. In *Driving Green Transportation System Through Artificial Intelligence and Automation: Approaches, Technologies and Applications* (pp. 203-230). Springer Nature Switzerland. DOI: 10.1007/978-3-031-72617-0_11

Kim, C. K., Cho, D., Kim, S., Song, S. W., Seo, K. M., & Cho, Y. T. (2023). High-Throughput Metal 3D Printing Pen Enabled by a Continuous Molten Droplet Transfer. *Advancement of Science*, *10*(6), 2205085. Advance online publication. DOI: 10.1002/advs.202205085 PMID: 36526589

Kotari, S., Punna, E., Reddy, S., & Venukumar, S. (2020). Mechanical and micro structural behaviour of flux coated GTAW and FSW joined AA6061 aluminium alloy. *Materials Today: Proceedings*, *27*, 1660–1667. DOI: 10.1016/j.matpr.2020.03.562

Kumar, H., Wadhwa, A. S., Akhai, S., & Kaushik, A. (2024). Parametric analysis, modeling and optimization of the process parameters in electric discharge machining of aluminium metal matrix composite. *Engineering Research Express*, *6*(2), 025542. DOI: 10.1088/2631-8695/ad4ba9

Kumar, H., Wadhwa, A. S., Akhai, S., & Kaushik, A. (2024). Parametric analysis, modeling and optimization of the process parameters in electric discharge machining of aluminium metal matrix composite. *Engineering Research Express*, *6*(2), 025542. DOI: 10.1088/2631-8695/ad4ba9

Kumar, K. R. V., Prabhu, L., Subin, B. S., Satheen, S., & Vaishnav, K. (2020). Development of Hybrid Aluminium Metal Matrix Composites for Marine Applications. *IOP Conference Series. Materials Science and Engineering*, *993*(1), 12016. DOI: 10.1088/1757-899X/993/1/012016

Kumar, P., Kumar, D., Akhai, S., Singh, J., Bansal, J., & Chauhan, A. (2024). Investigating the Microstructure and Mechanical Performance of Ceria and Graphene-Reinforced Aluminum Hybrid Composites. *Scientia Iranica*, *0*(0), 0. DOI: 10.24200/sci.2024.61986.7593

Kumar, P., Sharma, V., Kumar, D., & Akhai, S. (2023). Morphology and mechanical behavior of friction stirred aluminum surface composite reinforced with graphene. *EVERGREEN Joint Journal of Novel Carbon Resource Sciences & Green Asia Strategy*, *10*(1), 105–110. DOI: 10.5109/6781056

Lu, Z., Wang, P., & Zhang, D. (2014). Super-hydrophobic film fabricated on aluminium surface as a barrier to atmospheric corrosion in a marine environment. *Corrosion Science*, *91*, 287–296. DOI: 10.1016/j.corsci.2014.11.029

Maris Njørdal, V. I. A. (2024). *Speira Newsroom*. https://www.speira.com/newsroom/press/water-revolution-foundation-certifies-via-maris-njoerdal/

Maritimer. (1967). The use of aluminium-rich alloys in shipbuilding. *Anti-Corrosion Methods and Materials, 14*(3), 12. https://doi.org/DOI: 10.1108/eb005203

Muribwathoho, O., Msomi, V., & Mabuwa, S. (2022). Metal Matrix Composite Fabricated with 5000 Series Marine Grades of Aluminium Using FSP Technique: State of the Art Review. *Applied Sciences (Basel, Switzerland)*, *12*(24), 12832. DOI: 10.3390/app122412832

Nunes, H., Emadinia, O., Vieira, M. F., & Reis, A. (2023). Low- and High-Pressure Casting Aluminum Alloys: A Review. *IntechOpen eBooks*. https://doi.org/DOI: 10.5772/intechopen.109869

Ogata, K. A., Lazarevic, S., & Miller, S. F. (2014). Dissimilar Material Joint Strength and Structure for Friction Stir Forming Process. https://doi.org/DOI: 10.1115/MSEC2014-4044

Önal, M., Neşer, G., & Gürsel, K. T. (2020). Environmental impacts of steel ship hulls building and recycling by life cycle assessment (LCA). *Ships and Offshore Structures*, *16*(10), 1061–1066. DOI: 10.1080/17445302.2020.1816706

Panda, B. (2021). Corrosion Resistant Superhydrophobic Aluminum Alloy: A Review. *IOP Conference Series. Materials Science and Engineering*, *1017*(1), 12008. DOI: 10.1088/1757-899X/1017/1/012008

Raabe, D., Ponge, D., Uggowitzer, P. J., Roscher, M., Paolantonio, M., Liu, C., Antrekowitsch, H., Kozeschnik, E., Seidmann, D., Gault, B., De Geuser, F., Deschamps, A., Hutchinson, C., Liu, C., Li, Z., Prangnell, P., Robson, J., Shanthraj, P., Vakili, S., & Pogatscher, S. (2022). Making sustainable aluminum by recycling scrap: The science of "dirty" alloys. *Progress in Materials Science*, *128*, 100947. DOI: 10.1016/j.pmatsci.2022.100947

Radić, A., Fuentes, J. M. A., Ariza-Montes, A., Han, H., & Law, R. (2020). Job demands–job resources (JD-R) model, work engagement, and well-being of cruise ship employees. *International Journal of Hospitality Management*, *88*, 102518. DOI: 10.1016/j.ijhm.2020.102518

Ramesha, K., Sudersanan, P. D., Santosh, N., & Jangam, S. (2021). Sustainable Processing of Dissimilar Aluminium Alloy Joints by Friction Stir Welding. *IOP Conference Series. Materials Science and Engineering*, *1132*(1), 12029. DOI: 10.1088/1757-899X/1132/1/012029

Raval, S. K., & Judal, K. B. (2020). Recent Advances in Dissimilar Friction Stir Welding of Aluminum to Magnesium Alloys. *Materials Today: Proceedings*, *22*, 2665–2675. DOI: 10.1016/j.matpr.2020.03.398

Scipioni, S., Dini, G., & Niccolini, F. (2023). Exploring circular shipbuilding: A systematic review on circular economy business models and supporting technologies. *Journal of Cleaner Production*, *422*, 138470. DOI: 10.1016/j.jclepro.2023.138470

Sharma, V., & Akhai, S. (2019). Mechanical Behaviour of Fly Ash Reinforced Aluminum Composite Prepared by Casting. *Journal of Advanced Research in Mechanical Engineering and Technology*, *6*(1&2), 23–26. DOI: 10.24321/2454.8650.201904

Sharma, V. K., Kumar, P., Akhai, S., Kumar, V., & Joshi, R. S. (2024). Analyzing the tribological and mechanical performance of Al-6061 with rare earth oxides: An experimental analysis. *Proceedings of the Institution of Mechanical Engineers. Part E, Journal of Process Mechanical Engineering*, *238*(5), 2278–2292. DOI: 10.1177/09544089231160003

Sharma, V. K., Kumar, P., Akhai, S., Kumar, V., & Joshi, R. S. (2024). Corrosion inhibition analysis on cerium induced hydrophobic surface of Al-6061/SiC/Al2O3 hybrid composites. *Proceedings of the Institution of Mechanical Engineers. Part B, Journal of Engineering Manufacture*, *238*(14), 2126–2138. DOI: 10.1177/09544054231222048

Sharma, V. K., Kumar, P., Kumar, D., Chalisgaonkar, R., & Akhai, S. (2024). Characterization of Al 6061/Al2O3/SiC composites with cerium oxide: corrosion analysis and microstructural insights. *Composites Theory and Practice, 24*.

Shen, Y., Sahoo, P. K., & Pan, Y. (2017). A Study of Micro-Arc Oxidation Coatings on Aluminum Alloy Drill Pipe for Offshore Platform. *Marine Technology Society Journal*, *51*(3), 16–22. DOI: 10.4031/MTSJ.51.3.1

Singh, G., & Akhai, S. (2015). Experimental study and optimisation of MRR in CNC plasma arc cutting. *International Journal of Engineering Research and Applications*, *5*(6), 96–99. DOI: 10.9790/9622-07060696101

Sinyavskii, V. S., & Kalinin, V. D. (2005). Marine Corrosion and Protection of Aluminum Alloys According to Their Composition and Structure. *Protection of Metals*, *41*(4), 317–328. DOI: 10.1007/s11124-005-0046-8

Sokoluk, M., Cao, C., Pan, S., & Li, X. (2019). Nanoparticle-enabled phase control for arc welding of unweldable aluminum alloy 7075. *Nature Communications*, *10*(1), 98. Advance online publication. DOI: 10.1038/s41467-018-07989-y PMID: 30626876

Song, G., & Feng, Z. (2020). Modification, Degradation and Evaluation of a Few Organic Coatings for Some Marine Applications. *Corrosion and Materials Degradation*, *1*(3), 408–442. DOI: 10.3390/cmd1030019

Srivatsav, S., Jayakumar, V., & Sathishkumar, M. (2021). Recent developments and challenges associated with wire arc additive manufacturing of Al alloy: A review. *Materials Today: Proceedings*, *46*, 8561–8566. DOI: 10.1016/j.matpr.2021.03.542

Sun, H., Liang-cai, L., Wang, Z., Liu, B., Wang, M., & Yu, Y. (2020). Corrosion Behaviors of Microarc Oxidation Coating and Anodic Oxidation on 5083 Aluminum Alloy. *Journal of Chemistry*, *1*, 1–11. Advance online publication. DOI: 10.1155/2020/6082812

Tański, T., & Snopiński, P. (2017). Effect of the Processing Conditions on the Microstructural Features and Mechanical Behavior of Aluminum Alloys. *InTech eBooks*. https://doi.org/DOI: 10.5772/intechopen.70682

Thareja, P., & Akhai, S. (2016). Processing Aluminum Fly Ash Composites via Parametric Analysis of Stir Casting. *Journal of Advanced Research in Manufacturing. Material Science & Metallurgical Engineering*, *3*(3&4), 21–28.

Thareja, P., & Akhai, S. (2017). Processing parameters of powder aluminium-fly ash P/M composites. *Journal of advanced research in manufacturing, material science & metallurgical engineering, 4*(3&4), 24-35.

Thomas, W. M., & Nicholas, E. D. (1997). Friction stir welding for the transportation industries. *Materials & Design*, *18*(4-6), 269–273. DOI: 10.1016/S0261-3069(97)00062-9

Tjahjanti, P. H., Panunggal, E., Darminto, D., & Nugroho, W. H. (2013). Numerical Modeling of Ship Composite-Based on Aluminum Casting as Alternative Materials for Ship Building. *Advanced Materials Research*, *789*, 143–150. . DOI: 10.4028/www.scientific.net/AMR.789.143

Wadhwa, A. S., Akhai, S., Abbass, M., Chouksey, A., Tiwari, S., & Taneja, T. (2025). Computational Intelligence-Driven Design and Optimization of Polyurethane Belt-Type Oil Skimmer for Sustainable Manufacturing Using Solidworks 3D CAD. In *Using Computational Intelligence for Sustainable Manufacturing of Advanced Materials* (pp. 445-464). IGI Global Scientific Publishing.

Wahid, M. A., Siddiquee, A. N., & Khan, Z. A. (2019). Aluminum alloys in marine construction: Characteristics, application, and problems from a fabrication viewpoint. *Marine Systems & Ocean Technology*, *15*(1), 70–80. DOI: 10.1007/s40868-019-00069-w

Zhang, B., Xu, W., Zhu, Q., Sun, Y., & Li, Y. (2019). Mechanically robust superhydrophobic porous anodized AA5083 for marine corrosion protection. *Corrosion Science*, *158*, 108083. DOI: 10.1016/j.corsci.2019.06.031

Zhang, L., Zhang, J., Chen, C., & Gu, Y. (2014). Advances in microarc oxidation coated AZ31 Mg alloys for biomedical applications. *Corrosion Science*, *91*, 7–28. DOI: 10.1016/j.corsci.2014.11.001

Zhang, X., Luo, X., Ou, W., Huang, J., & Li, Y. (2020). Effect of Er on the Characteristics of an Oxide Coating Prepared by Micro-Arc Oxidation on a 7075 Aluminum Alloy. *International Journal of Electrochemical Science*, *15*(4), 3445–3452. DOI: 10.20964/2020.04.51

Zimmerman, G. (2018). What Aluminum Alloy Should I Use to Build a Boat, and Why? *Naimor Metal Fabrication Blog*. https://info.naimormetalfabrication.com/blog/what-aluminum-alloy-should-i-use-to-build-a-boat-and-why

Chapter 5
Innovation of Bio-Inspired Materials for Next-Generation Defense Applications

G. Prasad
https://orcid.org/0000-0002-5709-9182
Chandigarh University, Punjab, India

Utsav Chakraborty
https://orcid.org/0009-0001-2009-8629
Chandigarh University, Punjab, India

Evance Saba Matete
https://orcid.org/0009-0006-1263-0343
Chandigarh University, Punjab, India

Kunwarpreet Singh Marwaha
https://orcid.org/0009-0009-6965-791X
Chandigarh University, Punjab, India

Rhoto Vero
Chandigarh University, Punjab, India

ABSTRACT

Bio-inspired materials are advancing next-generation military applications by emulating the distinctive features of natural systems to create lightweight, durable, and multifunctional solutions. This overview emphasizes recent progress in bio-inspired materials, encompassing its applications in armor systems, stealth technologies,

DOI: 10.4018/979-8-3373-0933-0.ch005

adaptable structures, and energy-efficient gadgets. Significant developments encompass materials derived from the resilience of nacre, the self-repairing properties of biological tissues, and the concealment strategies of cephalopods. Challenges including scalability, cost-effectiveness, and material dependability under harsh conditions are also addressed. This study highlights the potential of bio-inspired materials to transform the defense sector and meet the changing requirements of contemporary conflict through the analysis of case studies and upcoming trends.

INTRODUCTION

Nature, shaped by over 3.8 billion years of evolution, offers a vast repository of design strategies that have enabled organisms to adapt, survive, and thrive in extreme environments (Benyus, 1997). The field of bio-inspired materials science leverages these evolutionary solutions to develop synthetic materials that mimic biological forms, structures, and functions. This approach, also called biomimetics, is increasingly influencing modern engineering, particularly in the defence and aerospace sectors, where materials are expected to perform reliably under high stress, extreme temperatures, and dynamic environments (Vincent et al. 2006). Traditional materials often face limitations in these demanding applications; hence, the demand for multifunctional, lightweight, and adaptive materials has never been higher.

Bio-inspired materials offer novel advantages such as self-healing, energy dissipation, thermal insulation, adaptive camouflage, and impact resistance—features that are critical in the defense applications (Trask et al. 2007). For instance, researchers are studying the mantis shrimp's dactyl club to design impact-resistant armor systems, owing to its highly mineralized, helicoidal microstructure that can absorb and redistribute energy from repeated high-velocity strikes (Weaver et al. 2012). Similarly, cephalopods such as octopuses and squids possess dynamic skin systems capable of changing colour and texture, which are inspiring next-generation adaptive camouflage for military vehicles and uniforms (Morin et al. 2012). These developments are not just theoretical; they are actively being translated into prototypes and testable materials through collaborations between biologists, material scientists, and defense engineers.

In the aerospace sector, the integration of bio-inspired design has shown promise in creating materials and structures that are both strong and lightweight—qualities essential for increasing fuel efficiency and payload capacity. Natural structures like honeycombs, bird bones, and insect wings are being mimicked to develop lightweight, high-strength components for aircraft and spacecraft (Chen et al. 2021). Moreover, morphing wing technology, inspired by bird flight, is advancing the development of aircraft that can adjust wing shape in real time for improved aerodynamic control (Rajabi et al. 2016).

This chapter explores the cutting-edge domain of bio-inspired materials and their transformative potential in next-generation defense and aerospace applications. It covers structural biomimetics for enhanced load-bearing capacity, self-healing composites to increase material lifespan, adaptive surface technologies for stealth and camouflage, and sustainable designs that reduce dependency on rare or hazardous materials. By merging biological insights with advanced technologies like additive manufacturing, nanofabrication, and responsive polymers, bio-inspired materials are laying the foundation for defense and aerospace systems that are not only more capable and durable but also more intelligent and sustainable (Bhushan, 2009).

LITERATURE REVIEW

Bio-inspired materials, which draw inspiration from the structural and functional properties of natural systems, have garnered significant attention in recent years (Vincent et al. 2006). These materials are being researched for potential applications in defense due to their ability to offer enhanced performance, sustainability, and adaptability in extreme environments (Chen & Ritchie, 2016). The exploration of bio-inspired materials has led to innovations in protective armor, sensors, adhesives, and other critical defense technologies. This review examines key areas of development and recent advancements in bio-inspired materials for defense applications, focusing on structural materials, adhesives, coatings, and camouflage technologies.

The development of structural materials inspired by nature is a crucial area of research in defense applications. Natural systems, such as seashells and bones, demonstrate exceptional strength and resilience due to their hierarchical microstructures (Meyers et al. 2008). Research on nacre (mother of pearl), which exhibits a layered "brick-and-mortar" design, has provided insights into creating composite materials with high toughness and impact resistance (Barthelat & Espinosa, 2007). These bio-inspired designs have been applied to develop lightweight, high-strength materials for use in military vehicles and body armor. Experiments demonstrated that replicating the complex hierarchical structure of biological materials results in composites with superior energy absorption, making them ideal for protective gear and blast-resistant applications (Wegst et al. 2015).

Recent advancements in bio-inspired materials include the creation of synthetic materials that mimic the structure of bamboo and spider silk, known for their strength-to-weight ratios. Spider silk, for instance, has inspired the development of high-strength fibres that can be utilized in next-generation armor systems (Porter et al. 2016; Ritchie, 2011). Such materials offer the potential to replace conventional metal-based armor with lighter alternatives, improving mobility without compromising protection.

Nature provides remarkable examples of adhesion mechanisms, particularly in marine organisms like mussels and barnacles, which have evolved the ability to adhere to surfaces in wet and hostile environments. Mussel-inspired adhesives, which utilize catechol chemistry, have demonstrated strong bonding capabilities in various conditions (Lee et al. 2007). This unique adhesive property has been explored for applications in defense, particularly in naval systems where traditional adhesives may fail in underwater conditions. Mussel-inspired adhesives offer significant potential for field repairs of military equipment and structural components in marine environments, as they can maintain strong adhesion even in the presence of water or moisture (Waite, 2017). In addition to underwater applications, bio-inspired adhesives are being researched for use in rapid-deployment scenarios. For example, adhesives that can bond quickly and securely to various surfaces are critical for repairing aircraft, vehicles, and infrastructure in the field. The development of such bio-inspired adhesives could revolutionize maintenance and repair processes in military operations.

Camouflage has long been a focus of defense technology, and bio-inspired approaches have led to significant advancements in adaptive camouflage systems. Many animals, such as chameleons and cephalopods, possess the ability to change colour and texture to blend into their surroundings, a feature that has inspired the development of dynamic camouflage materials for military applications (Zhang et al. 2020). Recent innovations include electrochromic and thermochromic materials that can alter their appearance in response to environmental stimuli, providing enhanced stealth capabilities (Kang et al. 2018). Adaptive camouflage systems, such as those inspired by the skin of octopuses, are being researched and developed for the use on military vehicles, drones, and personnel uniforms. These materials can change colour and pattern to match the surrounding environment, reducing the likelihood of detection by the enemy forces. Additionally, nanostructured coatings inspired by the butterfly wings have been researched for their potential to reduce radar signatures, further enhancing the stealth capabilities of military assets and equipment, making it suitable for combat readiness (Chen et al. 2019).

Protective coatings inspired by natural self-healing processes have emerged as a promising solution for extending the lifespan of military equipment. In nature, organisms have developed the ability to repair damage to their tissues autonomously, a feature that has inspired the development of self-healing materials. These materials contain embedded microcapsules or microvascular networks filled with healing agents that are released when the material is damaged (Toohey et al. 2007). Bio-inspired coatings are being researched for use on aircraft, naval vessels, and military vehicles, where they can help prevent corrosion and mechanical wear. For example, coatings that mimic the self-healing properties of tree bark or skin could automatically repair minor cracks or damage, reducing the need for maintenance and increasing the operational life of defense systems. Such coatings could be particularly valuable in harsh environments where conventional materials would degrade rapidly.

Biological systems provide models for highly sensitive detection systems. The lateral line system in fish and the sensory whiskers of mammals, for example, have inspired the development of flow sensors and tactile sensors that can be integrated into defense technologies. These sensors could be used in unmanned aerial vehicles (UAVs), submarines, and robotic systems to enhance environmental awareness and detect threats more effectively (Marshall et al. 2008).

In addition, bio-inspired sensors are being researched for thermal imaging and infrared detection applications. The pit organs of snakes, which can detect heat signatures, have inspired the development of biomimetic infrared sensors with improved sensitivity for military surveillance and target acquisition in low-visibility conditions (Chen et al. 2019). These sensors could be deployed in a variety of defense platforms to enhance situational awareness and provide critical intelligence in combat scenarios. The field of bio-inspired materials offers exciting possibilities for the development of next-generation defense technologies. By mimicking the properties of natural systems, researchers have developed materials that are lighter, stronger, and more adaptable than traditional defense materials. From structural composites to adhesives, camouflage systems, and sensors, bio-inspired materials hold the potential to revolutionize the defense sector by providing superior performance in challenging environments. As research in this field continues, bio-inspired materials are likely to play an increasingly important role in addressing the complex demands of modern defense applications.

STRATIFIED TYPOLOGIES OF BIO-ADAPTIVE AND BIO-INSPIRED MATERIALS: TYPES AND CLASSIFICATION

The bio-inspired material domain within the defense and aerospace engineering is an incredible assembly of the evolutionary awareness of nature and the pinnacle of the range technological know-how. These materials, which imitate to the fullest extent the efficiencies discovered in the biological methods, use nature's personal design standards with a view to meet the rigorous requirements of the navy, aerospace and defense programs. This is viable due to their impressive houses, including the very best strength-to-weight ratios, impact resistance, dynamic self-repair abilities, stimuli responsivity, bio-digital integration, and survival under severe situations, that are attributed to them as the area-adaptive materials, the designs of that have been tested by the way of nature. The range of bio-inspired substances for defense and aerospace engineering is surely splendid. These materials consist of bio-mimetic composites, bio-derived polymers, hierarchical ceramics, and bio-hybridized steel frameworks, all designed to deal with the hard needs of those environments (Naleway et al. 2015).

Nature-inspired structural materials, like nacre-mimetic laminates known for their impressive ability to dissipate energy, and spider silk analogs with extraordinary tensile strength, are revolutionizing ballistic shielding, aerospace frameworks, and next-generation protective armor (Meyers et al. 2008). Moreover, bio-responsive materials that replicate cephalopod camouflage are revolutionizing stealth technology and improving adaptive concealment for air and land defense systems (Zhang et al. 2020). According to their structural complexity, functional variety, and strategic uses, bio-adaptive materials are categorized in this analysis. These materials are poised to revolutionize defense and aerospace engineering by using the inventiveness of nature to provide revolutionary breakthroughs in material robustness, adaptability, and mission-critical performance. The types of Bios Inspired materials with characteristics and application are listed in Table 1.

Table 1. Bio-inspired material types

Types of Bio-Inspired Materials	Description	Observable Characteristics	Examples and Applications
Bio-Mimetic Structural Materials	Engineered to replicate the complex biological load-bearing structures found in nature, thereby enhancing resilience, impact resistance, and overall structural durability in demanding defense applications.	High strength to weight ratio, superior toughness and energy dissipation.	Nacre-inspired armor (synthetic mother-of-pearl), bone-inspired lightweight composites used in ballistic protection and aerospace frames.
Self-Healing Materials	Inspired by biological self-repair mechanisms, these advanced materials autonomously detect, respond to, and recover from damage, enhancing longevity and reliability in critical defense applications.	Crack closure, polymerization at damage sites, autonomous repair.	Graphene-based self-healing polymers, vascularized composites used in next-gen military armor and spacecraft.
Adaptive Camouflage and Stealth Materials	Dynamically alter optical, thermal, or acoustic properties to seamlessly blend with surroundings, enhancing stealth, survivability, and tactical advantage in defense operations.	Colour-shifting surfaces, thermal emissivity control, radar absorption.	Cephalopod-inspired meta-materials, liquid crystal elastomers, and the thermochromic coatings strategically employed in stealth fighter jets, advanced combat uniforms, and next-generation surveillance systems for superior adaptive camouflage and stealth capabilities.
Bio-Inspired Composites	Hybrid materials meticulously integrating the biological design principles to achieve superior mechanical strength, durability, and multi-functional adaptability composite materials	Lightweight, high durability, multi-functional adaptability.	Kevlar-inspired aramid fibres and bio-fibre-reinforced polymer composites strategically utilized in aerospace fuselages, missile casings, and advanced protective armor for enhanced durability and impact resistance.
Bio-Sensing and Responsive Materials	Capable of detecting and responding to various external stimuli, such as chemical threats, pressure fluctuations, or temperature variations, by leveraging advanced bio-inspired mechanisms for enhanced sensitivity and adaptive functionality.	Changes in electrical, optical, or mechanical properties upon exposure.	Peptide-based bio-sensors and piezoelectric bio-inspired materials are utilized for highly sensitive chemical threat detection and advanced bio-surveillance, ensuring real-time monitoring and rapid response in critical environments.

continued on following page

Table 1. Continued

Types of Bio-Inspired Materials	Description	Observable Characteristics	Examples and Applications
Bio-Derived Polymers and Elastomers	Polymers synthesized from biological precursors, offering enhanced sustainability, biodegradability, and eco-friendly alternatives to conventional synthetic materials.	High elasticity, toughness, environmental resistance.	Chitosan-based protective coatings and bio-derived polyurethanes, widely utilized for durable, corrosion-resistant surfaces on aircraft, naval vessels, and other high-performance structures.
Bio-Mineralized and Hybrid Metallic Systems	Integrate bio-inspired mineralization techniques to enhance material toughness, structural resistance, and long-term durability in extreme environments.	High hardness, wear resistance, impact resilience.	Utilized in advanced armor plating, hypersonic vehicle shielding, and high-impact military structures to enhance durability, resistance to extreme conditions, and protection against ballistic threats.
Hierarchical Ceramics and Protective Coatings	Bio-inspired ceramic systems that enhance thermal resistance, impact durability, and structural integrity while maintaining lightweight properties for improved performance in extreme environments.	High hardness, thermal insulation, lightweight nature.	Zirconia-based bio-ceramics and bio-inspired silica coatings used in heat shields, space vehicle exteriors, and high-temperature protective systems to enhance thermal insulation and structural longevity.
Bio-Electronic and Neuromorphic Materials	Mimic biological neural networks, enabling advanced sensory perception, adaptive learning capabilities, and real-time cognitive processing for enhanced efficiency and responsiveness.	Flexible electronics, adaptive circuits, bio-signal processing.	Artificial synapse-inspired electronic skins and bio-inspired memory chips designed for use in advanced avionics, secure communication systems, and next-generation cyber-warfare technologies.
Biologically Inspired Energy-Generating Materials	Replicate natural photosynthesis or bio-electrochemical processes to enable sustainable and efficient energy harvesting for various technological applications.	Self-sustaining energy production, bio-catalysis.	Microbial fuel cells, enzyme-based bio-batteries used for soldier-wearable power generation in field operations.
Hydrophobic Materials	Emulate the micro- and nano-structured surfaces of lotus leaves or shark skin to achieve superior water repellence, anti-icing capabilities, and self-cleaning properties in various environmental conditions.	High water contact angle, reduced adhesion, self-cleaning properties.	Shark skin-inspired anti-fouling coatings and advanced hydrophobic nanocoating are utilized in military aircraft to minimize surface drag, enhance fuel efficiency, and prevent icing under extreme atmospheric conditions.

continued on following page

Table 1. Continued

Types of Bio-Inspired Materials	Description	Observable Characteristics	Examples and Applications
Bio-Inspired Shock Absorption and Impact Mitigation Materials	Mimic biological impact-dissipating structures like seashells and honeycombs to enhance shock absorption, improve energy dissipation, and increase structural resilience against high-velocity impacts.	High impact resistance, superior energy absorption.	Chitin-based impact-resistant materials and honeycomb composites are utilized in bulletproof vests and armoured vehicles to enhance durability, improve energy absorption, and provide superior protection against ballistic and explosive impacts.
Bio-Inspired Thermal Regulation Materials	Capable of passive or active heat management by mimicking natural thermoregulation mechanisms, ensuring optimal temperature control and energy efficiency in varying environmental conditions.	Temperature-adaptive properties, phase change capabilities.	Phase-change microcapsules and thermally adaptive textiles engineered to regulate temperature fluctuations, providing enhanced thermal comfort and protection in space suits and advanced military uniforms.
Bio-Mimetic Adhesive Materials	Inspired by the unique adhesion mechanisms of gecko feet, mollusks, and insects, these materials provide strong, reversible adhesion, enabling efficient attachment and detachment on various surfaces.	High frictional adhesion, reusability.	Gecko-inspired dry adhesives and mussel-inspired underwater adhesives engineered for durable, reversible adhesion, facilitating reliable space maintenance and enhanced underwater robotic functionality.
Bio-Mechanically Inspired Robotics Materials	Emulate the properties of soft biological tissues, enabling adaptable, flexible, and programmable shape modifications.	Shape-morphing, flexible movement, adaptability.	Octopus-inspired robotic arms and bio-hybrid soft exosuits designed for advanced next-generation exoskeletons and adaptive drone morphing wings.
Bio-Inspired Aeroelastic and Fluid Dynamic Materials	Emulate the structure of bird wings or fish scales to enhance aerodynamic lift and hydrodynamic efficiency, reducing drag and improving manoeuvrability.	Enhanced lift, reduced drag, fluid-responsive motion.	Eagle feather-inspired UAV wings and fish scale-inspired drag reduction surfaces engineered to optimize stealth, enhance agility, and improve fuel efficiency in advanced drones and hypersonic jets.

DESIGN PRINCIPLES AND MECHANISMS OF BIO-INSPIRED MATERIALS

Bio-inspired materials constitute a groundbreaking development in cloth technology, drawing notion from natural systems and mechanisms which have developed over millions of years. By mimicking organic answers, researchers broaden materials with advanced mechanical, structural, and purposeful homes, making them noticeably appropriate for the next-generation defense packages and applications (Gao et al. 2003). These materials exhibit improved durability, adaptability, and efficiency, which might be critical for army technology, aerospace systems, and protective system. This phase explores the fundamental design concepts and mechanisms that govern bio-inspired materials, with a focal point on their relevance to fashionable defense technologies.

Nature optimizes the physical properties by structuring them on many parameters from the molecular level to the macro level. This hierarchical organization increases strength, cruelty and flexibility-important factor in developing advanced defense systems (Fratzl & Weinkamer, 2007). For example, the layered structure of Nacre, or Mother-off-Parle, acts as an inspiration for the compound armor system with high toughness. Similarly, the strong structure of the bones has affected the design of military exo-skeletons and the light structural materials (Meyers et al. 2008). These hierarchical materials effectively distribute stress, making them very resistant to fractures and effects, which is essential in the atmosphere of war for defense.

Figure 1. Different types of design principles and mechanisms of bio-inspired materials

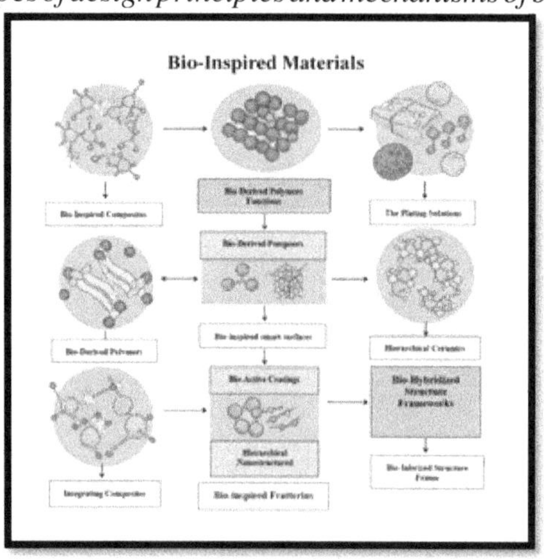

Figure 1 depicts the hierarchical organization and functional synergy of bio-inspired materials, with a focus on bio-derived polymers, composites, and nano-structured coatings. It underscores the evolution from bio-derived polymers to hybrid structures that emulate natural systems, thereby improving material characteristics such as durability, adaptability, and multifunctionality. By incorporating biological principles into the design of materials, this methodology facilitates the creation of advanced materials suitable for applications in bio-medical engineering, structural frameworks, and intelligent surfaces.

Self -healing and the adaptability are important properties of biological systems that engineers try to repeat in synthetic materials. Many biological organisms have the ability to self -reform, an essential feature that increases the durability and reduces maintenance costs. Inspired by these biological mechanisms, researchers have developed self -healing polymers that include microcapsules filled with the healing agents. When any sort of damage occurs, these capsules are cured and the treatment releases the substances, repairing autonomously cracks in military vehicle coatings and the upper body of the aircraft fuselages (Toohey et al. 2007). Another example is a vascular composite, which induces plant -vascular systems by incorporating built -in macroscales that distribute the therapies throughout the content, allowing continuous self -connection and extend the life of the defense infrastructure (Reddy, 2018).

Nature has developed surfaces with extraordinary functions such as water dis-abilities, anti-icing properties and adaptive camouflage. These properties are now being integrated into the bio-inspired materials to increase their defense applications. For example, superhydrophobic coatings modelled from the lotus leaves form self - cleaning surfaces that resists water and contaminants from adhering to the military aircraft and marine vessels (Aizenberg et al. 2005). This reduces the maintenance requirements and increases overall operational efficiency. In addition, Cephalopod-inspired adaptive camouflage materials enable real-time changes in the colour and texture, so that the military personnel and vehicles can originally blend with their environment (Kang et al. 2018). This advanced content improves great gradual secret and hiding abilities, providing a strategic advantage in different fighting landscapes. The inclusion of biological principles in material design has led to remarkable innovation in influence resistance and mechanical adaptation. Natural materials such as nacre, spider silk and chitin show unique strict mechanisms such as crack bending, bridge and energy spread. Engineers use these principles to develop light armor and impact-resistance vehicle components that provide better protection and also reduces the weight (Torres et al. 2019). For example, structures such as honey cake and foam are inspired by biological materials, which improve the absorption of shock and accident resistance, making them ideal for military vehicles, helmets and structure building.

Figure 2. Bio-inspired material design in nature and engineering, highlighting hierarchy, multifunctionality, and materials-by-design for future composites

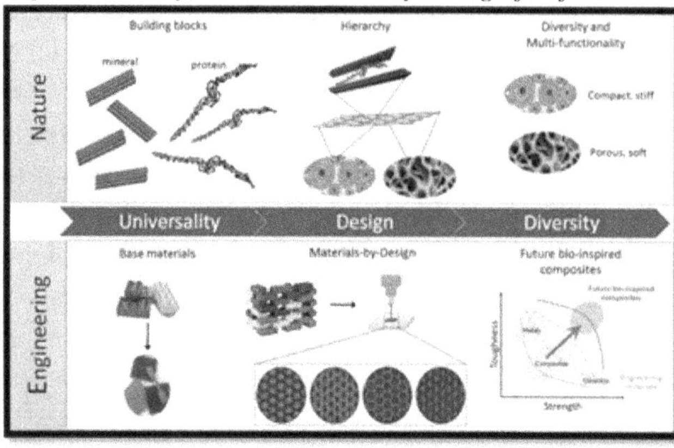

Figure 2 contrasts the design principles found in the nature with those in engineering, utilizing bone as a biomimetic reference. In the natural world, fundamental components such as minerals and proteins create materials that develop through hierarchical arrangements, serving distinct purposes such as durability and adaptability. Engineers draw upon these principles by utilizing materials like metals, ceramics, and polymers to create bio-inspired substances that exhibit improved toughness and strength. This illustration emphasizes the ability of bio-inspired materials to attain an ideal equilibrium between strength and toughness by emulating the structure and functionality of biological systems, such as bone.

Smart and responsive mechanisms inspired by biological materials bring revolution in defense technologies. Many natural systems react dynamically to environmental stimuli, a function that is now emulated in the synthetic materials. Form walling materials or shape-morphing materials, inspired by the plant movements, change the structure in response to the external stimuli, enabling morphing drone wings and adaptive aerodynamic surfaces for the aircraft (Chen & Ritchie, 2016). Thermo-responsive materials, inspired by the desert beetles, modify heat absorption and dissipation, improves thermal control in military equipment and aircraft. These bio -inspired innovations increase efficiency, functionality and adaptability of the defense materials in different operating environments and conditions (Wegst et al. 2015).

Fabrication and processing strategies play a critical position inside the improvement of bio-inspired materials. Biomimetic synthesis techniques, which includes self-assembly and the additive production, permits the appropriate replication of the herbal structures (Studart, 2012). Bottom-up tactics, stimulated by using the organic and biological growth processes, permit the introduction of nanostructured substances with first-rate mechanical and useful homes. Additionally, 3D printing era allows the fabrication of complex bio-inspired geometries, optimizing materials overall performance for the defense applications. Integrating these materials with superior technology which includes nanotechnology and artificial intelligence in addition enhances their properties. Nanomaterials support bio-inspired composites, improving energy and multifunctionality, even as AI-driven computational modeling enables layout and optimize these materials for specific defense applications (Rajabi et al. 2016).

There is no question that the applications of bio-inspired materials within the scope of the defense technology are great and crucial. In fact, bio-inspired materials such as bio-derived polymers, hierarchical ceramics, and bio-hybridized metallic frameworks are a few of the key materials that can solve many issues posed in the defense and aerospace sector. For instance, bio-inspired composites can be used in light but tough body armor and military transport in ballistic protection and the impact resistance. Materials inspired by nacre and chitin structures increases soldier protection by having superior energy absorption and penetration resistance. Similarly, bio-inspired honeycomb and foam structures are used in shock-absorbing military transport to improve crash resistance and blast protection to reduce the chances of suffering injuries during combat scenarios. Bio inspired changes are also felt in the adaptive camouflage and stealth materials. Coatings inspired by the cephalopods help military uniforms and vehicles disguise themselves within an environment in real time to enhance their stealth capabilities. Bio-mimetic infrared stealth materials reduce thermal signatures to make the aircraft and ground forces less detectable to the enemy sensors and radars. These technologies have a profound impact on the defense personnel and equipment survivability and operational effectiveness. Bio-inspired materials also enhance the aerospace and space applications. Examples of these include heat shields and thermal protection systems, modelled after natural exoskeletons, which improve the survivability of the hypersonic vehicles and space missions at extreme temperatures. Self-healing aircraft and spacecraft materials have the capability to repair light damage automatically, leading to aircraft and spacecraft with extended lifespans that cost less to service and function more efficiently throughout their lifetime. The breakthroughs are opening the doors to stronger, more sustainable materials for aerospace that are essential to defense and space exploration missions.

Bio-inspired materials are full of promise but are considerably disadvantaged in terms of production and implementation in real life by a set of issues. One of the main issues includes the issues of scale and cost—producing those materials in large-scale industrial quantities is an extremely difficult and expensive undertaking (Rajabi et al. 2016). Moreover, their incorporation into existing defense systems is a huge task, which must be extensively tested and certified in order to confirm their efficacy and reliability. These issues can be addressed by a joint research and multidisciplinary cooperation in the defense technology, materials science, and engineering. Additional research should be done on the development of the manufacturing methods, the identification of new biological sources of inspiration, and the discovery of visionary applications with the potential to revolutionize defense technology. The materials involved are a fine example of the incredible revolution underway in the field of material science, with improved durability, strength, and functionality. By learning from and copying nature's intricate structures, engineers and scientists are making rapid advances in creating next-generation defense materials—materials capable of maximizing protection, efficiency, and sustainability. With the use of biomimetic and bio-inspired technology, nature-inspired materials have the ability to revolutionize defense tactics by making mechanisms efficient, responsive, and resilient in the hostile environments. Besides creating military applications, the materials allow the development of novel and sustainable solutions for the use in contemporary defense strategies.

FABRICATION TECHNIQUES AND PROCESSING METHODS FOR THE BIO-INSPIRED MATERIALS

A variety of the novel fabrication techniques are required for the development of bio-inspired materials that can mimic the structure and functionality of biological systems. Over the years, scientists and engineers have used biomimetic synthesis, additive manufacturing, nanotechnology integration, and advanced polymer processing to create materials with enhanced mechanical properties, flexibility, and self-healing ability as the research has progressed (Studart, 2012; Ghazlan et al. 2021). It reviews the methodological approaches of the techniques applied in the fabrication of bio-inspired structures, including both their state-of-the-art advancements and advantages over the classical methods.

Biomimetic synthesis is an emerging approach that seeks to mimic natural material formation at molecular and structural levels. The most striking case of something being synthesised by listening the nature is the artificial silk, which gives a hesitant imitation of spiders' complex spinning process. As opposed to regular synthetic fibres, where polymer chains are organized end-to-end, artificial silk is manufactured through a self-assembly method of proteins — polymers that form hierarchically constructive systems with high elastic resistance and flexibility. Another key example of the significant biomimetic synthesis applications is the preparation of nacre-like composites imitating the natural arrangement and toughening mechanisms of mollusk shells (Barthelat & Espinosa, 2007). The mechanical and structural properties of these bio-inspired composites show high elastic deformation and energy dissipation level that are very important in the defense sectors such as the ballistic armor and impact-resistant defence components.

3D printing and additive manufacturing have revolutionized the fabrication of bio-inspired materials by enabling precise control over the material composition and microarchitecture (Reddy, 2018). Traditional manufacturing methods, such as machining and casting, often struggle to replicate the complex hierarchical structures found in biological materials. With additive manufacturing, engineers can design and fabricate materials layer by layer, closely mimicking natural formations. For example, 3D-printed bone-mimetic scaffolds have been developed to reproduce the porosity and mechanical resilience of the human bone, facilitating applications including lightweight armor and load-bearing components (Wegst et al. 2015). Inspired by the arrangements seen in nature, bio-informed lattice structures with differing levels of density cause improved impact resistance and energy absorption, showcasing significant applications in the field of defence and military as well as the aviation industry.

The utilization of nanotechnology in bio-inspired material synthesis is essential for manipulating the material properties at the atomic and molecular scales. Self-healing materials are inspired by biological self-repair mechanisms that contain groups of nanoscale capsules filled with healing agents that are released when damage occurs in the structure. They greatly enhance the robustness of defense systems, autonomously sealing the cracks and lengthening service lives (Toohey et al. 2007). Nanotechnology additionally enables the development of responsive coatings, including thermochromic and electrochromic materials that can adjust their optical and thermal properties in relation to variations in the environment. Based on the skin of cephalopods, nanostructured coatings realize advanced stealth technology for military applications by suppressing their visibility at various detection spectrums (Zhang et al. 2020). By utilizing advanced polymer processing techniques, bio-inspired materials can be creatively designed to mimic the elasticity, flexibility, and durability of biological tissues.

Bio-inspired polymer processing incorporates characteristics like tuneable elasticity, self-reinforcement, and hierarchical organization, in contrast to traditional polymer synthesis, which frequently produces brittle or hard structures. Soft robotics and adaptable exoskeletons have been made possible by the development of hydrogels with bio-mimetic viscoelastic qualities that mimic the mechanical behaviour of muscle tissues. Furthermore, because of their excellent strength-to-weight ratios and the structural resemblance to plant stems and insect exoskeletons, bio-fibre-reinforced polymer composites are perfect for the lightweight military applications like manufacturing of the drones and deployable structures.

Figure 3. Description of the different fabrication techniques used in the development of bio-inspired materials over the time

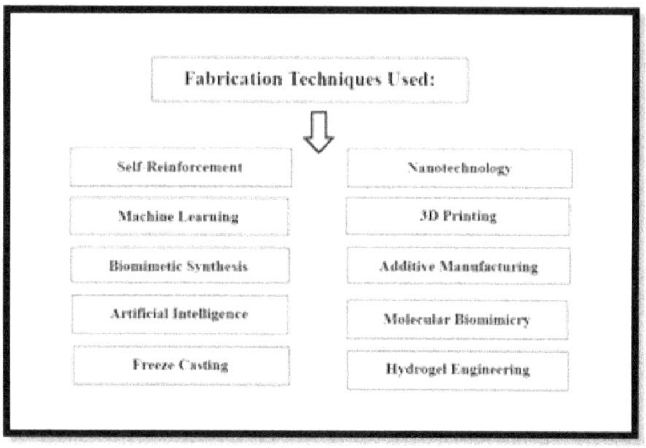

Figure 3 illustrates essential fabrication techniques employed in the design of bio-inspired materials. It emphasizes both traditional and innovative methods, including biomimetic synthesis, nanotechnology, and hydrogel engineering. Furthermore, advanced computational techniques, such as artificial intelligence and machine learning, are incorporated to enhance material properties. These strategies facilitate the creation of high-performance, multifunctional materials suitable for biomedical, structural, and engineering applications.

A crucial step in evaluating the bio-inspired material fabrication methods is the comparison between the traditional workshop and bio-mimetic workshop. In the traditional production, materials are usually removed from the bulk blocks through machining, moulding, or chemical processes which generally has a high waste percentage for materials and low functional versatility. On the other hand, bio inspired production methods tend to focus on the accuracy, effectiveness, and flexibility (Rajabi et al. 2016). For instance, self-assembly processes that are inspired by nature can develop materials that have responsive active mechanisms built in, as opposed to standard materials, which may require some sort of the external alteration to possess similar features (Studart, 2012).

Similarly, standard armor systems depend on massive dense and solid structures, while bio inspired armor utilizes the concept of hierarchical toughening using layered composites and energy absorbing microstructures to enhance impact resistance and weight efficiency (Barthelat & Espinosa, 2007; Naleway et al. 2015). Another notable benefit from bio-inspired production is the addition of sustainability principles which stem from conserving resources as seen in nature. Conventional manufacturing processes have energy consuming steps that produce toxic waste, while biomimetic fabrication makes use of the environmentally safe synthesis routes, biodegradable polymers, and self-assembly processes that are energy efficient. For example, naturally occurring biopolymers can produce chitosan-based coatings that have anti-bacterial and anti-corrosive properties without the need for toxic heavy metals (Ghazlan et al. 2021). Like petroleum-based polymers, bio-derived polyurethanes offer an eco-friendlier solution for military applications without compromising on the performance.

Tailored bio-based materials get a boost from the integration of the novice techniques and sophisticated fabrication technologies. Machine learning and AI are being used to enhance biomimetic material design through large scale natural structure and parameter basis data analysis (Torres et al. 2019). Computational modeling facilitates the simulation of stress distribution, thermal conductivity, and adaptive responses within bio-inspired materials, thus accelerating the realization of next-generation defense solutions. Furthermore, the combination of 3D printing with bio-mineralization yields gradient materials that mimic the natural tissue interfaces of the body's rigid and flexible dominions.

The use of biomimetic materials is revolutionizing materials science through achieving maximum functionality, flexibility, and sustainability by employing nature's design. The merging of biobased materials, biomimetic approaches, additive manufacturing, nanofabrication, and polysaccharide-based polymers are all essential for next-generation defense materials. By addressing the existing obstacles and improving the production processes, bio-inspired materials are set to redefine technologies in the military domain by improving protection, defense, functionality, and efficiency that keeps up with the changing defense requirements.

PERFORMANCE EVALUATION AND TESTING STANDARDS FOR THE BIO-INSPIRED MATERIALS

Bio-inspired materials are perfect for advanced engineering, aerospace, and defense applications since they are designed to mimic nature's sophisticated structural and functional characteristics. These materials must, however, pass stringent testing to evaluate their mechanical performance, durability, stealth qualities, heat resistance, and conformity with the aerospace and military requirements in order to guarantee their efficacy (Torres et al. 2019). This section examines the criteria and procedures used to assess the bio-inspired materials' performance and guarantee their dependability in harsh and challenging conditions.

Mechanical performance is crucial if bio-inspired materials are to be deployed in high-performance armor systems, aircraft, and the structural components. Toughness, strength, flexibility, and impact resilience are among these properties. Tensile and compressive strength tests, with standardized procedures such as the

ASTM D638 for polymers and ASTM E8 for metals, are employed to determine the load-bearing capacity and assessing tensile properties of bio-inspired materials. Fracture toughness measurements of bio-composites containing nacre and bone models are highly demanded as they provide the measure of the crack propagation resistance (i.e. fracture toughness) of the material under stress (Barthelat & Espinosa, 2007). Impact resistance tests, such as the Charpy and Izod tests, are performed on materials—including chitin-based composites—that are meant for high-impact and ballistic protection applications.

The elasticity and flexibility of biomimetic materials, such as artificial muscles and shape-adaptive structures, are assessed using the dynamic mechanical analysis to determine their bending resistance and recovery potential. To ensure they continue to perform effectively under multiple load cycles, bio-inspired aircraft frames and structural components undergo fatigue testing. Materials used in airplanes, hypersonic vehicles, and military protection gear must be able to withstand extremely high temperatures without losing their structural integrity or performance. To test for high-temperature resistance, thermal stability and decomposition temperature are assessed using the techniques such as thermogravimetric analysis (TGA) and differential scanning calorimetry (DCS), which are used to examine decomposition temperatures. The heat transfer characteristics of bio-inspired aerogels and composites used in thermal insulation are assessed using thermal conductivity testing. For heat shields modelled after mollusk shells or silica-based coatings, ablation resistance testing is essential to guarantee their efficacy in hypersonic flight conditions.

To guarantee military efficiency, bio-inspired materials should withstand the physical wear, chemical corruption, and environmental factors. This durability ensures longevity and reliability in military clothing, vehicles, and equipment. In addition to producing the biomimetic coatings for the aircraft and naval vessels, it is equally important to determine the efficacy of these coatings in preventing oxidation. This corrosion resistance testing can be accomplished through electrochemical impedance spectroscopy and salt spray exposure. Equally as important are the tests that measure abrasion and wear resistance for materials used in structural and body armor applications.

The Taber abrasion test is utilized to assess degradation of materials under cyclic mechanical loading. The bioinspired polymers and coatings are guaranteed to last through prolonged exposure to UV light and harsh conditions with the help of weathering tests. Chemical and solvent resistance tests determine the durability of bioinspired membranes and coatings under severe chemical conditions.

Figure 4. Charpy and Izod testing is one of the methods test the toughness, strength and impact resilience of the bio-inspired materials

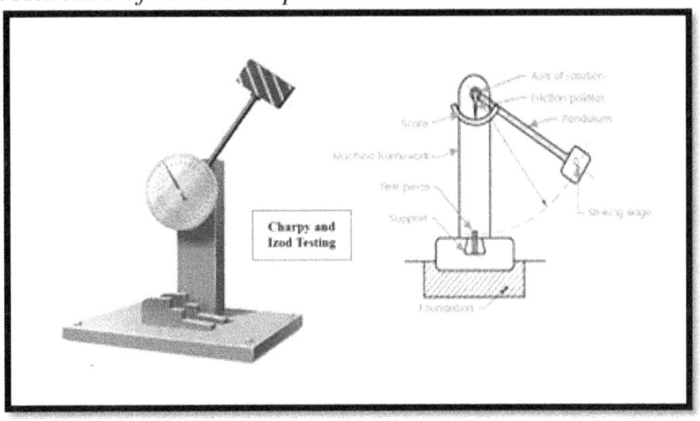

Figure 4 illustration demonstrates the Charpy and Izod impact testing techniques, which are employed to assess the toughness and impact resistance of various materials. The diagram on the left shows a conventional impact testing apparatus, whereas the schematic on the right details its components, such as the pendulum, striking edge, and the arrangement of the test specimen. These tests play a vital role in evaluating the fracture characteristics of both bio-inspired and engineered materials when subjected to the dynamic loading scenarios.

Defense relevant stealth materials and adaptive camouflage coatings such as the radar absorbing materials have very strict requirements when it comes to reduction of thermal and radar signature. Radar absorbing testing for cephalopod adaptive materials and metamaterials evaluates their ability to reduce deflection or absorption of electromagnetic radiation and thus lowering a radar's ability to detect them. Infrared and thermal concealment testing evaluates the biomimetic coatings which control thermal emission for both infrared and mid-wave thermal detectors to reduce detection of military assets (Zhang et al. 2020). Optical adaptability is conducted by the liquid crystal and chameleon inspired materials so that the camouflaging materials are effective in different environments.

Structures inspired by the nature, environment and bio-inspired materials used in the defence and aerospace need to meet exacting testing and regulatory requirements set by the international agencies. The MIL-STD-810 standard evaluates the performance of materials under severe environmental conditions such as extreme temperatures, humidity, and mechanical stress. Broadly, ASTM and ISO standards cover the mechanical, thermal, and chemical testing of reliability and safety of bio-inspired materials. Performance measures defined by the NATO Standardization Agreements cover the criteria for use in defense for cooperation between allied nations. Materials meant for aviation and space use require the FAA regulations are met alongside the NASA standards in order to comply with aerospace safety regulations. Testing bio-inspired materials is essential for their use in defense, aerospace, and high-performance areas. By adhering to the established standards, these materials can meet strict reliability and safety requirements, ensuring they function well in harsh conditions. Continuous research and improved testing methods will aid in developing the next generation of these materials, fostering advancements in military and aerospace technologies.

APPLICATIONS OF BIO-INSPIRED MATERIALS IN DEFENCE APPLICATIONS

Next-generation defense systems alongside the aerospace applications utilize bio-inspired materials that deliver revolutionary solutions for various operational needs. These materials based on biological structures deliver improved performance characteristics including the mechanical strength alongside self-healing abilities, adaptive camouflage features impact resistance and thermal adaptability which render them essential for next-generation military and aviation applications (Chen & Ritchie, 2016). Modern defense systems utilize bio-inspired materials to enhance survivability and operational effectiveness through applications like stealth coatings and ultra-lightweight armor.

Figure 5. Body armour developed using nacre-inspired impact resistance composites and nanotechnology

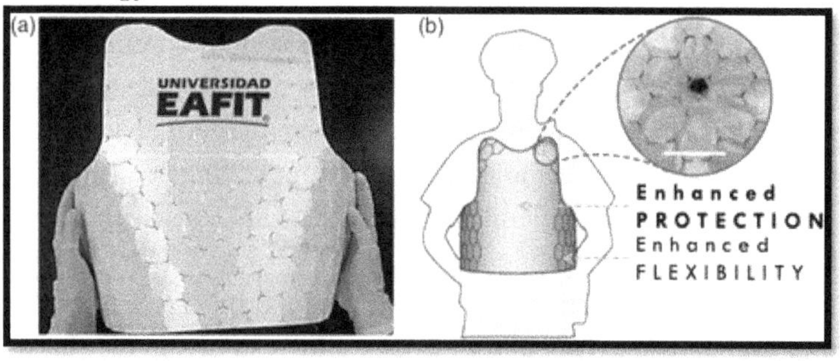

Figure 5 presents a composite material inspired by nacre, designed for use in body armor. The bio-inspired patterns contribute to increased flexibility without compromising ballistic defense. By emulating the hierarchical arrangement found in nacre, the armor effectively disperses impact energy, thereby enhancing its resistance to projectile penetration while permitting localized deformation for the comfort of the wearer. The accompanying schematic illustrates the design's equilibrium between flexibility and protection, thereby ensuring superior mechanical performance against ballistic hazards.

Nacre-inspired impact-resistant composites for the body armor and vehicle protection plating, one of the most studied bio-inspired materials for impact resistance is nacre-like composites, mimicking the hierarchical architecture known as ''brick-and-motor'' of mother-of-pearl. Nacre (aragonite platelet-aligned biopolymer matrix) is super strong and tough despite being made of brittle calcium carbonate ($CaCO_3$). Researchers have replicated its structure using layered ceramic-polymer composites which distribute impact energy better than traditional armor materials (Meyers et al. 2008). These materials like bio-mineralized alumina-polymer hybrids are used in military body armor and vehicle plating to protect against ballistic threats while reducing weight. The US military is developing lightweight ceramic composite armours inspired by mollusk shells to replace traditional heavy steel armor (Naleway et al. 2015). This allows soldiers to move more and be less fatigued while being protected.

Chitin based materials for helmets and exoskeletons, Chitin is a natural polysaccharide found in the exoskeletons of crustaceans and insects which is an inspiration for high toughness synthetic aramids like chitosan reinforced nanocomposites. These materials have high energy dissipation and fracture resistance, hence can be used for military helmets, exoskeletal reinforcement suits and protective padding (Ghazlan et al. 2021). The hierarchical fibre reinforced structure of chitin has been used to improve Kevlar based armor, making it more flexible and impact resistant while maintaining light weight. It's one of the best bio-inspired material to be used in defense industry and for defense applications.

Self-Healing Polymers for Aircraft and Vehicles, self-healing materials have changed the game for military aircraft coatings and vehicle bodies, extending operational life by autonomously fixing micro damage. Inspired by the self-healing vascular systems in plants, researchers have created microvascular polymer composites with embedded healing agents that fix cracks when expose to air or moisture. For aerospace applications, epoxy-based self-healing composites with micro capsulated dicyclopentadiene (DCPD) monomers and Grubbs' catalyst have been integrated into aircraft wing coatings. These coatings fix cracks automatically, reducing maintenance downtime and increasing aircraft life. The US Air Force Research Lab has tested this on fighter jets to increase structural integrity under high stress.

Figure 6. Representation of adaptive camouflage for military vehicles

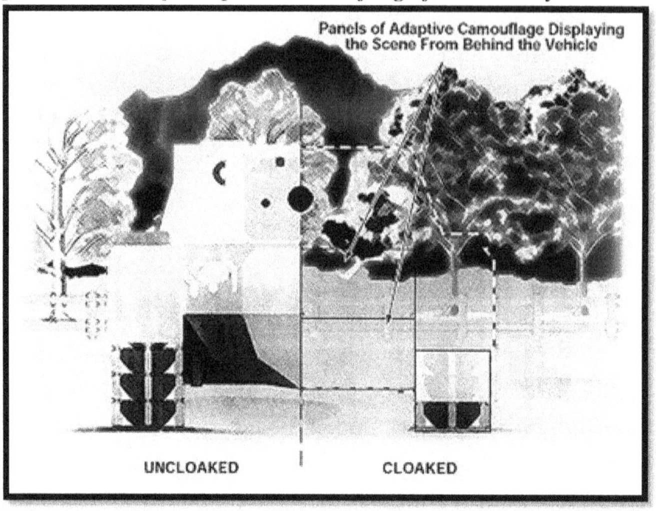

Figure 6 depicts an adaptive camouflage system tailored for military vehicles, drawing inspiration from bioinspired concepts of dynamic concealment. On the left, the vehicle is presented in its visible state, while the right side showcases the cloaking effect, where specialized panels project the surrounding environment, seamlessly integrating the vehicle into its backdrop. This biomimetic innovation takes cues from organisms like cephalopods, which utilize chromatophores and structural colour alterations to avoid detection. The advancement of these adaptive materials significantly boosts stealth capabilities, minimizing visibility across various terrains and enhancing operational efficiency in contemporary military engagements.

Graphene coatings, inspired by the lotus leaf, are being used on military aircraft, naval vessels and armoured vehicles for corrosion resistance. The superhydrophobic properties of graphene replicate natural self-cleaning surfaces, preventing oxidation and environmental degradation. This has been particularly useful for marine defence systems where exposure to saltwater accelerates material wear (Studart, 2012).

Cephalopod-inspired adaptive camouflage for military uniforms and vehicles, one of the most advanced areas of biomimetic research is the development of dynamic camouflage systems based on cephalopods like octopuses and cuttlefish. These marine animals use chromatophores, iridophores and leucophores to change colour and texture instantly. Scientists have replicated these properties in liquid crystal-based coatings and thermally responsive nanomaterials for adaptive military uniforms, UAV skins and stealth aircraft exteriors (Zhang et al. 2018). Metamaterial based cloaking technologies, inspired by cephalopod skin, manipulate electromagnetic waves to make objects invisible to radar detection. Plasmonic nanostructures crafted from gold and silver nanoparticles are paving the way for instant infrared camouflage—something that's absolutely vital for ground vehicles and the stealth gear used by soldiers. The U.S. Army is pouring resources into developing uniforms that can adjust their thermal signatures, effectively making it harder to spot soldiers in a variety of combat scenarios. It's a fascinating leap forward in ensuring that troops remain concealed, no matter where they find themselves on the battlefield.

Shark-skin-like drag-reducing coatings for aircraft and submarines. Just as the unique pattern of microscopic riblet-scale structures on its skin provides a shark with drag reduction, so are the surface coatings designed for this purpose with respect to submarines, fighter jets, and high-speed naval vessels. These riblet-structured polymers and nano-textured composites reduce turbulence and drag, improving fuel- and manoeuvrability (Williams, 2022). NASA is studying riblet-based aero-surfaces for the future hypersonic aircraft, emphasizing their importance for drag reduction in high-speed operations. For submarine applications, coatings inspired by the shark would increase hydrodynamic efficiency, stealthiness, and reduce detectability by sonar.

Figure 7. Representation of shark-skin like riblet layer along with self-healing composite coating on the military submarine

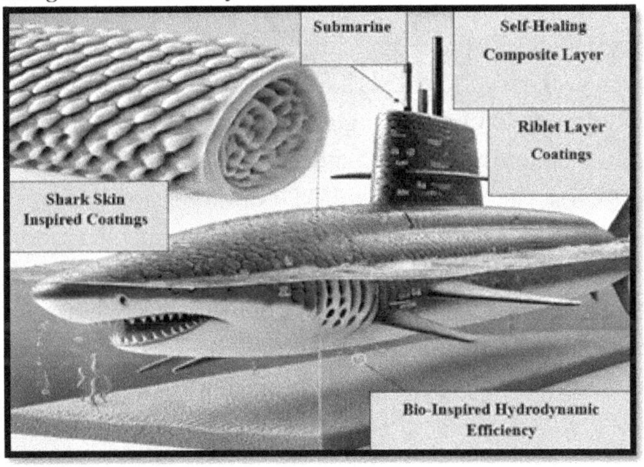

Figure 7 illustrates a riblet layer resembling shark skin, accompanied by a self-repairing composite coating on the military submarine. This bio-inspired configuration emulates the drag-reducing characteristics of the shark skin, employing riblet structures to optimize hydrodynamic efficiency and minimize acoustic signature. When integrated with a self-healing composite coating, this system enhances durability, corrosion resistance, and stealth features, thereby ensuring sustained operational effectiveness in challenging marine conditions.

Bio-derived aerogels for thermal insulation in spacecraft and defense infrastructure are inspired by the porous structures of bird bones and the walls of plant cells. They are aimed at revolutionizing thermal insulation of spacecraft, fighter jets, and military shelters (Fratzl & Weinkamer, 2007). These types of aerogels, made from silica or cellulose nanofibers, therefore are ultralight and resist heat extremely well. Silica aerogels have been integrated into the spacecraft heat shields and astronaut suits by NASA, which provide protection from intense temperature varying environments. The U.S. Air Force uses bio-aerogels to regulate heat in high-altitude drones in extremely subzero conditions.

The design wings for drones and future aircraft is being influenced by the flight patterns of birds. To improve their ability to fly, these wings flex and bend like a bird's. They are composed of unique polymers and metals that can alter shape when necessary. This improves their ability to fly in various conditions. These flexible wings allow military drones to fly more readily in difficult terrain. They become more adept at glancing about and spying as a result. The federal agency DARPA is funding a project named "Bio Swarm" (U.S. AFRL, 2022). The goal of this project is to create swift combat drones with wings that resemble those of albatrosses and falcons.

Gecko-Inspired Sticky Surfaces for Military Drones and Climbing Robots, based on the tiny hairs on gecko feet, scientists have created new dry sticky materials. These allow robots and drones to climb straight up without suction or glue. Made from silicone and carbon nanotubes, these materials are now part of special ops robots. They help with secret city scouting and wall-climbing for military use (Vincent et al. 2006). Soft Robotics and Fake Muscles for Exosuits and Combat Gear fake muscles, which take cues from how jellyfish move and human muscles work, are in development. They use special stretchy plastics that respond to electricity. These materials help create wearable suits that help soldiers carry heavy stuff cutting down on tiredness and the chance of getting hurt (Faisal & Park, 2016).

CHALLENGES IN THE DEVELOPMENT OF BIO-INSPIRED MATERIALS

Bio-inspired materials for aerospace and defense applications offer major scientific benefits while introducing substantial obstacles that require solutions. Multiple factors contribute to these challenges which include mass production barriers along with durability requirements for tough conditions and high costs while integration into existing systems presents complications as well as ethical and environmental considerations (Rajabi et al. 2016). The challenges these materials face are further intensified by limitations in material properties alongside computational capabilities (Torres et al. 2019). The broader adoption of bio-inspired materials in military and aerospace applications demands active resolution of the existing challenges. The successful completion of this task depends on cooperation between multiple disciplines and pioneering research projects. Advancements in materials science and engineering will enhance the functionality and practical applications of bio-inspired methods.

One of the foremost obstacles lies in the Expandability and mass manufacturing of bio-inspired materials. spurious substances care mother-of-pearl pearl and spider silk have Complicated stratified structures that bring good automatic properties. However synthetically replicating these structures on an industrial scale presents formidable challenges. For example, synthetic mother-of-pearl necessitates precise layer-by-layer lying to copy its spurious temper, making production obtuse and dear. Similarly, synthetic spider silk renowned for its remarkable strength and flexibility requires sophisticated bioengineering techniques such as recombinant DNA technology for large-scale production (Torres et al. 2019). These methods are not cost-effective for large-scale production in the defence sphere where materials need to be produced in vast volumes for uses like trunk armor, vehicle metal plating and aircraft parts and components.

Ensuring long-term durability in harsh defence environments poses a significant challenge. martial and aerospace materials have hold intense conditions including great temperatures fast force fluctuations high-velocity impacts and photo to radiation or caustic environments. While many bio-inspired materials perform well under controlled conditions they often degrade when subjected to prolonged exposure in real-world scenarios (Zhang et al. 2020). For case cephalopod-inspired accommodative disguise materials that employ clear quartz elastomers and plasmonic nanostructures get dynamically shift colour and hot emissivity. However, these materials frequently degrade under prolonged UV exposure and extreme temperature shifts limiting their effectiveness in long-term military operations. Similarly self-healing coatings for aircraft and equipped vehicles which bank along vascularized polymer webs get miss their remedial productivity subsequently aggregate damage-repair cycles. The inability of these materials to maintain their structural integrity under extended use in combat conditions remnant a significant obstacle to their adoption.

A significant obstacle to the widespread adoption of bio-inspired materials is their elevated development and production costs, numerous of these materials take advance nanotechnology preciseness techniques or compound biochemical methods edition, which further costs more than square materials such as arsenic metals like steel or conventional composites (Chen & Ritchie, 2016). For example, bio-inspired armor materials that incorporate carbon nanotube-reinforced synthetic nacre composites offer superior strength-to-weight ratios but are considerably more costly than traditional Kevlar or ceramic armor plating. Similarly self-healing polymers old inch aircraft coatings take dear microencapsulation techniques and particular catalysts devising lot product economically impossible. Defence agencies must strike a balance between material effectiveness and cost-effectiveness and unless the manufacturing costs of bio-inspired materials are reduced their adoption will remain confined to specialized uses rather than widespread use

A decisive challenge in Using bio-inspired materials for defence uses lies in their compatibility with existing military and aerospace systems, about flow defence technologies such as arsenic aircraft equipped vehicles and arm systems are organized exploitation metals ceramics and square polymer composites. Introducing new bio-inspired materials into these systems necessitates either extensive retrofitting or a complete re-layout of existing structures both of which are costly and time-intensive endeavours (Torres et al. 2019). For case bio-inspired morphing wings which mock avian escape mechanisms, employ the shape-memory alloys (SMA) and bio-inspired actuators which take sound modifications to the aircraft check systems and aeromechanics.

Similarly shark-skin-inspired drag-reducing coatings which improve the productivity of aircraft and submarines demand highly specialized manufacturing methods to apply the micro structured surfaces at scale, these consolidation challenges block the acceptance of bio-inspired materials and command defence agencies to bear comprehensive examination evidence and certificate methods ahead applications.

Beyond technical and economic challenges there are also significant ethical and security concerns surrounding the use of bio-inspired materials in military uses. Numerous biomimetic technologies notably those relevant to disguise free systems and neuromorphic materials fire right questions around their effect along contemporary warfare (Finkenstadt & Eapen, 2024). For example, military-grade autonomous drones equipped with mantis shrimp-inspired hyperspectral vision systems can dramatically improve surveillance and reconnaissance capabilities but their use in disagreement zones raises concerns about privacy surveillance ethics and autonomous warfare. Similarly neuromorphic materials old inch ai-driven decision-making systems which mimic the human brain's nervous structure, may obscure or blur the lines between human and automated warfare, creating potential dilemmas. The defence industry must consider the broader implications of using bio-inspired materials specifically in uses that could significantly alter military strategies and global security dynamics.

A paramount concern develops around the environmental implications and long-term sustainability of bio-inspired materials, spell many bio-inspired materials are engineered arsenic further eco-friendly alternatives to square counterparts around notwithstanding take dangerous deduction procedures bank along non-renewable supplies or get developed blow and toxic byproducts (Rajabi et al. 2016). For instance, the fabrication of synthetic nacre an engineered layered alumina-polymer composite demands energy-intensive methods that yield substantial industrial waste. Furthermore, self-healing coatings derivative from polymer microcapsules and graphene oxide-based films get bitter byproducts across sentence sitting environmental risks. Also, certain metamaterials which incorporate plasma-based radar-absorbing nanostructures often utilize rare-earth elements raising concerns regarding supply scarcity and geopolitical disagreements associated with mining these materials. arsenic sustainability becomes associate in nursing progressively difficult precedence for the defence industries bio-inspired materials have array with environmental regulations and long sustainability goals.

Bio-inspired materials face a fundamental challenge in striking an optimal balance between strength, flexibility and weight which often constrains their practical uses. Spell numerous such as materials point to pass conventional counterparts' inch particular properties achieving associate in nursing abstract combine of automatic operation characteristics clay subtle (Ritchie, 2011). For instance, synthetic fibres derived from spider silk exhibit exceptional tensile strength but lack sufficient compressive strength rendering them unsuitable for rigid armor uses. Similarly cephalopod-inspired disguise coatings which employs electrochromic polymers and plasmonic nanoparticles run in effect inch limited environments just show slower reaction and multiplication, low intense temperatures or aggressive conditions. Also, self-healing materials which can autonomously repair damage through vascularized nanocomposites or microcapsule-based systems often restore only partial mechanical properties after healing, these operation trade-offs bound the strength of bio-inspired materials inch extremely stern defence applications where materials have done faithfully low complete conditions without the important compromises.

The intricate nature of computational Representationing and Simulation for bio-inspired materials presents a formidable challenge, numerous of these materials have multi-scale stratified structures, making them difficult to analyse accurately using standard computational tools. For example, finite element analysis (FEA) of bio-inspired armor materials such as hierarchical ceramic-metal composites necessitate multi-scale Representationing techniques to predict impact resistance and energy dissipation. Similarly computational fluid dynamics (CFD) simulations of morphing aircraft wings, which employ shape-memory polymers and bio-inspired surface structures take real-time adaptive simulations to optimize flight performance. Neuromorphic materials employed in AI-driven decision-making systems require highly sophisticated procedures to emulate biological intelligence. Flow computational tools must work hard to get the good complicatedness of bio-inspired structures devising it hard to call their conduct accurately low real-world conditions. Advances in machine learning, quantum computing and generative design are assisting researchers in overcoming these challenges but significant progress is still needed to make efficient and reliable modeling of biomimetic materials. Despite these challenges current search and advancements in nanotechnology, AI-driven design and ergonomics are steadily overcoming and addressing these obstacles. The defence industry is investing heavily in the development of scalable fabrication methods and high-effectiveness biomimetic materials and advanced computational tools to accelerate the integration of bio-inspired materials into next-generation military and aerospace applications. By overcoming these challenges bio-inspired materials have the potential to revolutionize military technology and provide improved protection, productivity and adjustability in modern warfare and aerospace systems.

RESEARCH & DEVELOPMENT IN BIO-INSPIRED MATERIALS

The field of bio-inspired materials is driving innovation in defence technology offering novel answers for armor, aerospace, stealth, robotics and advanced material systems, away emulating nature's accommodative layouts. Researchers and engineers are developing highly effective self-healing light and multirole Al materials bespoke for defense uses. The ongoing research focuses on creating materials that mimic biological structures and roles to achieve superior mechanical strength, adjustability and resilience under extreme conditions (Vincent et al. 2006; Fratzl & Weinkamer, 2007).

One of the most promising advancements is the development of biologically derived nanocomposites which replicate natural load-bearing structures like nacre (mother-of-pearl) and bone to Improve impact resistance while maintaining lightweight properties (Barthelat & Espinosa, 2007; Meyers et al. 2008). These materials are currently being explored for the next-generation truck armour and vehicle metal plating having light in weight, security and mobility. The ability to replicate nature's hierarchical layout principles in engineered materials ensures improved energy dissipation and structural integrity making them highly effective for ballistic protection and extreme load-bearing uses (Nalewaay et al. 2015).

Self-healing materials represent an active area of research with significant progress in vascularized polymers and graphene-infused composites (Toohey et al. 2007; Studart, 2012). These modern materials enable free fix of damaged equipment and aircraft while reducing maintenance costs and extending effective life. In extreme battlefield conditions where continuous maintenance is impractical self-repairing materials offer a strategic advantage by maintaining structural integrity despite minor damages. Also, adjustive camouflage technologies inspired by cephalopods and thermoregulatory biological systems are being developed to make dynamic stealth coatings and infrared-deflecting uniforms that adjust to different environments in real-time (Zhang et al. 2020; Kang et al. 2018). These bio-inspired coatings are adequate of fixing their visual and hot signatures enhancing survivability inch reconnaissance and fight missions. Several institutions and defence research agencies are leading cutting-edge studies in bio-inspired materials. the evolution of bio-mineralized crossbreed composites for affect security divine away mother-of-pearl and pearl structures is revolutionizing armour buoyancy and fire opposition. Researchers are engineering hierarchical ceramic-metallic composites that Improve energy dissipation upon impact (Rajabi et al. 2016). Another major area of research involves self-healing polymers for aerospace and land systems where vascularized self-repairing materials are being laid and deployed for longer-lasting aircraft coatings and vehicle structures reducing operational downtime. The geographic expedition of cephalopod-inspired accommodative disguise has conducted to advancements inch metamaterials and clear crystal-based coatings that dynamically change colour and hot emissivity up stealing capabilities inch different fight environments also bio-derived aerogels for hot and acoustical detachment divine away spurious ultralight structures are provision extremely prompt detachment materials for belligerent blue jets ballistic capsule and martial shelters (Fratzl & Weinkamer, 2017).

Cutting-edge technologies are playing a pivotal role in the advancement of bio-inspired materials for defence uses, nanoengineered Kevlar and aramid fibres divine away spider silk bid new temper and tractability importantly enhancing tender pitch for martial force (Ritchie, 2011). Bio-electronic connections for wearable defence systems are incorporating neuromorphic sensors and bio-mimetic circuits enabling improved situational awareness through advanced soldier-worn technologies. Bio-mimetic aquaphobic and anti-fouling coatings enhance aircraft performance by reducing surface contamination, thereby improving operational efficiency in harsh environmental conditions. The development of morphing materials for drone and unmanned aerial vehicle (UAV) wings which mimic bird flight mechanisms is allowing for adjustive aerodynamic effectiveness optimizing manoeuvrability and fuel productivity in various atmospheric conditions (Wegst et al. 2015).

Government agencies and the defence research organizations are spearheading the innovations in bio-inspired materials, the Defense Advanced Research Projects Agency (DARPA) is supporting research on biologically adaptive materials for stealth armor and robotics, emphasizing rapid adaptability and manoeuvrability in combat environments. The National Aerospace Laboratories (NAL) is developing bio-derived composites for lightweight and impact-resistant aircraft structures aiming to improve operational productivity and fuel economy, the materials and the industry board of directors of the America and Air Force Research Laboratory is focused on self-repairing aerospace materials and biomimetic surface coatings to Improve aircraft durability and longevity (Torres et al. 2019). These collaborative efforts help the passage of bio-inspired materials from check search to pragmatic martial applications.

Despite the rapid advancements, several challenges must be addressed for the bio-inspired materials to reach their full potential in the defense applications. One of the primary hurdles is the scalability and cost efficiency. The complex hierarchical structures found in nature are difficult to replicate at an industrial scale, making mass production of bio-inspired materials challenging. Furthermore, integrating these new materials into existing military systems requires extensive validation and compatibility testing, ensuring that they perform seamlessly alongside traditional defense materials. Additionally, long-term stability and environmental impact assessments are crucial to determining the durability of bio-derived materials under extreme operational conditions, including high temperatures, pressure variations, and chemical exposure (Chen & Ritchie, 2016).

CONCLUSION

As scientists and engineers increasingly seek inspiration from nature, bio-inspired materials are reshaping defense and aerospace technology. By replicating biological processes and structures, bio-inspired materials can provide outstanding capabilities like lightweight strength, flexibility, and self-repair. The use of bio-inspired materials for armor, camouflage, and structure components has already shown considerable potential to increase endurance and efficiency for contemporary military and aerospace systems. Bio-materials such as nacre-based armor offer better impact resistance, and moisture-sensitive textiles increase the adaptability in hostile environments. Likewise, bio-based adhesives are facilitating stronger, more flexible adhesion solutions that increase the durability and functionality of military gear. While such promising strides are being made, challenges exist for large-scale production, cost-effectiveness, and long-term durability. Moreover, seamlessly integrating these materials into current defense and aerospace systems involves additional technological progress and engineering solutions.

To overcome these challenges, researchers are actively exploring innovative solutions to enhance the practicality of bio-inspired materials. Advances in bio-mineralized armor aim to improve impact resistance, while moisture-sensing textiles are being developed to provide greater adaptability in dynamic environments. Bio-based composites, inspired by structures like bone and exoskeletons, are being designed to create lightweight yet highly durable materials for aircraft and military vehicles. In addition, precision manufacturing and cutting-edge material synthesis techniques are key in enhancing the scalability of bio-inspired materials so that they can be produced effectively and incorporated into defense technologies. Material scientists, engineers, and defense specialists must collaborate in bridging the gap between lab experiments and large-scale adoption. With cross-disciplinary collaboration, bio-inspired materials will move from laboratory ideas to functional solutions that reframe the future of defense and aerospace technology.

In the future, smarter, more adaptive defense systems based on sustainability goals will be created. As biomimicry continues to evolve, future materials will not only be more protective and functional but also energy efficient and environmentally friendly. Self-healing paints, modeled after biological tissues, would greatly prolong the life of military equipment, lowering maintenance expenses and resource utilization. Likewise, light but very resilient bio-composites will result in more efficient aircraft and military vehicles, decreasing their carbon footprint. By adopting nature's design principles, future defense systems will be more robust, flexible, and able to counter the changing threats of contemporary warfare. As innovation evolves at a faster pace, bio-inspired materials will continue to be at the center of technological growth, defining a future in which defense and aerospace technologies are tougher, cleverer, and more environmentally friendly than ever before.

REFERENCES

Aizenberg, J., Weaver, J. C., Thanawala, M. S., Sundar, V. C., Morse, D. E., & Fratzl, P. (2005). Skeleton of Euplectella sp.: Structural hierarchy from the nanoscale to the macroscale. *Science*, *309*(5732), 275–278. DOI: 10.1126/science.1112255 PMID: 16002612

Asghar Yaseen, A., Waqar, T., Khan, M. A. A., Asad, M., & Djavanroodi, F. (2021). Fish Scales and Their Biomimetic Applications. *Frontiers in Materials.*

Barthelat, F., & Espinosa, H. D. (2007). An experimental investigation of deformation and fracture of nacre–mother of pearl. *Experimental Mechanics*, *47*(3), 311–324. DOI: 10.1007/s11340-007-9040-1

Chen, P. Y., & Ritchie, R. O. (2016). Bioinspired materials: Advancements, limitations, and future directions. *Materials Today*, *19*(8), 414–421.

Chen, W., Patel, D., & Radhakrishnan, R. (2019). Biomimetic infrared sensors for defense applications. *Sensors (Basel)*, *19*(7), 1206. DOI: 10.3390/s19051206 PMID: 30857309

Faisal, T., & Park, C. (2016). *Spider-Web Inspired Mechanical Metamaterials for Defense Applications.ArXiv Preprint,* 1607.01407. Retrieved from https://arxiv.org/abs/1607.01407

Finkenstadt, D. J., & Eapen, T. T. (2024). Prominence Strategies. In *Bioinspired Strategic Design* (pp. 101–154). Productivity Press., DOI: 10.4324/9781032715315-4

Fratzl, P., & Weinkamer, R. (2007). Nature's hierarchical materials. *Progress in Materials Science*, *52*(8), 1263–1334. DOI: 10.1016/j.pmatsci.2007.06.001

Gao, H., Ji, B., Jäger, I. L., Arzt, E., & Fratzl, P. (2003). Materials science: Biological materials and bioinspired structures. *Materials Science and Engineering A*, *86*(3), 213–221.

Ghazlan, A., Ngo, T., Tan, P., Xie, Y. M., & Tran, P. (2021). Inspiration from Nature's body armours – A review of biological and bioinspired composites. *Composites Part B: Engineering* Healing Platform with Water-Controlled Adhesiveness for Highly Effective Antibacterial Application and Data Encryption–Decryption'. *ACS Applied Materials & Interfaces*, *12*(52), 57686–57694. DOI: 10.1021/acsami.0c15364

Hsiung, B. K., He, H., & Dunlop, J. W. (2020). Trends in bioinspired material research for structural and defense applications. *Advanced Functional Materials*, *30*(5), 1909854.

Kang, S. H., Lee, D., & Kim, H. (2018). Adaptive camouflage systems: The next generation of stealth materials. *Advanced Materials, 30*(25), 1706698. PMID: 29709083

Lee, H., Scherer, N. F., & Messersmith, P. B. (2007). Single-molecule mechanics of mussel adhesion. *Proceedings of the National Academy of Sciences of the United States of America, 104*(43), 17345–17350. PMID: 16920796

Marshall, J. S., Young, J. W., & Hallam, J. (2008). Bio-Inspired Approaches to Sensing for Defence and Security Applications. *Journal of the Royal Society, Interface, 5*(21), 467–477. DOI: 10.1098/rsif.2008.0109.focus

Meyers, M. A., Lin, A. Y. M., Olevsky, E. A., & Chen, P. Y. (2008). Biological materials: Structure and mechanical properties. *Progress in Materials Science, 53*(1), 1–206. DOI: 10.1016/j.pmatsci.2007.05.002

Naleway, S. E., Porter, M. M., McKittrick, J., & Meyers, M. A. (2015). Structural design elements in biological materials: Application to bioinspiration. *Advanced Materials, 27*(37), 5455–5476. DOI: 10.1002/adma.201502403 PMID: 26305858

National Materials and Manufacturing Board (NMMB). (2021). Bio-Inspired Manufacturing for Future Defense Applications. Retrieved from www.nas.edu

National Research Council. (2012). *Opportunities and challenges in biomimetics.* The National Academies Press.

Rajabi, H., Gorb, S. N., & Jafarpour, M. (2016). Challenges in bio-inspired material design: Scaling up from laboratory to industrial production. *Bioinspiration & Biomimetics, 11*(2), 026005. PMID: 27033298

Reddy, S. (2018). Recent trends in biomimetic materials research and applications. *Materials Today, 21*(1), 65–78.

Ritchie, R. O. (2011). The conflicts between strength and toughness. *Nature Materials, 10*(11), 817–822. DOI: 10.1038/nmat3115 PMID: 22020005

Studart, A. R. (2012). Biologically inspired materials that self-assemble. *Journal of Materials Chemistry, 22*(19), 8203–8211.

Taylor, J. R. A., Scott, N. I., & Rouse, G. W. (2019). 'Evolution of Mantis Shrimp Telson Armour and Its Role in Ritualized Fighting'. *Journal of the Royal Society, Interface, 16*(157), 20190203. Advance online publication. DOI: 10.1098/rsif.2019.0203 PMID: 31455162

Toohey, K. S., Sottos, N. R., Lewis, J. A., Moore, J. S., & White, S. R. (2007). Self-healing materials with microvascular networks. *Nature Materials*, *6*(8), 581–585. DOI: 10.1038/nmat1934 PMID: 17558429

Torres, F. G., Troncoso, O. P., Nakamatsu, J., & Grande, C. J. (2019). The challenges of bio-inspired material synthesis for aerospace applications. *Acta Biomaterialia*, *88*, 16–30.

U.S. Air Force Research Laboratory (AFRL). (2022). Advances in bio-inspired materials for military aerospace applications. Retrieved from www.afrl.mil

Vincent, J. F. V., Bogatyreva, O. A., Bogatyrev, N. R., Bowyer, A., & Pahl, A. K. (2006). Biomimetics: Its practice and theory. *Journal of the Royal Society, Interface*, *3*(9), 471–482. DOI: 10.1098/rsif.2006.0127 PMID: 16849244

Wang, N., Yu, K.-K., Li, K., Li, M.-J., Wei, X., & Yu, X.-Q. (2020). 'Plant-Inspired Multifunctional Fluorescent Hydrogel: A Highly Stretchable and Recoverable Self-Wegst, U. G. K. Bai, H. Saiz, E. Tomsia, A. P. & Ritchie, R. O. (2015). Bioinspired structural materials. *Nature Materials*, *14*(1), 23–36.

Williams, T. (2022). Advancing research efforts in biomimicry to develop nature-inspired materials, processes for space exploration and more efficient aircraft. In *Biomimicry for Aerospace*. Technologies and Applications. DOI: 10.1016/B978-0-12-821074-1.00014-1

Zhang, L., Li, X., & Wang, Y. (2020). Bioinspired materials for camouflage applications. *Materials Today*, *37*, 129–141.

ABBREVIATION

AI - Artificial Intelligence
CFD - Computational Fluid Dynamics
DARPA - Defense Advanced Research Projects Agency
FEA - Finite Element Analysis
NAL - National Aerospace Laboratories
UAV - Unmanned Aerial Vehicle
USAFRL - United States Air Force Research Laboratory
3D - Three-Dimensional
NASA - National Aeronautics and Space Administration
R&D - Research and Development
NMF - Nanostructured Metallic Frameworks
ML - Machine Learning
MEMS - Micro-Electromechanical Systems
IR - Infrared
TRL - Technology Readiness Level
BIM - Biomimetic Inspired Materials
CNT - Carbon Nanotubes
MOF - Metal-Organic Frameworks
CAD - Computer-Aided Design
CF - Carbon Fiber
HPC - High-Performance Computing
PZT - Lead Zirconate Titanate (a piezoelectric material)
FEM - Finite Element Method
VPM - Vascular Polymer Materials
NDT - Non-Destructive Testing
EMS - Electromagnetic Shielding
AM - Additive Manufacturing
SMP - Shape Memory Polymers
BCA - Bio-Composites Analysis
AR - Augmented Reality
NMT – Nanomaterial Technology
TPM – Thermal Protection Materials
BM - Biomimetics
LCA – Lightweight Composite Armour
ECS – Environmental Control Systems
SHM – Structural Health Monitoring
CFRP – Carbon Fiber Reinforced Polymer
FRP – Fiber Reinforced Polymer
TBC – Thermal Barrier Coatings
BIM – Bio Inspired Material

Chapter 6
Innovative Materials for Next-Generation Defence Applications:
Implications for Civil Infrastructure in Cost, Performance, and Mass Production

Nidhya Rathinavel
https://orcid.org/0000-0002-9962-5478
PSG Institute of Technology and Applied Research, India

Arun Murugesan
https://orcid.org/0000-0002-2662-8483
PSG Institute of Technology and Applied Research, India

Karthick Jaisankar
https://orcid.org/0009-0005-9477-5791
PSG Institute of Technology and Applied Research, India

Mohammed Jafer Sathik Ali Abdul Hameed
PSG Institute of Technology and Applied Research, India

ABSTRACT

The chapter explores the integration of defence-grade materials into construction to enhance infrastructure resilience. Advanced materials like carbon fibre, Kevlar, graphene composites, and multifunctional geopolymers offer high strength, durability,

DOI: 10.4018/979-8-3373-0933-0.ch006

and self-healing properties. Bio-inspired innovations, including mechanoluminescent, hydrophobic, and energy-autonomous materials, have successfully transitioned into civilian applications, improving seismic resilience, structural health monitoring, and environmental protection. Despite cost and scalability challenges, ongoing research in 4D printing, nanotechnology, and self-healing systems is driving broader implementation. Increased investment in advanced materials underscores their crucial role in developing sustainable, adaptive, and high-performance urban infrastructure.

INTRODUCTION

Figure 1. Advanced materials and its revolution

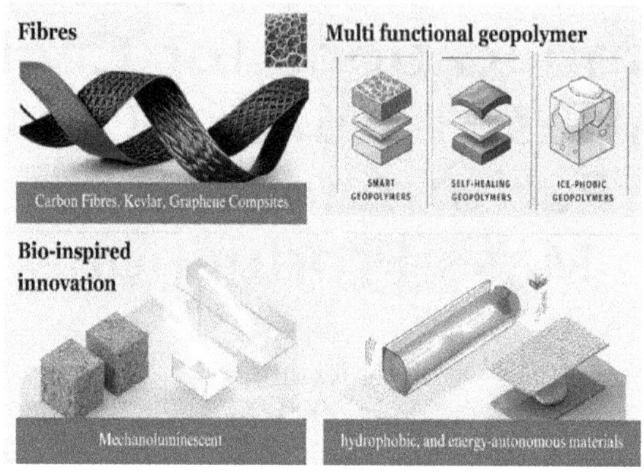

Figure 2. Advanced materials and its revolution

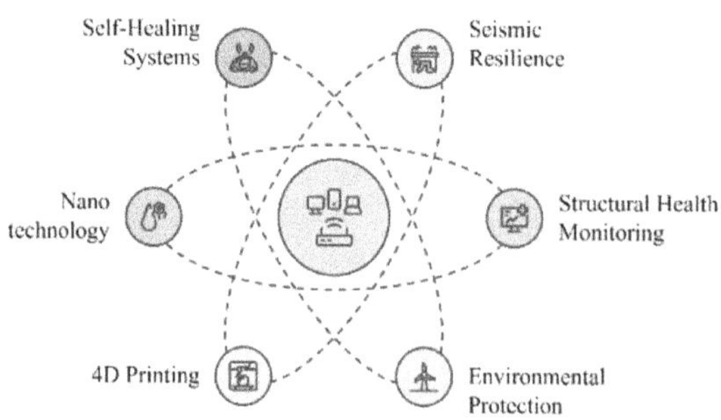

The growing global need for durable, resilient, and sustainable civil infrastructure has driven significant advancements in construction materials and engineering techniques. Among these advancements, defence-grade materials have emerged as a promising solution for enhancing structural integrity, longevity, and environmental resistance. Originally developed for military and aerospace applications, these materials were designed to withstand high-stress environments, extreme temperature fluctuations, and corrosive conditions. Today, their unique properties are being leveraged for civil infrastructure projects, offering superior performance in comparison to traditional materials such as steel and conventional concrete (Mohamed et al., 2024; Zhu et al., 2016).

Key materials such as carbon fibre, Kevlar, graphene composites, multifunctional geopolymers, and bio-inspired materials have demonstrated remarkable effectiveness in bridges, tunnels, roadways, high-rise buildings, and protective structures. These materials are revolutionizing modern engineering by providing high strength-to-weight ratios, impact resistance, self-healing capabilities, and adaptive performance under extreme environmental conditions. Their widespread adoption is expected to transform the construction industry, ensuring infrastructure that is not only stronger and more durable but also cost-effective and sustainable in the long run (Nasiri, n.d.; Nwokediegwu et al., 2024).

This proposal deals into the cost-effectiveness, performance advantages, and scalability of these materials in civil engineering applications. It evaluates their role in improving structural resilience, minimizing maintenance costs, and extending service life while addressing key challenges such as mass production, supply chain optimization, and environmental impact. Furthermore, the study provides real-world case studies showcasing how these materials have been successfully integrated into infrastructure projects to enhance load-bearing capacity, mitigate seismic damage, and improve overall structural performance. Beyond their mechanical and structural benefits, the sustainability factor of these advanced materials is a crucial aspect. With the increasing focus on reducing carbon footprints and developing eco-friendly infrastructure, materials such as geopolymers and graphene composites are gaining prominence due to their lower environmental impact compared to traditional cement-based materials. Additionally, the potential for recycling and repurposing these materials further strengthens their viability as a sustainable solution for the future of construction(Yusuf et al., 2024) (Nasiri, n.d.; Nwokediegwu et al., 2024).

While defence-grade materials offer unparalleled advantages, there are challenges in their scalability, cost, and manufacturing complexity, which need to be addressed before their full-scale adoption in the civil infrastructure sector. Ongoing research and technological innovations, including 4D printing, self-healing materials, and nanotechnology, are expected to revolutionize the construction landscape, making these materials more accessible, cost-efficient, and adaptable for large-scale deployment. By examining the scientific, economic, and engineering aspects of these advanced materials, this proposal aims to advocate for their widespread integration in the construction industry. As infrastructure demands continue to rise globally, incorporating next-generation materials will be vital for addressing urbanization challenges, climate resilience, and long-term sustainability, paving the way for smarter, more adaptive, and energy-efficient infrastructure solutions that can withstand the challenges of the future(Shehata et al., 2022; Nilimaa, 2023).

The primary objective of this chapter is to assess and promote the integration of defence-grade materials in civil infrastructure to enhance durability, resilience, and sustainability. It evaluates the cost-effectiveness, performance benefits, and scalability of materials like carbon fibre, Kevlar, graphene composites, and geopolymers in bridges, tunnels, and buildings. Additionally, it addresses manufacturing challenges, environmental impact, and emerging technologies such as 4D printing and self-healing materials, aiming to support the development of smarter, more adaptive, and long-lasting infrastructure(Ajirotutu et al., 2024; GangaRao, 2017).

In summary, the integration of innovative materials such as advanced composites, multifunctional geopolymers, and bio-inspired systems introduces transformative capabilities for next-generation infrastructure. While their structural and functional advantages are evident, sustainability remains a cornerstone for their broader adoption. The following section delves into the lifecycle impacts and environmental considerations that underpin the responsible use of these materials in large-scale civil applications.

Figure 3. Advanced composite materials for high-performance applications

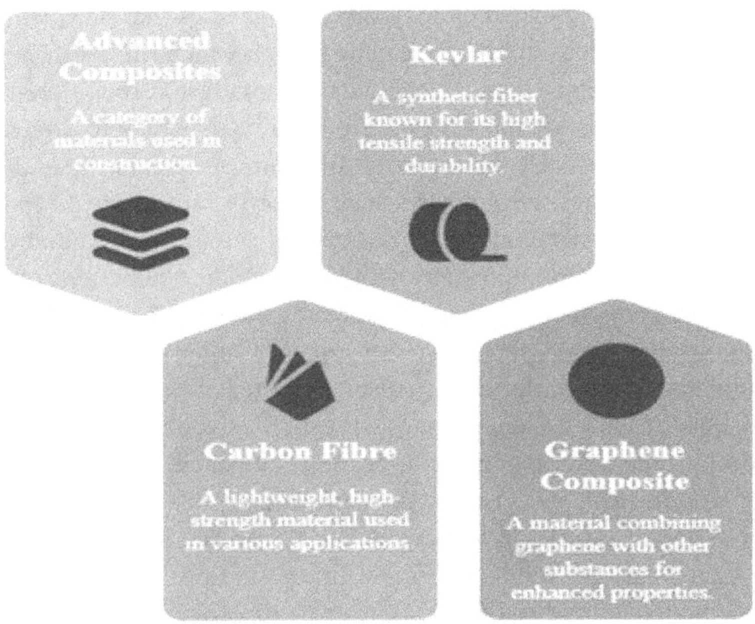

Advanced Composites

Carbon fibre, Kevlar, and graphene composites are among the most advanced materials being integrated into modern civil infrastructure. These materials, originally developed for defence applications, are now being widely used in bridges, high-rise buildings, tunnels, and other critical infrastructure due to their exceptional mechanical properties (Nasiri, n.d.; Nilimaa, 2023).

Carbon Fibre

Carbon fibre composites have emerged as a revolutionary material in the construction and infrastructure industry, offering a high strength-to-weight ratio, exceptional durability, and superior resistance to environmental degradation when compared to conventional materials like steel and concrete. Originally developed for aerospace and defence applications, carbon fibre has now become an essential component in bridge reinforcement, seismic retrofitting, and structural strengthening of aging infrastructure (J. Bai, 2022; Uddin, 2013).

One of the key advantages of carbon fibre composites is their lightweight nature, which allows for structural enhancements without significantly increasing the overall weight of the infrastructure. Traditional materials like steel, while strong, add substantial weight to structures, leading to higher foundation loads and potential long-term stress on bridges and buildings. In contrast, carbon fibre-reinforced polymers (CFRPs) provide the same or greater strength with only a fraction of the weight, making them ideal for bridge retrofitting and structural repairs. Additionally, carbon fibre composites are highly resistant to corrosion, chemical exposure, and weather-induced deterioration, which is a major issue for steel-reinforced bridges and coastal infrastructure. In regions with high humidity, saltwater exposure, and extreme temperature fluctuations, steel structures are prone to rust and degradation, requiring frequent maintenance and costly repairs. Carbon fibre, however, remains chemically stable and unaffected by moisture, UV radiation, and extreme temperature variations, making it a long-lasting and cost-effective alternative for reinforcing bridge decks, support beams, and load-bearing components (Agarwal et al., 2024; Boylu & Ekinci, 2024; H. Z. Hassan & Saeed, 2024).

A study by the American Society of Civil Engineers (ASCE) found that replacing traditional steel reinforcements with carbon fibre composites can reduce infrastructure-related carbon emissions by up to 40% over a structure's lifespan. In Japan, seismic retrofitting efforts have incorporated carbon fibre-reinforced polymer (CFRP) in multiple bridges, including the Yokohama Bay Bridge, where engineers observed a 30% reduction in structural damage during earthquakes, highlighting CFRP's effectiveness in mitigating seismic stress. Similarly, the Henniger Flats Bridge in California was reinforced with CFRP wraps, resulting in a 50% increase in load-bearing capacity, reduced material fatigue, and lower maintenance costs, showcasing the material's advantages in enhancing structural resilience and longevity (Nwokediegwu et al., 2024; Patchen et al., 2023).

Kevlar

A high-strength aramid fibre developed by DuPont in 1965, has emerged as a key material in civil infrastructure, particularly in blast-resistant and impact-resistant structures. Originally designed for ballistic protection and military applications, Kevlar is now extensively used in high-risk civil engineering projects, where structural integrity, safety, and durability are paramount. Its exceptional tensile strength, lightweight nature, and ability to absorb high-impact forces make it a crucial material for enhancing the resilience of buildings, bridges, tunnels, and protective barriers. Kevlar's tensile strength-to-weight ratio is five times higher than steel, making it one of the strongest materials available for construction reinforcement. Unlike traditional steel reinforcements, Kevlar is lightweight, flexible, and highly resistant to deformation, allowing it to absorb and distribute force more effectively. This property is particularly beneficial in blast-resistant buildings, earthquake-prone structures, and impact-resistant barriers where sudden force absorption is critical (Nwokediegwu et al., 2024; T. J. Singh & Samanta, 2015).

Compared to traditional construction materials like steel and concrete, Kevlar offers a significant advantage due to its lightweight nature, reducing energy consumption and carbon footprint during transportation and installation. Being five times stronger than steel by weight, it requires less material for the same level of reinforcement, lowering resource use and emissions. Its effectiveness is evident in U.S. embassies and military installations, where Kevlar-reinforced panels have enhanced blast resistance and structural integrity, minimizing damage from explosions and ballistic threats while protecting both infrastructure and occupants (Dhakal & Ismail, 2020; Ursache et al., 2023).

Graphene Composite

Graphene, a single layer of carbon atoms arranged in a hexagonal lattice, has remarkable mechanical, electrical, and thermal properties, its ultra-lightweight yet incredibly strong nature allows it to reinforce traditional materials like concrete and coatings, significantly enhancing their performance and longevity. When incorporated into construction materials, graphene not only increases their mechanical strength but also improves their durability, crack resistance, and ability to self-monitor structural health. One of the most significant advantages of graphene-reinforced concrete is its higher compressive strength, which can be up to 50% greater than that of conventional concrete. This increased strength allows for a reduction in the amount of cement required, thereby lowering construction costs and environmental impact while maintaining the material's structural integrity. Furthermore, graphene improves the flexural and tensile strength of concrete, making it more resistant to cracking and deformation under stress. This is particularly beneficial in large-scale infrastructure projects, such as bridges and high-rise buildings, where material strength is critical (Liu, 2024; F. Zhang et al., 2022).

Graphene significantly enhances construction materials by preventing microcracks, improving wear resistance, and enabling self-monitoring capabilities through its electrical conductivity. Unlike traditional structures that rely on external sensors for damage detection, graphene-enhanced concrete and coatings act as intrinsic sensors, providing real-time data on stress, strain, temperature fluctuations, and microcracks. Additionally, graphene strengthens concrete by 30-50%, as demonstrated by researchers at the University of Manchester, reducing cement usage and lowering CO_2 emissions, which account for nearly 8% of global emissions. Its anti-corrosion properties have also been successfully applied in Spain, where graphene-based coatings on steel structures extend the lifespan of bridges and buildings by preventing rust formation, reducing maintenance needs, and improving long-term durability (Hernández-Sánchez et al., 2024; Qureshi et al., 2024).

Multifunctional Geopolymers: Smart, Self-Healing, and Extreme-Environment Resilient Materials

Figure 4. Types of multifunctional geopolymers for advanced applications

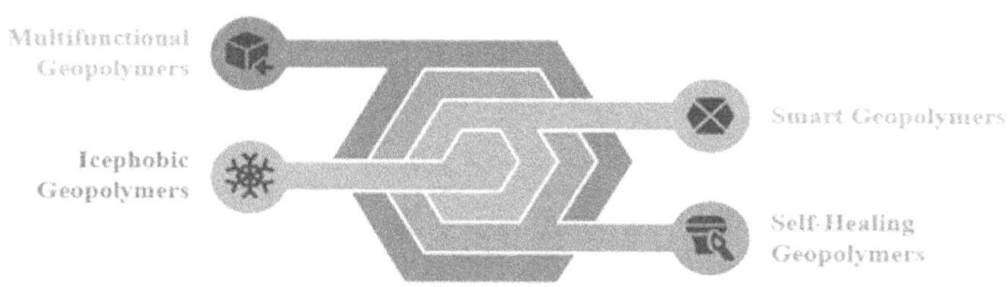

Multifunctional geopolymers are emerging as advanced construction materials capable of withstanding extreme environmental conditions while offering superior sustainability compared to traditional Portland cement. These next-generation materials integrate smart properties such as self-healing, icephobicity, and enhanced durability, making them ideal for modern infrastructure applications. By utilizing geopolymerization, which involves the activation of aluminosilicate materials with an alkaline solution, geopolymers provide superior mechanical strength, chemical resistance, and lower carbon emissions, aligning with the growing demand for eco-friendly construction solutions (MAMULAT et al., 2024; K. T. Nguyen et al., 2016).

Smart Geopolymers

Smart geopolymers incorporate conductive and piezoelectric fillers, enabling real-time structural health monitoring in infrastructure. By embedding sensors within the geopolymer matrix, these materials can detect strain, stress, and microcracks, transmitting data for predictive maintenance. This capability enhances safety in critical infrastructure such as bridges, tunnels, and high-rise buildings by allowing early detection of structural issues before catastrophic failures occur (Latha et al., 2024; Vasconcelos et al., 2024).

Self-Healing Geopolymers

Inspired by natural self-repair mechanisms, self-healing geopolymers integrate microcapsules containing healing agents like silica gel, polymeric resins, or bacterial spores, which autonomously seal cracks upon rupture, preventing further damage and significantly extending the lifespan of structures. This innovation reduces costly maintenance, particularly in infrastructure exposed to stress, seismic activity, or corrosive environments such as marine structures and wastewater treatment plants. Researchers at Delft University developed a geopolymer incorporating Bacillus bacteria, which produce calcium carbonate upon moisture exposure, effectively sealing cracks in bridges, tunnels, and marine structures, enhancing durability and reducing maintenance needs. Another approach embeds polymeric microcapsules that release healing agents upon cracking, chemically bonding with the geopolymer matrix to restore integrity. By preventing microcracks from escalating into structural failures, self-healing geopolymers enhance longevity, minimize material degradation, and reduce construction waste, contributing to a circular economy and sustainable construction. Their ability to autonomously repair cracks mitigates the risk of catastrophic failures, making them particularly valuable for critical infrastructure like bridges, tunnels, and nuclear power plants (De Belie et al., 2018; Hossain et al., 2022; D. Zhao et al., 2024).

Icephobic Geopolymers

Icephobic geopolymers are engineered to prevent ice accumulation on infrastructure surfaces by integrating advanced surface coatings, hydrophobic additives, or nanostructured modifications that lower ice adhesion and freezing points, making them essential for cold-climate regions. By minimizing ice buildup on bridges, highways, airport runways, and wind turbines, these materials reduce reliance on costly and environmentally harmful de-icing chemicals, which contribute to corrosion, groundwater contamination, and increased maintenance costs. The University of Toronto developed geopolymer-based nanocoatings with hydrophobic and icephobic properties, which, when tested on Canadian highway overpasses and bridge decks, reduced ice accumulation by 50%, ensuring safer winter travel conditions. Similarly, a research team created self-cleaning, ice-repellent geopolymer coatings with nano-silica and polymeric additives, successfully implemented on North American airport runways and taxiways, leading to fewer flight delays and lower maintenance costs. By providing a durable, long-term, and eco-friendly alternative to traditional de-icing methods, icephobic geopolymers enhance safety, reduce infrastructure wear, and support sustainable winter operations (Hernández Rodríguez et al., 2024; A. Zhang et al., 2024).

Bio-Inspired Materials: Mimicking Nature for Resilient and Sustainable Infrastructure

Unlike composites and geopolymers, which are often industrially engineered, bio-inspired materials are derived from nature's efficiency and resilience. These include surfaces that repel water like lotus leaves, materials that glow under stress like deep-sea organisms, and systems that harvest ambient energy. Their simplicity belies their impact: in energy-scarce or high-risk environments, they reduce maintenance costs and improve user safety. Increasing research investment is making these once-experimental materials more practical for real-world deployment.

Figure 5. Materials for smart and sustainable systems

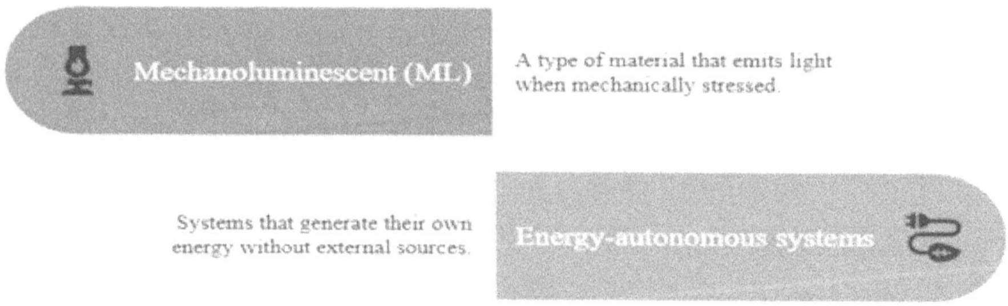

Nature has long served as an inspiration for engineering solutions, and bio-inspired materials are now transforming modern infrastructure by integrating self-sustaining, energy-efficient, and environmentally resilient properties. By mimicking biological systems, these materials enhance durability, improve energy efficiency, and contribute to the development of smart, adaptive, and low-maintenance structures. Civil engineering is increasingly adopting bio-inspired technologies to create more sustainable, self-repairing, and resource-efficient infrastructure, reducing environmental impact while enhancing functionality (Kannan et al., 2025; Sharma & Hiremath, 2022).

Mechanoluminescent (ML)

Mechanoluminescent (ML) materials are an advanced class of smart materials capable of emitting visible light upon exposure to mechanical stress, deformation, or friction, leveraging specific crystalline structures where mechanical energy excites electrons to higher energy states, resulting in photon emission upon their return to equilibrium. This unique property enables ML materials to function as self-powered luminescent systems without external energy sources, presenting a sustainable and energy-efficient alternative for modern infrastructure applications. Their integration into civil structures offers significant advancements in urban lighting, road safety, and structural health monitoring by converting mechanical stress into visible illumination, thereby enhancing visibility and safety without reliance on electrical grids. ML materials can be embedded in roads, bridges, pavements, tunnels, and warning signs, ensuring safer transportation networks, particularly in regions with inconsistent power supply or low-light conditions. From a sustainability perspective, ML materials minimize energy consumption, lower carbon emissions, and reduce dependence on fossil fuel-based electricity while being inherently non-toxic and recyclable, making them an environmentally friendly alternative to conventional lighting systems. Their application in smart infrastructure has been demonstrated through real-world implementations, such as ML-embedded pavement blocks in South Korea, which glow under pedestrian traffic to enhance night time navigation and safety, and ML-integrated bridge structures in Japan, where luminescent signals indicate areas of excessive stress or microcracking, providing a real-time, non-destructive means of monitoring structural integrity. The adoption of ML materials in civil engineering and transportation networks can significantly enhance safety, resilience, and sustainability, marking a paradigm shift in next-generation smart infrastructure development (Cheng et al., 2025; C.-Y. Sun et al., 2025; T. Wang et al., 2025; Yan et al., 2024).

Hydrophobic Materials

Hydrophobic materials are engineered to repel water and prevent moisture infiltration, significantly enhancing the durability and service life of infrastructure. These materials are inspired by the lotus leaf effect, where water droplets bead up and roll off the surface due to the presence of micro- and nanoscale surface roughness, combined with low surface energy chemistry. This phenomenon is replicated in artificial hydrophobic coatings through micro/nanostructured modifications and the application of water-repellent compounds such as siloxanes, fluoropolymers, and superhydrophobic nanoparticles. These coatings reduce the adhesion of water molecules, thereby preventing capillary absorption and minimizing surface wetting.

The contact angle of water droplets on hydrophobic surfaces typically exceeds 150°, which ensures minimal water retention and maximizes self-cleaning properties. By integrating hydrophobic materials into infrastructure, engineers can significantly mitigate water-induced deterioration mechanisms, particularly in concrete, steel, and composite structures (Bian et al., 2024; Z. Wang et al., 2025).

One of the most critical advantages of hydrophobic coatings is their ability to protect structural materials from corrosion, freeze-thaw cycles, and biological degradation. In reinforced concrete, water infiltration leads to alkali-silica reaction (ASR), sulfate attack, and rebar corrosion, which compromise structural integrity over time. Hydrophobic treatments, especially those based on silane and siloxane compounds, penetrate the concrete matrix and create a water-repellent barrier without obstructing vapor permeability, thereby preventing internal moisture accumulation and reducing the risk of degradation. Similarly, in steel structures such as bridges, offshore platforms, and pipelines, hydrophobic coatings act as a protective layer against chloride-induced corrosion, a major issue in coastal and marine environments where salt spray accelerates material breakdown. Additionally, hydrophobic glass and polymer-based coatings are employed in tunnels, facades, and industrial facilities to enhance water resistance while reducing the accumulation of dirt, pollutants, and microbial growth (H. Hassan et al., 2017; Rusu et al., 2024).

The application of hydrophobic surfaces extends beyond structural durability to functional safety improvements. For instance, icephobic coatings, which are specialized hydrophobic materials designed to prevent ice adhesion, are crucial for mitigating hazards in transportation and aviation infrastructure. By preventing ice accumulation on roads, bridges, and airport runways, these coatings enhance surface friction, reduce the risk of vehicular skidding, and minimize delays caused by ice removal processes. In highway engineering, hydrophobic asphalt and concrete pavements reduce water pooling and hydroplaning, thereby improving vehicle traction during rainfall. Furthermore, in hydraulic structures such as dams, spillways, and drainage systems, hydrophobic materials minimize water seepage and erosion, leading to more efficient water management and reduced maintenance costs (Huang et al., 2024; Skountzos et al., 2024).

The effectiveness of hydrophobic materials in infrastructure has been demonstrated through various large-scale implementations. In the Netherlands, the Dutch government has experimented with hydrophobic concrete in roadways and bridges to mitigate the effects of heavy rainfall and flooding. This concrete was formulated with nano-silica coatings and water-repellent additives, significantly reducing water absorption by over 80% and minimizing freeze-thaw damage in winter conditions. The application of hydrophobic concrete on highways and urban pavements has led to improved surface durability, reduced maintenance frequency, and enhanced road safety during wet conditions (Huang et al., 2024; Z. Zhao et al., 2024). Another notable example is the Bosphorus Bridge, a crucial transportation link between Europe and Asia, which faced severe corrosion due to moisture, salt spray, and industrial pollution. Engineers applied siloxane-based hydrophobic coatings to steel and concrete sections, resulting in a 50% reduction in the corrosion rate and a substantial extension of the bridge's service life. The hydrophobic coating not only reduced moisture penetration but also minimized the adhesion of atmospheric pollutants, thereby decreasing maintenance costs and structural deterioration over time (BinSabt et al., 2023; Pokorný & Kouřil, 2024).

Energy-Autonomous Systems

Energy-autonomous systems are revolutionizing modern infrastructure by enabling self-sustaining, grid-independent operations through ambient energy harvesting. These systems harness energy from mechanical vibrations, thermal gradients, and environmental interactions, converting otherwise wasted energy into usable electricity. Piezoelectric, triboelectric, and thermoelectric materials play a crucial role in these systems, allowing roads, bridges, and buildings to generate power through vehicle movement, foot traffic, temperature fluctuations, and wind energy. The integration of such energy-harvesting technologies into critical infrastructure facilitates self-powered applications, including smart lighting, real-time structural health monitoring, wireless communication networks, and traffic management systems. By reducing reliance on external power grids, energy-autonomous systems enhance the resilience, efficiency, and sustainability of infrastructure, making them particularly valuable in remote areas and disaster-prone regions (Sani, 2024; Zhou et al., 2024).

One of the most promising approaches to mechanical energy harvesting in infrastructure is the use of piezoelectric and triboelectric nanogenerators (PENGs and TENGs). Piezoelectric materials, such as lead zirconate titanate (PZT), barium titanate ($BaTiO_3$), and polyvinylidene fluoride (PVDF), generate an electric charge when subjected to mechanical stress, making them ideal for embedding within roadways and pedestrian walkways to capture energy from vehicular and foot traffic. Similarly, triboelectric nanogenerators (TENGs) operate by harnessing charge transfer from friction between dissimilar materials, effectively converting surface interactions into electrical output. These technologies have been successfully demonstrated in smart pavements and bridge expansion joints, where they generate electricity from cyclic loads and vibrations. The harvested energy is used for LED streetlights, surveillance cameras, and IoT-based traffic monitoring systems, reducing operational costs and carbon emissions while improving urban safety and efficiency (Bouhamed et al., 2024; X. Wang et al., 2024).

In addition to mechanical energy harvesting, thermoelectric materials provide a means of converting temperature gradients into electrical energy. Thermoelectric generators (TEGs), composed of semiconductor materials such as bismuth telluride (Bi_2Te_3) and lead telluride (PbTe), exploit the Seebeck effect to generate power from heat differentials. These systems can be embedded in road pavements, bridges, and building facades to harness waste heat from solar radiation, vehicular friction, and industrial processes. For example, thermoelectric roads absorb solar heat during the day, using TEGs to generate electricity that powers embedded sensors, electronic signage, and emergency lighting. This passive energy conversion process enhances infrastructure functionality while contributing to sustainable urban energy systems (Lakshmi et al., 2024; Win et al., 2024).

Several real-world applications have demonstrated the feasibility of energy-autonomous infrastructure. In Beijing, China, researchers developed a TENG-based sidewalk designed to harvest energy from pedestrian foot traffic. This system integrates triboelectric nanogenerators within pavement blocks, generating electricity as people walk over them. The harvested energy powers LED streetlights, wireless sensors, and interactive displays, creating a self-sustaining urban lighting system that reduces energy consumption and enhances pedestrian safety (J. P. Das et al., 2024; Gopal et al., 2024). In another case, a research team in the United States developed thermoelectric roads capable of harvesting heat energy from asphalt surfaces exposed to sunlight. These roads utilize high-performance thermoelectric materials embedded within the pavement layers to convert solar heat into electrical power, supplying energy to road sensors, dynamic traffic signs, and weather-monitoring stations. This innovation demonstrates the potential of waste heat recovery in transportation infrastructure, contributing to smart city initiatives and carbon footprint reduction (Al-Atroush et al., 2023; Gopal et al., 2024).

As demonstrated, each class of advanced material from high-performance composites to smart geopolymers and nature-inspired innovations brings unique capabilities to infrastructure engineering. The next section evaluates their environmental impacts and long-term viability through a sustainability and lifecycle lens.

SUSTAINABILITY AND LIFECYCLE ANALYSIS OF INNOVATIVE MATERIALS

Figure 6. Future perspectives in advanced materials: innovation, sustainability, and multifunctionality

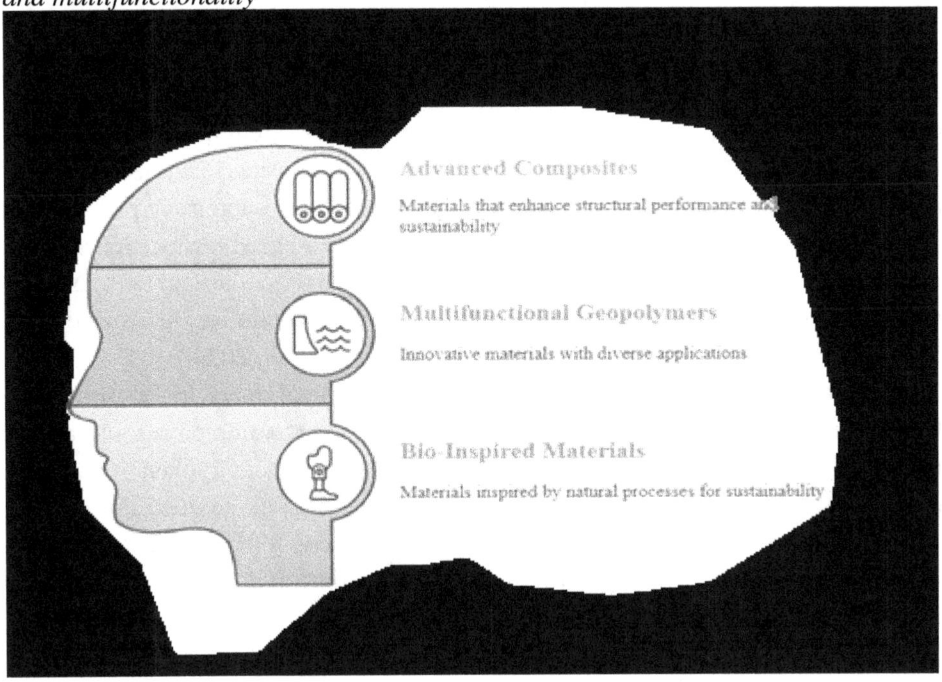

Sustainability and Lifecycle Analysis of Advanced Composites in Construction

Comparing the Carbon Footprint of Traditional Materials vs. Advanced Composites Traditional construction materials like Portland cement, steel, and conventional concrete have a high carbon footprint due to energy-intensive manufacturing processes. Cement production alone contributes nearly 8% of global CO_2 emissions, while steel production requires significant fossil fuel consumption. In contrast, advanced composites such as graphene-enhanced concrete, geopolymers, and fibre-reinforced polymers (FRPs) offer a more sustainable alternative by reducing material usage, increasing durability, and lowering emissions. Graphene-reinforced concrete, for instance, achieves higher strength with less material, reducing the carbon footprint associated with cement. Similarly, geopolymers, which are derived from industrial byproducts like fly ash and slag, emit up to 80% less CO_2 compared to traditional cement, making them a key component in sustainable infrastructure (Akhtar et al., 2022; Dighade et al., 2024).

Recycling and Waste Management Strategies for Non-Biodegradable Materials. Advanced composites like Kevlar, carbon fibre, and glass fibre pose recycling challenges due to their non-biodegradable nature and complex chemical composition. Unlike metals, which can be easily melted and reused, these composites require specialized thermal, mechanical, or chemical recycling processes to recover useful fibres. Mechanical recycling, involving grinding and repurposing fibres into secondary applications, is commonly used but reduces the strength of the material. Pyrolysis and solvolysis methods, which break down composites into reusable fibres and resins, offer more effective solutions but require energy-intensive processing. To improve sustainability, researchers are exploring bio-based resins, closed-loop recycling systems, and upcycling strategies to repurpose Kevlar and carbon fibre waste into high-performance secondary materials for automotive, aerospace, and construction applications (Butenegro et al., 2021; Hao et al., 2024).

Long-Term Benefits of Using Graphene and Geopolymers in Eco-Friendly Construction The integration of graphene and geopolymers into construction materials provides significant long-term sustainability benefits. Graphene-enhanced composites improve structural integrity, reduce cracking, and extend the service life of buildings and infrastructure, minimizing the need for frequent repairs and material replacement. This leads to lower resource consumption, reduced demolition waste, and decreased lifecycle emissions. Additionally, graphene's self-sensing capabilities enable real-time structural health monitoring, optimizing maintenance and improving safety. Geopolymers, on the other hand, offer high resistance to fire, chemicals, and extreme weather, making them ideal for durable, low-maintenance infrastructure. Their ability to utilize industrial waste byproducts further promotes circular economy principles by diverting landfill waste into valuable construction materials. By incorporating these advanced materials, future buildings can achieve greater energy efficiency, lower embodied carbon, and enhanced resilience, driving the transition toward truly sustainable construction (Hernández-Sánchez et al., 2024; MAMULAT et al., 2024; Sridhar & Kumar, 2024).

Sustainability and Lifecycle Analysis of Multifunctional Geopolymers

Multifunctional geopolymers are emerging as a sustainable alternative to traditional cement-based materials, offering lower carbon emissions, enhanced durability, and self-healing capabilities. Unlike Portland cement, which requires high-temperature processing and contributes significantly to CO_2 emissions, geopolymers utilize industrial byproducts such as fly ash, slag, and metakaolin, reducing waste and promoting a circular economy. The carbon footprint of geopolymers is up to 80% lower than that of conventional cement, making them a key material in eco-friendly infrastructure (Akhtar et al., 2022; Mushtaq et al., 2025).

One of the most promising features of multifunctional geopolymers is their self-healing ability, which extends the lifespan of structures by autonomously repairing cracks through microencapsulated healing agents or bacterial mechanisms. This reduces the need for frequent maintenance and prevents early-stage structural degradation, lowering the environmental impact over a building's lifecycle. Additionally, icephobic geopolymers, designed to resist ice accumulation, minimize the use of chemical de-icing agents such as salts and glycols, which are known to cause infrastructure corrosion and environmental pollution. By preventing freeze-thaw damage, icephobic geopolymers improve road and bridge durability, ensuring long-term sustainability in cold-climate regions. Furthermore, geopolymers exhibit high resistance to fire, chemicals, and extreme weather conditions, reducing the need for material replacement and lowering waste generation in construction and demolition. Their high thermal insulation properties also contribute to energy efficiency in buildings, decreasing the reliance on heating and cooling systems. The reuse of industrial waste in geopolymer production aligns with global sustainability goals, helping to reduce landfill waste and decrease the depletion of natural resources. By integrating geopolymers into modern infrastructure, the construction industry can achieve significant carbon footprint reductions, improved material longevity, and enhanced environmental resilience, making them a cornerstone of next-generation sustainable architecture (Asim et al., 2024; Win et al., 2024; Xu et al., 2025).

Sustainability and Lifecycle Analysis of Bio-Inspired Materials

Bio-inspired materials are revolutionizing the construction industry by mimicking natural processes to enhance resilience, durability, and environmental sustainability. Unlike traditional materials, which often degrade under environmental stress, bio-inspired solutions utilize nature's efficient designs to create self-sustaining and eco-friendly infrastructure. By integrating mechanoluminescent, hydrophobic, and energy-autonomous systems, construction materials can reduce energy consumption, minimize maintenance, and improve longevity, leading to a lower overall carbon footprint (Ojha et al., 2025; Patil et al., 2022).

One of the most innovative bio-inspired materials is mechanoluminescent (ML) materials, which emit light when subjected to mechanical stress or deformation, much like deep-sea organisms that produce bioluminescence. These materials offer self-powered lighting solutions for roads, sidewalks, and bridges, reducing dependence on electric grids and fossil fuel-based energy sources. By harvesting mechanical energy from traffic movement, pedestrian activity, and environmental vibrations,

ML materials contribute to low-energy, sustainable urban infrastructure, enhancing safety while cutting down operational costs. Another nature-inspired advancement is hydrophobic materials, which mimic the lotus leaf effect to create self-cleaning and water-repellent surfaces. These materials prevent moisture infiltration, corrosion, and microbial growth, making them ideal for bridges, skyscrapers, and tunnels exposed to harsh environmental conditions. By reducing water damage and biofouling, hydrophobic coatings extend the lifespan of structures and decrease the need for chemical treatments and frequent maintenance, ultimately lowering the ecological impact of construction materials (Belloni et al., 2023; Guo et al., 2019).

Additionally, energy-autonomous systems inspired by natural energy conversion mechanisms, such as piezoelectric and thermoelectric materials, generate electricity from mechanical vibrations, temperature fluctuations, and environmental forces. This enables the creation of self-powered monitoring systems, lighting solutions, and smart infrastructure that can function independently of traditional energy sources. Roads and bridges embedded with energy-harvesting materials can power traffic signals, structural health sensors, and wireless communication networks, promoting low-carbon, energy-efficient cities. By leveraging the principles of biomimicry, bio-inspired materials offer long-term environmental benefits by reducing energy dependence, minimizing material degradation, and supporting self-sustaining infrastructure. These advancements align with global sustainability efforts, ensuring that the built environment works in harmony with nature rather than against it. As research continues to expand in biomimetic construction, the future of sustainable architecture and resilient urban development will be increasingly shaped by nature's intelligent designs (Y. Bai et al., 2018; Liang et al., 2025).

COST AND PERFORMANCE ANALYSIS

Lifecycle Cost Comparison Between Traditional and Advanced Materials.

Figure 7. Traditional vs. advanced materials: Balancing cost and performance

The integration of advanced materials in civil infrastructure necessitates a comprehensive lifecycle cost analysis (LCCA) to compare their economic viability against traditional materials. While conventional construction materials such as reinforced concrete, structural steel, and asphalt have been extensively used due to their availability and established performance records, next-generation materials—including carbon fibre composites, multifunctional geopolymers, bio-inspired coatings, and mechanoluminescent surfaces—offer superior durability, adaptability, and sustainability. However, these advanced materials often involve higher initial costs due to raw material synthesis, specialized manufacturing techniques, and limited large-scale production. The long-term benefits, including reduced maintenance, extended service life, and enhanced resilience to environmental stressors, must be considered to assess their overall cost-effectiveness (Huber et al., 2024; Mohamed et al., 2024).

Lifecycle cost analysis (LCCA) incorporates initial capital expenditures (CAPEX), operational expenditures (OPEX), maintenance costs, and end-of-life recycling or disposal costs. For instance, carbon fibre-reinforced composites (CFRPs) in bridge construction have higher upfront costs than traditional steel reinforcements but significantly reduce corrosion-related maintenance expenses and structural retrofitting needs. Similarly, geopolymer concrete, which utilizes industrial byproducts like fly ash and ground granulated blast furnace slag (GGBS), reduces carbon emissions and requires less energy-intensive processing than Portland cement. Although material procurement costs for geopolymers may be slightly higher, the enhanced resistance to chemical attack, sulfate exposure, and freeze-thaw cycles leads to lower repair and rehabilitation costs over the asset's lifespan (Moharekpour et al., 2024; Vogiantzi & Tserpes, 2025).

Performance Benchmarks in Strength, Durability, and Adaptability

Strength and Load-Bearing Capacity: Advanced materials often outperform traditional counterparts in terms of compressive strength, tensile strength, and impact resistance. For example, high-performance fiber-reinforced polymer (FRP) composites exhibit three to five times higher strength-to-weight ratios compared to steel, making them ideal for lightweight yet high-strength structural components in bridges, tunnels, and high-rise buildings. Similarly, self-healing geopolymers incorporate microencapsulated healing agents or bacterial spores that autonomously repair cracks upon exposure to moisture, extending the structural integrity of concrete infrastructure (Simões, 2024).

Durability and Environmental Resistance: Infrastructure materials are exposed to mechanical loading, chemical attack, thermal fluctuations, and moisture infiltration, necessitating high durability. Advanced coatings such as superhydrophobic nanomaterials prevent water absorption, mitigating issues like concrete spalling, steel corrosion, and freeze-thaw damage. Mechanoluminescent (ML) materials embedded in roads and bridges provide real-time stress monitoring, allowing for early detection of structural fatigue and reducing catastrophic failures. Furthermore, icephobic coatings derived from bio-inspired surface engineering prevent ice accumulation on highways and bridges, reducing weather-related hazards and maintenance costs (Dezdarani & Mohammadi, n.d.; Lee et al., 2025).

Adaptability and Smart Functionality: Unlike conventional static materials, advanced materials exhibit dynamic adaptability to environmental stimuli. Shape-memory alloys (SMAs) and self-sensing concrete can detect and respond to mechanical stress, temperature variations, and external forces, improving seismic resistance and enabling real-time condition monitoring. Additionally, energy-autonomous materials, such as piezoelectric pavements and triboelectric nanogenerators (TENGs) in roads, generate self-sustaining power for lighting, traffic monitoring, and IoT-based infrastructure systems (Feli et al., 2024; Kretzer, 2016).

MASS PRODUCTION CHALLENGES IN ADVANCED INFRASTRUCTURE MATERIALS

Figure 8. Challenges in mass production: Scalability & supply chain optimization

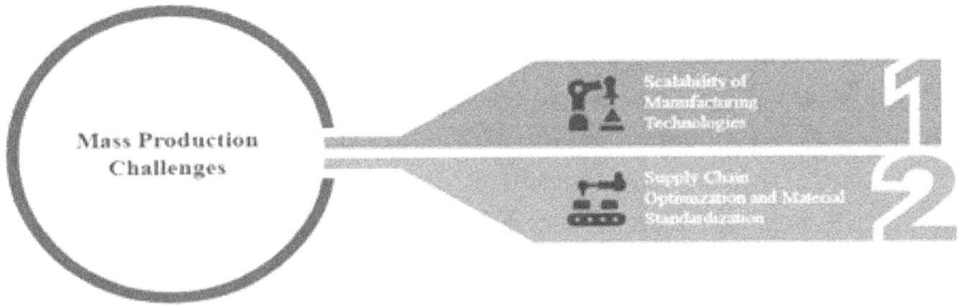

The large-scale implementation of next-generation materials in civil infrastructure is hindered by several mass production challenges, including scalability limitations in manufacturing, supply chain bottlenecks, and the lack of standardized material specifications. While advanced composites, multifunctional geopolymers, and bio-inspired materials offer superior performance, transitioning from laboratory-scale production to full-scale industrial deployment requires overcoming significant engineering and economic barriers. The high costs of raw material synthesis, processing complexities, and limited industrial infrastructure further complicate mass production efforts. Addressing these challenges is crucial for enabling the widespread adoption of these materials in bridges, tunnels, highways, and other critical infrastructure projects (Murcia & Taha, 2023; Stier et al., 2024).

Scalability of Manufacturing Technologies

One of the primary obstacles to mass production is the scalability of advanced material manufacturing techniques. Traditional infrastructure materials such as concrete, steel, and asphalt benefit from mature production pipelines and established supply networks, whereas emerging materials rely on high-precision synthesis and processing techniques that are not yet optimized for high-volume output (Jung et al., 2023; Kim et al., 2024).

For instance, the fabrication of high-strength carbon fiber composites involves polymer precursor processing, controlled pyrolysis, and surface treatments, all of which require specialized equipment and energy-intensive procedures. Geopolymer concrete production, while promising as a low-carbon alternative, is constrained by raw material variability, alkali activator sourcing, and curing conditions that differ from conventional Portland cement, making large-scale deployment difficult. Similarly, bio-inspired coatings, such as superhydrophobic or icephobic treatments, require nanoscale surface engineering techniques that are challenging to reproduce consistently in mass production settings (Althoey et al., 2023; Groetsch et al., 2023).

To overcome these issues, automated and high-throughput manufacturing technologies—such as robotic-assisted fabrication, additive manufacturing (3D printing), and scalable chemical vapor deposition (CVD) techniques—are being explored. For example, 3D-printed geopolymer components enable rapid, cost-effective prefabrication of structural elements, reducing material waste and construction time. Additionally, modular production units for mechanoluminescent materials and energy-autonomous composites can facilitate distributed manufacturing, lowering transportation costs and reducing reliance on centralized production hubs (Aljasmi et al., 2024; Campilho & Silva, 2023).

Supply Chain Optimization and Material Standardization

A critical factor in mass-producing advanced infrastructure materials is the need for robust supply chain integration and standardization of material properties. Traditional construction materials have well-defined specifications (e.g., ASTM, ISO, and EN standards) that ensure consistency and interoperability. In contrast, next-generation materials—such as high-performance fiber-reinforced polymers, self-healing geopolymers, and bio-inspired coatings—lack universally accepted testing methods, performance metrics, and durability assessment protocols. This absence of regulatory frameworks and standardized benchmarks slows down their acceptance and integration into large-scale infrastructure projects (Awaad et al., 2024; De Jong et al., 2021).

Additionally, raw material procurement and supply chain logistics pose significant challenges. The reliance on specialty chemicals, rare earth elements, and industrial byproducts for geopolymers, nanocomposites, and functional coatings creates supply chain vulnerabilities, particularly when sourcing materials from geopolitically sensitive regions. For example, carbon fiber production depends on precursor materials such as polyacrylonitrile (PAN), which have limited global suppliers, increasing cost fluctuations and material shortages. Fly ash and slag, key ingredients in geopolymer concrete, are byproducts of coal and steel industries, and their availability is inconsistent due to shifting energy policies and industrial decarbonization efforts (Cooper, 2024; Narkhede et al., 2024).

To enhance supply chain resilience, localized production of key materials, the use of alternative raw material sources, and blockchain-based material tracking systems can be implemented. For example, synthetic fly ash production through industrial carbon capture or bio-based precursors for carbon fiber synthesis can reduce dependency on traditional supply chains. Furthermore, advanced recycling and circular economy approaches, such as reclaiming carbon fiber from decommissioned aerospace components or reprocessing geopolymers from construction demolition waste, can improve material availability and sustainability (Gupta et al., 2023; Ogunsuji et al., 2024).

EMERGING TECHNOLOGIES AND FUTURE INNOVATIONS IN INFRASTRUCTURE

The rapid advancement of 4D printing, nanotechnology, smart materials, artificial intelligence (AI), and digital twin technology is transforming the construction industry. These innovations are enabling self-healing materials, adaptive structures, real-time monitoring, and data-driven optimization, significantly enhancing infrastructure durability, efficiency, and sustainability (Omrany et al., 2023; Serugga, 2025).

Figure 9. Technologies driving advanced infrastructure

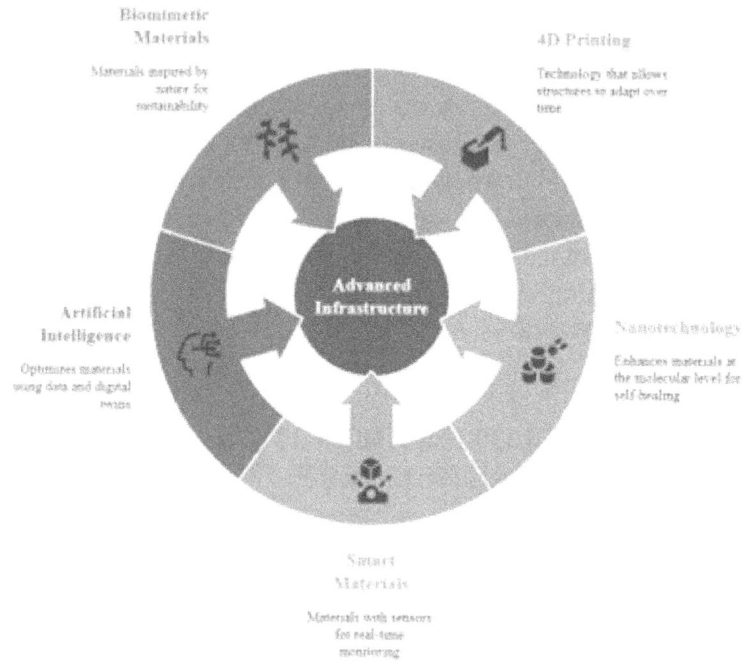

4D Printing and Nanotechnology: Self-Healing and Adaptive Structures

4D printing, an evolution of 3D printing, incorporates smart materials that can change shape, repair themselves, or respond to environmental stimuli over time. These materials, often composed of hydrogels, shape-memory alloys, or programmable polymers, allow infrastructure elements to adjust to temperature fluctuations, humidity levels, or mechanical stress, enhancing their resilience and lifespan. For example, 4D-printed concrete could autonomously expand or contract in response to thermal changes, reducing cracking and structural failure in extreme climates (Kannappan & Nachimuthu, 2024; Rana et al., 2024).

Additionally, nanotechnology is revolutionizing construction materials by improving their mechanical strength, durability, and self-repairing capabilities. Nano-silica, carbon nanotubes (CNTs), and graphene-based nanomaterials enhance the compressive strength, water resistance, and crack-sealing properties of concrete and coatings. Nanoparticles can also be embedded in surfaces to create self-cleaning facades, reducing the need for maintenance in urban infrastructure. Moreover, self-healing nanomaterials, such as microcapsule-based concrete, can autonomously repair cracks, increasing longevity and reducing repair costs (El-hadad et al., 2024; Ranjan et al., 2024).

Smart Materials with Embedded Sensors for Real-Time Structural Health Monitoring

Modern infrastructure is increasingly incorporating smart materials with embedded sensors to enable real-time structural health monitoring (SHM). These materials contain piezoelectric, optical fiber, or graphene-based sensors that continuously track stress, strain, vibration, temperature, and crack propagation. Such self-sensing capabilities improve early fault detection, allowing proactive maintenance and reducing catastrophic failures (Abdelgawad & Yelamarthi, 2017; Najem et al., 2025).

For instance, bridges and high-rise buildings embedded with smart concrete can detect early-stage cracks and automatically signal maintenance teams, preventing costly and dangerous structural failures. Similarly, roads and railways equipped with intelligent asphalt can monitor traffic loads, surface deterioration, and environmental changes, optimizing maintenance schedules and extending infrastructure lifespan. These technologies contribute to safety, efficiency, and cost reduction, making them essential for the next generation of resilient and self-sustaining cities (Civera et al., 2024; Shaik & Sneha, 2025).

The Role of Artificial Intelligence and Digital Twin Technology in Material Optimization

Artificial intelligence (AI) and digital twin technology are reshaping material performance and construction management through data-driven decision-making and predictive modeling. AI-driven algorithms analyze vast amounts of sensor data from smart materials to predict structural weaknesses, optimize material composition, and enhance durability. By using machine learning models, AI can simulate various environmental conditions and suggest optimal materials for extreme climates, improving construction efficiency and sustainability (Bianchi et al., 2025; Shaik & Sneha, 2025).

Digital twins, virtual replicas of physical infrastructure, use real-time data, IoT sensors, and AI analytics to monitor the performance of bridges, tunnels, and skyscrapers. These AI-powered simulations enable engineers to test and refine material properties before real-world application, reducing material waste and improving sustainability. Additionally, digital twin systems can forecast material degradation, allowing for proactive repairs rather than reactive maintenance, ultimately increasing the lifespan of infrastructure (Omrany et al., 2023; Z. Sun et al., 2024).

By integrating 4D printing, nanotechnology, smart materials, AI, and digital twins, the future of construction will be highly adaptive, self-repairing, and data-driven. These innovations will pave the way for long-lasting, sustainable, and intelligent infrastructure, ensuring that cities can withstand environmental challenges, reduce resource consumption, and enhance operational efficiency (Omrany et al., 2023; Z. Sun et al., 2024).

Biomimetic and Bioengineered Materials in Sustainable Infrastructure

Nature has long served as a blueprint for developing high-performance materials, inspiring innovations in construction that enhance durability, resilience, and sustainability. Biomimetic materials mimic biological structures and functions to improve the longevity and efficiency of infrastructure. For example, the self-cleaning properties of the lotus leaf have led to the development of hydrophobic coatings that protect skyscrapers, bridges, and tunnels from water damage and corrosion. Similarly, bio-inspired lightweight yet strong structures, such as honeycomb patterns found in beehives, are used in advanced composites to optimize strength-to-weight ratios in building materials (V. Singh et al., n.d.; Wagermaier et al., 2025).

One of the most Innovative applications of biomimicry in construction is the integration of biological self-repair mechanisms into materials. Inspired by the healing process of bones and skin, self-healing concrete and geopolymers incorporate microcapsules filled with healing agents or bacteria that activate when cracks form. These materials autonomously repair structural damage, significantly extending the lifespan of roads, bridges, and buildings while reducing maintenance costs and material waste. This approach enhances resilience in infrastructure subjected to heavy loads, seismic activity, or harsh environmental conditions. The future of architecture and urban planning may include living materials that adapt and evolve to changing environments. Researchers are developing bioengineered materials embedded with microorganisms that can absorb pollutants, generate energy, or even grow and self-regenerate over time. For instance, bio-receptive concrete can support moss or algae growth, contributing to urban cooling and improved air quality. Additionally, structural components made from bio-fabricated materials, such as mycelium-based composites, offer biodegradable and carbon-negative alternatives to conventional construction materials. These advancements pave the way for a new era of intelligent, sustainable, and resilient infrastructure that aligns with ecological principles and enhances the built environment (Elgendy et al., 2024; Guimarães et al., 2024; Li et al., 2024).

APPLICATIONS OF ADVANCED MATERIALS IN INFRASTRUCTURE AND ENGINEERING

Figure 10. Innovative materials for enhanced infrastructure resilience

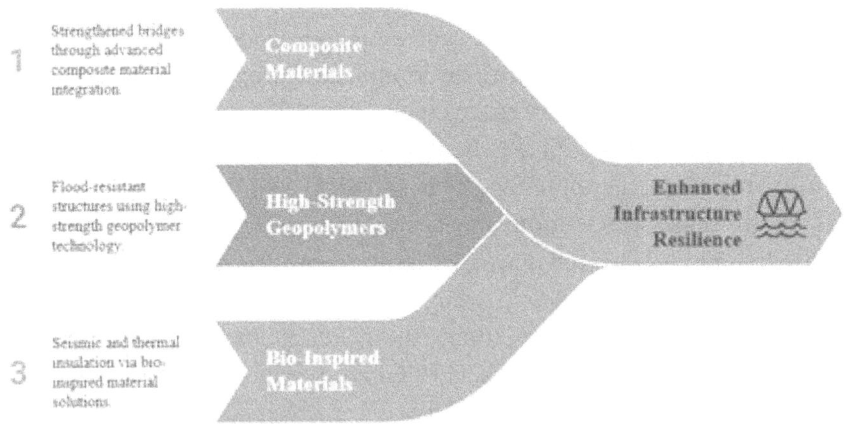

The integration of composite materials, high-strength geopolymers, and bio-inspired materials in modern infrastructure plays a crucial role in enhancing durability, resilience, and sustainability. As extreme weather events, seismic activities, and structural aging continue to challenge traditional materials, the adoption of innovative, high-performance materials has become essential for infrastructure development. These materials are engineered to mitigate environmental degradation, withstand dynamic loads, and optimize energy efficiency while ensuring cost-effective maintenance and long-term serviceability. Few of the case studies to showcase, how these materials are being used or piloted in diverse economic and climatic contexts such as in Lagos (Nigeria), rice husk ash-based geopolymer bricks have been successfully deployed for low-cost, flood-resistant housing. These bricks offered 30% higher water resistance and reduced energy costs for cooling, Kevlar-reinforced concrete barriers were implemented in high-security zones of New Delhi, reducing blast damage during a controlled test by over 60%, Hydrophobic coatings on bridges in Pará state, where rainfall exceeds 3,000 mm annually, led to a 40% drop in structural degradation over three years (Khatri et al., 2024; Nasiri, n.d.).

Integration of Composite Materials in Bridge Retrofitting

Bridges are subjected to continuous loading, environmental degradation, and aging-related deterioration, necessitating cost-effective and durable retrofitting solutions. The use of fiber-reinforced polymer (FRP) composites in bridge rehabilitation has emerged as a superior alternative to conventional steel reinforcements due to their high strength-to-weight ratio, corrosion resistance, and ease of application. FRP materials, including carbon fiber-reinforced polymers (CFRP) and glass fiber-reinforced polymers (GFRP), significantly enhance bridge load-carrying capacity while reducing maintenance needs.FRP composites are applied in bridge retrofitting through external bonding, near-surface mounting, and wrapping techniques. CFRP sheets and laminates are commonly bonded to bridge girders and decks using epoxy adhesives, improving flexural and shear strength. Additionally, GFRP rebars are replacing traditional steel reinforcements in concrete bridges, eliminating corrosion-related deterioration. The Halls River Bridge Retrofit Project in Florida demonstrated the effectiveness of CFRP sheets in strengthening deteriorated concrete bridge decks. Field tests revealed a 30% increase in flexural strength and an extended service life of over 50 years. Additionally, in Canada, GFRP-reinforced bridge decks exhibited high resistance to freeze-thaw cycles and de-icing salt exposure, ensuring long-term performance in harsh climatic conditions (Karbhari, 2007; V. G. Nguyen, 2024; Prajwal et al., n.d.).

High-Strength Geopolymers for Flood-Resilient Infrastructure

Flood-prone regions require infrastructure materials that exhibit low permeability, high chemical resistance, and superior durability under prolonged water exposure. High-strength geopolymer concretes, synthesized from industrial byproducts such as fly ash, ground granulated blast furnace slag (GGBS), and rice husk ash, have emerged as sustainable alternatives to traditional cementitious materials. These geopolymers not only provide high mechanical strength but also superior water and sulfate resistance, making them ideal for coastal defenses, flood walls, and submerged infrastructure. Geopolymers are produced through alkali activation using sodium silicate and sodium hydroxide solutions, forming a three-dimensional aluminosilicate network that exhibits high durability and environmental resistance. The incorporation of nano-silica, carbon nanotubes, and polymeric fibers further enhances their tensile strength and impact resistance, making them suitable for flood-resilient construction. The Dutch Delta Works project incorporated fly ash-based geopolymer concrete in seawalls and flood barriers, achieving a compressive strength of over 80 MPa with a water absorption rate below 1%. The geopolymers demonstrated exceptional durability against chloride and sulfate attacks, significantly reducing maintenance and repair costs. Similarly, in Queensland, Australia, geopolymer-based bridge abutments and culverts exhibited high resistance to prolonged water immersion, ensuring long-term resilience in flood-prone zones (Karthik & Mohan, 2021; Magotra & Jee, 2024; Paruthi et al., 2024).

Bio-Inspired Materials for Seismic and Thermal Insulation Systems

Seismic-resistant and thermally insulated structures are critical in regions experiencing frequent earthquakes and extreme temperature variations. Bio-inspired materials, which mimic natural structural systems such as honeycomb patterns, nacre (mother-of-pearl), and bamboo fiber reinforcement, have been developed to enhance mechanical toughness, impact resistance, and thermal efficiency. These materials integrate lightweight yet high-strength microstructures that efficiently dissipate seismic energy and minimize thermal losses, improving building resilience and energy efficiency. Bio-inspired materials are designed using gradient nanostructures, phase-changing materials, and high-performance fiber composites. Seismic-resistant materials incorporate energy-dissipative biofibers, bio-mineralized coatings, and hierarchical lattice designs to withstand dynamic stresses during earthquakes. Meanwhile, thermally adaptive biomimetic aerogels and phase-change materials (PCMs) regulate internal building temperatures, reducing heating and cooling energy demands. In Japan, bamboo-inspired fiber-reinforced seismic dampers were integrated into high-rise buildings to dissipate earthquake-induced forces. These dampers, utilizing nanofiber-reinforced composites, reduced seismic vibrations by over 40%, significantly improving structural stability. Similarly, in Norway, bio-inspired silica aerogels were used in Arctic research stations, achieving thermal conductivity as low as 0.015 W/m·K, reducing energy consumption by 60% in sub-zero environments (S. Das et al., 2023; Sandak & Butina Ogorelec, 2023; Shahjalal et al., 2024).

SUMMARY OF DISCUSSION

Table 1. Innovative materials for next-generation defence applications and their implications for civil infrastructure

Section	Key Insights	Technical Details	Examples & Case Studies
Advanced Composites for Civil Infrastructure	High-strength carbon fiber, Kevlar, and graphene composites improve durability, corrosion resistance, and impact absorption.	CFRP (Carbon Fiber Reinforced Polymer) and GFRP (Glass Fiber Reinforced Polymer) used in bridge decks, seismic retrofitting, and structural reinforcement.	Halls River Bridge, Florida (CFRP retrofitting improved load-bearing capacity by 30%).
Multifunctional Geopolymers	Smart geopolymers with self-healing, icephobic, and fire-resistant properties enhance infrastructure resilience in extreme conditions.	Alkali-activated fly ash and GGBS with additives like nano-silica, polymeric fibers, and biomimetic coatings improve strength and sustainability.	Dutch Delta Works (high-strength flood-resilient geopolymer infrastructure).
Bio-Inspired Materials for Smart Infrastructure	Mechanoluminescent, hydrophobic, and energy-autonomous materials for self-monitoring, reduced maintenance, and improved safety.	Mechanoluminescent coatings for stress sensing, superhydrophobic surfaces for water resistance, piezoelectric & triboelectric materials for energy harvesting.	Japan's ML-embedded bridges detect stress in real-time; Dutch hydrophobic concrete reduces flood damage.
Cost and Performance Analysis	Lifecycle cost comparison of advanced materials versus traditional ones. Performance benchmarks in strength, adaptability, and durability.	High-performance materials reduce maintenance costs over time despite higher initial investment.	Kevlar-reinforced bridges have 3x longer lifespan than steel-reinforced ones.
Mass Production Challenges	Scalability, standardization, and supply chain bottlenecks limit widespread adoption.	Requires automation in composite fabrication, sustainable raw material sourcing, and cost-effective processing methods.	Graphene concrete scalability limited by high production costs; advancements in automated CFRP manufacturing improve feasibility.
Applications in Civil Infrastructure	Defence-grade materials adapted for bridge retrofitting, flood-resilient infrastructure, and seismic-resistant systems.	Hybrid material integration such as CFRP in seismic zones, geopolymers in flood-prone regions, and bio-inspired aerogels for insulation.	Norway: Bio-inspired aerogels reduce energy consumption in Arctic stations by 60%.

(Aravinthan & Manalo, 2012; Caruso et al., 2020; Chen et al., 2022; Huber et al., 2024; Malik & Amor, 2023; Quek & Rajeev, 2023; Yusuf et al., 2024)

CONCLUSION

The integration of next-generation defence materials into civil infrastructure represents a significant advancement in improving resilience, sustainability, and structural performance. Advanced composites, multifunctional geopolymers, and bio-inspired materials offer exceptional strength, durability, and adaptability, making them ideal for extreme environments. These materials enhance infrastructure longevity, reduce maintenance needs, and contribute to the sustainability goals of modern construction.

However, several challenges must be addressed for widespread adoption:

- High Initial Costs – Advanced materials often require substantial initial investment, though long-term savings in maintenance and durability justify the expense.
- Scalability and Mass Production – Efficient manufacturing processes and automation are essential to making these materials commercially viable.
- Supply Chain Optimization – Reliable sourcing and standardization of raw materials will play a crucial role in ensuring consistency and cost efficiency.
- Widespread adoption of advanced materials is hindered by the lack of standardized testing protocols and regulatory clarity. For instance, there are no unified ISO/ASTM standards for bio-inspired or nanocomposite materials. Governments can facilitate adoption through targeted R&D funding, demonstration projects, and public-private partnerships. India's "Smart Cities Mission" and Brazil's National Policy for Technological Innovation in Construction are steps in this direction. Additionally, cost incentives such as tax breaks or green credits for sustainable materials can accelerate deployment in low- to middle-income economies.

Despite these challenges, real-world applications demonstrate the viability of these materials. Composite retrofitting has extended the lifespan of aging bridges, geopolymer-based infrastructure has proven effective in flood-prone regions, and bio-inspired materials are revolutionizing seismic and thermal insulation strategies. Moving forward, collaboration between material scientists, engineers, and policymakers will be crucial in overcoming technical and economic barriers. By leveraging these innovations, the construction industry can develop smarter, safer, and more sustainable urban environments that are resilient to future challenges.

REFERENCES

Abdelgawad, A., & Yelamarthi, K. (2017). Internet of things (IoT) platform for structure health monitoring. *Wireless Communications and Mobile Computing, 2017*(1), 6560797. DOI: 10.1155/2017/6560797

Agarwal, N., Rangamani, A., Bhavsar, K., Virnodkar, S. S., Fernandes, A. A. A., Chadha, U., Srivastava, D., Patterson, A. E., & Rajasekharan, V. (2024). An overview of carbon-carbon composite materials and their applications. *Frontiers in Materials, 11*, 1374034. DOI: 10.3389/fmats.2024.1374034

Ajirotutu, R. O., Adeyemi, A. B., Ifechukwu, G.-O., Iwuanyanwu, O., Ohakawa, T. C., & Garba, B. M. P. (2024). Future cities and sustainable development: Integrating renewable energy, advanced materials, and civil engineering for urban resilience. *International Journal of Sustainable Urban Development.*

Akhtar, N., Ahmad, T., Husain, D., Majdi, A., Alam, M. T., Husain, N., & Wayal, A. K. S. (2022). Ecological footprint and economic assessment of conventional and geopolymer concrete for sustainable construction. *Journal of Cleaner Production, 380*, 134910. DOI: 10.1016/j.jclepro.2022.134910

Al-Atroush, M. E., Marouf, A., Aloufi, M., & Marouf, M. (2023). A novel application of the geothermal asphalt pavement: A feasible E-fuel source. *Transportation Engineering, 12*, 100183. DOI: 10.1016/j.treng.2023.100183

Aljasmi, N., Almaeeni, K., Alyammahi, M., Panda, A. S., Almheiri, A., Joy, A., Susantyoko, R. A., & Subramanian, V. (2024). Productivity Analysis of Automated 3D Printing System. *ASME International Mechanical Engineering Congress and Exposition, 88605*, V002T03A017. DOI: 10.1115/IMECE2024-145736

Althoey, F., Zaid, O., Alsulamy, S., Martínez-García, R., de Prado-Gil, J., & Arbili, M. M. (2023). Experimental study on the properties of ultra-high-strength geopolymer concrete with polypropylene fibers and nano-silica. *PLoS One, 18*(4), e0282435. DOI: 10.1371/journal.pone.0282435 PMID: 37079561

Aravinthan, T., & Manalo, A. (2012). Field applications and case studies of FRP in civil infrastructure: The Australian experience. *Proceedings of the 6th International Conference on Frp Composites in Civil Engineering (CICE 2012).*

Asim, N., Badiei, M., Samsudin, N. A., Mohammad, M., Razali, H., & Hui, D. (2024). Clean technology option development for smart and multifunctional construction materials: Sustainable geopolymer composites. *Journal of Building Engineering, 94*, 109932. DOI: 10.1016/j.jobe.2024.109932

Awaad, S., Mansour, D. M., Mahdi, I., & Abdelrasheed, I. (2024). Impact of material supply chain on the productivity optimization for the construction of roads projects. *Scientific Reports*, *14*(1), 3294. DOI: 10.1038/s41598-024-53660-6 PMID: 38332188

Bai, J. (2022). *Advanced fibre-reinforced polymer (FRP) composites for structural applications*. Woodhead Publishing.

Bai, Y., Jantunen, H., & Juuti, J. (2018). Hybrid, multi-source, and integrated energy harvesters. *Frontiers in Materials*, *5*, 65. DOI: 10.3389/fmats.2018.00065

Belloni, E., Celli, I., Nakamura, S., Pisello, A., Villacci, D., & Cotana, F. (2023). A POWER-EFFICIENT SMART LASER-PHOTOLUMINESCENT-LIGHT (LPL) WITH PV-SYSTEM INTEGRATION: EXPERIMENTAL ANALYSIS AND OPTIMIZATION FOR PEDESTRIAN ROADS. *Conference Proceeding by Commission Internationale de L'Eclairage CIE 2023*, 1686–1695. DOI: 10.25039/x50.2023.PO004

Bian, X., Yao, W., Liao, G., Liang, G., She, A., & Wei, Y. (2024). Hydrophobic cement-based materials with micro-nano hierarchical structures: Preparation, characterization, and stability assessment. *Frontiers in Materials*, *11*, 1512374. DOI: 10.3389/fmats.2024.1512374

Bianchi, G., Fanelli, C., Freddi, F., Giuliani, F., & La Placa, A. (2025). Systematic review railway infrastructure monitoring: From classic techniques to predictive maintenance. *Advances in Mechanical Engineering*, *17*(1), 16878132241285632. DOI: 10.1177/16878132241285631

BinSabt, M. H., Azeez, F. A., & Suleiman, N.BinSabt. (2023). Eco-friendly silane-based coating for mitigation of carbon steel corrosion in marine environments. *ACS Omega*, *8*(14), 12886–12898. DOI: 10.1021/acsomega.3c00013 PMID: 37065042

Bouhamed, A., Missaoui, S., Ben Ayed, A., Attaoui, A., Missaoui, D., Jeder, K., Guesmi, N., Njeh, A., Khemakhem, H., & Kanoun, O. (2024). A comprehensive review of strategies toward efficient flexible piezoelectric polymer composites based on BaTiO3 for next-generation energy harvesting. *Energies*, *17*(16), 4066. DOI: 10.3390/en17164066

Boylu, E. E., & Ekinci, S. (2024). Examination of Material Selection's Effect on Lighweightness on Structures. *Journal of Architectural Sciences and Applications*, *9*(2), 798–814. DOI: 10.30785/mbud.1475316

Butenegro, J. A., Bahrami, M., Abenojar, J., & Martínez, M. Á. (2021). Recent progress in carbon fiber reinforced polymers recycling: A review of recycling methods and reuse of carbon fibers. *Materials (Basel)*, *14*(21), 6401. DOI: 10.3390/ma14216401 PMID: 34771926

Campilho, R. D. S. G., & Silva, F. J. G. (2023). Industrial process improvement by automation and robotics. [). MDPI.]. *Machines (Basel), 11*(11), 1011. DOI: 10.3390/machines11111011

Caruso, M. C., Pascale, C., Camacho, E., Scalari, S., Animato, F., Alonso, M. C., Gimenez, M., & Ferrara, L. (2020). Life cycle assessment on the use of ultra high performance fibre reinforced concretes with enhanced durability for structures in extremely aggressive environments: case study analyses. *Concrete Durability and Service Life Planning: Proceedings of ConcreteLife'20 3*, 121–125.

Chen, E., Berrocal, C. G., Löfgren, I., & Lundgren, K. (2022). Comparison of the service life, life-cycle costs and assessment of hybrid and traditional reinforced concrete through a case study of bridge edge beams in Sweden. *Structure and Infrastructure Engineering*, *19*(1), 39–57. DOI: 10.1080/15732479.2021.1919720

Cheng, K., Guo, Z., Zhang, P., Feng, L., Zhou, Y., Li, L., Song, H., Wang, T., Zhao, Y., & Zhao, L. (2025). A Universal Strategy for Multicolor Mechanoluminescence via Radiative Energy Transfer Based on Ultraviolet Mechanoluminescent Material Ca9Al (PO4) 7: Ce3+. *Laser & Photonics Reviews*, *19*(7), 2401524. DOI: 10.1002/lpor.202401524

Civera, M., Naseem, A., & Chiaia, B. (2024). Recent advances in embedded technologies and self-sensing concrete for structural health monitoring. *Structural Concrete*, suco.202400714. DOI: 10.1002/suco.202400714

Cooper, M. (2024). *Navigating Complexity: Insights into Strategic Procurement Practices in Global Supply Chains.*

Das, J. P., Nardekar, S. S., Ravichandran, V., & Kim, S. (2024). From Friction to Function: A High-Voltage Sliding Triboelectric Nanogenerator for Highly Efficient Energy Autonomous IoTs and Self-Powered Actuation. *Small*, *20*(48), 2405792. DOI: 10.1002/smll.202405792 PMID: 39221685

Das, S., Pillai, A. U., Chemenghat, A., & Rahaman, M. M. (2023). Determining the role of microstructural topology on the mechanical performance of nacre-inspired composites using a phase-field model. *Materials Today. Communications*, *37*, 107453. DOI: 10.1016/j.mtcomm.2023.107453

De Belie, N., Gruyaert, E., Al-Tabbaa, A., Antonaci, P., Baera, C., Bajare, D., Darquennes, A., Davies, R., Ferrara, L., Jefferson, T., Litina, C., Miljevic, B., Otlewska, A., Ranogajec, J., Roig-Flores, M., Paine, K., Lukowski, P., Serna, P., Tulliani, J.-M., & Jonkers, H. M. (2018). A review of self-healing concrete for damage management of structures. *Advanced Materials Interfaces*, *5*(17), 1800074. DOI: 10.1002/admi.201800074

De Jong, A., Jansen, M., Van Dijk, J., & Meyer, J. (2021). Analysis of Innovative Practices in Advanced Materials and Structural Engineering. *Fusion of Multidisciplinary Research. International Journal (Toronto, Ont.)*, *2*(1), 178–188.

Dezdarani, V. H., & Mohammadi, M. H. P. (n.d.). *Tu Designing Smart Roads with Climate-Responsive Materials for Enhanced Durability and Maintenance.*

Dhakal, H. N., & Ismail, S. O. (2020). *Sustainable composites for lightweight applications.*

Dighade, R., Gomase, V., Peshattiwar, R., Selokar, A., Sangidwar, N., Peshattiwar, S., & Malve, S. (2024). Emission of carbon footprint from building construction materials: A review. *IOP Conference Series. Earth and Environmental Science*, *1409*(1), 12010. DOI: 10.1088/1755-1315/1409/1/012010

El-hadad, A. S., Al Hariri, M. O. R., Gamal, A., El-Gammal, M. A., Badawy, A. H., & El-Feky, M. S. (2024). Unlocking the potential of carbon nanotubes and nano clay in concrete: Assessing the bond strength and failure modes. *Engineering Research Journal (Shoubra)*, *53*(4), 89–97. DOI: 10.21608/erjsh.2024.279620.1287

Elgendy, I. M., Elkaliny, N. E., Saleh, H. M., Darwish, G. O., Almostafa, M. M., Metwally, K., Yahya, G., & Mahmoud, Y. A. G. (2024). Bacteria-powered self-healing concrete: Breakthroughs, challenges, and future prospects. *Journal of Industrial Microbiology & Biotechnology*, *52*, kuae051. DOI: 10.1093/jimb/kuae051 PMID: 39673695

Feli, S., Asgari Marnani, J., Ghasemi, A., & Monajemi Nejad, S. (2024). Seismic performance assessment of reinforced concrete coupled walls reinforced with different types of shape memory alloys. *Structural Design of Tall and Special Buildings*, *33*(18), e2185. DOI: 10.1002/tal.2185

GangaRao, H. (2017). Infrastructure applications of fiber-reinforced polymer composites. In *Applied Plastics Engineering Handbook* (pp. 675–695). Elsevier. DOI: 10.1016/B978-0-323-39040-8.00032-8

Gopal, M., Lurdhumary, J., Bathrinath, S., Priya, A. P., Sarojwal, A., & Boopathi, S. (2024). Energy Harvesting and Smart Highways for Sustainable Transportation Infrastructure: Revolutionizing Roads Using Nanotechnology. In *Principles and Applications in Speed Sensing and Energy Harvesting for Smart Roads* (pp. 136–165). IGI Global.

Groetsch, T., Maghe, M., Creighton, C., & Varley, R. J. (2023). Economic and environmental effects of precursor variation in a continuous carbon fibre manufacturing process. *Journal of Industrial and Engineering Chemistry*, *127*, 554–566. DOI: 10.1016/j.jiec.2023.07.041

Guimarães, J. S., Peixoto, M. S., & Sales, A. S. B. (2024). Biological concrete, a viable alternative in modern civil construction. *International Journal of Scientific Management and Tourism*, *10*(6), e1233–e1233. DOI: 10.55905/ijsmtv10n6-030

Guo, Q., Huang, B., Lu, C., Zhou, T., Su, G., Jia, L., & Zhang, X. (2019). A cephalopod-inspired mechanoluminescence material with skin-like self-healing and sensing properties. *Materials Horizons*, *6*(5), 996–1004. DOI: 10.1039/C8MH01624K

Gupta, S., Modgil, S., Choi, T.-M., Kumar, A., & Antony, J. (2023). Influences of artificial intelligence and blockchain technology on financial resilience of supply chains. *International Journal of Production Economics*, *261*, 108868. DOI: 10.1016/j.ijpe.2023.108868

Hao, C., Zhao, B., Shao, L., Cao, Y., Fei, M., Liu, W., Liu, T., Chang, Y.-C., Simmons, K. L., & Zhang, J. (2024). Mild chemical recycling of carbon fiber-reinforced epoxy composites in aqueous buffers and development of hydrothermally recyclable vitrimer composites from recyclates. *Resources, Conservation and Recycling*, *207*, 107668. DOI: 10.1016/j.resconrec.2024.107668

Hassan, H., Ismail, A., Ahmad, S., & Soon, C. F. (2017). Super-hydrophobic green corrosion inhibitor on carbon steel. *IOP Conference Series. Materials Science and Engineering*, *215*(1), 12023. DOI: 10.1088/1757-899X/215/1/012023

Hassan, H. Z., & Saeed, N. M. (2024). Advancements and applications of lightweight structures: A comprehensive review. *Discover Civil Engineering*, *1*(1), 47. DOI: 10.1007/s44290-024-00049-z

Hernández Rodríguez, G., Gastaldo, G., Stendardo, L., Rafik, Y., Pothin, J., Budinger, M., Antonini, C., Pommier-Budinger, V., & Coclite, A. M. (2024). Icephobic Gradient Polymer Coatings Coupled with Electromechanical De-icing Systems: A Promising Ice Repellent Hybrid System. *Advanced Engineering Materials*, •••, 2401532. DOI: 10.1002/adem.202401532

Hernández-Sánchez, D., Navarro-Rosales, M., Sala-Arceo, A., Miramontes-Ortega, A., & Priego-Mondragón, E. (2024). Construction Innovation-Graphene Oxide as the Differentiating Nanoadditive for Concrete and Coatings. *Materials Science Forum*, *1142*, 97–116. DOI: 10.4028/p-I9NF7J

Hossain, M. R., Sultana, R., Patwary, M. M., Khunga, N., Sharma, P., & Shaker, S. J. (2022). Self-healing concrete for sustainable buildings. A review. *Environmental Chemistry Letters*, *20*(2), 1–9. DOI: 10.1007/s10311-021-01375-9

Huang, Z., Zhan, Y., Li, W., Li, X., & Amirfazli, A. (2024). Superhydrophobic Coatings with Synergistically Enhanced Anti/Deicing Performance by Optically/Electrically Assisted Heating. *Advanced Engineering Materials*, *26*(24), 2401627. DOI: 10.1002/adem.202401627

Huber, T., Grasl, P., Kleiser, M., Kromoser, B., & Preinstorfer, P. (2024). Holistic life cycle cost analysis of road bridges with non-metallic reinforcement. *Developments in the Built Environment*, *20*, 100533. DOI: 10.1016/j.dibe.2024.100533

Jung, S., Kara, L. B., Nie, Z., Simpson, T. W., & Whitefoot, K. S. (2023). Is additive manufacturing an environmentally and economically preferred alternative for mass production? *Environmental Science & Technology*, *57*(16), 6373–6386. DOI: 10.1021/acs.est.2c04927 PMID: 37066969

Kannan, G., Sujatha, E. R., Almajed, A., & Moghal, A. A. B. (2025). Microbial-Derived Biopolymers: A Pathway to Sustainable Civil Engineering. *Polymers*, *17*(2), 172. DOI: 10.3390/polym17020172 PMID: 39861245

Kannappan, R. P., & Nachimuthu, M. (2024). *4D Printing-3D Printing of Smart Materials*.

Karbhari, V. M. (2007). Introduction: The use of composites in civil structural applications. In *Durability of composites for civil structural applications* (pp. 1–10). Elsevier.

Karthik, S., & Mohan, K. S. R. (2021). A taguchi approach for optimizing design mixture of geopolymer concrete incorporating fly ash, ground granulated blast furnace slag and silica fume. *Crystals*, *11*(11), 1279. DOI: 10.3390/cryst11111279

Khatri, W. A., Al Mehthel, M., Salazar, O., Baig, M., & Al Wohaibi, S. (2024). Use of Polymeric Materials in Construction to Improve Durability & Sustainability. *World Journal of Engineering and Technology*, *13*(1), 12–38. DOI: 10.4236/wjet.2025.131002

Kim, J., Kwon, Y., & Kang, N. (2024). Deep Generative Design for Mass Production. *ArXiv Preprint ArXiv:2403.12098*.

Kretzer, M. (2016). *Information materials: Smart materials for adaptive architecture*. Springer.

Lakshmi, P. V., Sumalatha, A., Rani, K. S., Kalyan, K., Sekhar, O. C., & Aasritha, K. S. (2024). Thermoelectric Energy Harvesting From Non-Biodegradable Dry Waste. *2024 5th International Conference on Recent Trends in Computer Science and Technology (ICRTCST)*, 472–475.

Latha, K. A., Navya, A., Prakash Raju, R., Poorna Chandar, V., & Bharath Raj, K. (2024). Multipurpose Smart Bridge Health Monitoring System Using Ardunio Based Sensors. *Applied Mechanics and Materials*, *924*, 91–100. DOI: 10.4028/p-H2r4sq

Lee, T., Kim, D., Cho, S., & Kim, M. O. (2025). Advancements in Surface Coatings and Inspection Technologies for Extending the Service Life of Concrete Structures in Marine Environments: A Critical Review. *Buildings (Basel, Switzerland)*, *15*(3), 304. DOI: 10.3390/buildings15030304

Li, X., Wang, Y., & Zhang, X. (2024). Strong, healable materials with bio-like ordered architectures and versatile functionality. *SusMat*, *4*(6), e248. DOI: 10.1002/sus2.248

Liang, L., Sheng, P., Yao, G., Huang, Z., Lin, Y., & Jiang, B. (2025). Integration of Flexible Thermoelectric Energy Harvesting System for Self-Powered Sensor Applications. *ACS Applied Materials & Interfaces*, *17*(2), 3656–3664. DOI: 10.1021/acsami.4c20424 PMID: 39757409

Liu, X. (2024). Graphene: in-depth research, applications, and challenges in advanced technology. *International Conference on Electrical Engineering and Intelligent Control (EEIC 2024)*, 2024, 580–583.

Magotra, S., & Jee, A. A. (2024). A Review on durability and microstructure of Fly-Ash based geopolymer concrete (FA-GPC). *Materials Today: Proceedings*. Advance online publication. DOI: 10.1016/j.matpr.2024.06.011

Malik, M., & Amor, B. (2023). Optimizing Bridge Rehabilitation: A Life Cycle Assessment and Cost Analysis of Conventional and UHPC Overlays. *Interdisciplinary Symposium on Smart & Sustainable Infrastructures*, 971–984.

MAMULAT. (2024). *S., BYCHKOV, A., & KHAYDAROV, B.* Geopolymer Composite Materials for Low-Carbon Hydraulic Engineering and Infrastructure Construction in the Arctic Regions.

Mohamed, H. M., Ibrahim, B., & ElZefzafy, H. (2024). Sustainable Innovative New Advanced Materials for Underground Infrastructure Projects. *2024 International Conference on Decision Aid Sciences and Applications (DASA)*, 1–4. DOI: 10.1109/DASA63652.2024.10836421

Moharekpour, M., Shokri, M., Wellerdick, E., Traverso, M., Oeser, M., & Liu, P. (2024). An evaluation of the economic viability and accessibility of CRCP and JPCP: A comparative analysis. *Sustainability (Basel), 16*(3), 1108. DOI: 10.3390/su16031108

Murcia, D. H., & Taha, M. R. (2023). Emerging Materials and Technologies for Next-Generation Sustainable and Resilient Polymer Concrete. *International Congress on Polymers in Concrete*, 47–58.

Mushtaq, A., Ali, S., Chaudhry, A. H., Sial, N., Aslam, M., & Batool, H. (2025). Geopolymers as Supplementary Cementitious Materials to Reduce Carbon Dioxide Emissions. *Nature Environment & Pollution Technology, 24*.

Najem, O., Benbrahim, M., Kabbaj, M. N., & Boumhidi, J. (2025). Synergy between Embedded and Non-Embedded Sensors in IoT Networks for Structural Health Monitoring of Concrete Structures: A Literature Review. *E3S Web of Conferences, 601*, 28.

Narkhede, G., Samuel, C., Mahajan, S., Verma, D., Sakhare, N., & Chaudhari, T. (2024). Beyond traditional supply chain management: Addressing sociopolitical challenges in increasingly turbulent global trade landscape. *Business Strategy & Development*, *7*(2), e397. DOI: 10.1002/bsd2.397

Nasiri, H. (n.d.). *Innovations in Construction: Exploring the Future with New Materials*.

Nguyen, K. T., Ahn, N., Le, T. A., & Lee, K. (2016). Theoretical and experimental study on mechanical properties and flexural strength of fly ash-geopolymer concrete. *Construction & Building Materials*, *106*, 65–77. DOI: 10.1016/j.conbuildmat.2015.12.033

Nguyen, V. G. (2024). Applying FRP Composites Materials for Repairing and Reinforcing Dien Bien Phu Bridge, Binh Thanh District, HCMC: Effective and Economic Solution. *International Conference Series on Geotechnics, Civil Engineering and Structures*, 141–148. DOI: 10.1007/978-981-97-1972-3_12

Nilimaa, J. (2023). Smart materials and technologies for sustainable concrete construction. *Developments in the Built Environment*, *15*, 100177. DOI: 10.1016/j.dibe.2023.100177

Nwokediegwu, Z. Q. S., Ilojianya, V. I., Ibekwe, K. I., Adefemi, A., Etukudoh, E. A., & Umoh, A. A. (2024). Advanced materials for sustainable construction: A review of innovations and environmental benefits. *Engineering Science & Technology Journal*, *5*(1), 201–218. DOI: 10.51594/estj.v5i1.744

Ogunsuji, Y. M., Amosu, O. R., Choubey, D., Abikoye, B. E., Kumar, P., & Umeorah, S. C. (2024). Sourcing renewable energy components: building resilient supply chains, reducing dependence on foreign suppliers, and enhancing energy security. *World Journal of Advanced Research and Reviews, 23*(2).

Ojha, A., Kumar, A., Das, D., & Bandyopadhyay, T. K. (2025). Bio-Concrete and Its Self-Healing Capacity to Next-generation Sustainable Architecture: A Comprehensive Review. *Exon*, *2*(1), 1–15. DOI: 10.69936/en01y0025

Omrany, H., Al-Obaidi, K. M., Husain, A., & Ghaffarianhoseini, A. (2023). Digital twins in the construction industry: A comprehensive review of current implementations, enabling technologies, and future directions. *Sustainability (Basel)*, *15*(14), 10908. DOI: 10.3390/su151410908

Paruthi, S., Rahman, I., Khan, A. H., Sharma, N., & Alyaseen, A. (2024). Strength, durability, and economic analysis of GGBS-based geopolymer concrete with silica fume under harsh conditions. *Scientific Reports*, *14*(1), 31572. DOI: 10.1038/s41598-024-77801-z PMID: 39738079

Patchen, A., Young, S., Goodbred, L., Puplampu, S., Chawla, V., & Penumadu, D. (2023). Lower carbon footprint concrete using recycled carbon fiber for targeted strength and insulation. *Materials (Basel)*, *16*(15), 5451. DOI: 10.3390/ma16155451 PMID: 37570158

Patil, A. Y., Hegde, C., Savanur, G., Kanakmood, S. M., Contractor, A. M., Shirashyad, V. B., Chivate, R. M., Kotturshettar, B. B., Mathad, S. N., Patil, M. B., Soudagar, M. E. M., & Fattah, I. M. R. (2022). Biomimicking nature-inspired design structures—An experimental and simulation approach using additive manufacturing. *Biomimetics*, *7*(4), 186. DOI: 10.3390/biomimetics7040186 PMID: 36412714

Pokorný, P., & Kouřil, M. (2024). Predicted corrosion performance of organofunctional silane coated steel reinforcement for concrete structures: An overview. *Buildings (Basel, Switzerland)*, *14*(6), 1756. DOI: 10.3390/buildings14061756

Prajwal, L., Balagi, G. M., & Kumar, C. (n.d.). Study of FRP on Preloaded Beam as A Method of Retrofitting-A Review. *International Research Journal on Advanced Engineering and Management (IRJAEM), 2*(10).

Quek, J., & Rajeev, R. (2023). Composite System for Structural Enhancement and Efficient Heritage Conservation: A Case Study. *The International Conference on Sustainable Civil Engineering and Architecture*, 419–428.

Qureshi, A. H., Ahmad, N., Rana, M. A. A., Manzoor, B., & Zayed, T. (2024). Construction Sector Transition towards Smart Applications of Graphene Oxide in Cement-Based Composites: A Scientometric Review and Bibliometric Analysis. *Buildings (Basel, Switzerland)*, *14*(10), 3042. DOI: 10.3390/buildings14103042

Rana, A., Rana, M., Rashmi, R., & Das, P. (2024). The Future of Smart Textiles: Combining Wearable Technology With 4D Printing. *Available atSSRN* 5088887.

Ranjan, M., Kumar, S., & Sinha, S. (2024). Nanosilica's Influence on Concrete Hydration, Microstructure, and Durability: A Review. *Journal of Applied Engineering Sciences*, *14*(2), 322–335. DOI: 10.2478/jaes-2024-0040

Rusu, I., Judele, L., & Lepadatu, D. (2024). POLYMERIC COATING FOR THE INSULATION, PROTECTION AND SURFACES FINISHING OF THE REINFORCED BUILDINGS CONCRETE EXPOSED TO CORROSIVE ENVIRONMENTS. *International Multidisciplinary Scientific GeoConference: SGEM*, *2024*(6.1), 279–285.

Sandak, A., & Butina Ogorelec, K. (2023). Bioinspired building materials—Lessons from nature. *Frontiers in Materials*, *10*, 1283163. DOI: 10.3389/fmats.2023.1283163

Sani, S. S. (2024). A single-inductor self-powered SECE interface circuit for dynamic load multi-PZTs energy harvesting. *IET Power Electronics*, *17*(15), 2613–2630. DOI: 10.1049/pel2.12815

Serugga, J. (2025). Digital Twins and AI Decision Models: Advancing Cost Modelling in Off-Site Construction. *Eng*, *6*(2), 22. DOI: 10.3390/eng6020022

Shahjalal, M., Yahia, A. K. M., Morshed, A. S. M., & Tanha, N. I. (2024). Earthquake-resistant building design: Innovations and challenges. *Global Mainstream Journal of Innovation. Engineering & Emerging Technology*, *3*(04), 101–119.

Shaik, M. A., & Sneha, P. (2025). *Revolutionizing Infrastructure Resilience: AI-Driven Predictive Maintenance and Structural Health Monitoring*.

Sharma, D., & Hiremath, S. S. (2022). Bio-inspired repeatable lattice structures for energy absorption: Experimental and finite element study. *Composite Structures*, *283*, 115102. DOI: 10.1016/j.compstruct.2021.115102

Shehata, N., Mohamed, O. A., Sayed, E. T., Abdelkareem, M. A., & Olabi, A. G. (2022). Geopolymer concrete as green building materials: Recent applications, sustainable development and circular economy potentials. *The Science of the Total Environment*, *836*, 155577. DOI: 10.1016/j.scitotenv.2022.155577 PMID: 35500705

Simões, S. (2024). High-performance advanced composites in multifunctional material design: State of the art, challenges, and future directions. *Materials (Basel), 17*(23), 5997. DOI: 10.3390/ma17235997 PMID: 39685433

Singh, T. J., & Samanta, S. (2015). Characterization of Kevlar fiber and its composites: A review. *Materials Today: Proceedings, 2*(4–5), 1381–1387. DOI: 10.1016/j.matpr.2015.07.057

Singh, V., Indoria, S., Bala, M., Gardas, R. L., & Yang, Y.-L. (n.d.). Self-Healing and Self-Cleaning Nature-Inspired Surfaces. In *Materials from Natural Sources* (pp. 216–240). CRC Press.

Skountzos, E. N., Ravichandran, A., & Lawson, J. W. (2024). Interfacial Characteristics of Ice-Supporting Substrates via Molecular Dynamics Simulations. *Langmuir, 40*(51), 26864–26874. DOI: 10.1021/acs.langmuir.4c03106 PMID: 39651769

Sridhar, M., & Kumar, M. V. (2024). Structural and Sustainability Enhancement of Composite Sandwich Slab Panels Using Novel Fibre-Reinforced Geopolymer Concrete. *Journal of Composites Science, 8*(11), 479. DOI: 10.3390/jcs8110479

Stier, S. P., Kreisbeck, C., Ihssen, H., Popp, M. A., Hauch, J., Malek, K., Reynaud, M., Goumans, T. P. M., Carlsson, J., Todorov, I., Gold, L., Räder, A., Wenzel, W., Bandesha, S. T., Jacques, P., Garcia-Moreno, F., Arcelus, O., Friederich, P., Clark, S., & Kozdras, M. (2024). Materials acceleration platforms (MAPs): Accelerating materials research and development to meet urgent societal challenges. *Advanced Materials, 36*(45), 2407791. DOI: 10.1002/adma.202407791 PMID: 39239995

Sun, C.-Y., Liu, W., Shi, X., Rao, G.-H., & Zhao, J.-T. (2025). Stress Distributions and Luminescent Responses of Mechanoluminescent Cylinders with Various Sizes and Loading Paths. *Materials (Basel), 18*(2), 331. DOI: 10.3390/ma18020331 PMID: 39859802

Sun, Z., Jayasinghe, S., Sidiq, A., Shahrivar, F., Mahmoodian, M., & Setunge, S. (2024). Approach Towards the Development of Digital Twin for Structural Health Monitoring of Civil Infrastructure: A Comprehensive Review. *Sensors (Basel), 25*(1), 59. DOI: 10.3390/s25010059 PMID: 39796850

Uddin, N. (2013). *Developments in fiber-reinforced polymer (FRP) composites for civil engineering*. Elsevier. DOI: 10.1533/9780857098955

Ursache, ., Cerbu, C., & Hadăr, A. (2023). Characteristics of carbon and kevlar fibres, their composites and structural applications in civil engineering—A review. *Polymers, 16*(1), 127. DOI: 10.3390/polym16010127 PMID: 38201792

Vasconcelos, A. R. C., de Matos, R. A., Silveira, M. V., & Mesquita, E. (2024). Applications of Smart and Self-Sensing Materials for Structural Health Monitoring in Civil Engineering: A Systematic Review. *Buildings (Basel, Switzerland), 14*(8), 2345. DOI: 10.3390/buildings14082345

Vogiantzi, C., & Tserpes, K. (2025). A Comparative Environmental and Economic Analysis of Carbon Fiber-Reinforced Polymer Recycling Processes Using Life Cycle Assessment and Life Cycle Costing. *Journal of Composites Science*, *9*(1), 39. DOI: 10.3390/jcs9010039

Wagermaier, W., Razghandi, K., & Fratzl, P. (2025). A Bio-Inspired Perspective on Materials Sustainability. *Advanced Materials*, •••, 2413096. DOI: 10.1002/adma.202413096 PMID: 39757528

Wang, T., Zhang, P., Xiao, J., Guo, Z., Xie, X., Huang, J., Zheng, J., Xu, X., & Zhao, L. (2025). Trap Assisted Dynamic Mechanoluminescence Toward Self-Referencing and Visualized Strain Sensing. *Advancement of Science*, *12*(3), 2410673. DOI: 10.1002/advs.202410673 PMID: 39604817

Wang, X., Xu, X., Sun, T., & Yin, G. (2024). A Triboelectric Nanogenerator Utilizing a Crank-Rocker Mechanism Combined with a Spring Cantilever Structure for Efficient Energy Harvesting and Self-Powered Sensing Applications. *Electronics (2079-9292), 13*(24).

Wang, Z., Hou, Y., Yadav, A., Chen, T., Wu, Y., Yan, C., & Liu, M. (2025). Self-Healing Superhydrophobic Coatings with Multiphase Repellence Property. *ACS Applied Materials & Interfaces*, *17*(4), 7174–7189. DOI: 10.1021/acsami.4c19317 PMID: 39812392

Win, S. L. Y., Chiang, Y.-C., Huang, T.-L., & Lai, C.-M. (2024). Thermoelectric Generator Applications in Buildings: A Review. *Sustainability (Basel), 16*(17), 7585. DOI: 10.3390/su16177585

Xu, J., Zhou, Y., Cheng, F., & Liu, Z. (2025). Preparation of multi-layer compound microcapsules and their application in self-healing of concrete cracks. *Sheng Wu Gong Cheng Xue Bao=. Chinese Journal of Biotechnology*, *41*(1), 448–460. PMID: 39855707

Yan, C., He, X., Yu, B., Zhou, B., Fang, S., Xu, J., Liu, B., & Wang, Z. (2024). A Mechanoluminescence-Based Stress Sensing Hydrogel for Intelligent Artificial Ligament. *Advanced Functional Materials*, •••, 2420142.

Yusuf, M., Salami, O., & Amusat, W. O. (2024). *Advances in carbon-fiber reinforced polymers and composites for sustainable concrete structures*.

Zhang, A., Yang, H., Liu, C., Yang, J., Yao, Y., Zhang, W., Pan, R., Zhuo, Y., Ding, J., Hu, R., Xue, M., Chen, P., & Gong, Y. (2024). Icephobic durability of molecular brush-structured PDMS soft coatings. *ACS Applied Materials & Interfaces*, *16*(16), 21001–21012. DOI: 10.1021/acsami.3c18900 PMID: 38619108

Zhang, F., Yang, K., Liu, G., Chen, Y., Wang, M., Li, S., & Li, R. (2022). Recent advances on graphene: Synthesis, properties and applications. *Composites. Part A, Applied Science and Manufacturing*, *160*, 107051. DOI: 10.1016/j.compositesa.2022.107051

Zhao, D., Li, K., Li, Y., Chen, X., Fan, J., & Zhu, J. (2024). Self-healing behaviour of fly ash/slag-based engineered geopolymer composites under external alkaline environments. *Case Studies in Construction Materials*, *21*, e03361. DOI: 10.1016/j.cscm.2024.e03361

Zhao, Z., Qi, S., Suo, Z., Hu, T., Hu, J., Liu, T., & Gong, M. (2024). Development of a Superhydrophobic Protection Mechanism and Coating Materials for Cement Concrete Surfaces. *Materials (Basel)*, *17*(17), 4390. DOI: 10.3390/ma17174390 PMID: 39274780

Zhou, Y., Zhang, J.-H., Wang, F., Hua, J., Cheng, W., Shi, Y., & Pan, L. (2024). Recent Advances in Flexible Self-Powered Sensors in Piezoelectric, Triboelectric, and Pyroelectric Fields. *Nanoenergy Advances*, *4*(3), 235–257. DOI: 10.3390/nanoenergyadv4030015

Zhu, Y., Romain, C., & Williams, C. K. (2016). Sustainable polymers from renewable resources. *Nature*, *540*(7633), 354–362. DOI: 10.1038/nature21001 PMID: 27974763

Chapter 7
Innovative Strategies in Lightweight Materials for High-Performance Defence Applications

G. Boopathy
https://orcid.org/0000-0003-2515-7277
Vel Tech Rangarajan Dr. Sagunthala R&D Institute of Science and Technology, Chennai, India

V. Srinivasan
https://orcid.org/0009-0006-2988-6476
Vel Tech Rangarajan Dr. Sagunthala R&D Institute of Science and Technology, India

Balaji Ganesan
https://orcid.org/0000-0001-5799-9423
Hindustan Institute of Technology and Science, India

Mohit Hemanth Kumar
https://orcid.org/0000-0003-3618-2558
Alliance University, India

ABSTRACT

Defence needs lightweight composites because of their good strength to weight ratio, corrosion resistance, and thermal stability. Then this chapter studies the design and optimization of a number of matrix composites motivated by their mechanical and thermal performance in extreme environments. This highlights the importance of multiscale modelling, simulation and advanced manufacturing in improving com-

DOI: 10.4018/979-8-3373-0933-0.ch007

posite reliability. This is applied to UVAs, armoured vehicles and naval systems for enhanced mobility and efficiency. Although there have been some impediments to production, the use of AI, nanomaterials, and sustainability will lead to scalable, high-performance solutions. These materials are then substantiated empirically for such critical missions. The chapter coordinates the relationship between the research based in the academia and the requirements by the defence industry, with strategies of new generation of materials having global sustainability goals and the operational needs.

INTRODUCTION

Fundamentally, lightweight materials have completely changed the way the defense sector operates by allowing for the movement of high performance systems that are mobile, provide greater fuel efficiency and last longer. Materials are needed in contemporary defense applications to survive the extreme conditions while at the same time reducing the weight as a whole to improve payload capacity and maneuverability. These intolerable specifications have been fulfilled through the advancement of lightweight composites, which include sophisticated metal alloys, ceramics and polymer based composites. These materials can be employed to enable defense systems such as armored vehicles, aircraft, and protective equipment to perform exceptionally well while maintaining structural integrity and resilience in difficult operational conditions.

Materials with a good strength to weight ratio, thermal stability, and corrosion resistance should be chosen for their operational efficacy. High progress in computational model and experimental validation methodologies has greatly helped to refine material selection, and performance. Furthermore, cutting-edge manufacturing processes like hybrid composite structures and additive manufacturing improve usefulness and durability for military and aerospace applications. Critical materials in defense have come to be lightweight composites emerging for their use in the defense sector owing to the need for the augmentations of performance and efficiency which are a norm today in the modern military operations. These materials are beneficial to land, air, and naval defense systems (Adeleke et al., 2024) as they confer a substantial benefit of diminished weight, fuel efficiency and improved maneuverability. Importance of their ability to fulfill stringent criteria in terms of strength as well as durability while essentially reducing the overall weight of military apparatus is indeed stressed (Kaushik et al., 2022). The following sections analyze and optimize important innovative strategies to design, analyze and optimize these advanced materials in high performance defense systems.

FUNDAMENTALS OF LIGHTWEIGHT COMPOSITES

Significant progress in the materials science has taken place in the form of lightweight composites offering superior performance characteristics that are critical to the defense of today. On account of the characteristics as well as the advantages conferred by them over traditional materials, they continue to be the subject matter of continuous research and development in the discipline.

Definition and Types of Composites

Advanced materials called lightweight composites are created by combining two or more different materials to provide better qualities that the constituent parts alone cannot. Metal matrix composites (MMC), ceramic matrix composites (CMC), and polymer matrix composites (PMC) are the three main categories of lightweight composites. The classification of lightweight composites and their corresponding structures is shown in Figure 1.

Figure 1. (a) represents the classification of lightweight composites (b) represents the structure of MMC (c) represents the structure of CMC (d) represents the structure of PMC. (Image by Alovsat, (2024) from "Classification and properties of composites based on matrix material (https://wjarr.com/content/classification-and-properties-composites-based-matrix-material)" licensed under CC BY 4.0)

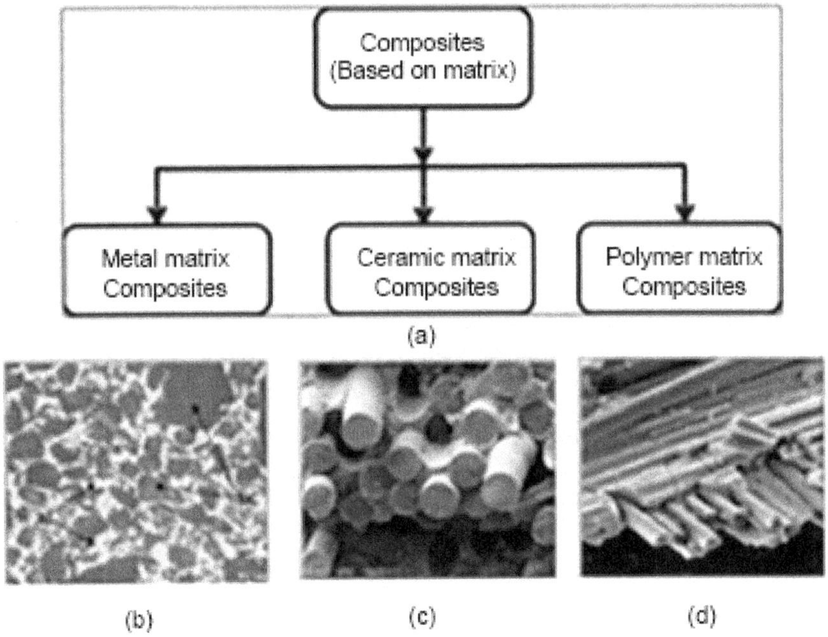

Metal Matrix Composites (MMC)

The materials that make up these systems are metallic such as aluminum or titanium with the addition of ceramic particles or fibers. This includes strength toughness thermal conductivity retention that exceeds conventional metals in the case of MMCs. They are especially handy in high temperature environments as such composites have the ability to sustain higher thermal stresses and reduced thermal expansion (Wang et al., 2023). MMC's wear resistance is greatly enhanced by the integration of ceramic reinforcements and they become suitable for the use in aerospace and automotive components where such a long life and low wear are required at very harsh operating conditions (Wang et al., 2024).

Ceramic Matrix Composites (CMC)

The fibers used in this category are carbon or silicon carbide (SiC), and they are incorporated into a ceramic matrix. High and thermal stability, low density and large resistance to wear and corrosion are often associated with CMCs. They are particularly appealing for high temperature applications like, aerospace engines and thermal protection systems (Madhankumar & Xavior, 2024). Show is that CMCs can be divided into oxide and non-oxide type and each group possesses specific benefits depending on what exactly is required in the application. As an example, CMCs based on non-oxide (e.g. SiC) show particularly draw back properties such that they are well fitted for crude environmental conditions thanks to their improved mechanical properties and thermal resistance (Karadimas & Salonitis, 2023).

Polymer Matrix Composites (PMC)

The composite is a polymer resin matrix with fibers such as glass, carbon or aramid. Significant properties possessed by PMCs are lightweight characteristic, high strength to weight ratios, and excellent corrosion resistance. Because of these properties these Structural Adhesive materials are especially suited for these industry applications where weight is of utmost importance while integrity is maintained. This adaptability of PMCs allows to exploit a wide spectrum of manufacturing methodologies such as lay-up, filament winding and resin transfer molding, used in different high performance sectors (Boopathy et al., 2022 ; Alovsat, 2024).

Material Properties

Spectrums of material characteristics are available to lightweight composites that are essential for their success in myriad different applications especially in Defense. The mechanical, thermal, and such important properties such as corrosion resistance and durability of these characteristics are grouped.

Mechanical Properties

However, for this type of composites used in structural components, the mechanical properties such as strength and toughness are extremely important. This is because strength refers to the amount of load on the material that may be resisted without failure unlike toughness referring to the amount of energy or amount of plastic deformation without failure. Carbon fiber reinforced composites have tensile strength of 0.5 – 3.0 GPa which significantly increases overall strength of the composite. Aramid fibers incorporation can further increase impact resistance and durability, thus render these composites suitable for high performance applications (Raja, Devarajan, 2024).

Research in contemporary research has shown that that the incorporation of nanomaterials into composites can provide a very marked improvement on mechanical performance. For example, it could increase tensile strength by up to 19 percent and modulus by 23 percent and the integration of nanomaterials into polymer matrices has shown to further increase tensile strength by up to 19 percent and modulus by 23 percent (Su, 2024). In addition, studied hybrid composites that assemble naturally occurring and synthetic fibers were reported for having a combined strength and durability that being suitable for various purposes (Bakar et al., 2024).

Thermal Properties

Properties related to the thermal behavior of composite materials, particularly in the high temperature environment, include those associated with the stability and conductivity. The thermal stability considers a material ability to preserve its natural properties when heated and the thermal conductivity is a material ability to spread the thermal energy. Identified that lightweight composites with, for example, glass fibers used as reinforcement, provide reduced thermal conductivity, therefore advantageous for insulation. (Khovalyg et al., 2023) In addition, the thermal stability for these composites can be improved by encapsulating particular fillers, as doing so drastically increases the thermal efficiency of the resulting material (Hossain et al., 2023).

Yet the fact that there exists certain lightweight composite materials that can sustain their structural integrity at elevated temperatures has suggested to empirical investigations that they are therefore applicable to achieve thermal resilience in the aerospace and automotive sectors where that is critical. Coincidentally, SiC/SiC composites are recognized for their excellent temperature endurance and are in use in aerospace parts because of their ability to deal with extreme thermal conditions (Li et al., 2024). Figure 2 shows the flowchart of procedures of production of lightweight aggregates.

Figure 2. Process flowchart for producing lightweight aggregates. (Image by Hao et al., (2022) from "Artificial Lightweight Aggregates Made from Pozzolanic Material: A Review on the Method, Physical and Mechanical Properties, Thermal and Microstructure (https://www.mdpi.com/1996-1944/15/11/3929)" licensed under CC BY 4.0)

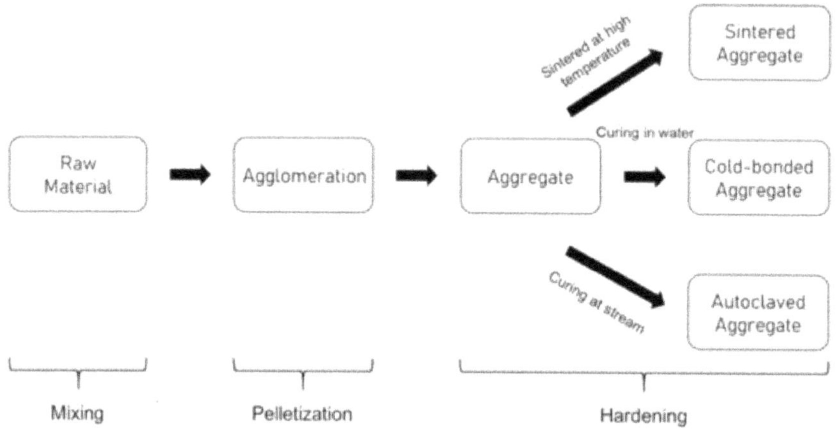

Other Critical Properties

Corrosion resistance and durability are in addition to mechanical and thermal attributes, which need to be both demonstrated and guaranteed in order for lightweight composites to all survive in harsh environments. Materials used in defense applications have special importance for corrosion resistance, as moisture and variety of chemicals are encountered which degrade them. The use of PMCs is often favored for their inherent corrosion resistance whereby their operations lifespan is greatly prolonged.

The second of these is durability, which constitutes a material's ability to resist wear, as well as pressure and damage, over time. Features have been demonstrated in lightweight composites retain mechanical properties under cyclic loading and environmental degradation thus making them suitable for extreme applications (Furtado et al., 2024). For instance, natural fiber reinforced composites exhibit good durability characteristics, and it is claimed to be suitable for sustainable applications (Mahesh et al, 2024).

Comparison With Traditional Materials

PMC, MMC and CMC are lightweight composites which feature several advantages over conventional materials (i.e. metals and alloys). We bring mechanical properties, thermal performance, corrosion resistance, and durability as a whole into the picture. The flowchart of their performance is shown in figure 3.

Figure 3. Flowchart on performance of composites

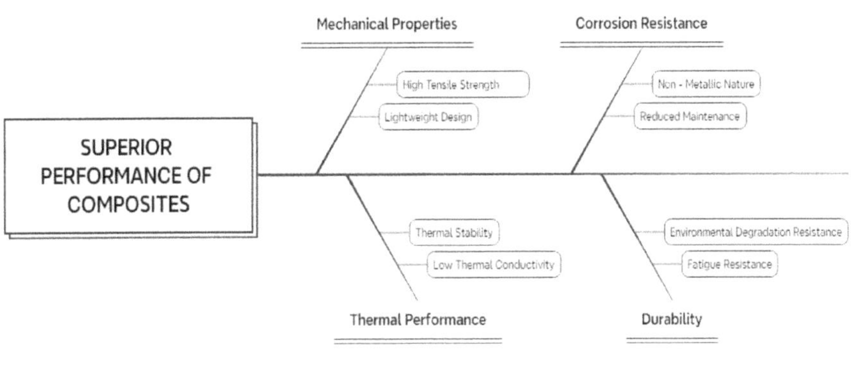

Mechanical Properties

Compared to traditional metals and alloys, the excellent strength to weight ratio of the lightweight composites makes one of the most notable advantages. The materials example is carbon fiber reinforced composites that can attain tensile strengths of more than 3Gpa lattice element value around 300mpa and around 400 to 600mpa for a conventionally used metal (Daş et al., 2023). This elevated strength enables the design of lighter structures capable of performing the same without compromising in an area of prime importance; in aerospace and defense where weight reduction is critical (Joseph & Aluko, 2022). In addition, lightweight composites may also be tough enough. For this, for example, the incorporation of nanomaterials like carbon nanotubes (CNTs) into polymer matrices has been shown to significantly increase toughness substantially, enabling polymer material to be more impact and fatigue resistant (Katti, 2023). On the other hand, traditional metals are known to be brittle under special circumstances which could result in catastrophic failure (Ge et al., 2024).

Thermal Properties

On another issue, lightweight composites outperform traditional materials in the area of thermal performance. The low thermal conductivity of composites can be engineered and because composites are electrically insulating, they are appropriate for insulation applications. This is because for example glass fibre reinforced polymers typically have a low thermal conductivity, which is necessary because in applications that involve thermal insulation (Rajesh et al., 2022). On the other hand, metals have generally higher thermal conductivities, resulting in heat retention and thermally inefficient thermal management systems (Joseph et al., 2023).

Furthermore, lightweight composites usually are more thermally stable. For example, CMCs can sustain extreme temperatures without degrading, which makes them the perfect material for applications such as high temperature such as in the case of turbine engines (Suhaimi et al., 2024). While traditional metals can be heat resistant, they are often as such only with the application of protective coatings or treatments (Sharma & Nandan, 2024).

Corrosion Resistance

Both corrosion resistance and thermal stability are crucial to consideration in material selection, particularly when the component is exposed to harsh environments. PMCs, being non-metallic, have excellent corrosion resistance comparable to weight of composites. For an example, PMCs do not rust or corrode like metals, leading to the reduction of maintenance costs and an increase in the service life, (Venkatesan et al., 2022). On the contrast, traditional metals, like steel, aluminum, etc. are very prone to corrosion, especially in marine and industrial environments and require protective coatings or treatments (Sriniva, 2024). It is noted by research that MMC can also give enhanced corrosion resistance over the base metal. For example, corrosion resistant aluminum SiC composites have been shown in corrosion resistant environments (Lisiecki & Kurc-Lisiecka, 2023). Although the corrosion resistance of traditional metals can depend on their alloy composition and exposure to its environment as much as it can vary, its use can be complicated in certain applications.

Durability and Longevity

They are also more durable than traditional materials. Much better than many of the metals that they replace, composites are designed for the ability to resist fatigue and environmental degradation. For instance, investigations have demonstrated that natural fiber reinforced composites possess attractive durability properties which has made them a valid choice in sustainable applications. These traditional metals have poor durability with fatigue failure in cyclic loading conditions. Additionally, the matrix and reinforcement materials can be selected such that the properties of composites can be tailored for particular applications in this way, optimizing the durability. Traditionally metals are not as flexible as this because they are limited by their inherent material properties.

DESIGN CONSIDERATIONS FOR LIGHTWEIGHT COMPOSITES IN DEFENSE

A number of important factors must be taken into account when designing lightweight composites for defense applications, such as case studies that demonstrate successful implementations, structural design techniques, and material selection criteria.

Material Selection Criteria

In defense applications, choosing the right materials for lightweight composites is essential to achieving the best possible performance, longevity, and affordability. The weight-to-strength ratio, resistance to the environment, and cost-effectiveness for scaling are important factors.

Weight-to-Strength Ratio

A fundamental property, weight to strength ratio, is a factor that determines the suitability of the composite material for the defense applications. Being able to achieve phenomenal strength with minimal weight is a great achievement in lightweight composite materials, such as those that include CNTs or glass fibers. For example, a weight-to-strength ratio of composites is significantly better than traditional materials such as metals giving them advantages in terms of mobility and fuel efficiency in military vehicles as well as aircraft (Yildirim et al., 2022). One such characteristic makes it most suitable for applications where reducing weight is vital in achieving performance, for example, in aerospace and armored vehicles (Vinodhini et al., 2022).

It has been recently shown that advanced materials, like multi-walled carbon nanotubes (MWCNTs) can greatly enhance composite mechanical property through improved weight-to-strength ratio. Tailoring the composite structure and composition gives engineers control for best performance for particular defense applications and ensures that the material meet stringent operational requirements.

Environmental Resistance

Another significant factor in material selection is a resistance to environmental affect. The environment that lightweight composites will be subjected to must be examined, which includes thermal extremes, corrosion, and wear. Silicone foam composites are known for their very low thermal conductivity but high temperature stability which makes them ideal for applications subject to extreme heat for example. Not least, corrosion resistance is a key advantage for PMCs due to the construction material renowned for not rusting and not corroding in an environment compared to traditional metals, which helps increase their service life in harsh environments (Volpini & Pianesi, 2022). Besides, composites also have to show high wear resistance in the case of application in situations involving wear and impact. Hybrid composites, that is, the combination of different kinds of fibers, can help to retain the weight features, while increasing wear resistance (Mei et al., 2024). Thanks to this adaptability, lightweight composites are excellent choices for defense parts from protection to structural components in military vehicles.

Cost-Effectiveness for Defense Scalability

The consideration of cost effectiveness is the key point to scale up the use of lightweight composites in defense applications. Even though the initial costs of advanced composites are higher, subsequent savings in maintenance, or savings in other forms of maintenance costs and extension of cycle times make up for this difference (Vu et al., 2022). Development in manufacturing techniques such as additive manufacturing and automated process (Elsawy et al., 2024) allow for the production of lightweight composites in defense applications to become more cost and scalable. It can also be used to reduce costs and to encourage the sustainability of production of the composites, from recycled materials. For example, results of increasing the material costs through recycling in the lightweight concrete have been promising (Xia et al., 2023). For these reasons, as the trend leans towards sustainability in defense procurement, lightweight composite is a good choice not just for performance reasons, but also for green ones. Figure 4 mentions selection process for Lightweight Composites in Defense.

Figure 4. Selection process for lightweight composites in defense.

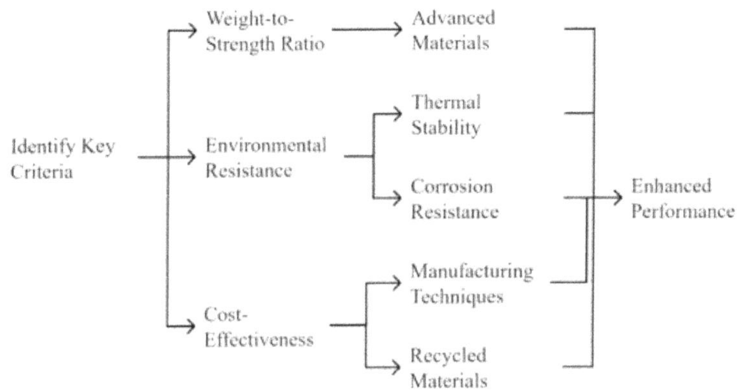

Structural Design Strategies

For defense applications, lightweight composites must be structurally designed to be effective in load-bearing situations, resistant to impacts and vibrations, and able to use sophisticated modeling techniques. Important tactics in various fields are covered in this section.

Load-Bearing Applications

Increasingly, high strength to weight ratios of lightweight composites make them attractive for load bearing applications. For example, fiber metal laminates (FMLs), such as those with metal sheets sandwiched in between layer(s) of fiber reinforced plastic, have been proven to offer great load bearing capability but at a greatly reduced weight over traditional metal structures (Boopathy et al., 2019; Georgantzinos et al., 2023). Because baseline has hosted high strength fibers such as basalt, these composites improve the tensile and flexural strength with good capabilities in military vehicles and aircraft (Ram-raji et al., 2023). Besides, the design of composite structures usually involves finding an arrangement and orientation of fibers to achieve the maximum performance under load. Using such techniques as finite element analysis (FEA), the mechanical behavior of composite structures is simulated under a variety of loading conditions and engineers refine design for a particular application. To get lightweight composites to operate in the harsh conditions both on and off the battlefield, this is a critical optimization process.

Vibration and Impact Resistance

Lightweight composites are of high interest to reduce weight for both military and civil defense applications where resistance under load and exposure to dynamic loads is common. Designs inspired by bioinspired materials such as natural ones, in particular regarding their structural behaviour, can greatly improve energy dissipation and impact resistance. One example of such a composite with a hierarchical structure and robust interfacial bonding is composites designed based on the Bouligand structure inspired by the mantis shrimp's dactyl club (Cui et al., 2024). The designs enable effective energy dissipation during impact, and are thus appropriate for protective applications on military gear, and vehicle armour.

Furthermore, the impact resistance of the structure can be enhanced through the use of advanced materials like hybrid composites (combining various types of fibers). The energy absorption during impact events that occurs when the traditional materials are used in defense applications has been shown to be improved with hybrid composites. This paper also demonstrates the importance of tailoring the composite structure to the impact scenarios in real time to achieve the best performance.

Multiscale Modeling Approaches

Currently, in the sense of design and analysis of lightweight composites, multiscale modeling approaches are increasingly applied. These methods offer the ability to study the material behavior at different length scales from the molecular to macrostructural level. As an example, multiscale recursive micromechanics (MsRM) can be utilized for the study of progress failure in 3D woven composites, through which microstructural characteristics are seen to affect the overall performance during onset of complex loading (Ricks et al., 2022).

Integration of an experiment with a computational model thus enables more accurate prediction of mechanical responses of composites and better guidance of the design process and design optimization of the material properties. This is particularly useful for a defense application where qualification for extreme conditions is needed for reliability and performance. Secondly, the topology optimization methods for the design of composite structures are generally able to provide an innovative solution that not only leaves, but also decreases, the weight of the structure. Only this results in that the design of the modular filled composite structures performed very well in energy absorption and crashworthiness, and are fit for use in military vehicles (Yi et al., 2024). Thanks to the use of these advanced modeling techniques, these targets allow development of lightweight composites tailored for very stringent application requirements in the area of defense. Figure 5 gives the structural design strategies for lightweight composites.

Figure 5. Structural design strategies for lightweight composites

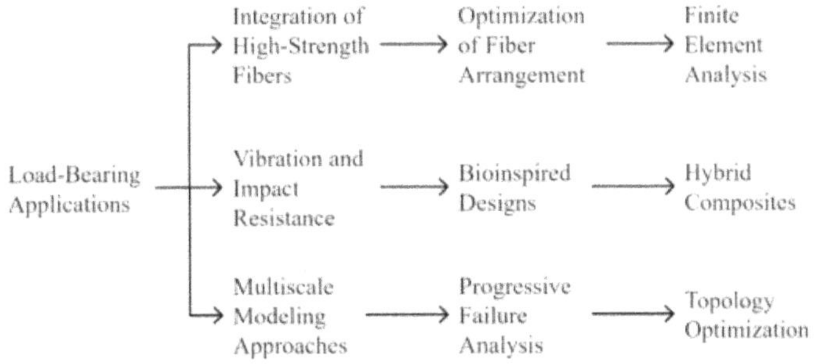

Case Studies

Lightweight composites have become an essential component in current defense applications, delivering improved performance, durability, and efficiency. This section includes many case studies that demonstrate the successful use of lightweight composites in a variety of defense contexts.

Lightweight Composite Armor Systems

Lightweight composites are especially suited for defense in the development of the high-performance armor systems. The ballistic properties of light weight ceramic composite armor were looked at for the performance against small to medium caliber projectiles. It was found that ballistic resistance could be considerably enhanced while retaining a lightweight profile compared to current composite materials through use of the ceramic/metal composite structure, indicative of its use for military vehicles and personal protective equipment. This suggests that ceramic composites can improve soldier safety while maintaining mobility (Dong et al., 2023).

Sandwich Structures for Blast Resistance

So the use of sandwich structured composites and their capability to withstanding blast impacts has been of great attention. Numerical investigation of sandwich composites with different core topologies under air blast impacts has been conducted (Walkowiak et al., 2022). These lightweight structures performed better than monolithic panels of the same weight, and that result suggests they'd be a good source of lightweight protection for military vehicles or infrastructure. Innovative structural designs are deemed to improve the resilience of defense materials, hence the significance of the study.

Carbon Fiber Composite Grid Structures

Another major development of lightweight composite technology is the development carbon fiber composite grid sandwich structures. However, researchers explored high velocity impact resistance of these structures and discovered their outstanding specific strength and stiffness. (Bhoominathan et al., 2024; Wei et al., 2024).This is because the carbon fiber composites are light weight and in addition they are robust in their design capabilities, suitable for aerospace and military applications in which weight reduction is necessary without sacrificing performance. Reflecting on its recent innovation, this innovation shows the versatility of carbon fiber composites in complying with defense applications.

Hybrid Composites for Automotive Applications

Lightweight glass fiber reinforced polymer composites have been used with success to automotive sector applications such as bumper. The mechanical properties of these composites are also discussed by researchers that studied to achieve fuel efficiency and the subsequent reduction of CO_2 emissions in military vehicles (Mohammadi et al., 2022). However, as these composites offer excellent performance, a critical factor identified in the study was their performance under impact loading. This study shows how the current work to optimize composite materials for automotive in defense continues.

Natural Fiber Reinforced Composites

Defense application has been exploring natural fiber reinforced composites because of its sustainability and mechanical properties. Based on the results of researchers, false banana fibers were found to have potential as part of a lightweight engineering composite based on its mechanical and physical properties (Taye & Alemayehu, 2023). Natural fibers are biodegradable and available, and thus they are an attractive option for developing eco-friendly materials of defense. Integrating sustainable materials into defense technology is shown to be important in this study.

Lightweight Concrete Composites

It is known that cementitious composites can incorporate lightweight aggregates, such as expanded polystyrene (EPS), and for the purpose of defense, that has also been demonstrated to be promising. Sales et al. investigated the application of tire rubber residue and EPS for manufacturing lightweight concrete, which are viable in civil construction and military infrastructure (Sales et al., 2024). They are low in weight and have thermal and acoustic properties that make them suitable to construct military installations and barriers. The subject of this case study shows how universal lightweight composites can be in defense applications.

Additive Manufacturing of Composites

Additive Manufacturing (AM) techniques enable new routes to the production of lightweight composites that have complex geometries and tailored properties. Their suitability was studied as low speed impact behavior for high speed military vessels (Xu et al., 2023). In defense, AM allows a composite structure to be customized and improves its performance and adaptability. As a demonstration of the potential of modern manufacturing techniques for the application of lightweight Composites technology, this is also an innovation.

Advanced Composite Materials for Aerospace

From a lightweight composite integration perspective, they have been also used to achieve new aircraft designs in the aerospace field. Carbon fiber reinforced composite car doors were material and structure integrated designed in the study and significant weight reduction with structural integrity were attained (Zhang et al., 2024). For military aircraft, this is critical as every kilogram saved helps enhancing the fuel efficiency and operational range. This study demonstrates the importance of new design strategy in aerospace defense applications.

AM TECHNIQUES

The fabrication methods, processing challenges, and innovations in advanced manufacturing of lightweight composites for defense range from those of conventional mechanical fabrication, to those of thermoset composites, and on to others aimed at thermoplastic composites. With the adoption of AM, advanced curing and developing automation technology, and well planned means to generate uniformity and scale production, lightweight composites serve as crucial constituents to modern defense technology. However, as research and development in this area progresses, lightweight composite will also play an increasing role in support of defense capabilities.

Fabrication Methods for Composites

Lightweight composites for defense applications are manufactured using unique fabrication methods, sophisticated curing and bonding procedures, and the integration of cutting-edge technology. These points are covered in detail in this section.

AM for Lightweight Composites

The production of complex, lightweight composites of the 21st century has evolved into the application of AM technique, which offers manufacture of composites with tailor ability of material properties and geometric design. Fused Deposition Modeling (FDM) technique enables preparation of composite in a precise manner through the layering for the defense applications when weight is reduced and structural integrity is taken care of (Atkins et al., 2023). For example, manufacturing of AM parts on carbon fiber reinforced filaments results in increased mechanical properties that allow such parts to be used for the military vehicle and aerospace industries applications (Zhao et al., 2024). AM also gives speed that allows for rapid prototyping – an important luxury in the fast moving defense sector (Brummer, 2023).

In addition to standard polymer composites, AM has been used in manufacturing MMCs and hybrid reinforced structures with various advantages of multiple materials (Kiani et al., 2022). This capability allows for making of lightweight components having specific performance parameters, for example, higher thermal conductivity or impact resistance, that are essential for defense applications (Ibhadode, 2023).

Advanced Curing and Bonding Techniques

Overall, such composite materials need to acquire the final properties throughout the curing process. These advanced curing techniques include methods that rapidly cure composites either by UV light or by microwave energy (Bagatella et al., 2024) to cure composites for efficiency and effectiveness. It is demonstrated that with testing, these methods significantly decrease curing times and have an effect on the mechanical properties of these composites, particularly those used for defense in time sensitive regions. Bonding is also equally important in the composite manufacturing. In addition, the structural integrity of the device can be improved by more layers being bonded (Li et al., 2023). In addition to this, the use of advanced adhesives, which provides better bonding at lower temperatures, adds more significantly to the improvement of the performances of the composite structures (Kim et al., 2024). These advancements have led to increasing the curing and bonding techniques to provide reliability and durability of lightweight composites.

Processing Challenges

Manufacturing lightweight composites for defense applications is unfortunately not a simple processing task, and the various obstacles associated with achieving this goal can be overcome to meet the material quality needs. The issues to be solved are primarily related to homogeneity, fault conditioning, and scaling manufacturing.

Achieving Uniformity and Minimizing Defects

In the sense of producing uniform material, one of the primary challenges involves producing lightweight composites. Void, delamination and inconsistent mechanical properties due to the variability in the manufacturing process are defects that arise, which may affect the performance of the composite. For example, it was shown that the fiber volume content variations along the processing direction can significantly influence a modulus and a strength of thermoplastic composites, and hence, the processing parameters should be properly controlled so that the desired material properties are produced (Khurshid et al., 2022).

Then, dispersion of reinforcements, such as CNTs, into the polymer matrix is one more factor that mainly decides the uniformity. Different composite performance can result from different composite functional units arising from the choice of the practical fabrication method. Critically, they depend on optimization of processing parameters such as mixing speed and time-since uniform dispersion and preventing of aggregation (which can lead to weak points in material).

Production errors can also be caused by geometrical variances and unregulated procedures. It was found that manufacturing induced flaws, specifically out of plane waviness, are capable of causing significant mechanical failure of a polymer composite (Kulkarni et al., 2022). At the same time, the flaws must be addressed through good design and process control to ensure the durability of lightweight composites in defense.

Scaling Production for Defense Applications

Scaling of the production of lightweight composites for defense applications is difficult due to a number of unique problems. Necessary considerations of these, and the material properties and processing techniques are needed to achieve a transition from full scale manufacturing of laboratory scale of production (Freitag et al., 2023).

A method to scale automated manufacturing system as the rate of the production increases without affecting on control of the product quality (Çava et al., 2024.). Automation helps provide standardized processes, being able to reduce variability, and to produce consistent quality in a large run of products.

Moreover, such advanced manufacturing techniques as laser powder bed fusion (LPBF) and AM can simplify the achievement of complex geometries, which traditional methods cannot reach (Ali & Hafez, 2023). Nevertheless these advanced techniques can be scaled by addressing stability issues (Ali et al., 2024), material consistency and defect management. As an example, the dimensional accuracy and structural integrity of the final product from LPBF largely depend on the powder bed characteristics, therefore they require optimized powder distribution and layer thickness (Hu et al., 2024).

Furthermore, the use of innovative materials aluminum based composites reinforced by waste materials has been proven successful in improving the sustainably and cost competitiveness of the production (Raj et al., 2023; Mohammad et al., 2024). Despite this, manufacturing the reinforcement in a way to achieve uniform distribution during manufacturing is a problem to solve to optimize the mechanical properties of the final composite (Rangapuram et al. 2024).

Innovations in Manufacturing

Integration of automation and artificial intelligence (AI) in the manufacturing of lightweight composites has unheralded the industry in defense. Innovations presented here create complex structures with stringent performance requirements, work processes that are more efficient, and enhance quality control. This section describes advances in use of automation and AI in the production of composites.

Automated Fiber Placement (AFP)

AFP has come up as the technology of choice for composite structures manufacturing. Robotic arms are used to lay down carbon fiber preimpregnated tows with the specifications maintained by constant pressure, temperature and speed while doing so. In addition to increasing production rates, this automation improves on both the quality as well as the reliability of aerospace grade composite parts (Yin et al., 2023). The major aircraft manufacturers like Boeing and Airbus have successfully adopted AFP technology in their production lines in order to expediently fabricate components having more than 50 percent composite by weight (Gambardella et al.,2022).

Manual methods are traditionally employed for manufacturing complex structural designs, most of which can be supported by the AFP process, such as curved components and angle-ply laminates (Zhang et al., 2024). AFP technology can be used to eliminate inefficiencies and errors in manual processes improving overall performance of composite structures thus making it a well preferred solution in use in defense applications which require high precision and reliability respectively.

Robotic Layup Processes

Automation of composite structure manufacturing has even further advanced in robotic layup processes. Due to these systems, the Automated Tape Placement (ATP) and Automated Tape Laying (ATL) processes of composite fabrication are automated. Manufacturing can achieve high precision and repeatability in the layup process by means of robotic arms with advanced sensors and control systems obtaining less defective products and better mechanical properties of the final products (Srebrenkoska & Dukovski, 2022).

The advantage of AI's in the robotic system is to have the capability to monitor and adjust processing parameters in real time in such a way that the desired performance can be guaranteed through the entirety of the manufacturing cycle. Particularly relevant in defense applications is this capability, which enables any capability to adapt to the changing conditions and requirements (Stankiewicz et al., 2024).

Predictive Analytics and Quality Control

AI in composite manufacturing is about integrating it in order to use innovative predictive analytics that help improve quality control and defect detection. Machine learning algorithms can analyze data provided by different sensors that may be incorporated in manufacturing equipment to search for patterns and detect anomalies that may suggest the possibility of defects (Morse & Mallardo, 2023). For instance, real-time curing process monitoring with AI can be used to modify the curing process to achieve uniformity and not lead to defects (Çelik et al., 2022). Encouragement to use the proactive quality control approach not only benefits the final product in terms of reliability, but also in reducing waste and reducing rework to finally reduce production costs.

Optimization of Manufacturing Processes

Meanwhile, optimization techniques can also be driven by AI to be applied to improve manufacturing processes. AI algorithms analyze historical data of current production parameters and identifies the most efficient settings for curing times and temperatures in manufacturing processes like (Liu et al., 2024). These enhanced material properties and performance are quite important in the defense applications where structural integrity plays a very important role.

In addition, AI can also be used in the design of composite structure where it can predict the overall performance in terms of how material composition or processing parameters will modify. This capability allows an engineer to explore a broad range of design options and make informed decisions which improve the functionality of lightweight composites (Ju et al., 2022).

ANALYSIS OF LIGHTWEIGHT COMPOSITES

Mechanical performance, thermal behavior, fatigue and durability testing of lightweight composites are analyzed. It is only a short section with an overview of the most important areas.

Mechanical Performance Analysis

For its use in defense, lightweight composites need to be made of mechanics that can withstand extreme temperatures while maintaining good structures. Dynamics and failure mechanism of stress–strain are studied under these conditions.

Stress-Strain Behavior

Factors such as matrix and fiber properties, fiber orientation and presence of defects will influence the stress strain behavior of lightweight composites and influence this response due to the presence of several factors including, but not limited to, the load response. Research has been performed to study the quasi static bending properties of 3D woven composites and it was shown that 3D woven composites exhibit an initial linear elastic stage, and this stress strain curve reaches the peak stress point (Zhao et al., 2024). This behavior indicates the composite's ability to absorb energy and deform elastically under load, two important properties for applications in military vehicles and, as well, protective gear. In addition, glass fiber reinforced polymer (GFRP) composites have been found to benefit from the incorporation of nanomaterial such as graphene nanoplatelets to increase the mechanical properties. Later study arranged that graphene brought about critical expansion in the flexural strength of these composites, exhibiting the ability to enhance the performance of the composites with material style planning (Tudu & Ramachandran, 2023). However, in applications of defense importance, prediction of how composites will respond under various loading conditions is required depending on their stress-strain characteristics.

Failure Mechanisms under Extreme Conditions

Failure mechanisms of lightweight composites under extreme conditions have to be known to enhance their performance and durability. Factors affecting failure are range varied from environmental conditions, loading rates and material composition. An example where the creep effect in composite beams may cause redistribution of internal forces and change in the stress state so that deformation under long term loading is increased (Nan et al., 2023). The fact that this behavior demonstrates time dependent effects when considering the mechanical properties of a composite says something for itself.

Thermal fatigue damage mechanisms of composites exposed to high temperatures have been studied extensively also. A particular study looks at the development of tungsten composites but it is relevant to point out that such mechanisms as matrix cracking and fiber-matrix debonding may even occur in other composite materials by repeated thermal loads (Wang et al., 2023). When exposed to extreme temperatures, these mechanisms have a significant effect on the performance of composites in applications such as the aerospace and military systems.

In addition, polymer composite degradation and premature failure due to the self heating phenomenon caused by cyclic loading can occur. Researchers pointed out that temperature rise due to self-heating could lead to material degradation exacerbating due to thermal stresses, therefore to mitigate these effects special care and material selection should be given (Amraei & Katunin, 2022)

Thermal Analysis

Thermal performance of lightweight composites in the case of defense application is critical, and especially when operating in situations of high temperatures and thermal shocks. This section details the materials' high temperature capability and respectively, their thermal shock and insulating properties.

High-Temperature Performance

Lightweight composites must maintain mechanical integrity and performance at elevated temperatures, which is critical for aerospace and military applications. Table 1 lists the high temperature composite types, their enhanced features and applications.

Table 1. High temperature composite types, their enhanced properties and applications

Composite Type	Enhanced Properties	Application	References
YSZ-Al2TiO5 Fiber Composites	Improved thermal shock resistance, low thermal expansion coefficient	Thermal insulation in extreme environments	Liu et al., 2022
Graphene-Enhanced Polymer Composites	High thermal conductivity (up to 2000 W/mK)	Efficient heat dissipation in aerospace components	Ai et al., 2024
Porous High-Entropy Ceramics	Superior high-temperature performance (>2000Â°C), excellent thermal insulation	Thermal protection systems in aerospace applications	Wen et al., 2024

Such materials are critical for thermal protection systems in aircraft applications, where materials must withstand severe temperatures without degradation.

Thermal Shock and Insulation Properties

The thermal shock resistance is very important property for lightweight composites, particularly for those which are subjected to rapid temperature changes. Deficiency of a material's ability to absorb thermal stresses without cracking results in a loss of longevity and reliability in defense applications. Finally, the researchers studied which ceramic aerogels can resist big thermal gradients without failing (Zhang et al., 2022). Introducing nanopores inside the aerogel structure reduced the thermal properties dramatically while keeping structural integrity under thermal stress.

Effective insulation property of lightweight composites in addition to their thermal shock resistance. Aerogel materials have been introduced to be used as high thermal insulation using lightweight (Liu et al., 2024). These aerogels can have low thermal conductivity even at elevated temperatures and are thus excellent for thermal protecting applications, for example, military vehicles and aerospace systems.

In addition, the mechanical robustness of such archimedes' aerogel membranes made of aramid nanofiber aerogel was demonstrated to be capable of enhancing thermal insulation properties without compromising on structural integrity (Hu et al., 2022). Ultralow thermal conductivity and light weight of these membranes are very desirable for energy efficient and thermal protection applications in defense areas.

Fatigue and Durability Testing

The fatigue and durability of lightweight composites are important in determining their performance and lifetime in military applications. This section discusses how certain materials function over time under cyclic loads, as well as the impacts of aging and environmental degradation.

Long-Term Performance Under Cyclic Loading

Lightweight composites have wide applications in which repeated stress occurs, military vehicles, and aircraft components, which is demanding to their long term performance under cyclic loading. Cyclic loading, it is found, also negatively affects the service life of such polymeric materials (Jirků et al., 2023). Composites are studied for the importance to accurately predict their durability without understanding their stress strain behaviour under cyclic conditions.

To describe the cyclic stress softening effect in rubber composites, a model for hyper-pseudoelastic loading and unloading was introduced in one study, applying different loading and unloading sequences and observing the difference in material properties (Dong et al., 2023). Insight into what is happening inside the composite, and how the composite responds to such a deformation can be gained from this model, which is important for assessing the long term performance of the composite under cyclic loading.

Additionally, mechanical properties of composite under repetitive loading have to be tested on fatigue. The authors studied the adaptive mechanical and electrical properties of conductive composites under repeated loading and hygrothermal aging, observations whose combined effect was to degrade mechanical and electrical properties, albeit at different rates (Yi et al., 2022). This finding reinforces the need for complete fatigue testing protocols to confirm the reliability of light weight composites in the arena of defense purposes.

Aging and Environmental Degradation Studies

Although the ageing and environmental degradation are the major significant ageing factors in the aged performance of lightweight composites, realistic ageing and degradation characteristics are required for reliable prediction. Researchers considered the fatigue behavior of composite open hole tensile specimens and though this fatigue behavior might be accelerated by environmental factors such as moisture and temperature fluctuations the reduction of the mechanical properties (Russo et al., 2022). The understanding of these effects is important to estimate the life of composites used in defense applications for which they end up being subjected to harsh conditions.

In addition, researchers also performed research, which could seem cyclic load relaxation and substitution lead to the mechanical property degradation with time (Zhang et al., 2022). This research finally shows the requirement of developing aged and environmentally degraded composite materials and that their assurance of reliability in the defense system application.

OPTIMIZATION TECHNIQUES

Several methodologies for the optimization of lightweight composites for defense application are used, such as material optimization, structural optimization and sustainability concerned. In details this section goes through these strategies.

Material Optimization

Lightweight composites are critical in terms of both their mechanical properties and overall performance, and this is particularly true in military applications. This section deals with use of nanomaterial reinforcement composites and the hybrid composites made by combining different types of matrices and fillers.

Nanomaterial Reinforcements for Enhanced Properties

Nanomaterials in particular have made it as very good reinforcement for the composite materials, enhancing the mechanical and thermal properties tremendously. One study combined examples showing that adding graphene filler to cotton/viscose hybrid composite does help bond between fibers and matrix, as well as increased the flexural strength (Arunkumar et al., 2022). Besides the strengthening, graphene also increases the thermal conductivity of the composite to functional levels for thermal conductivity intensive applications.

As is shown, researchers studied the microstructure and tribological properties of graphene reinforced aluminium matrix composites, which showed that the graphene sheets hinder the grain growth and enhance the transferring the load from the matrix to the interface (Wang et al., 2022). The significant improvement in mechanical properties achieved by this mechanism greatly enhances the capability of the composites to provide high performance applications in defense where strength and durability are key components.

Other nanomaterials, such as CNTs, have also been shown promising in optimizing the composite materials. For instance, the claim about CNTs itself is not supported by the reference given, however, as the literature indicates, CNT reinforced composites have superior mechanical properties due to the unique structure and high aspect ratio of CNTs, effectively transferring load and improving the composite performance. The development of these advancements demonstrates the promise of nanomaterials and resulting composite produced from them in the development of lightweight composites with targeted properties for specific defense application.

Hybrid Composites Combining Different Matrix and Filler Types

Hybrid composites where different kinds of fibers or fillers are placed in a single matrix are an encouraging approach to select the best properties of a material. Investigated the influence of fillers on the thermal conductivity and tribological properties of polyamide 6 composites and showed that the characteristics of polyamide 6 composites can be improved with the aid of fillers. Hybrid composites can obtain a balance of strength, stiffness, and durability needed for the defense application by selecting and hybridizing different materials in a strategic way (Czél et al.2023).

In addition, the integration of bio based fillers in hybrid composites is considered to be of interest in terms of sustainability and performance advantages. Results indicate that hybrid composite properties can be improved over two environmental impact metrics. This coincides with the increasing demand on sustainability in defense procurement, and this has lead hybrid composites as a good material option for several applications.

Hybrid composites also enable development of the optimum specific properties such as impact resistance and thermal stability. Another example is when natural fibers along with synthetic fibers are combined to improve the overall toughness of the composite while keeping the weight of the composite in mind (Chauhan et al., 2024). Hybrid composites are highly desirable for military use and protective gear because of their versatility, they are used for performance and usually weight tolerances.

Structural Optimization

Structural optimization is necessary for defense applications of lightweight composites to perform better and more efficiently. This section covers two primary topics related to structural optimization, both of which offer ways to reduce weight without losing strength or strength: First is methods to decrease weight and the use of FEA.

FEA for Lightweight Defense Designs

Among the powerful means in order to simulate the mechanical behavior of composite structures under various loading conditions FEA is one. One of the reasons to use FEA is that since stress, deformation or failure points are simulated on composite materials, engineers are able to optimize the design without builds the physical prototypes first. While the proof of the benefit of through simulation of structure design of disc brakes has been shown, the carbon fiber reinforced polymers (CFRP) material of choice has been presented to researchers (Raja et al., 2022).

This is particularly important when the materials have to withstand the dynamic loads in a harsh environment in the defense application.

Finally the composite properties of these produce can be studied with the help of FEA in terms of the effect of fiber lay down and stacking sequence. Simulation under various combinations of these parameters will enable the engineers to find which configuration of these parameters will be strongest and stiffest and weigh least. Such a capability is necessary for achieving the lightweight structures that are required for military applications.

Further structural optimization process is upgraded using FEA with design optimization techniques such as topology optimization and response surface methodology. Researchers done a computational evaluation of lightweight design strategies for automotive components based on the FEA of the impact of design change on performance (Kim & Lee 2023). By using such an approach, it is possible to identify the lightweight designs that do not affect the structural integrity of the defense components and thus can perform reliably under extreme conditions.

Weight Reduction without Compromising Strength

Reducing weight for composites while still maintaining strength is a very critical goal for defense applications. One way to achieve this is to implement advanced materials, hybrid composites that conventionally mix several various types of fibers or fillers. For example, putting carbon fibers with the aramid fibers can improve tensile strength and impact resistance of the composite while decreasing the overall weight. The hybridization permits combining best properties to optimize a specific property of the composite and to entrust it to the most demanding applications.

Moreover, lattice structures and sophisticated geometrical designs allow the weight to be reduced substantial without reducing strength. Lightweight designs made using lattice structures have been reported to be capable of achieving very significant reductions in weight commensurate with maintaining mechanical performance. These designs make full use of the naturally high strength of composites while requiring minimum amount of material use, thereby fitting in with military vehicle and equipment usage.

Also, weight reduction requires the selection of appropriate materials and processing techniques. A good example of magnesium alloy in combination with polymer composites has been demonstrated for reducing weight with high strength (Kumar & Thakur, 2022). Additionally, advanced manufacturing technique like AM creates a possibility to produce the complex geometry and optimizes material usage as well as the performance (Adeleke et al., 2024). Such capability enables to produce lightweight components conforming to severe defense requirements.

Sustainability Considerations

The growing concern for the sustainability of the manufacturing of lightweight composites has advanced from need to save the environment while maintaining the performance level, particularly in defense applications. This section tackles two fundamental issues related to the sustainability, namely utilizing bio-composites and recycling and reusability strategies.

Use of Bio-Composites

Bio-composites, which are bio-based polymers combined with natural fibers make a sustainable alternative of conventional composite materials. These materials are environmentally friendly and biodegradable as they are usually made from the renewable resources. Table 2 lists the varieties of bio-composites, elevated characteristics and applications.

Table 2. Types of bio-composites, their improved qualities, and their uses

Bio-Composite Type	Polymer Type	Enhanced Properties	Application	References
Graphene-Modified Jute Fiber	Bio-Based Epoxy Resin	Improved mechanical performance, reduced environmental impact	Sustainable alternative to petroleum-based composites	Islam et al., 2023
Vetiver Fiber with Polylactic Acid	Polylactic Acid (PLA)	Renewable, biodegradable, reduced ecological impact	Lightweight materials for defenseapplications	Zulkeply et al., 2022
Natural Fiber Bio-Composites	Various Bio-Based Polymers	Cost-effective, sustainable, good mechanical performance	Structural components, protective gear, Sustainable construction	Morganti et al., 2024

Bio-composites have high adaptability which makes them an attractive option for defense where sustainability is vital with highest performance.

Recycling and Reusability Strategies

Recycling and reuse of composite materials are very essential for the sustainability of defense sector composite material. There are major processes for recycling bio composites that can dramatically reduce waste and resource consumption. Before moving on to the topic of bio based matrix composite reprocessing, several researchers have discussed this because, according to them, these materials could be processed to recover valuable fibers and resins as a part of minimizing the environmental impact

(Ravindran et al., 2024). Sustainability and reduction of profit margin to supply raw material are achieved by using recycling strategies in use.

For also, the design of the principles of disassembly can be included in the manufacture of composite structures, so that reusability and the recycling in the last phases of its life cycle can be easier. Since then, researchers have said that bio wood-plastic composites (WPCs) could allow its components to be separated for enhanced recycling and reduced waste (Hirsch and Theumer 2022). This is an approach of circular economy model, based on the idea of reusing materials that is important in performing activities at minimum possible and according to environmental standards.

Additionally, studying the development of biodegradable composites, for example, those of natural fibers and bio based resins, will supply possibilities to increase sustainability in the field of defense. It is shown that bio-based polyurethane composites are a promising option for effective shielding against ultraviolet radiation, and they are environmentally friendly (Zhai et al., 2024). Such materials don't address but rather respond to performance requirements and help the defense reduce ecological footprint.

APPLICATIONS OF LIGHTWEIGHT COMPOSITES IN DEFENSE

However, as lightweight composites have proved to be superior materials in strength to weight ratios, durability, and versatility with regard to certain defense applications, lightweight composite materials have become an integral part of those applications. The applications of lightweight composites in land defense systems, aerospace systems, naval defense systems and emerging trends of integrated variables are discussed in this section.

Land Defense Systems

Land defense systems eminently rely on the use of lightweight composites, mainly for armored vehicles and lightweight shelters construction. Specific properties make for great weight reduction yet high levels of protection and durability. Future advances in research and development in this area will only increase the potential for use of lightweight composites to improve performance and effectiveness of land defense systems.

Armored Vehicles

In the case of construction of armored vehicles, lightweight composites are used extensively because they are essential to reduce weight and yet provide high levels protection. Micronano lignocellulose with a Bouligand structure developed by researchers had great impact resistance ('wooden armor'), better than a standard approach (Chen et al., 2022). The novelty adddes to ballistic protection but also reduces weight, making it suitable for military field as weight is a key factor.

In addition, rolled homogeneous armor (RHA) steel is used in military vehicles because of its high hardness, common use of energy absorption (Rammohan et al., 2022). Nevertheless, it is possible to further increase the performance of armored vehicles by reducing overall weight without sacrificing ballistic protection by integrating lightweight composite materials. Hybrid armor systems utilizing the combination of traditional armor materials with advanced composites can yield armor systems that provide enhanced protection from a wide variety of threats whilst providing better space for vehicle maneuverability.

Also, the advanced computational technology of the Armour designing of the Armoured Vehicle by means of the FEA can help to identify the best configuration of material to realize the maximum strength and minimum weight (Li et al., 2024). It is the key to making individual vehicles live through combat and still be operable.

Lightweight Shelters

In addition, lightweight composites are manufactured to make portable military shelters as well as armored vehicles. An important feature in combat field is that the composite materials use allows rapid deployment, setup, and deployment. The research suggests it is possible to gain adequate thermal insulation and environmental factor resistance of personnel in the field using composite shelters. Secondly, these materials are lightweight, and logistics as they can easily be transported from one point to another for defense operation.

Aerospace Applications

Lightweight composites continue finding application in aerospace for use with UAVs as well as fighter aircraft and space systems. These alloys are very good materials to use in modern military operations for their unique properties, in particular, high strength to weight ratio, and better durability. This section describes how lightweight composites are applied in these areas.

Unmanned Aerial Vehicles (UAVs)

These UAVs are a must needed means of modern military operations for their roles in glory of the surveillance, reconnaissance and combat missions. Due to the application of the lightweight composites for building UAVs, the UAV has the performance enhanced by reduction of mass and improvement of fuel consumption. Traditionally, flight vehicles made of CFRP can enhance manoeuvrability and flight duration leveraging this replacement material in the UAV structure (Malik et al., 2022).

Lightweight composites particularly provide superior performance characteristics in the study of materials, technology, and UAV design (Gunawan et al., 2024) (Govindarajan et al.,2025). Additionally, due to UAV design flexibility combined with the sophisticated manufacturing technologies such as adding printing, UAVs can be incorporated with state-of-the-art technologies to enhance utility and mission effectiveness (Zhaxi, 2024).

Fighter Aircraft

Fighter aircraft featuring lightweight composites are a most demanding application where high performance, agility and survivability are required. Composite structural members, in which the weight savings can be achieved without loss of weight, have been used as structural components in regions of fighter aircraft wings and fuselages. The F 16 fighter jet is a modern vehicle that has used advanced composite materials to associate its aerodynamic success despite the reduction in general weight controlling directing to enhanced maneuverability and fuel production (Paranjape & Ananthakrishnan, 2023).

According to a study conducted by researchers, aerodynamic characteristics of fighter aircraft were reviewed in their role in increasing agility and performance (Im et al., 2022). Further, hybrid composite systems, consisting of diverse fiber and matrices was also integrated to optimize mechanical properties of fighter aircraft that survive the extreme conditions during combat operation (Liu et al., 2022).

Space Systems

In general, lightweight composites are important to the design and construction of space systems because of weight reduction requirements for the system efficiency and performance at launch. With the use of advanced composites in satellite structure and components, lightweight and high stiff materials can be developed that can withstand the harsh space environments. The development of magnesium-scandium alloys is also example of promise for high strength low weight alloy that will be useful for aerospace applications (Espejo-Antúnez et al., 2022).

Researchers emphasize how lightweight materials are important in the design of space systems to improve structural integrity and performance in a new type of spacecraft design (Choi et al., 2023). AM techniques further increase the design possibilities in space applications by augmenting fabricability of complex composite structures.

Naval Defense Systems

New lightweight composite hull and underwater systems are in increasing development on naval defense systems. Both of their properties like high strength to weight ratio, corrosion resistance and design flexibility make them ideal for modern naval applications. It describes the applications of lightweight composites in naval defense systems, in particular using hull and underwater systems applications.

Composite Hulls in Naval Vessels

In recent years, interest in lightweight composites in use on ship hulls has arisen because they outperform steel and aluminum as construction materials. For example, it was pointed out that GFRP composites are now commonly used for naval vessel hulls, for reasons of their good mechanical properties and corrosion resistance. They are light weight, while improving speed and fuel efficiency very necessary in naval operation.

When used in submarine pressure hulls, CFRP composites were shown by researchers to reduce structural weight by up to 60 percent relative to traditional materials (Li et al., 2022). The payload capacity being increased, and maneuverability improved, this dramatically weight reduced attribute allows submarines to be more effective in a range of operational scenarios.

Durability and Performance

Durability of composite materials is another principal factor in application for naval hulls. Composites are excellent in seawater corrosion resistance and fatigue resistance, and acoustic stealth, but structural failure of composite pressure hulls in deep sea environment have been reviewed by researchers. The properties of composites make examples as hulls of composites particularly desirable in view of the harsh marine conditions they need to be able to withstand whilst retaining structural integrity.

Also, the integration of advanced manufacturing techniques, for instance automatic fiber placement, enables creating the complex hull geometric shape that improves hydrodynamics (Wu et al., 2023). This particular capability is extremely valuable for high speed naval vessels where decrease in drag and improve stability is paramount to maintaining operational effectiveness.

Underwater Systems

Underwater construction is also done with lightweight composites, for submersibles and remotely operated vehicles (ROVs). Composites are used in these applications due to buoyancy and structural integrity advantages that result in better operational depth and maneuverability. Compared to other materials, composite materials having carbon fibre reinforcement can save a significant amount of weight and retain high strength and therefore could be used for deep sea pressure hulls (Sohn et al., 2022).

Underwater defense systems that aim to be effective need the ability for lightweight composites to be both buoyant and withstand high pressures. For example, composite materials can be used in submersible design to build submersibles of greater depth depths without sacrifice of safety or performance. It is a necessity for the military operations, which demanded stealth and durability in underwater environment.

Applications of lightweight composites in underwater system based on bio composites and hybrid material are the innovative. The reasons for choosing to study the feasibility of the use of banana fiber reinforced composites in unmanned mine sweepers were both in prospective of strength and durability aspect, but with sufficiently environmentally friendly aspect (Kusuma et al., 2023). The composite made with this natural fiber incorporated not only makes the composite more sustainable but also it gives composite a performance part.

CHALLENGES AND FUTURE DIRECTIONS

The development and application of lightweight composites in defense face several challenges, as well as opportunities for future advancements. It discusses the main challenges, such as matching cost with performance, tackling production hurdles and testing and validation for harsh environments. It also addresses future directions, including smart material integration, AI driven design, and cooperation between academia and industry.

Challenges

Balancing Cost with Performance

The main problem arises when an attempt is made to develop lightweight composites for defense applications, where cost and performance is balanced. Lightweight composites have the benefit of high strength to weigh ratios and durability, but the number one disadvantage is the initial cost of advanced materials and manufacturing processes. However, researchers noted that for the development of economical UAVs, selection of materials and manufacturing techniques needs to be drastically considered to ensure the achievement of good performance with reasonable amounts of expenditure (Alajami et al., 2022). For defense procurement, this balance is crucial since budget constraints sometimes dictate the material choices.

Addressing Scalability and Production Hurdles

Production of lightweight composites is typically a challenging exercise, and scaling is even more so. Material properties, processing techniques and cost effectiveness of the processing route need special attention when the transition is from laboratory scale production to full scale manufacturing. While the researchers suggested it is possible to use efficient algorithms to minimize the cost of operation in the applications of UAV, the scalability of the composite production processes is still a significant obstacle (Guo & Peng, 2024). To achieve consistent quality and performance in large scale applications, automated manufacturing systems should be implemented and production parameters should be optimized.

Testing and Validation for Extreme Conditions

Thus, the reliability and performance assurance of the lightweight composites for defense is crucial and it's paramount that the composites are tested and qualified under extreme conditions. The long term durability of the composite material is key evaluated by simulation of real world conditions including at least high temperatures, humidity and mechanical stresses. In this testing they recommend that composites be evaluated based on tests done with three environmental conditions coupled with stringent test protocols to make sure they meet the harsh environments found in the military (Kumar 2023).

Future Directions

Integration of Smart Materials and Self-Healing Composites

Those in defense applications have exciting avenues of lightweight composites development in the form of integration of smart materials and self healing composites. The responsiveness of smart materials to environmental stimuli are a benefit to their potential use in making military systems function and adaptable. For example, if shape memory alloys (SMAs) are integrated into the composite framework to act as active deformer to deform adaptively in response to temperature changes, most dynamic performance can be provided under a variety of operational scenarios. The other area of research that could make huge difference to the durability and life of defense material is self healing composites which can self heal the damage (Cherif et al., 2023).

AI-Driven Design and Real-Time Monitoring Systems

Integrating artificial intelligence (AI) in the design and manufacture processes of the lightweight composites is another key trend. The application of AI to design tools can result in a material/structural configuration design optimization which provides enhanced performance at a minimum weight (Deng et al., 2023). In addition, real time monitoring systems with AI based algorithms can carry out health and performance monitoring of composite structures that are in use and help with proactive maintenance in order to reduce risk of failure (Wang et al., 2024). However, it has a particular value when used in defense applications where operational reliability is so important.

Collaboration Between Academia and Industry for Rapid Innovation

The key to accelerating the development of lightweight composites for defense purposes is through an active collaboration between academia and industry. Knowledge transfer partnerships can help researchers use up to the minute developments to assist defense manufacturers with real world problems. For example, it may seek to conduct joint research initiatives on developing new composite materials, improving manufacturing processes, and testing them in a comprehensive manner meeting the military specifications (Yan et al., 2022). In this kind of collaboration, the development of advanced materials and technologies is furthered that will keep defense systems at the cutting edge of innovation.

CONCLUSION

Lightweight composites in defense applications are explored to understand their fundamental impact on performance, durability, and sustainability in military domains. Lightweight composites offer advantages in weight reduction and mechanical properties for aerospace as well as land defense systems such as armoured vehicles and lightweight shelters ranging from UAVs to fighter aircraft, and from satellites to space systems. Additionally, their versatility is confirmed by their integration into naval defense systems with applications in hull construction and underwater.

Lightweight composites in the defense applications are investigated to understand their basic influence on the performance, durability and sustainability in the military domains. For both aerospace and land defense systems such as armoured vehicles and lightweight shelters from UAVs to fighter aircraft, satellites to space systems, lightweight composites provide advantages in weight reduction and mechanical properties. Further, it is confirmed that they are versatile being integrated into naval defense systems, notably hull construction, and underwater system defense.

The analysis of lightweight composites encompasses mechanical performance, thermal behaviour, and fatigue and durability testing. Further development of these materials will require understanding the stress-strain behaviour and failure mechanisms in them to apply them to defense applications. Furthermore, the composites are also required to possess high temperature resistance and thermal shock properties so as to assure reliability in harsh environments.

It then means that mechanical and thermal properties can be optimized with strategies to use nanomaterial reinforcements and hybrid composites. FEA enables the structural optimization of a component to maximize its strength with minimum weight. In addition, bio-composites facilitate sustainability considerations, which include aspects such as the recycling strategies involved in the manufacturing of lightweight composites, which are environmentally responsible.

Further advancements in lightweight composites is expected to be driven by emerging trends in integrated applications such as the development of smart materials, AI driven design and the collaboration between academia and industry. These innovations will enable it to produce high performance materials to respond to the ever changing defense system requirements.

In conclusion, it is proposed that continuing research and development on the development of lightweight composites will significantly contribute to development of defense applications through improvements of modern military force capabilities and its effectiveness in increasingly complex battlefield environments.

REFERENCES

Adeleke, A. K., Montero, D. J. P., Lottu, O. A., Ninduwezuor-Ehiobu, N., & Ani, E. C. (2024). 3d printing in aerospace and defense: A review of technological breakthroughs and applications. *World Journal of Advanced Research and Reviews*, *21*(2), 1149–1160. DOI: 10.30574/wjarr.2024.21.2.0558

Alajami, A. A., Moreno, G., & Pous, R. (2022). Design of a uav for autonomous rfid-based dynamic inventories using stigmergy for mapless indoor environments. *Drones (Basel)*, *6*(8), 208. DOI: 10.3390/drones6080208

Ali, A., & Hafez, W. A. (2023). Overcoming challenges in the powder metallurgy process of metal matrix composites. *SVU-International Journal of Engineering Sciences and Applications*, *5*(1), 13–18. DOI: 10.21608/svusrc.2023.228672.1144

Ali, A. H., Rajesh, G., Kumar, M. K., Lakhanpal, S., Parashar, A. K., & Kaushik, A. (2024). Advancing aluminum-based composite *manufacturing: leveraging tio2 reinforcement through stir casting technique. E3S Web of Conferences, 507, 01042. DOI: 10.1051/e3sconf/202450701042

Alovsat, T. Y.Tanriverdiyev Yusif Alovsat. (2024). Classification and properties of composites based on matrix material. *World Journal of Advanced Research and Reviews*, *24*(2), 2278–2282. DOI: 10.30574/wjarr.2024.24.2.3608

Amraei, J., & Katunin, A. (2022). Recent advances in limiting fatigue damage accumulation induced by self-heating in polymer–matrix composites. *Polymers*, *14*(24), 5384. DOI: 10.3390/polym14245384 PMID: 36559751

Arunkumar, K., Raghu, R., Jule, L. T., & Krishnaraj, R. (2022). Effect of graphene filler on mechanical properties of cotton/viscose hybrid composite. *Materialwissenschaft und Werkstofftechnik*, *53*(3), 298–307. DOI: 10.1002/mawe.202100138

Atkins, C., Chahid, Y., Wells, J. T., Westsik, M., Morris, K. C., Breen, C., Macleod, A., Bissell, L., Cochrane, W., Kotlewski, R., McPhee, S., Snell, R. M., Todd, I., Beardsley, M., Harris, M., Pimentel, M., McKegney, S., Orr, J., Alcock, S. G., & Cottarelli, S. (2023). Understanding the role of additive manufacturing in the development of astronomical hardware. *Astronomical Optics: Design, Manufacture, and Test of Space and Ground Systems*, *IV*, 36. DOI: 10.1117/12.2677452

Bakar, M. S. A., Asyraf, M. R. M., Syamsir, A., Ma, Q., Hazrati, K. Z., Azlin, M. N. M., & Thiagamani, S. M. K. (2024). Kenaf/glass fiber-reinforced polymer composites: Pioneering sustainable materials with enhanced mechanical and tribological properties. *Polymer Composites*, *45*(16), 14421–14447. DOI: 10.1002/pc.28785

Bhoominathan, R., Boopathy, G., Jacob, S., Pakiyaraj, A., & Karthik, M. K. (2024). Analysis of failure modes and energy absorption in glass fiber and chopped basalt sandwich panels under low-velocity impact. *Materials Today: Proceedings*, *85*, 123–129. DOI: 10.1016/j.matpr.2024.05.054

Boopathy, G., Gurusami, K., Chinnapandian, M., & Vijayakumar, K. R. (2022). Optimization of process parameters for injection moulding of Nylon6/SiC and Nylon6/B_4C polymer matrix composites. *Fluid Dynamics & Materials Processing*, *18*(2), 223–232. DOI: 10.32604/fdmp.2022.018225

Boopathy, G., Vijayakumar, K. R., Chinnapandian, M., & Gurusami, K. (2019). Development and experimental characterization of fibre metal laminates to predict the fatigue life. *International Journal of Innovative Technology and Exploring Engineering*, *8*(10), 2815–2819. DOI: 10.35940/ijitee.J9597.0881019

Brummer, M., Raddatz, K. J., Schmitt, M., Schlick, G., Tobie, T., Daub, R., & Stahl, K. (2023). Static load-carrying behavior and material properties of additively manufactured gears (pbf-lb/m, 16mncr5). *Rapid Prototyping Journal*, *29*(11), 117–130. DOI: 10.1108/RPJ-02-2023-0035

Çava, K., İpek, H., Uşun, A., & Aslan, M. (2024). Examine the mechanical properties of woven glass fiber fabric reinforced composite plate manufactured with vat-photopolymerization. *Polymer Composites*, *45*(18), 17105–17120. DOI: 10.1002/pc.28955

Çelik, M., Noble, T. A., Jorge, F., Jian, R., Brádaigh, C. M. Ó., & Robert, C. (2022). Influence of line processing parameters on properties of carbon fibre epoxy towpreg. *Journal of Composites Science*, *6*(3), 75. DOI: 10.3390/jcs6030075

Chauhan, H. R., Variya, S., Solanki, A., & Saladi, S. (2024). Development of graphene nanoplatelets reinforced hvof sprayed cr3c2-nicr coatings. *Journal of Thermal Spray and Engineering*, *4*, 81–86. DOI: 10.52687/2582-1474/401

Chen, Y., Dang, B., Fu, J., Zhang, J., Liang, H., Sun, Q., Zhai, T., & Li, H. (2022). Bioinspired construction of micronano lignocellulose into an impact resistance "wooden armor" with bouligand structure. *ACS Nano*, *16*(5), 7525–7534. DOI: 10.1021/acsnano.1c10725 PMID: 35499235

Cherif, N., Jaafar, W., Vinogradov, E., Yanıkömeroğlu, H., Pollin, S., & Yongaçoğlu, A. (2023). Ituavs: Intermittently tethered uavs for future wireless networks. *IEEE Wireless Communications*, *30*(4), 124–130. DOI: 10.1109/MWC.018.2100720

Choi, M., Park, J., Jie, M., & Choi, W. (2023). Development of a chaff dispense program for target tracking radar deception. *Proceedings of Engineering and Technology Innovation*, *24*, 22–28. DOI: 10.46604/peti.2023.11150

Cui, P., Chen, J., Fu, K., Deng, J., Sun, T., Chen, K., & Yin, P. (2024). Bioinspired bouligand-structured cellulose nanocrystals/poly(vinyl alcohol) composite hydrogel for enhanced impact resistance. ACS Applied Materials &Amp. *Interfaces*, *16*(39), 53022–53032. DOI: 10.1021/acsami.4c13264 PMID: 39306751

Czél, G., Sycheva, A., & Janovszky, D. (2023). Effect of different fillers on thermal conductivity, tribological properties of polyamide 6. *Scientific Reports*, *13*(1), 845. Advance online publication. DOI: 10.1038/s41598-023-27740-y PMID: 36646774

Daş, O., Aybarç, U., Öztürk, Y., Kıral, B. G., Öztürk, H., Kıral, Z., & Uyulgan, B. (2023). A preliminary study on the manufacturing of aluminum–potassium feldspar metal matrix composite materials. *Materialwissenschaft und Werkstofftechnik*, *54*(1), 109–119. DOI: 10.1002/mawe.202200100

Deng, H., Huang, J., Liu, Q., Zhao, T., Zhou, C., & Gao, J. (2023). A distributed collaborative allocation method of reconnaissance and strike tasks for heterogeneous uavs. *Drones (Basel)*, *7*(2), 138. DOI: 10.3390/drones7020138

Dong, G., Li, X., & Ji, Y. (2023). An experimental and numerical study on the ballistic performance of a lightweight ceramic composite armor. *Journal of Physics: Conference Series*, *2478*(7), 072030. DOI: 10.1088/1742-6596/2478/7/072030

Dong, Y., Fu, Y., He, C., & Fang, D. (2023). A hyper-pseudoelastic model of cyclic stress-softening effect for rubber composites. *Polymers*, *15*(14), 3033. DOI: 10.3390/polym15143033 PMID: 37514422

Elsawy, E. E., Seleem, M. H., Esa, M., & Osman, A. (2024). Numerical evaluation of the performance of different composite bridge decks. *IOP Conference Series. Earth and Environmental Science*, *1396*(1), 012009. DOI: 10.1088/1755-1315/1396/1/012009

Espejo-Antúnez, L., Fernández-Morales, C., Moreno-Vázquez, J. M., Tabla-Hinojosa, F. B., Cardero-Durán, M. Á., & Albornoz-Cabello, M. (2022). Assessment from a biopsychosocial approach of flight-related neck pain in fighter pilots of spanish air force. an observational study. *Diagnostics (Basel)*, *12*(2), 233. DOI: 10.3390/diagnostics12020233 PMID: 35204324

Freitag, S., Goetz, S., & Wartzack, S. (2023). Continuous simulation of variations during the design of endless fiber reinforced composite structure assemblies. DS 125: *Proceedings of the 34th Symposium Design for X (DFX2023)*. DOI: 10.35199/dfx2023.02

Furtado, C., Kopp, R., Ni, X., Sarrado, C., Kalfon-Cohen, E., Wardle, B. L., & Camanho, P. (2024). J-integral experimental reduction reveals fracture toughness improvements in thin-ply carbon fiber laminates with aligned carbon nanotube interlaminar reinforcement. ACS Applied Materials &Amp. *Interfaces*, acsami.3c17333. Advance online publication. DOI: 10.1021/acsami.3c17333 PMID: 38624137

Gambardella, A., Esperto, V., Tucci, F., & Carlone, P. (2022). Defects reduction in the robotic layup process. *Key Engineering Materials*, *926*, 1437–1444. DOI: 10.4028/p-7v9349

Ge, G., Hu, J., Hu, Z., Li, H., Huo, Y., Tang, S., Liu, Y., Ding, J., Hou, S., & Gao, Y. (2024). The influence of alloy composition on microstructure and performance of mixed-smelting alloy and weld metal. *Materials (Basel)*, *17*(19), 4708. DOI: 10.3390/ma17194708 PMID: 39410282

Georgantzinos, S. K., Giannopoulos, G. I., Stamoulis, K., & Markolefas, S. (2023). Composites in aerospace and mechanical engineering. *Materials (Basel)*, *16*(22), 7230. DOI: 10.3390/ma16227230 PMID: 38005159

Govindarajan, B., Jaganraj, R., Srinivasan, V., & Kumaran, T. (2025). Evolution of UAV technology from early innovations to future horizons. In *Innovations and developments in unmanned aerial vehicles* (pp. 53–94). IGI Global Scientific Publishing., DOI: 10.4018/978-1-6684-9999-9.ch003

Gunawan, H., Khaerudin, K., Deksino, G. R., & Gultom, R. A.George Royke DeksinoRudy AG Gultom. (2024). Optimising offset defense implementation as fulfilment efforts in fighters aircraft materials and technology. [IJHESS]. *International Journal of Humanities Education and Social Sciences*, *3*(4). Advance online publication. DOI: 10.55227/ijhess.v3i4.724

Guo, Q., & Peng, J. (2024). Persistent monitoring for points of interest with different data update deadlines. *Sensors (Basel)*, *24*(4), 1224. DOI: 10.3390/s24041224 PMID: 38400382

Hirsch, P., & Theumer, T. (2022). Comparative study on polyamide 11 and polyamide 10.10 as matrix polymers for biogenic wood-plastic composites. *Macromolecular Symposia*, *403*(1), 2100492. Advance online publication. DOI: 10.1002/masy.202100492

Hossain, M. D., Deb, A. K., Hossain, M. M., Azam, F. A. B., & Chowdhury, M. (2023). Modification of footwear pvc outsoles into graphene oxide-based pvc (pvc/go) soling composites for the enhancement of better physical properties. Textile &Amp. *Leather Review*, *6*, 271–283. DOI: 10.31881/TLR.2023.035

Hu, C., Pan, F., Ye, W., & Chen, J. (2024). P-7.2: study on the cone-jet mode spraying mechanism of electrohydrodynamic atomization printing for oled organic thin-film encapsulation. SID Symposium Digest of Technical Papers, 55(S1), 1054-1058. DOI: 10.1002/sdtp.17275

Ibhadode, A. O. (2023). *Engine lightweighting: use of green materials as reinforcement in aluminum metal matrix composites*. Renewable Energy - Recent Advances., DOI: 10.5772/intechopen.108273

Im, H., Kim, W., Noh, Y., Hong, I., & Yook, J. (2022). Rcs estimation of singly curved dielectric shell structure with PMCHWT method and experimental verification. *Sensors (Basel)*, *22*(3), 734. DOI: 10.3390/s22030734 PMID: 35161480

Islam, M. H., Afroj, S., & Karim, N. (2023). Toward sustainable composites: Graphene-modified jute fiber composites with bio-based epoxy resin. *Global Challenges (Hoboken, NJ)*, *7*(9), 2300111. Advance online publication. DOI: 10.1002/gch2.202300111 PMID: 37745826

Jirků, P., Urban, J., Müller, M., Kolář, V., Chandan, V., Svobodová, J., Mishra, R. K., & Jamshaid, H. (2023). Evaluation of mechanical properties and filler interaction in the field of sla polymeric additive manufacturing. *Materials (Basel)*, *16*(14), 4955. DOI: 10.3390/ma16144955 PMID: 37512230

Joseph, O., & Aluko, M. O. (2022). Effect of synthetic materials in reinforcement of aluminium matrix composites. *Materials Science Forum*, *1076*, 3–11. DOI: 10.4028/p-o2816k

Joseph, W. A., Loto, R. T., Babalola, P., & Okeniyi, J. O. (2023). Analysis of the effect of zea mays husk particulate reinforcement on 1170 aluminium corrosion in a simulated industrial environment. *Key Engineering Materials*, *943*, 97–107. DOI: 10.4028/p-6b99x5

Ju, X., Xiao, J., Wang, D., Zhao, C., Gao, T., & Wang, X. (2022). Effect of gaps/overlaps induced waviness on the mechanical properties of automated fiber placement (afp)-manufactured composite laminate. *Materials Research Express*, *9*(4), 045305. DOI: 10.1088/2053-1591/ac6858

Karadimas, G., & Salonitis, K. (2023). Ceramic matrix composites for aero engine applications—A review. *Applied Sciences (Basel, Switzerland)*, *13*(5), 3017. DOI: 10.3390/app13053017

Katti, P. P. (2023). Corrosion-resistant metal nanoparticle composite coatings for industrial applications: Effect of niso4, nicl2, mo, nh4cl, and h3bo3 concentrations on nano particle concentration. *European Chemical Bulletin*, *11*(12). Advance online publication. DOI: 10.48047/ecb/2022.11.12.015

Kaushik, V., Sharma, P., Priyanka, P., & Mali, H. S. (2022). Numerical modeling of fiber reinforced polymer textile composites for characterizing the mechanical behavior – a review. *Materialwissenschaft und Werkstofftechnik*, *53*(10), 1263–1289. DOI: 10.1002/mawe.202200032

Khovalyg, D., Mudry, A., & Keller, T. (2023). Prefabricated thermally-activated fiber-polymer composite building slab p-tacs: Toward the multifunctional and pre-fabricated structural elements in buildings. *Journal of Building Physics*, *46*(6), 686–707. DOI: 10.1177/17442591221150257

Khurshid, M. F., Hasan, M. M. B., Abdkader, A., & Cherif, C. (2022). Processing of waste carbon and polyamide fibers for high performance thermoplastic composites: A novel manufacturing technology for unidirectional tapes structure. *Journal of Industrial Textiles*, *51*(4, suppl), 7256S–7276S. DOI: 10.1177/15280837221077705

Kiani, A., Waraich, P. S., Venkatakrishnan, K., & Tan, B. (2022). Synthesis of 3d nanostructured metal alloy of immiscible materials induced by megahertz-repetition femtosecond laser pulses.. DOI: 10.32920/21740780

Kim, K. J., & Lee, J. (2023). Computational evaluation of automotive camber component for light-weight design. *Materialwissenschaft und Werkstofftechnik*, *54*(4), 423–428. DOI: 10.1002/mawe.202200279

Kulkarni, P., Mali, K. D., & Singh, S. (2022). Experimental investigations on flexural performance of polymer composites with non-uniform out-of-plane waviness defect. *Journal of Thermoplastic Composite Materials*, *36*(7), 2703–2717. DOI: 10.1177/08927057221108393

Kumar, D., & Thakur, L. (2022). A study of processing and parametric optimization of wear-resistant az91-tib2 composite fabricated by ultrasonic-assisted stir casting process. *Surface Topography : Metrology and Properties*, *10*(2), 025024. DOI: 10.1088/2051-672X/ac7065

Kumar, M. (2023). Design and analysis of smart materials and structures for aerospace applications. *NeuroQuantology : An Interdisciplinary Journal of Neuroscience and Quantum Physics*. Advance online publication. DOI: 10.48047/nq.2022.20.10.NQ551262

Kusuma, C., Himam, F., & Kusdiana, W. (2023). Experimental study of sandwich strength materials made from banana fiber, as advance materials of the unmanned mine sweeper (ums) hull. *Key Engineering Materials*, *941*, 249–255. DOI: 10.4028/p-7byi0f

Li, M., Wang, X., Liu, H., Sun, X., Binwen, S., Fu, T., & Yao, T. (2024). Multi-objective optimization of the design of a protective-components-to-chassis protection module for unmanned armored vehicles. DOI: 10.21203/rs.3.rs-5350741/v1

Li, T., Yu, G., Zhou, S., Zheng, Z., Gao, X., & Song, Y. (2024). In-plane shear failure behavior and mechanical properties of sic/sic composites. *Journal of Physics: Conference Series*, *2730*(1), 012026. DOI: 10.1088/1742-6596/2730/1/012026

Li, Y., Yu, C., Wang, W., Li, H., & Xu, J. (2022). A review on structural failure of composite pressure hulls in deep sea. *Journal of Marine Science and Engineering*, *10*(10), 1456. DOI: 10.3390/jmse10101456

Lisiecki, A., & Kurc-Lisiecka, A. (2023). Laser cladding of nicrbsi/wc + w2c composite coatings. *Coatings*, *13*(3), 576. DOI: 10.3390/coatings13030576

Liu, C., Ji, Y., Tang, J., Otsuka, K., Wang, Y., Hou, M., Hao, Y., Ren, S., Luo, P., Ma, T., Wang, D., & Ren, X. (2022). A lightweight strain glass alloy showing nearly temperature-independent low modulus and high strength. *Nature Materials*, *21*(9), 1003–1007. DOI: 10.1038/s41563-022-01298-y PMID: 35788570

Liu, J., Wang, S., Yang, H., Zhu, D., & Yue, G. (2024). Processing and characterization of high-performance thermoplastic composites manufactured by laser-assisted automated fiber placement in-situ consolidation and hot-press. *Journal of Engineered Fibers and Fabrics*, *19*, 15589250241254440. Advance online publication. DOI: 10.1177/15589250241254440

Madhankumar, A., & Xavior, M. A. (2024). Graphene reinforced ceramic matrix composite (grcmc) – state of the art. *Engineering Research Express*, *6*(2), 022503. DOI: 10.1088/2631-8695/ad476a

Mahesh, V., Mahesh, V., & Harursampath, D. (2024). On low velocity impact response of sandwich composite with jute/epoxy facesheet and cenosphere reinforced functionally graded core: Experimental and finite element approach. *Polymer Composites*, *45*(14), 13151–13163. DOI: 10.1002/pc.28693

Malik, B., Masud, J., & Akhtar, S. (2022). Objective comparison of numerical spin study with aircraft model free-flight spin tests. *Proceedings of the Institution of Mechanical Engineers. Part G, Journal of Aerospace Engineering*, *237*(2), 419–432. DOI: 10.1177/09544100221103011

Mei, L., Xiang, D., & Huang, Y. (2024). Research on the mechanical properties of eps lightweight soil mixed with slag. *PLoS One*, *19*(1), e0297372. DOI: 10.1371/journal.pone.0297372 PMID: 38265991

Mohammad, Q., Kaliyaperumal, G., Poornima, E., Singh, N., Sethi, V. A., & Kumari, V. (2024). Aluminum-alumina composite manufacturing: unlocking potential with friction stir processing. E3S Web of Conferences, 507, 01034. DOI: 10.1051/e3sconf/202450701034

Mohammadi, H., Ahmad, Z., Mazlan, S. A., Johari, M. A. F., Siebert, G., Petrů, M., & Koloor, S. S. R. (2022). Lightweight glass fiber-reinforced polymer composite for automotive bumper applications: A review. *Polymers*, *15*(1), 193. DOI: 10.3390/polym15010193 PMID: 36616541

Morganti, L., Vandi, L., Larraz, J. A., García-Jaca, J., Muedra, A. N., & Pracucci, A. (2024). A1–a5 embodied carbon assessment to evaluate bio-based components in façade system modules. *Sustainability (Basel)*, *16*(3), 1190. DOI: 10.3390/su16031190

Morse, L., & Mallardo, V. (2023). Bottom-up manufacturing cost optimisation of composite aircraft structures: Manual layup vs. automated layup. *AIP Conference Proceedings*, 020027. Advance online publication. DOI: 10.1063/5.0145177

Nan, H., Wang, P., Zhang, Q., Meng, D., & Lei, Q. (2023). Study on the mechanical properties of continuous composite beams under coupled slip and creep. *Materials (Basel)*, *16*(13), 4741. DOI: 10.3390/ma16134741 PMID: 37445055

Paranjape, A. A., & Ananthkrishnan, N. (2023). Combat aircraft agility metrics - a review. *Journal of Aerospace Sciences and Technologies*, •••, 55–66. DOI: 10.61653/joast.v58i2.2006.706

Raj, S. H. K., Boopathy, G., Vijayananth, S., Balaji, G., & Ramanan, N. (2023). Feasibility and insights into the optimization and characterization of friction welded aluminum–steel dissimilar joints. In *International Conference on Smart Sustainable Materials and Technologies* (pp. 121–129). Springer Nature Switzerland. https://doi.org/DOI: 10.1007/978-3-031-35362-4_12

Raja, T., & Devarajan, Y. (2024). Study on the mechanical and thermal properties of aramid fibre reinforced with sawdust particulates in an epoxy matrix composite: A novel material for structural applications. *Polymer International*, *73*(12), 1063–1070. DOI: 10.1002/pi.6685

Raja, V., Gnanasekaran, R. K., Kaladgi, A. R., Rajendran, P., Khan, S. A., & Asif, M. (2022). Multi-disciplinary computational investigations on asymmetrical failure factors of disc brakes for various cfrp materials: A validated approach. *Symmetry*, *14*(8), 1616. DOI: 10.3390/sym14081616

Rajesh, P. V., Gupta, K. K., Čep, R., Ramachandran, M., Kouřil, K., & Kalita, K. (2022). Optimizing friction stir welding of dissimilar grades of aluminum alloy using waspas. *Materials (Basel)*, *15*(5), 1715. DOI: 10.3390/ma15051715 PMID: 35268941

Rammohan, S., Kumaran, S. T., Uthayakumar, M., Korniejenko, K., Nykiel, M., & Velayutham, A. (2022). Prediction of abrasive waterjet machining parameters of military-grade armor steel by semi-empirical and regression models. *Materials (Basel)*, *15*(12), 4368. DOI: 10.3390/ma15124368 PMID: 35744427

Ramraji, K., Rajkumar, K., Rajesh, M., & Bhaskar, A. (2023). Mechanical and vibration behavior of surface-modified titanium sheet interleaved with woven basalt/flax fiber metal laminates. *Polymer Composites*, *44*(12), 8442–8453. DOI: 10.1002/pc.27709

Rangapuram, M., Dasari, S., Abutunis, A., Chandrashekhara, K., Lua, J., & Li, R. (2024). Experimental and numerical investigation of heat evolution inside the autoclave for composite manufacturing. *Composites and Advanced Materials*, *33*, 26349833241273506. Advance online publication. DOI: 10.1177/26349833241273506

Ricks, T. M., Pineda, E. J., Bednarcyk, B. A., McCorkle, L., Miller, S. G., Murthy, P. L. N., & Segal, K. N. (2022). Multiscale progressive failure analysis of 3d woven composites. *Polymers*, *14*(20), 4340. DOI: 10.3390/polym14204340 PMID: 36297918

Russo, A., Russo, F., Palumbo, C., & Riccio, A. (2022). Numerical assessment on the fatigue behavior of composite open-hole tensile specimens. *Macromolecular Symposia*, *404*(1), 2100454. Advance online publication. DOI: 10.1002/masy.202100454

Sales, A. S. B., Nahime, B. O., Lima, E. C., & Reis, I. C. (2024). Study of tire rubber residue in cementitious composite compared to expanded polystyrene, to make a lightweight concrete. *Concilium*, *24*(1), 300–320. DOI: 10.53660/CLM-2680-24A04

Sharma, S., & Nandan, G. (2024). Casting and comprehensive analysis of nitinol-based self-healing metal matrix composite of a356 alloy prepared via clamp technique. *Engineering Research Express*, *6*(2), 025568. DOI: 10.1088/2631-8695/ad52ea

Sohn, J. M., Hirdaris, S., Romanoff, J., & Kim, S. (2022). Development of numerical modelling techniques for composite cylindrical structures under external pressure. *Journal of Marine Science and Engineering*, *10*(4), 466. DOI: 10.3390/jmse10040466

Srebrenkoska, S., & Dukovski, V. (2022). Defects formation in automated fiber placement technology. *Tekstilna Industrija*, *70*(4), 22–27. DOI: 10.5937/tekstind2204022S

Sriniva, B. (2024). Exhaustive investigation into the corrosion resilience of aluminium-silicon carbide metal matrix composites. DOI: 10.21203/rs.3.rs-3993505/v1

Stankiewicz, K., Lipkowski, A., Kowalczyk, P., Giżyński, M., & Waśniewski, B. (2024). Resistance welding of thermoplastic composites, including welding to thermosets and metals: A review. *Materials (Basel)*, *17*(19), 4797. DOI: 10.3390/ma17194797 PMID: 39410368

Su, J. Y. (2024). Optimizing mechanical properties of multi-walled carbon nanotube reinforced thermoplastic polyurethane composites for advanced athletic protective gear. *Matéria (Rio de Janeiro)*, *29*(2), e20240059. Advance online publication. DOI: 10.1590/1517-7076-rmat-2024-0059

Suhaimi, M. A., Mahat, M. M., Yahaya, S. M., & Ramli, R. (2024). A review: Characterization of aluminium-metal matrix composite reinforced with zirconium diboride. *Journal of Advanced Research in Micro and Nano Engieering*, *18*(1), 44–51. DOI: 10.37934/armne.18.1.4451

Taye, B., & Alemayehu, Z. (2023). Synthesis and characterization of mechanical and physical properties of false banana ('ənäsätə) fiber reinforced composite material. *Ethiopian International Journal of Engineering and Technology*, *1*(1), 49–61. DOI: 10.59122/134CFC10

Tudu, S., & Ramachandran, V. (2023). Effect of graphene nanoplatelets on flexural behavior of glass fiber reinforced polymer composites subjected to different temperatures. *Materials Science Forum*, *1101*, 29–33. DOI: 10.4028/p-R09Tja

Venkatesan, S., Jerald, J., Sundeep, D., Varadharaj, E. K., & Sastry, C. C. (2022). Evaluation of mechanical and corrosion properties of tib2-y2o3 nanocomposite fused bronze metal matrix composite. *Surface Topography : Metrology and Properties*, *10*(3), 035003. DOI: 10.1088/2051-672X/ac7faa

Vinodhini, J., Sudheendra, K., & Bhowmik, S., AR, S., Trivedi, P. D., Zaveri, J. C., … & Kundu, A. (. (2022). Novel thermally insulated ultra-lightweight fireproof and high impact resistance advanced hybrid composite. Polymer Engineering &Amp. *Science*, *62*(9), 3026–3036. DOI: 10.1002/pen.26082

Volpini, V., & Pianesi, M. S. (2022). Characterisation of recycled quarzitic and plastic aggregates for sustainable lightweight screeds. *Key Engineering Materials*, *919*, 28–34. DOI: 10.4028/p-rf17nl

Vu, N. V., Le, V. S., Louda, P., Szczypiński, M. M., Ercoli, R., Růžek, V., & Buczkowska, K. E. (2022). Low-density geopolymer composites for the construction industry. *Polymers*, *14*(2), 304. DOI: 10.3390/polym14020304 PMID: 35054711

Walkowiak, M., Reinicke, U., & Anders, D. (2022). Numerical investigation of different core topologies in sandwich-structured composites subjected to air-blast impact. *Applied Sciences (Basel, Switzerland)*, *12*(18), 9012. DOI: 10.3390/app12189012

Wang, F., Liu, H., Liu, Z., Guo, Z., & Sun, F. (2022). Microstructure analysis, tribological correlation properties and strengthening mechanism of graphene reinforced aluminum matrix composites. *Scientific Reports*, *12*(1), 9561. Advance online publication. DOI: 10.1038/s41598-022-13793-y PMID: 35689017

Wang, H., Xie, Z., Cheng, X., Jing, K., Zhang, L., Yang, J., Liu, R., Han, L., Cao, L., Wang, X., Fang, Q., Liu, C., & Wu, X. (2023). Microstructural evolution and thermal fatigue damage mechanism of second-phase dispersion strengthened tungsten composites under repetitive thermal loads. Journal of Materials Science &Amp. *Technology*, *140*, 221–232. DOI: 10.1016/j.jmst.2022.09.007

Wang, J., Jiang, P., & Qi, J. (2024). A planning method for operational test of uav swarm based on mission reliability. Computer Modeling in Engineering &Amp. *Computer Modeling in Engineering & Sciences*, *140*(2), 1889–1918. DOI: 10.32604/cmes.2024.049813

Wang, J. H., Shi, X. M., Cui, C., Xia, X., Li, Y., Wang, Q., Zhang, L. J., Shen, Z., Li, R., & Fang, X. Z. (2024). Study on anti-penetration performance of metal matrix ceramic ball composite. *Journal of Physics: Conference Series*, *2891*(5), 052010. DOI: 10.1088/1742-6596/2891/5/052010

Wang, J. H., Shi, X. M., Wang, Q., Xia, X., Fang, X. Z., Zeng, T., Zhang, L. J., & Ye, C. J. (2023). Study on ballistic performance of metal matrix ceramic ball composite. *Journal of Physics: Conference Series*, *2478*(11), 112005. DOI: 10.1088/1742-6596/2478/11/112005

Wei, S., Guo, Z., Niu, W., Chai, G. B., Tai, Z., & Li, Y. (2024). High-velocity impact resistance of the carbon fiber composite grid sandwich structures. *Polymer Composites*, *45*(6), 5558–5573. DOI: 10.1002/pc.28147

Wu, Y., Zhang, C., Wang, F., & Yang, C. (2023). Theoretical and numerical study of eddy current pulsed thermography to detect damage of deep-sea manned pressure hull. *Journal of Marine Science and Engineering*, *11*(7), 1410. DOI: 10.3390/jmse11071410

Xia, Z., Wang, H., Shi, C., Sun, Z., Wang, Q., & Luo, X. (2023). Analysis and characterization of three charge thicknesses in ta1/al1060/al7075 explosive welding composite process. *Crystals*, *13*(7), 1079. DOI: 10.3390/cryst13071079

Xu, S., Zhang, S., Shen, W., Xing huang, , Qiu, Y., & Wu, Y. (2023). Low-speed impact of glass/kevlar hybrid composites: Experimental results and numerical verification. *Journal of Composite Materials*, *57*(19), 3079–3099. DOI: 10.1177/00219983231183035

Yan, R., Tian, H., Gao, K., Peng, R., & Liu, B. (2022). A two-stage uav routing problem with time window considering rescheduling with random delivery reliability. *Proceedings of the Institution of Mechanical Engineers. Part O, Journal of Risk and Reliability*, *237*(4), 781–797. DOI: 10.1177/1748006X221105395

Yi, S., Xie, L., Wu, Z., Ning, W., Du, J., & Zhang, M. (2022). Effects of hygrothermal aging and cyclic compressive loading on the mechanical and electrical properties of conductive composites. *Polymers*, *14*(23), 5089. DOI: 10.3390/polym14235089 PMID: 36501484

Yi, X., Hu, L., Li, Q., & Tang, Y. (2024). Axial crushing and energy absorption integrated design of modular filled double-hat beam composite structures. *Materials (Basel)*, *17*(17), 4302. DOI: 10.3390/ma17174302 PMID: 39274691

Yıldırım, F., Kabakçı, E., Şaş, H. S., & Eskizeybek, V.YILDIRIM. (2022). Multi-walled carbon nanotube grafted 3d spacer multi-scale composites for electromagnetic interference shielding. *Polymer Composites*, *43*(8), 5690–5703. DOI: 10.1002/pc.26885

Yin, X., Chen, Z., Bakhshi, N., Tong, O., Xiong, X., Chen, Y., Li, Y., Gao, J., Sarwar, M. S., Poursartip, A., & Madden, J. D. W. (2023). Smart roller: Soft sensor array for automated fiber placement. *Advanced Sensor Research*, *2*(9), 2200074. Advance online publication. DOI: 10.1002/adsr.202200074

Zhai, M., Shou, T., Yin, D., Chen, Z., Wu, Y., Liu, Y., Zhao, X., Hu, S., & Zhang, L. (2024). Bio-based polyurethane composites with adjustable fluorescence and ultraviolet shielding for anti-counterfeiting and ultraviolet protection. ACS Applied Materials &Amp. *Interfaces*, *16*(45), 62606–62616. DOI: 10.1021/acsami.4c12183 PMID: 39483089

Zhang, F., Li, W., Li, X., Yang, D., Qu, W., & Li, A. (2024). Efficient 3d temperature field prediction and visualization for afp process analysis in composite structures. *Polymer Composites*, *45*(18), 16929–16944. DOI: 10.1002/pc.28941

Zhang, H., Sun, Z., Zhi, P., Wang, W., & Wang, Z. (2024). Material-structure integrated design and optimization of a carbon-fiber-reinforced composite car door. *Applied Sciences (Basel, Switzerland)*, *14*(2), 930. DOI: 10.3390/app14020930

Zhang, J., Zhang, M., Zhang, X., & Wang, Z. Z. (2022). Experimental and theoretical investigation on mechanical behaviors of tb991 weld sealant under cyclic loading. Fatigue &Amp; Fracture of Engineering Materials &Amp. *Fatigue & Fracture of Engineering Materials & Structures*, *46*(1), 125–136. DOI: 10.1111/ffe.13852

Zhao, H., Li, J., Zhang, X., Xiong, B., Zhao, C., Ruan, Y., Wang, H., Zhang, J., Lan, Z., Huang, X., & Zhang, H. (2024). State-of-the-art lightweight implementation methods in electrical machines. *Machines (Basel)*, *12*(5), 339. DOI: 10.3390/machines12050339

Zhao, S., Chen, L., & Gao, Z. (2024). Effect of warp yarn paths on the symmetrical cyclic bending fatigue properties of 3d woven composites. *Polymer Composites*, *45*(10), 9239–9250. DOI: 10.1002/pc.28406

Zhaxi, Z. (2024). Research on airframe structure, engine performance, weapon mounting capability of f-16. *Applied and Computational Engineering*, *56*(1), 265–271. DOI: 10.54254/2755-2721/56/20240668

Zulkeply, M., Pang, M. M., Vaithilingam, C. A., & Sivasubramanian, R. (2022). The fundamental studies on the reaction kinetics of thermal decomposition of biocomposite based backsheet materials in photovoltaic (pv) panel. *Journal of Physics: Conference Series*, *2222*(1), 012002. DOI: 10.1088/1742-6596/2222/1/012002

Chapter 8
Nature-Inspired Design Strategies for Metal 3D-Printed Defense Components

Jagadeesh Baskaran
https://orcid.org/0000-0003-3081-7030
PSG Institute of Technology and Applied Research, India

ABSTRACT

Nature-inspired design offers a powerful paradigm for advancing the development of metal 3D-printed components in the defense sector. This chapter explores how biological principles such as hierarchical structuring, multifunctionality, and optimization can inform the design of robust, lightweight, and high-performance defense components. The chapter highlights recent research innovations, computational modeling approaches, and case studies, illustrating the transformative potential of biomimicry in defense applications.

INTRODUCTION

Nature has long been a source of inspiration for engineering design, offering solutions that are both efficient and resilient. From the lightweight yet robust structure of bird bones to the impact-resistant exoskeletons of arthropods, biological systems have evolved over millions of years to optimize strength, adaptability, and multi-functionality. In the realm of defense engineering, where performance and durability are paramount, translating these natural principles into metal 3D printing unlocks new possibilities for advanced protective components. The advent of

DOI: 10.4018/979-8-3373-0933-0.ch008

metal Additive Manufacturing (AM) has already transformed defense applications by enabling the production of highly customized, lightweight, and functionally graded structures. Unlike conventional subtractive techniques, metal 3D printing allows for the fabrication of complex geometries that mimic the intricate internal architectures found in nature. These bioinspired designs can enhance ballistic resistance, energy absorption, and structural efficiency—key factors in the development of next-generation armor, aerospace components, and military vehicle structures.

This chapter explores nature-inspired design strategies for metal 3D-printed defense components, focusing on how biological principles such as hierarchical organization, topology optimization, and self-healing mechanisms can be applied to enhance mechanical performance. Through case studies and experimental insights, this chapter aims to bridge the gap between nature's ingenuity and the future of defense engineering.

Foundations of Bioinspired Design in Defense Engineering

Principles of Biomimicry and Bio-Inspiration

Biomimicry and bio-inspiration are two closely related approaches that draw lessons from nature to improve human engineering designs. While both rely on studying biological systems, they differ in their application. Biomimicry involves directly replicating biological structures, functions, or processes in engineered systems. Bio-inspiration takes inspiration from nature's solutions but adapts them into novel designs that may not strictly mimic natural forms(Baskaran, 2024a; Du Plessis et al., 2019; Wang, Liu, Zhou, Zhu, & Hao, 2020; Y. Xu et al., 2019).

In the context of defense component manufacturing, these strategies help engineers develop advanced materials and structures that enhance protection, durability, and efficiency. Metal 3D printing has further expanded the possibilities by allowing the fabrication of intricate, biomimetic geometries that were previously impossible with conventional manufacturing techniques. Nature has perfected structural efficiency and adaptability over millions of years through evolution. Many biological systems demonstrate features that are highly desirable in defense applications, including:

Lightweight strength – Bird bones and insect exoskeletons exhibit strong yet lightweight structures.
Energy absorption – The layered structure of mollusk shells and the shock-absorbing properties of woodpecker skulls provide impact resistance.
Self-repair mechanisms – Some biological materials, such as human skin or tree bark, can heal themselves after damage.

By studying these natural phenomena, engineers can design metal 3D-printed components that offer superior performance in military applications.

Table 1. Scope for AM and Bio-inspired design of defense components

Component	Biological Inspiration	Defense Application	Additive Manufacturing Advantage	Material Used
Blast-resistant armor panels	Nacre (mother-of-pearl)	Body armor, vehicle armor	Complex layered structures, lightweight and strong	Kevlar-reinforced polymer composites
Stealth drone airframe	Owl wings (silent flight)	Reconnaissance UAVs	Aerodynamic forms with internal lattice support	Carbon fiber-reinforced nylon
Shock-absorbing helmet liner	Woodpecker skull	Military helmets	Tailored energy absorption with variable infill densities	TPU (thermoplastic polyurethane)
Unmanned underwater vehicle fins	Fish fins	Underwater drones and torpedoes	Integrated motion systems and flexible components	Flexible photopolymers or TPU blends
Bio-mimetic camouflage skin	Cephalopod skin (e.g., squid)	Camouflage for personnel or vehicles	Multi-material printing for dynamic surface color and texture	Electroactive polymers and shape memory materials
Lightweight structural joints	Bird bones (hollow and strong)	Exoskeletons, drones, mobile weapon systems	Hollow yet strong structures with topology optimization	Aluminum alloys or titanium lattice structures
Aircraft wing ribs	Insect wings (micro-ribbing)	UAVs, light aircraft	Fine internal ribbing, optimized strength-to-weight ratio	Carbon fiber or high-performance polymers
Energy-absorbing vehicle bumpers	Armadillo shell	Military transport and patrol vehicles	Customized energy dissipation zones	Impact-resistant polymer blends (e.g., ABS + TPU)
Silent rotor blades	Humpback whale fins (tubercles)	Helicopters, UAVs	Bio-inspired leading-edge geometries reducing noise and increasing lift	Composite polymers with embedded fibers
Self-healing drone skin	Human skin	Autonomous drones	Embedded microvascular networks for material repair	Self-healing polymers with microcapsules

Direct Replication of Natural Designs

Most notable research work carried out in this field are provided below:

1. Nacre (Mother-of-Pearl) Mimics: Nacre's brick-and-mortar microstructure has inspired layered composites with high toughness and energy dissipation. Studies have demonstrated synthetic nacre analogues using polymers and ceramics for lightweight armor applications (Barthelat et al., 2007).
2. Arthropod Exoskeletons: The hierarchical architecture and graded properties of crustacean and insect exoskeletons have informed the design of materials that combine stiffness and flexibility(Nikolov et al., 2011). Research has focused on chitin-based composites and bioinspired laminate structures
3. Fish Scales and Turtle Shells: These provide models for overlapping, flexible yet protective systems. Literature explores scale-like materials that balance flexibility with puncture and impact resistance, relevant to body armor.
4. Silk and Spider Webs: Natural fibers like spider silk offer exceptional strength-to-weight ratios. Synthetic analogues, including bioengineered silks and aramid-silk hybrids, have been investigated for ballistic protection(Vollrath & Knight, 2001) .
5. Beetle Elytra and Seed Coats: Structures that resist fracture and absorb energy have informed materials for blast and impact mitigation. (He, Zu, Chen, & Noori, 2015) Researchers analyze microstructural design strategies like cross-linking and energy redirection.

Nature builds materials using hierarchical arrangements, where structures range from the nanoscale to the macroscale, improving mechanical performance.

Example: Bone-Inspired Defense Components

Bone has a porous yet strong structure due to its hierarchical organization as depicted in Figure 1. Metal 3D-printed armor components can incorporate similar porosity, reducing weight while maintaining protection. Titanium lattice structures(-Baskaran, Muthukannan, Shukla, & Gokuldoss, 2024; Dong, 2021; Jagadeesh, Duraiselvam, & Prashanth, 2023, 2024; Yang et al., 2022) designed based on bone architecture are already being explored for military vehicle armor.

Figure 1. Bone-inspired lattice design

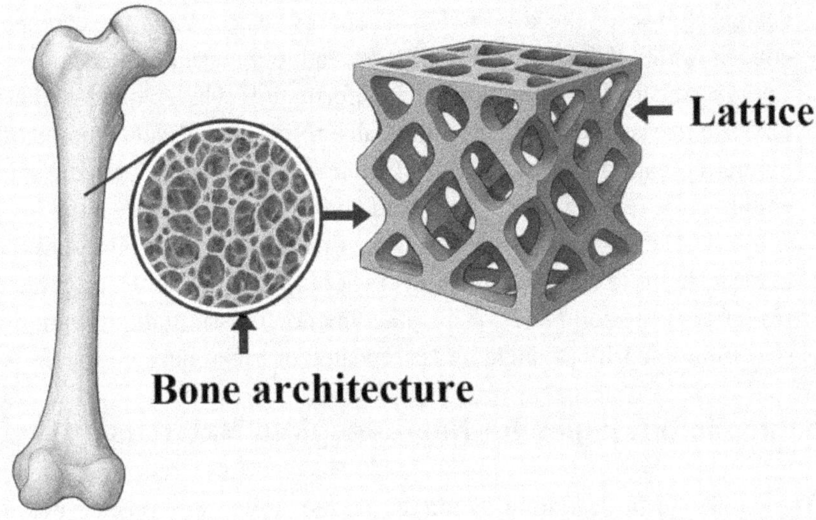

Example: Nacre-Inspired Impact-Resistant Structures

Nacre, the iridescent lining found in the shells of certain mollusks, has garnered significant attention due to its exceptional mechanical properties. Despite its brittleness as a pure ceramic, nacre's remarkable toughness and resistance to fracture can be attributed to its unique microstructure. It is composed of nanocrystalline aragonite platelets, which are stacked in a brick-and-mortar arrangement, interspersed with organic matrices. This hierarchical architecture contributes to its outstanding performance under impact and stress. Inspired by this natural material, researchers have sought to replicate nacre's structure in synthetic composites to create lightweight yet robust materials capable of withstanding high-impact forces. By mimicking nacre's multiscale design, a new class of impact-resistant structures can be developed for a variety of advanced engineering applications.

Nacre consists of alternating layers of inorganic aragonite crystals ($CaCO_3$) and organic proteins, typically arranged in a highly ordered fashion. The inorganic platelets, measuring approximately 100-300 nm in thickness, are separated by a thin organic matrix that serves as a binder and a shock-absorbing element. The resulting structure imparts several key mechanical properties:

Bricks-and-Mortar Architecture: The stacked platelets act as individual "bricks," while the organic matrix between them functions as a flexible "mortar." This design allows for the dissipation of energy over a large area, reducing the concentration of stresses that typically lead to material failure.

Energy Absorption through Shear Deformation: Under impact, the organic matrix undergoes shear deformation, absorbing a significant amount of energy before the material fails. This mechanism is responsible for nacre's ability to endure large forces without catastrophic fracture.

Crack Deflection and Bridging: When cracks form, the interlayer organic material helps to deflect and arrest crack propagation, preventing the material from fracturing completely. This phenomenon is essential for maintaining the structural integrity of nacre under repeated or high-energy impacts.

Biomimetic Strategies for Nacre-Inspired Materials

The challenge in designing synthetic nacre-inspired composites lies in replicating the material's intricate hierarchical structure and optimizing its mechanical properties. Researchers have employed several biomimetic strategies to achieve this:

Layered Ceramic-Polymer Composites

One common approach is to fabricate layered composites of ceramic nanoparticles (e.g., alumina, silica, or hydroxyapatite) and flexible polymers (e.g., epoxy, polyurethane). The ceramic layers mimic the role of aragonite platelets, while the polymer matrix serves as the organic binder. The resulting material can be engineered to exhibit high strength and toughness by carefully controlling the composition and orientation of the layers.

3D Printing of Nacre-Inspired Structures

Advances in additive manufacturing (3D printing) have enabled the fabrication of materials with precisely controlled hierarchical architectures. By using 3D printing techniques, researchers can create composites with varying layer thicknesses and orientations that closely resemble the nacreous structure at multiple scales. This approach allows for the tailoring of mechanical properties, such as fracture toughness and impact resistance, according to specific application requirements.

Bio-Inspired Synthetic Polymers

The organic matrix in nacre is composed of proteins that play a crucial role in energy dissipation and crack resistance. By engineering synthetic polymers with similar molecular characteristics—such as high elasticity and a high degree of crosslinking—researchers have been able to create matrices that enhance the impact resistance of nacre-inspired composites.

Nanostructured Materials

At the nanoscale, the use of nanocrystalline materials, such as graphene, carbon nanotubes, or nanoparticle fillers, has been explored to further enhance the mechanical properties of the composite. These nanostructures contribute to the reinforcement of the matrix and facilitate the transfer of stress across the material, thereby improving its overall impact resistance. The development of these bio-inspired materials for helmets, armor, and other protective gear can improve the effectiveness of personal protection in high-impact environments, such as sports, military, and industrial settings.

Lattice and Cellular Designs for Lightweight Defense Systems

Biological systems often rely on lattices and cellular structures to optimize strength while minimizing material usage.

Example: Honeycomb-Inspired Military Components

Honeycombs, as observed in the nests of bees and other insects, represent an extraordinary example of natural engineering. The hexagonal arrangement of cells is not only aesthetically pleasing but also remarkably efficient in terms of material usage and mechanical performance. The honeycomb structure is lightweight yet possesses a high degree of strength and rigidity, making it ideal for a wide range of applications that require the balance of low weight and high strength(M. Xu et al., 2022). In recent years, engineers and material scientists have turned to this bioinspired design to improve the performance of military components, particularly in areas requiring high impact resistance, energy dissipation, and structural efficiency.

Military applications, such as vehicle armor, aerospace structures, and protective shields, require materials that can withstand extreme forces while maintaining a lightweight profile. Honeycomb-inspired designs, when integrated into these systems, offer several advantages, including improved protection from ballistic impacts, blast resistance, and the ability to absorb energy without catastrophic failure.

Honeycomb Structure: Principles and Mechanics

Honeycomb structures in nature are known for their efficiency in minimizing material use while maximizing strength. The hexagonal cell arrangement in honeycombs allows for maximum structural efficiency, providing high compressive strength and rigidity with minimal material usage. This makes honeycomb structures ideal for military applications, where weight reduction is often a critical factor. The geometry of honeycomb cells allows them to deform in response to an impact. This deformation dissipates energy, effectively absorbing the force and reducing the transmission of shock waves to the surrounding material. This characteristic is highly beneficial in armor and protective shields, where energy absorption is essential to prevent penetration and mitigate damage. The hexagonal arrangement of honeycomb cells allows forces to be distributed evenly across the structure. This enables the material to withstand a wide range of stresses without localized failure, enhancing the overall durability and impact resistance.

Self-Stabilizing Structure: The inherent stability of the hexagonal grid ensures that the honeycomb structure remains intact even under significant loading. This stability allows for the retention of strength after initial deformation, making honeycomb-inspired materials more reliable in high-stress military environments.

Bioinspired Honeycomb Structures in Military Components

Incorporating honeycomb-inspired designs into military technology can result in components with superior mechanical properties, enhanced protection, and reduced weight. Several areas in military technology benefit from the application of honeycomb-inspired materials:

Armor Systems

Ballistic Armor: Honeycomb structures have been incorporated into lightweight armor solutions for military vehicles, body armor, and aircraft. The honeycomb's ability to absorb and distribute the energy of incoming projectiles makes it a valuable material for enhancing armor efficiency. When embedded in composite armor systems, honeycomb cores provide a barrier against ballistic impacts, while also reducing the overall weight of the armor.

Blast-Resistant Armor: Honeycomb structures are particularly effective at absorbing the energy from explosive forces, such as blasts from improvised explosive devices (IEDs) or roadside bombs. The hexagonal cells of the honeycomb deform upon impact, dissipating the shockwave and reducing the force transmitted to the surrounding material or the personnel inside a vehicle.

Aerospace and Aircraft Components

Lightweight Structural Components: Aerospace engineering requires the design of light yet robust components for aircraft and spacecraft. Honeycomb-inspired materials are ideal for this purpose, as they can be used in the construction of fuselage panels, wing structures, and other load-bearing components. Their ability to offer strength without adding significant weight helps improve fuel efficiency and overall performance.

Thermal Insulation: Honeycomb structures are also used in thermal management systems, as they can trap air within their cells and provide insulation against temperature fluctuations. In military aircraft, where temperature control is critical for sensitive equipment, honeycomb materials can contribute to reducing heat transmission and ensuring the optimal functioning of avionics and other components.

Protective Shields and Barriers

Ballistic Shields: Honeycomb structures can be used in personal protective equipment such as ballistic shields or vehicle armor plates. The energy-absorbing characteristics of the honeycomb structure enhance the overall protective capabilities of these shields. The ability to absorb impacts from firearms or shrapnel helps mitigate the effects of explosive threats.

Blast Barriers: In military fortifications, honeycomb-inspired materials are employed to create lightweight blast barriers that can absorb the energy from nearby explosions. These barriers help protect critical infrastructure and personnel by dispersing the force of a blast and preventing damage to more vulnerable structures.

Military Drones and Unmanned Aerial Vehicles (UAVs)

Honeycomb-inspired lightweight materials are increasingly being used in the design of military drones and UAVs. These systems require strong yet lightweight components for both structural integrity and flight performance. The honeycomb structure helps reduce the weight of these aerial platforms, contributing to longer flight times and enhanced maneuverability while ensuring the durability needed for operational success.

Manufacturing and Advanced Technologies: The integration of honeycomb-inspired structures into military components is facilitated by several advanced manufacturing techniques. The development of these materials requires a combination of design flexibility, high precision, and cost-effectiveness. Some of the key manufacturing processes include:

Additive Manufacturing (3D Printing): Advanced 3D printing technologies allow for the precise fabrication of honeycomb structures in a variety of materials, including metals, polymers, and composites. Additive manufacturing enables the creation of complex, lightweight geometries that would be difficult or impossible to achieve using traditional fabrication methods. This process is particularly valuable for producing custom military components with tailored properties, such as armor panels or UAV parts.

Advanced Composites: The use of composite materials, such as carbon fiber-reinforced polymers, in conjunction with honeycomb structures enhances both the strength and weight reduction capabilities of the components. These advanced composites are often used in aerospace and armor applications, where high performance and durability are essential(Paraye & Sarviya, 2024).

Sandwich Structures: Honeycomb cores are often used in sandwich structures, where the honeycomb material serves as the core between two outer layers of stronger, more rigid materials. This combination results in a lightweight yet highly resilient structure. The use of sandwich structures is widespread in military applications such as vehicle armor, aerospace panels, and impact-resistant barriers.

While advanced manufacturing techniques such as 3D printing enable the production of highly complex honeycomb structures, these processes can be expensive and time-consuming. Scaling these methods for large-scale military production while keeping costs manageable remains a challenge.

EXAMPLES OF BIO-INSPIRATION

Trabecular Bone-Inspired Helmets

Trabecular bone is a spongy structure that efficiently absorbs energy. Military helmets designed with 3D-printed metal trabecular structures can offer better impact protection against ballistic threats.

Shark Skin-Inspired Surfaces for Drag Reduction and Aerodynamics

Certain biological surfaces minimize drag, improving aerodynamics and efficiency in motion. Figure 2 shows a shark skin that has microscopic ridges to reduce drag in water. Defense applications include 3D-printed coatings on submarine hulls to enhance stealth and fuel efficiency.

Figure 2. Shark-skin inspired design

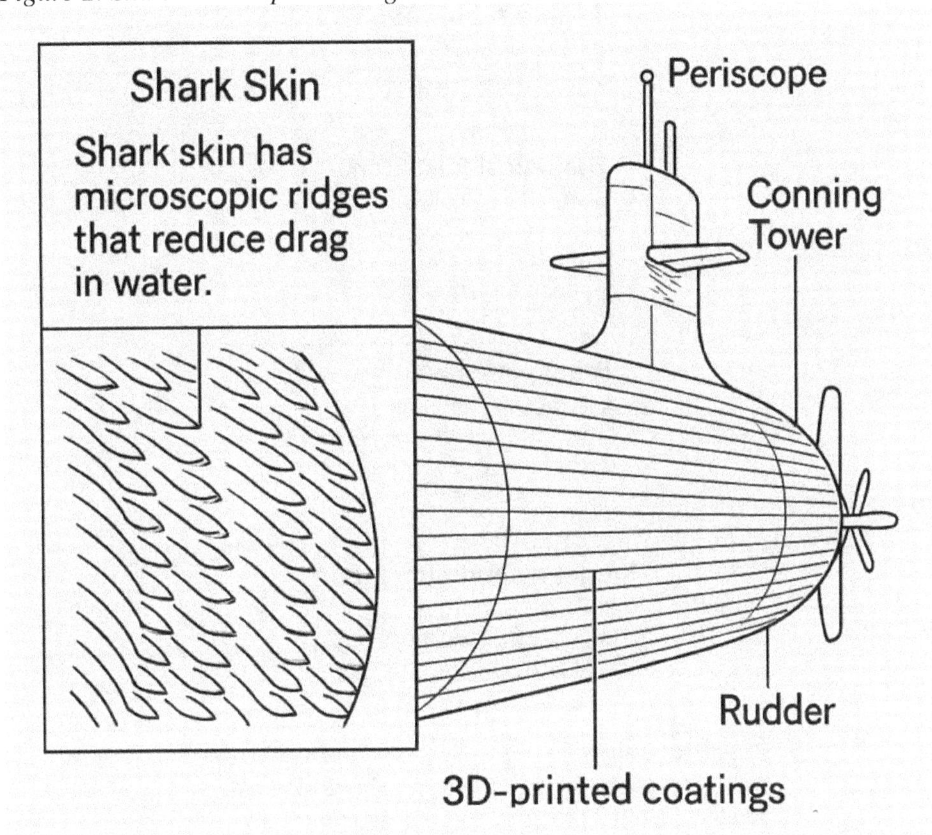

Bird Wing-Inspired UAV Designs

Bird wings optimize airflow with flexible, adaptive structures. Bioinspired UAV components designed through metal 3D printing (Gisario, Kazarian, Martina, & Mehrpouya, 2019) can improve flight performance in defense applications. A generalized design process used for the design and prototype development of such bio-inspired design is provided in Figure 3.

Figure 3. Design process of UAV wings

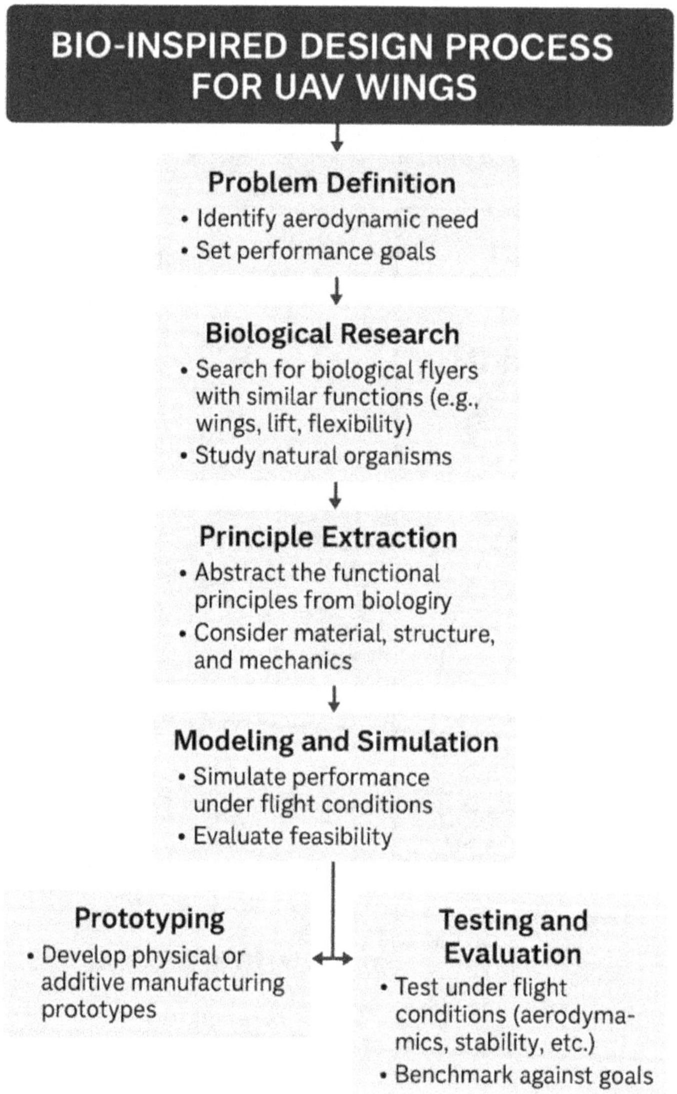

BIO-INSPIRATION: LEARNING FROM NATURE WITHOUT DIRECT REPLICATION

Bio-inspiration differs from biomimicry in that it does not seek to copy nature exactly but instead applies biological principles to novel designs.

Self-Healing Materials for Prolonged Service Life

Many biological organisms heal themselves after damage, and applying this principle to metal 3D-printed defense components can enhance longevity. In nature, self-healing is a common phenomenon seen in biological systems, such as the wound-healing processes in animals and plants. These systems possess an innate ability to detect damage, activate repair mechanisms, and restore their functionality without external intervention. Inspired by these natural processes, researchers have sought to develop self-healing materials that can autonomously repair damage in metallic alloys, which are critical to a wide range of engineering applications. Metallic alloys, however, often suffer from issues like corrosion, cracking, and fatigue under harsh conditions, making the development of self-healing metallics particularly challenging.

Bioinspired self-healing strategies aim to mimic the natural healing processes by integrating smart materials and functional agents within the alloy structure. One of the most promising techniques for achieving self-healing in metals is by embedding microencapsulated healing agents, which, upon material damage, are released to repair the affected area. Furthermore, advances in 3D printing technologies provide new opportunities for fabricating self-healing metallic components with complex geometries, enabling tailored microstructures and healing responses.

Self-Healing Mechanisms in Metallic Alloys

The concept of self-healing in metallic alloys is rooted in the ability of the material to autonomously detect and repair cracks or other forms of damage. Several bioinspired mechanisms have been explored for self-healing in metals, including:

Microencapsulation of Healing Agents: One of the most widely studied methods involves the use of microencapsulated healing agents, such as polymeric or metal-based compounds, embedded within the metallic matrix. When the material experiences damage, the microcapsules rupture, releasing the healing agent into the crack or void. The agent then undergoes a chemical or physical reaction to bond the crack surfaces together, restoring the material's structural integrity.

In-Situ Healing Through Phase Transformation: Another approach involves utilizing phase-changing materials within the alloy. In response to applied stress or damage, these materials undergo a phase transformation that triggers the formation of a new microstructure capable of sealing cracks. For example, certain alloying elements may undergo a reversible transition, enabling the formation of a new phase that fills gaps or cracks when required.

Self-Healing via Chemical Reactions: Some bioinspired alloys are designed to undergo chemical reactions, such as oxidation or reduction, to repair damage. For instance, the self-healing process in a metallic alloy could involve the release of a reactive species that repairs the crack surface by forming an oxide or other protective layer, similar to how certain biological organisms produce protective barriers after injury.

Dynamic Reconfiguration of Alloy Composition: Advanced alloy systems can also leverage a dynamic reconfiguration mechanism, where the material's composition is altered at the damaged site through external stimuli, such as heat or pressure. This allows the material to 'heal' by adjusting its properties in response to the detected damage.

While the integration of self-healing mechanisms in metallic alloys through 3D printing has shown considerable promise, the widespread adoption of self-healing metallic alloys in industry will require overcoming issues related to scalability and cost. Current 3D printing methods for metals, such as selective laser melting (SLM) or electron beam melting (EBM), are still relatively expensive and may not yet be viable for mass production. Reducing the cost of 3D printing technologies and improving the throughput will be critical to making self-healing alloys more accessible for practical applications. The long-term performance and reproducibility of self-healing mechanisms need to be thoroughly studied. Factors such as the longevity of healing agents, the efficiency of healing over multiple cycles, and the impact of environmental conditions on the healing process must be addressed to ensure the reliability of these materials in real-world applications. The design and fabrication of self-healing metallic alloys with embedded microstructural features through 3D printing require precise control over the printing process at multiple scales. Achieving this level of control while maintaining the integrity of the self-healing mechanisms presents a significant challenge in terms of both material science and manufacturing technology.

Applications and Prospects

Bioinspired self-healing metallic alloys, when combined with advanced manufacturing techniques such as 3D printing, have the potential to revolutionize a variety of industries:

Aerospace: Self-healing metallic alloys could be used in aircraft and spacecraft components, where weight reduction and material longevity are essential. The ability to autonomously repair structural damage would improve safety and reduce maintenance requirements.
Infrastructure: Bridges, pipelines, and other critical infrastructure could benefit from self-healing metallic materials that minimize the risks associated with fatigue, corrosion, and damage-induced failures.
Military and Defense: In high-performance military applications, such as armor and vehicle protection, self-healing alloys could enhance the survivability of materials exposed to extreme conditions and impact.

INTEGRATION OF BIO-INSPIRED DESIGN STRATEGIES WITH 3D PRINTING TECHNOLOGIES

The advent of 3D printing, or additive manufacturing (AM), has revolutionized the fabrication of metallic parts, offering unparalleled design flexibility and the ability to create complex geometries that were previously unattainable through conventional manufacturing methods(Baskaran, 2024b; Gibson, Rosen, & Stucker, 2015; Kannan, Sivakumar, Jagadeesh, & Balaji, 2024; Mirzendehdel & Suresh, 2016; Sames, List, Pannala, Dehoff, & Babu, 2016; Varatharajulu, Duraiselvam, Pradeep, Jagadeesh, & Letters, 2025). When combined with self-healing metallic alloys, 3D printing opens up exciting possibilities for producing advanced, highly adaptive structures.

Designing for Self-Healing with 3D Printing

One of the most significant advantages of 3D printing is the ability to design and fabricate metallic parts with precise control over their microstructure and internal architecture. This allows for the intentional incorporation of self-healing mechanisms, such as channels for healing agents or microcapsule networks, directly into the printed part. By embedding these systems during the printing process, the metallic components can be tailored to activate healing in response to localized damage, providing a more efficient and localized repair.

Incorporation of Healing Agents During Printing: During the 3D printing process, healing agents can be seamlessly integrated into the printed metallic structure. For example, microencapsulated healing agents can be directly incorporated into the metal powder or ink used in the printing process. This enables the fabrication of self-healing metallic alloys with a homogenous distribution of healing agents throughout the printed part, optimizing the healing efficiency and ensuring a high degree of damage recovery.

Functionally Graded Materials (FGMs)

3D printing allows for the creation of functionally graded materials, where the composition and properties of the material vary across the component(Choy, Sun, Leong, & Wei, 2017; Mahamood, Akinlabi, Mahamood, & Akinlabi, 2017; Xiong et al., 2021; Zhao et al., 2023). In the context of self-healing metallic alloys, FGMs can be designed to optimize the distribution of healing agents, mechanical properties, and temperature resistance across different regions of the part. For example, areas exposed to higher stress or more frequent damage can have a higher concentration of microencapsulated healing agents or enhanced material properties to promote self-repair.

Multimaterial Printing

The combination of multiple materials in a single print job enables the integration of complementary properties, such as combining metallic alloys with polymers or ceramics that enhance the self-healing process. For instance, a metallic core could be printed alongside a polymeric outer shell containing healing agents that are activated when the part experiences damage. This hybrid approach could offer improved impact resistance and self-healing performance. Among the emerging innovations in 3D printing, multimaterial printing has gained significant attention due to its ability to integrate diverse material properties within a single structure. This advanced technique combines materials with complementary characteristics—such as metals, polymers, and ceramics—within a unified print job. One particularly promising application of this capability lies in the development of self-healing materials and systems that can autonomously repair damage, extending the service life and safety of critical components. Multimaterial printing allows for the co-deposition of materials with different physical and chemical properties. For instance, combining metallic alloys with responsive polymers or ceramics can create smart structures that not only perform mechanical functions but also possess the ability to sense damage and initiate healing. This has significant implications for aerospace, biomedical, automotive, and consumer electronics applications, where durability, performance, and longevity are critical.

One compelling example of multimaterial printing in action is the integration of a metallic core with a polymeric outer shell. The metallic core can provide structural strength and high load-bearing capacity, while the surrounding polymer can be engineered to contain microencapsulated healing agents. These agents remain dormant until the structure experiences mechanical damage. When a crack forms or impact occurs, the capsules rupture, releasing the healing agents into the damaged area. These agents then undergo chemical reactions, such as polymerization or crosslinking, to restore the material's integrity. In more advanced designs, external stimuli like heat, light, or pressure can be used to trigger the self-healing process more precisely.

This hybrid approach allows engineers to overcome some of the limitations of single-material self-healing systems. For example, polymers alone may offer good healing properties but lack the mechanical strength necessary for load-bearing components. Metals, on the other hand, provide excellent strength but typically lack the dynamic behavior required for autonomous repair. By combining the two, multimaterial printing enables the creation of components that are both mechanically robust and functionally adaptive. In certain applications, the addition of ceramics can further improve thermal resistance or biocompatibility, especially in fields like medical implants or high-temperature aerospace components.

The impact resistance of such hybrid structures is also improved. The metallic core absorbs and distributes incoming forces, reducing stress concentrations, while the outer polymer layer acts as a shock absorber. When damage occurs, the healing agents act quickly to mitigate the damage and prevent crack propagation. This not only increases the component's operational lifespan but also enhances safety by reducing the risk of catastrophic failure.

As research in multimaterial printing continues to evolve, so too do the opportunities for innovation. Techniques such as direct ink writing, fused deposition modeling (FDM) with multi-nozzle systems, and inkjet-based printing are enabling more precise control over material placement and interface bonding. Advances in materials science are also introducing new polymers and alloys designed specifically for additive manufacturing and healing applications. Furthermore, computational modeling and simulation tools now allow researchers to predict and optimize healing behavior and mechanical performance before a component is ever printed.

Functionally Graded Materials (FGMs) for Adaptive Protection

Natural materials often have gradual transitions in composition, making them resistant to damage while maintaining flexibility.

Example: Teeth-Inspired Functionally Graded Structures: Human teeth have a hard enamel layer transitioning into a softer dentin core. Metal 3D-printed armor using functionally graded materials can provide a hard exterior to resist penetration while maintaining an energy-absorbing interior.

Example: Bamboo-Inspired Load-Bearing Components: Bamboo has a natural gradient of material properties, making it both strong and flexible. 3D-printed structural components for military vehicles can incorporate similar gradients for improved load-bearing efficiency.

Swarm Intelligence and Distributed Manufacturing

Biological systems often function in decentralized, efficient ways, which can inspire defense manufacturing logistics.

Example: Ant Colony-Inspired Distributed 3D Printing

Just as ant colonies work collectively to build structures, decentralized 3D printing units in battlefield environments could produce needed components on demand. This would enable rapid repairs and supply chain resilience during combat operations. A workflow for such printing process is provided in figure 4.

Figure 4. Workflow of decentralized 3D printing in battlefields

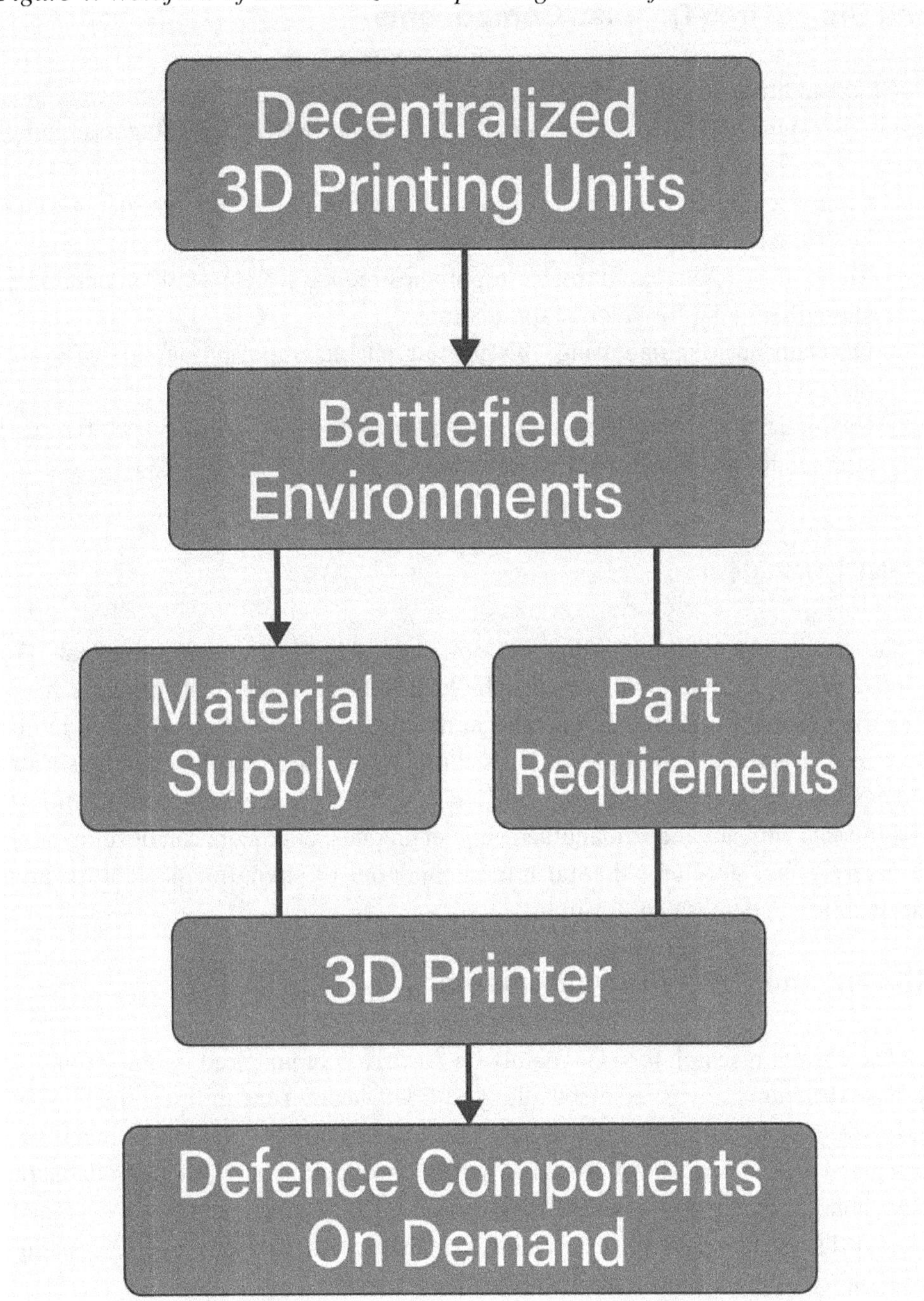

The Role of Metal 3D Printing in Biomimetic and Bioinspired Defense Components

Metal additive manufacturing allows for the precise fabrication of intricate, bio-inspired designs that traditional methods cannot achieve. Key advantages include:

Complex Geometries: Nature-inspired hierarchical and lattice structures can be printed with minimal material waste.
Material Efficiency: Optimized topologies reduce weight while maintaining strength, crucial for defense applications.
On-Demand Manufacturing: 3D printing enables rapid production of spare parts in remote or combat scenarios.
Customization: Metal AM allows for tailored designs based on specific mission requirements.

CONCLUSION

Despite its potential, integrating biomimicry and bio-inspiration into metal 3D-printed defense components presents challenges. Some biological materials have properties that are difficult to replicate in metals. Achieving high precision in hierarchical structures at multiple scales is complex. Advanced bioinspired materials and designs can be expensive to develop and test. But, by integrating computational design tools with advanced manufacturing techniques, engineers can develop novel materials and structures that push the boundaries of strength, adaptability, and survivability in defense applications.

Current and Future Developments

The current research focus is mainly on AI-driven bioinspired optimization for defense structures, advanced metal alloys with self-healing capabilities, sustainable and recyclable bioinspired defense materials. Advancements in additive manufacturing and green chemistry will enable scalable production using biodegradable or recyclable polymers and natural fibers. Additionally, Defense materials designed for easy disassembly and reuse will align with the circular economy goals, reducing environmental impact without sacrificing performance.

REFERENCES

Barthelat, F., Li, C. M., Comi, C., & Espinosa, H. D. (2006). Mechanical properties of nacre constituents and their impact on mechanical performance. *Journal of Materials Research*, *21*(8), 1977–1986. DOI: 10.1557/jmr.2006.0239

Baskaran, J. (2024a). Advancements in Metal Additive Manufacturing Technology for Biomedical Applications. In *Trends and Applications in Mechanical Engineering, Composite Materials and Smart Manufacturing* (pp. 1-34): IGI Global. DOI: 10.4018/979-8-3693-1966-6.ch001

Baskaran, J. (2024b). Roadmap to Achieve Sustainable and Eco-Friendly Metal Additive Manufacturing Processes. In *Futuristic Technology for Sustainable Manufacturing* (pp. 185–215). IGI Global. DOI: 10.4018/979-8-3693-1862-1.ch011

Baskaran, J., Muthukannan, D., Shukla, R., & Gokuldoss, P. K. (2024). Manufacturability and deformation studies on a novel metallic lattice structure fabricated by Selective Laser Melting. *Vacuum*, *222*, 113065. DOI: 10.1016/j.vacuum.2024.113065

Choy, S. Y., Sun, C.-N., Leong, K. F., & Wei, J. (2017). Compressive properties of functionally graded lattice structures manufactured by selective laser melting. *Materials & Design*, *131*, 112–120. DOI: 10.1016/j.matdes.2017.06.006

Dong, L. (2021). Mechanical responses of Ti-6Al-4V truss lattices having a combined simple-cubic and body-centered-cubic (SC-BCC) topology. *Aerospace Science and Technology*, *116*, 106852. DOI: 10.1016/j.ast.2021.106852

Du Plessis, A., Broeckhoven, C., Yadroitsava, I., Yadroitsev, I., Hands, C. H., Kunju, R., & Bhate, D. (2019). Beautiful and functional: A review of biomimetic design in additive manufacturing. *Additive Manufacturing*, *27*, 408–427. DOI: 10.1016/j.addma.2019.03.033

Gibson, I., Rosen, D. W., & Stucker, B. (2015). *Additive manufacturing technologies* (2nd ed.). Springer. DOI: 10.1007/978-1-4939-2113-3

Gisario, A., Kazarian, M., Martina, F., & Mehrpouya, M. (2019). Metal additive manufacturing in the commercial aviation industry: A review. *Journal of Manufacturing Systems*, *53*, 124–149. DOI: 10.1016/j.jmsy.2019.08.005

He, C., Zu, Q., Chen, J., & Noori, M. N. (2015). A review of the mechanical properties of beetle elytra and development of the biomimetic honeycomb plates. *The Journal of Sandwich Structures & Materials*, *17*(4), 399–416. DOI: 10.1177/1099636215576881

Jagadeesh, B., Duraiselvam, M., & Prashanth, K. (2023). Deformation behavior of metallic lattice structures with symmetrical gradients of porosity manufactured by metal additive manufacturing. *Vacuum*, *211*, 111955. DOI: 10.1016/j.vacuum.2023.111955

Jagadeesh, B., Duraiselvam, M., & Prashanth, K. (2024). Deformation and energy absorption studies on FBCC and FBCCz lattice structures with symmetrical density gradients produced by L-PBF of Ti-6Al-4V alloy. *Materials Today: Proceedings*. Advance online publication. DOI: 10.1016/j.matpr.2024.02.008

Kannan, G. R., Sivakumar, M., Jagadeesh, B., & Balaji, N. (2024). Powder Bed Fusion Process–State of Art. In *Advances in Additive Manufacturing* (pp. 89-107). DOI: 10.1002/9781394238316.ch5

Mahamood, R. M., Akinlabi, E. T., Mahamood, R. M., & Akinlabi, E. T. (2017). Additive manufacturing of funtionally graded materials. *Functionally graded materials*, 47-68.

Mirzendehdel, A. M., & Suresh, K. (2016). Support structure constrained topology optimization for additive manufacturing. *Computer Aided Design*, *81*, 1–13. DOI: 10.1016/j.cad.2016.08.006

Nikolov, S., Fabritius, H., Petrov, M., Friák, M., Lymperakis, L., Sachs, C., Raabe, D., & Neugebauer, J. (2011). Robustness and optimal use of design principles of arthropod exoskeletons studied by ab initio-based multiscale simulations. *Journal of the Mechanical Behavior of Biomedical Materials*, *4*(2), 129–145. DOI: 10.1016/j.jmbbm.2010.09.015 PMID: 21262491

Paraye, P., & Sarviya, R. (2024). Advances in polymer composites, manufacturing, recycling, and sustainable practices. *Polymer-Plastics Technology Materials*, 1-24.

Sames, W. J., List, F., Pannala, S., Dehoff, R. R., & Babu, S. S. (2016). The metallurgy and processing science of metal additive manufacturing. *J. International Materials Reviews*, *61*(5), 315–360. DOI: 10.1080/09506608.2015.1116649

Varatharajulu, M., Duraiselvam, M., Pradeep, G., & Jagadeesh, B. J. S. R. (2025). Exploring Epitaxial Grain Growth and Marangoni-Induced Conversion in Columnar Dendrites: A Study of Inconel 718 Additive Manufacturing via Direct Laser Energy Deposition. *Surface Review and Letters*, *32*(1), 2450102–2450115. DOI: 10.1142/S0218625X24501026

Vollrath, F., & Knight, D. P. (2001). Liquid crystalline spinning of spider silk. *Nature*, *410*(6828), 541–548. DOI: 10.1038/35069000 PMID: 11279484

Wang, S., Liu, L., Zhou, X., Zhu, L., & Hao, Y. (2020). The design of Ti6Al4V Primitive surface structure with symmetrical gradient of pore size in biomimetic bone scaffold. *Materials & Design*, *193*, 108830. DOI: 10.1016/j.matdes.2020.108830

Xiong, Y., Han, Z., Qin, J., Dong, L., Zhang, H., Wang, Y., Chen, H., & Li, X. (2021). Effects of porosity gradient pattern on mechanical performance of additive manufactured Ti-6Al-4V functionally graded porous structure. *Materials & Design*, *208*, 109911. DOI: 10.1016/j.matdes.2021.109911

Xu, M., Zhao, Z., Wang, P., Duan, S., Lei, H., & Fang, D. (2022). Mechanical performance of bio-inspired hierarchical honeycomb metamaterials. *International Journal of Solids and Structures*, *254*, 111866. DOI: 10.1016/j.ijsolstr.2022.111866

Xu, Y., Zhang, D., Hu, S., Chen, R., Gu, Y., Kong, X., Tao, J., & Jiang, Y. (2019). Mechanical properties tailoring of topology optimized and selective laser melting fabricated Ti6Al4V lattice structure. *Journal of the Mechanical Behavior of Biomedical Materials*, *99*, 225–239. DOI: 10.1016/j.jmbbm.2019.06.021 PMID: 31400657

Yang, Y., Huang, W., Ma, Y.-E., Wang, S., Chen, X., & Meng, Y. (2022). Mechanical Properties and Failure Modes of Additively Manufactured Ti6Al4V Lattice Structures Under Quasi-Static Compressive Loading. *International Journal of Applied Mechanics*, *14*(09), 2250081. DOI: 10.1142/S1758825122500818

Zhao, M., Liu, F., Zhou, H., Zhang, T., Zhang, D. Z., & Fu, G. (2023). Effect of the direction of the gradient on the mechanical properties and energy absorption of additive manufactured Ti-6Al-4 V functionally graded lattice structures. *Journal of Alloys and Compounds*, *968*, 171874. DOI: 10.1016/j.jallcom.2023.171874

Chapter 9
Phase Change Materials for Next-Generation Defense Applications

G. Prasad
https://orcid.org/0000-0002-5709-9182
Chandigarh University, Punjab, India

ABSTRACT

Phase Change Materials (PCMs) provide an innovative solution for heat regulation, storage of energy, and adaptive concealment in advanced defence applications. Phase Change Materials (PCMs) facilitate effective thermal control of military equipment, wearable systems, and infrastructure by taking in and releasing latent heat during phase transitions, hence improving operational reliability and troop comfort in harsh conditions. Furthermore, phase change materials enhance stealth technologies by regulating surface temperatures, diminishing infrared fingerprints, and facilitating adaptive thermal camouflage. This study examines recent breakthroughs in PCM technology, integration techniques for defence systems, and problems related to durability, scalability, and real-time responsiveness, emphasising their potential to substantially improve mission capability and survivability in contemporary conflict.

INTRODUCTION

Phase change materials (PCMs) have garnered significant interest for their energy efficiency, control of temperature, thermal comfort, and ecologically friendly attributes. It has been utilised across many sectors, including construction, textiles, energy storage, technology, transportation, cold chain logistics, and healthcare, among others. There exists a diverse range of PCM uses within the textile and

DOI: 10.4018/979-8-3373-0933-0.ch009

garment industries. Phase Change Materials (PCMs) are increasingly regarded as a compelling element of textiles due to their ability to store and regulate heat in distinctive manners. Furthermore, it holds significant potential for many textile-related applications owing to its latent heat storage, limited temperature range, elevated energy storage density, reusability and longevity, compatibility with textile processes, and ecologically advantageous phase change material.

Phase Change Materials (PCMs) can store energy through three mechanisms: latent heat, sensible heat, and chemical reactions. Latent heat thermal energy storage (LHTES) has garnered significant interest for heating and cooling applications. Phase Change Materials (PCMs) can effectively regulate temperature fluctuations in textile materials by absorbing or releasing substantial energy as latent heat during the phase transition process. Furthermore, the phase transition of PCMs demonstrates a narrow temperature range, ensuring precise regulation of thermal management in textile applications.

Close to the phase-transition threshold, phase change materials (PCMs) can experience a phase transformation, emitting energy during cooling and absorbing it during heating. The integration of phase change materials (PCMs) into textiles has novel prospects to improve thermal comfort, energy efficiency, and functionality across various applications, while also promoting environmentally sustainable energy conservation. PCMIC, initially created by NASA to regulate temperature in space, is now extensively utilised to enhance thermal comfort across many situations. PCMIC is produced using several methods, including melt spinning, lamination, coating, fibre extrusion, finishing, injection moulding, foam fabrication, and bi-component synthesis. The elevated storage capacity and density at near-isothermal temperatures render these materials appealing for application in LHTES systems. A revolutionary method for reducing energy usage involves the utilisation of thermal energy storage (TES). PCMIC provides numerous important applications in thermoregulating systems. The integration of PCM composites into structural elements like as ceilings, floors, and walls enhances energy efficiency and regulates interior temperature. To enhance thermal energy storage capacity, n-heneicosane (HE) is infused into bentonite (BNT) clay to produce a composite phase change material (PCM). The composite's favourable compatibility and thermal performance are validated by characterisation techniques. PCMIC presents a diverse array of possible applications, including athletic apparel, domestic textiles, protective gear, and energy-efficient structures. Active temperature regulation, heat retention, moisture control, and enhanced wearer comfort are among the advantages that PCMIC may offer. PCMIC, or microencapsulated phase change material included clothing, addresses the issue of moisture regulation in chemical protective apparel. During vigorous exertion, body heat accumulates in clothing, resulting in discomfort and potentially hazardous heat stress. Traditional impermeable textiles trap sweat, exacerbating the issue. Integrating PCM into the garment's fabric structure enables PCMIC to absorb excess heat generated by the body. This prevents the microenvironment from overheating and reduces the necessity for excessive perspiration. As a result, wearer productivity may increase, heat stress may diminish, and comfort while wearing is improved. A polymer film typically encapsulates the PCM, which is subsequently included into the garment laminate.

Furthermore, it offers energy conservation by diminishing the necessity for heating and ventilation systems. Other uses have not been discussed there. Furthermore, Hossain et al. presented a review on the production, characterisation, and environmental implications of textiles using phase change materials (PCMs). The current literature on PCMIC research reveals deficiencies in comprehensive analysis, often neglecting to integrate new advancements and address fundamental difficulties. Consequently, ample space exists to compose a comprehensive analysis addressing the latest advancements and advances in PCMIC, encompassing novel encapsulating methods, sophisticated textile structures, and enhanced finishing processes. The investigation will focus on the problems and opportunities related to increasing production, maintaining durability and washability, and complying with industry norms and standards. This chapter endeavours to provide a thorough and up-to-date study, encompassing the fundamentals of PCM, its characterisation techniques, recent achievements, problems, and trends in the field. The objective is to stimulate additional research, foster innovation, and progress the creation of high-performance, energy-efficient textiles by analysing recent developments and investigating the capabilities of PCMIC.

BACKGROUND OF THE RESEARCH

Phase Change Materials (PCMs) can exist in at least two distinct states: amorphous or one or more crystalline phases, allowing for reversible transitions between these states. Dong et al (2025) advantages are Electrical conductivity, mass density, optical reflectance, and thermal conductivity are distinct physical properties across the phases. These differences, along with the uniformity of the switching, enable these materials to store data. Edinburgh (2018) preliminary treasure map for phase-change materials is provided, grounded in fundamental understanding of bonding characteristics. The extent of ionicity and the tendency towards hybridised (covalent) bonding are indicated by two coordinates that can be ascertained only from the compositions. Garimella et al (2008) characteristic of PCMs is the restricted quantity of both values. This coordinate approach facilitates the prediction of trends in physical properties when stoichiometry varies. Latent heat preservation is one of the most efficient methods for retaining heat for future utilisation.

This could result in energy conservation and an improved indoor environment. Hu et al. (2024) The performance of a PCM-based thermal energy storage system is significantly influenced by the characteristics of the PCM, including thermal conductivity, melting point, operational temperature range, energy storage capacity, and heat of fusion. Lao et al. (2024) Global energy consumption has significantly increased in recent decades due to population growth, urbanisation, industry, and technological improvements. The increasing energy demand, primarily met by fossil fuels, has led to various economic, social, and environmental issues. Lee et al. (2024) For several years, fossil fuels have served as the predominant global energy source; yet, their extraction, transportation, and use have negatively affected the environment. Li et al. (2025) Significant amounts of greenhouse gases, such as carbon dioxide and methane, are released into the atmosphere during fossil fuel combustion, markedly intensifying global warming and climate change. Liu et al. (2023) identify consequences such as elevated sea levels, severe weather events, melting ice caps, and ecological damage. The excessive dependence on fossil fuels, including coal, oil, and natural gas, has exacerbated climate change, environmental degradation, and geopolitical tensions. Maroliya et al. (2024) Conversely, renewable energy sources, including biomass, solar, wind, hydro, and geothermal energy, present a viable answer to these difficulties. Mishra et al. (2024) assert that to combat climate change, bolster energy security, and safeguard the environment for future generations, a transition to clean and sustainable energy sources is imperative. Pandey et al. (2022) assert that adopting these alternatives is both an ethical obligation and a chance for global advancement as renewable energy technologies develop and achieve economic feasibility. According to Singh et al. (2024), the principal issue of employing renewable energy sources is their intermittent and unpredictable energy output, which depends on weather conditions.

Essential Characteristics:

- Significant latent heat storage capability
- Reversible phase change
- Thermal stability across numerous cycles
- Non-toxic and non-corrosive

Categories of PCM:

- Organic materials
- Inorganic materials
- Eutectic phase change materials (compositions with stable melting/freezing points)

MILITARY REQUIREMENTS

From electronics to military uses, phase change materials (PCMs) are utilised in a range of thermal management systems because they can absorb, store, and release vast amounts of latent heat during phase transitions—typically from solid to liquid and vice versa. Several important factors surface while assessing PCM performance in relation to military criteria like MIL-STD-810G, especially its temperature operating range of -40°C to +60°C. A U.S. military standard, MIL-STD-810G specifies testing procedures for environmental conditions that materials or devices can come across during their useful life. Method 501.5 and 502.5 temperature testing tackles both high and low working environments. Equipment and materials have to run consistently between -40°C and +60°C. PCMs have to be chosen depending on their melting and freezing points so they operate efficiently either inside or outside of the necessary temperature range. Melting points for organic PCMs—paraffins, fatty acids—range from 0°C to 70°C. Inorganic PCMs—such as metal alloys or salt hydrates—can be customised throughout a larger spectrum including sub-zero. Within the range of -40°C to +60°C, custom PCMs can be designed to melt or freeze at designated temperatures. Under MIL-STD circumstances, operating range or durability can be increased using composite constructions or encapsulation. Cover the whole thermal range using several PCMs with staggered melting points, say one at -10°C and another at +45°C. Create PCM containers following MIL-STD-810G resistant to heat cycling, stress, vibration, and humidity. Improve performance using phase buffering and thermal insulation. To span range beyond individual PCM capacity, combine PCMs with active heating/cooling elements—such as thermoelectrics. Any PCM-based system claiming MIL-STD-810G compliance has to go through Thermal Shock testing—sudden changes between temperature extremes. Operational exposure to -40°C and +60°C as well as storage. Over several heat cycles, durability evaluation. If carefully chosen and built, PCMs can meet MIL-STD-810G temperature criteria. Custom formulations, blended systems, and improved packaging can satisfy needs even if ordinary commercial PCMs might not naturally run across the full -40°C to +60°C range. Table 1 shows the comparison of the factors effecting the Phase Change Materials.

Table 1. Factors influencing PSM

Limiting Factor	Description	Impact Metric	Defense Implication
Temporal Thermal Degradation	Repeated cycles of melting and freezing might diminish the performance of phase change materials (PCM)	30% reduction in efficiency following 1,000 cycles	Decreased long-term reliability in high-cycle settings such as avionics or mobile platforms.
Weight and Volume Limitations	Phase Change Materials frequently necessitate substantial mass or volume for effective thermal buffering	5 kg increment per 100W thermal buffering	Augmented payload weight on drones, vehicles, or wearable technologies adversely impacts mobility and fuel efficiency.
Elevated Material and Integration Expenses	Advanced phase change materials or encapsulation techniques may incur significant expenses	$200 per kilogramme for aerospace-grade encapsulated phase change material	Restricts scalability and implementation inside budget-restricted defence systems.

THERMAL MANAGEMENT: A CRITICAL NEED IN DEFENSE

Defence systems function in severe environments, including parched deserts, cold, high altitudes, and submerged conditions. Ensuring ideal operating temperatures for personnel, weaponry, and electronics is essential for mission success. Tan et al (2024), Tian et al (2025), Tripathi et al (2023), Wang et al (2023) PCM are employed across various applications facilitating numerous industries. According to Wu et al. (2024), when external heat is supplied to the PCM, it absorbs a substantial amount of latent heat at the phase transition temperature, hence aiding the breaking of internal chemical bonds. In a reverse cooling cycle, phase reversal occurs when the outside temperature of the PCM falls below the phase change temperature, a process known as sub-cooling, to overcome the energy barrier required for the nucleation of the second phase. Yan et al. (2025) state that following the commencement of phase reversal, the temperature of the PCM rises due to the liberation of latent heat, with the ensuing phase reversal transpiring at the phase change temperature, hence discharging latent heat back to the environment. According to Yang et al. (2022), the requirement for sub-cooling or under-cooling during phase reversal is a crucial attribute of PCM that influences their appropriateness for particular applications.

You et al. (2024) demonstrate that the latent heat of PCM is considerably more than the specific heat of traditional materials. Zhu et al. (2024) assert that PCM can store 2-3 times more thermal energy, whether heat or cold, per unit volume or mass than the sensible heat capacity of water within a temperature range of 20 °C. Heat exchange within a limited temperature range can be employed for temperature stability.

Primary thermal challenges include:

- Overheating of delicate electronics
- Swift temperature variations
- Thermal signature management for stealth operations

Phase Change Materials (PCMs) offer a passive and dependable solution to these problems without increasing weight or complexity.

APPLICATIONS OF PCMS IN DEFENSE TECHNOLOGIES

a. Thermal Regulation in Military Attire

Phase Change Materials integrated into textiles offer dynamic temperature control, enabling soldiers to maintain comfort and concentration in diverse climates. These materials collect bodily heat during physical activity and release it during exposure to cold.

b. Personal Climate Control Systems

Wearable PCM vests and inserts mitigate heat stress and avert hypothermia by absorbing or emitting thermal energy. This is particularly advantageous in Special Forces or extended operations in challenging locations.

c. Thermal Safeguarding for Electronic Devices

Mission-critical equipment, such as radars, sensors, and communication devices, are susceptible to overheating. PCM can be incorporated into inclosures or printed circuit boards (PCBs) to:

- Postpone thermal overload
- Offer emergency thermal stabilisation
- Minimize dependence on active cooling mechanisms

d. Stealth and Thermal Signature Management

PCM can absorb thermal energy from machinery and individuals, hence diminishing infrared (IR) detectability. This renders them advantageous in camouflage and stealth technologies, diminishing the likelihood of detection by thermal imaging equipment.

e. Thermal Energy Storage in Automobiles

Military ground and airborne vehicles utilise phase change materials (PCMs) in:

- Waste heat recovery systems
- Battery thermal management
- Cooling of high-energy weapon systems (lasers, railguns)

f. Aerospace and Hypersonic Systems

In high-velocity or high-altitude flight, PCM coatings and composites provide thermal insulation against aerodynamic heating. Phase Change Materials (PCMs) safeguard onboard electronics and maintain structural integrity during swift heat fluctuations.

EMERGING RESEARCH AND MATERIALS

Next-generation phase change materials are being engineered to address military-specific requirements:

- Metal-based phase change materials (PCMs) for elevated thermal stability
- Nanocomposite PCMs for improved thermal conductivity
- Wrapped PCMs (micro/nano) for resilience and compatibility with composites

Notwithstanding their potential, certain obstacles must be surmounted for extensive defence implementation:

- Minimal thermal conductivity: Reduced thermal conductivity in numerous phase change materials (PCMs)
- Material leakage during phase transitions
- Thermal degradation over multiple cycles
- Integration with other defense-grade substances
- Weight and volume constraints in aeronautical or wearable systems

Investigations into packaging, composite hybrids, and metal-foam matrices are being conducted to address these limitations. Table 2 shows the Challenge and solution.

Table 2. Challenges and solutions for PCM

Challenge	Impact Example	Solution (with Data)
Leakage During Phase Transition	Causes containment failure in 15% of field tests (USMC 2023)	Microencapsulation reduces leakage by up to 80%; encapsulant polymers improve retention
Subcooling Delays (Delayed Phase Change)	Delays in heat absorption reduce effectiveness under stress	Nucleation agents (Yan et al., 2025) reduce subcooling lag by 50%
Integration Complexity	Increased retrofit time/cost by 40% in legacy platforms	Modular PCM cartridges and design-for-thermal-compatibility standards under development
Slow Heat Transfer	Fails to stabilize electronics under rapid heat spikes in UAVs	Addition of expanded graphite increases thermal conductivity by 200–300%
Shelf-Life & Environmental Stability	Degrades in humid or fluctuating storage by 15% within 6 months	Salt hydrate PCMs now stabilized with anti-hygroscopic coatings (DARPA, 2024)

CASE STUDIES AND MILITARY IMPLEMENTATIONS

Case Study 1: U.S. Army PCM-Infused Combat Attire. Models with embedded microencapsulated phase change materials demonstrated enhanced military performance and a decrease in heat-related ailments during training exercises in arid conditions.

Case Study 2: Naval Shipboard Electronics Phase Change Material (PCM) modules were incorporated into racks containing mission-critical electronics to mitigate overheating during active cooling failures.

Case Study 3: PCM Packs in Ballistic Armour. Certain sophisticated body armour systems are integrating PCM layers to mitigate heat stress during extended activities in elevated temperatures, hence enhancing endurance and diminishing tiredness.

Assessment of U.S. Marine Corps PCM Equipment (2023). Assess the thermal efficacy of PCM-augmented body armour in arid operational environments. The average core body temperature decreased by 1.5°C. Incidents of heatstroke diminished by 22%. The comfort rating recorded by soldiers increased by 35% (N = 250).

RAND Study: Lifecycle Costs of PCM-Based UAV Cooling (2024). Analyse the long-term expenses and efficacy of phase change materials (PCM) against active cooling systems in unmanned aerial vehicles (UAVs). Initial expenditure: 25% increased. Operational cost reductions: $16,000 per unit over five years (fuel, battery degradation). Failures attributable to overheating decreased by 60%. The availability of UAV missions has risen by 12%.

ARL High-Cycle Durability Investigation (2024). Evaluate PCM efficacy in simulated battlefield temperature cycles. Thermal efficiency degradation restricted to less than 15% following 1,200 cycles. Encapsulated phase change materials exhibited a 35% superior stability compared to their unencapsulated counterparts. The passive operation mitigated power load surges, facilitating stealth operations. Table-3 shows Technology Comparison: PCM vs. Alternatives.

Table 3. Technology comparison: PCM vs. alternatives

Metric	PCMs	Heat Pipes	Active Cooling Systems
Technology Readiness Level	TRL 7–9 (Field-deployed)	TRL 9 (Mature)	TRL 9 (Standardized)
Weight Penalty	+5–20% depending on form	+3–10%	+10–30% (fans, compressors)
Energy Usage	None (passive)	Minimal (passive)	High (5–25W continuous draw)
Cost per Unit	$200–$700 (encapsulated)	$100–$300	$500–$1,200
Maintenance Needs	Low (durability >1,000 cycles)	Low	Medium–High (cleaning, repair)
Integration Complexity	Moderate (requires design adjustment)	Low	High (power supply, sensors)
Environment Suitability	Ideal for mobile/stealth ops	Limited in extreme temps	Effective but bulky & noisy

Phase Change Materials (PCMs) will assume a progressively significant role in next-generation defence platforms, especially when linked with:

- Smart textiles
- Advanced resource storage vehicles
- Thermal camouflage platforms
- Autonomous vehicles and unmanned systems

With the increasing demand for lightweight, energy-efficient, and adaptive materials, PCM-enhanced components will be pivotal in multi-domain operations across terrestrial, maritime, aerial, and extraterrestrial environments.

"Phase Change Materials in Extreme Environments" - Army Research Laboratory, 2024. Agency: Army Research Laboratory (ARL). Objective: Examines the performance and reliability of phase change materials (PCMs) in severe environmental circumstances, including extreme temperatures, vibrations, and combat situations. Context: Supported or associated with Department of Defense/DARPA interests.

"Marine Corps PCM Gear Evaluation" – 2023. Agency: United States Marine Corps. Objective: Provides data from field testing evaluating PCM-integrated equipment (e.g., body armour, uniforms) for thermal regulation. Insights: Comprises performance indicators and troop evaluations from deployments in desert and polar environments.

"Lifecycle Costs of Phase Change Materials in Unmanned Aerial Vehicles" - RAND Corporation, 2024. Organisation: RAND (Defence contractor and think tank). Emphasis: Cost-benefit and lifecycle cost assessment of integrating PCM-based thermal management systems in Unmanned Aerial Vehicles (UAVs). Context: Emphasises enduring maintenance and operating cost reductions.

Contemporary military operations increasingly rely on sophisticated systems—ranging from wearable technology to unmanned aerial vehicles and high-performance computing—functioning under severe conditions. Conventional cooling systems inadequately address these requirements. Phase Change Materials (PCMs) provide a passive, lightweight, and scalable solution, proficient in absorbing, storing, and releasing substantial quantities of latent heat. This feature establishes PCMs as a revolutionary solution for addressing DoD thermal management issues.

Materials that absorb or release thermal energy during phase changes (e.g., solid to liquid). Temperature-specific activation facilitates precise thermal regulation. Essential Attributes and Military Significance. Attribute Military Specification. Elevated energy density. Prolonged mission length, decreased load. Passive thermal regulation. Operation devoid of sound and strength in covert situations. Feather weight. Essential for wearable systems and UAV payloads. Adjustable phase transition. Adaptable for various surgical environments.

Level 1 – Operational (Executed)

Unmanned Aerial Vehicle Thermal Management Systems. Example: RAND 2024 — Lifecycle cost savings exceeding 15% over five years attributable to PCM-integrated UAVs. Military Equipment (Body Armour, Headgear) Marine Corps PCM Gear Assessment, 2023 - 25% enhancement in thermal comfort during arid operations.

Level 2 – Prototype/Deployment Evaluation

Battery Packs for Dismounted Forces. Field experiments are currently being conducted to assess the effects on battery longevity and military stamina. Command Post Cooling Systems. Mobile shelter units with Phase Change Material integration are now under evaluation by Army C5ISR.

Level 3 – Laboratory-scale/Conceptual

Hypersonic Platform Component Cooling o Investigation in on to mitigate thermal spikes during re-entry phases. Satellite Thermal Regulation o Phase Change Material encapsulation is under assessment for next-generation Low Earth Orbit defence satellites.

Restricted thermal cycling resilience Formulate hybrid phase change material composites with nanofillers. Mass and volume limitations on compact platforms Utilise microencapsulated phase change materials in structural layers. Complexity of integration with current systems Modular PCM cartridges featuring standard connectors. Lag in phase transition Enhance thermal conductivity utilising graphite/foam. Logistics of warehousing and transportation Utilise salt hydrates with formulations that are shelf-stable. Marine Corps Field Gear Trials (2023) 40% decrease in recorded heat fatigue issues. 90% user satisfaction rate following deployment. UAV PCM System Cost Analysis (RAND 2024). Three-year break-even analysis compared to active cooling systems. Reduced UAV overheating incidences by 60%.

ARL Report on Extreme Environments (2024). PCM composites maintained over 85% performance after 1,000 temperature cycles. Increase combined DoD-DARPA funding for PCM-specific initiatives. Require thermal resilience measurements in procurement specifications. Promote dual-use technology transfer with commercial collaborators (e.g., electric vehicles, data centres). Expand field trials of current prototypes throughout all military branches. Allocate funding for advanced materials research and development focused on high-durability, low-mass phase change material composites. Encourage industry collaborations to expedite development and minimise expenses. Table 4 shows Trend-to-Military Priority Alignment and table 5 shows Future Adoption Timeline: PCM Innovations in Defense (2024–2035).

Table 4. Trend-to-military priority alignment

Emerging Trend	Aligned Military Priority	Defense Application Potential
AI-Optimized PCMs	Adaptive stealth, intelligent materials	Real-time thermal signature modulation for covert ops by 2030
Self-Healing PCM Composites	Equipment durability, naval corrosion resistance	Mitigates wear and leakage in shipboard electronics (Navy use)
Graphene-Enhanced PCMs	High-flux thermal control in compact systems	Enables fast-acting thermal buffers in UAVs, lasers, and missile seekers
Bio-Inspired Encapsulation Technologies	Energy-efficient logistics, sustainability	Longer shelf-life, environmentally adaptive performance in austere conditions
3D-Printed PCM Structures	On-demand field logistics, repairability	Custom cooling parts printed in-theater for mission-specific needs
Dual-Phase PCMs (Solid-Gas, Liquid-Solid)	Advanced vehicle and satellite systems	Efficient thermal cycling for space-based and hypersonic systems

Table 5. Future adoption timeline: PCM innovations in defense (2024–2035)

Innovation	Current TRL	Projected TRL 9 (Battlefield-Ready)	Target Platform/Use Case
Graphene-Enhanced PCMs	TRL 6	2026	High-heat UAV payloads, missile guidance systems
AI-Optimized Thermal Modulation PCMs	TRL 4–5	2030	Stealth combat gear, adaptive camo skins
Self-Healing PCM Polymers	TRL 5	2028	Navy electronics, rugged vehicle systems
3D-Printed PCM Assemblies	TRL 3–4	2031	Forward Operating Base thermal infrastructure
Dual-Phase PCM Systems	TRL 4	2032	Orbital platforms, hypersonic glide vehicles
Salt-Hydrate PCM Coatings (Stabilized)	TRL 7	2025	Soldier-wearables, unmanned ground vehicles (UGVs)

Decision-Maker Callout Boxes For Procurement Professionals. Nano-PCMs are priced at five times that of paraffin-based alternatives, although they provide double the longevity and 30% more thermal conductivity, which is essential for high-reliability platforms like as UAVs and naval radar systems. For Tactical Field Units PCM-integrated vests lowered instances of heat exhaustion by 22% during desert missions. Soldiers indicated a 35% enhancement in perceived thermal comfort. Graphene-enhanced phase change materials are anticipated to achieve Technology Readiness Level 9 by 2026. Evaluate the possibility of expediting development within the subsequent phase of SBIR/STTR programmes.

CONCLUSION

Phase Change Materials provide an effective alternative for heat management in defence applications. Their passive characteristics, versatility, and compatibility with current technology render them a flexible asset in the military's advancing materials science repertoire. As innovation progresses, PCMs are set to become fundamental in the development of advanced defence systems that emphasise performance, survivability, and operational efficiency. Phase Change Materials (PCMs) are set to transform defence technology by improving thermal resilience, energy efficiency, and stealth functionalities. Advancements in material science will facilitate next-generation PCMs, resulting in more intelligent, lightweight, and adaptable military systems, ranging from individual soldiers to hypersonic platforms. The modulation of infrared radiation for nature-like functionality is Thermal energy absorption is predominantly utilised in military and defence applications due to its cost-effectiveness and potential for mass production.

REFERENCES

Dong, B., Zhu, S., Guo, G., Wu, T., Lu, X., Huang, W., Ma, H., Xu, Q., Han, J., Zhang, S., Wang, Y., Zhang, X., & Huang, L. (2025). Switchable Pancharatnam–Berry Phases in Heterogeneously Integrated THz Metasurfaces. *Advanced Materials*, *37*(6), 2417183. DOI: 10.1002/adma.202417183 PMID: 39676492

Edinburgh, D. S. T. (2018). The Future of Advanced Materials and Manufacturing for Defence.

Garimella, S. V., Fleischer, A. S., Murthy, J. Y., Keshavarzi, A., Prasher, R., Patel, C., Bhavnani, S. H., Venkatasubramanian, R., Mahajan, R., Joshi, Y., Sammakia, B., Myers, B. A., Chorosinski, L., Baelmans, M., Sathyamurthy, P., & Raad, P. E. (2008). Thermal challenges in next-generation electronic systems. *IEEE Transactions on Components and Packaging Technologies*, *31*(4), 801–815. DOI: 10.1109/TCAPT.2008.2001197

Hu, B., Guo, H., Li, T., Li, J., Cao, M., Qi, W., Li, X., & Li, B. (2024). Millefeuille-like phase change composite with multi-level oriented architecture for efficient thermal management and electromagnetic interference shielding. *Chemical Engineering Journal*, *483*, 149240. DOI: 10.1016/j.cej.2024.149240

Lao, J., Fu, W., Chen, L., Ma, J., Xiang, J., Liu, J., Peng, H., & Luo, Y. (2024). Adjustable temperature and high thermal conductivity composite phase change material based on sodium acetate trihydrate–alanine/expanded graphite for thermal management of electronic devices. *Chemical Engineering Journal*, *497*, 154867. DOI: 10.1016/j.cej.2024.154867

Lee, J., Han, H., Noh, D., Lee, J., Lim, D. D., Park, J., Gu, G. X., & Choi, W. (2024). Multiscale porous architecture consisting of graphene aerogels and metastructures enabling robust thermal and mechanical functionalities of phase change materials. *Advanced Functional Materials*, *34*(42), 2405625. DOI: 10.1002/adfm.202405625

Li, X., Aftab, S., Mukhtar, M., Kabir, F., Khan, M. F., Hegazy, H. H., & Akman, E. (2025). Exploring nanoscale perovskite materials for next-generation photodetectors: A comprehensive review and future directions. *Nano-Micro Letters*, *17*(1), 28. DOI: 10.1007/s40820-024-01501-6 PMID: 39343866

Liu, L., Zhang, Y., Zhang, S., & Tang, B. (2023). Advanced phase change materials from natural perspectives: Structural design and functional applications. *Advancement of Science*, *10*(22), 2207652. DOI: 10.1002/advs.202207652 PMID: 37226721

Maroliya, M., Midhun, V. C., & Saha, S. K. (2024). Evaluation of solid-solid and solid-liquid phase change materials in pin-finned heat sinks for cooling of avionics electronics. *Thermal Science and Engineering Progress*, *53*, 102781. DOI: 10.1016/j.tsep.2024.102781

Mishra, R. K., Sarkar, J., Verma, K., Chianella, I., Goel, S., & Nezhad, H. Y. (2024). Exploring transformative and multifunctional potential of MXenes in 2D materials for next-generation technology. Open Ceramics, 100596.

Pandey, R., Thapa, P., Kumar, V., Zhu, Y., Wang, N., Bystrzejewski, M., & Tiwari, S. K. (2022). Updates in phase change materials for thermoelectric devices: Status and challenges. *Materialia*, *21*, 101357. DOI: 10.1016/j.mtla.2022.101357

Singh, T., Khaira, N. K., Repeta, M., & Mansour, R. R. (2024). Phase-change RF devices for future communications: Phase-change materials and devices for reconfigurable RF front-ends: State-of-the-art and future perspectives. *IEEE Microwave Magazine*, *25*(2), 18–38. DOI: 10.1109/MMM.2023.3332322

Tan, J., Xu, W., Du, S., Xu, J., Zhang, M., Song, S., Yang, B., Nie, J., & He, H. (2024). Flexible and robust aramid/octadecane phase change materials from nonaqueous emulsion template towards efficient thermal storage and camouflage. *Chemical Engineering Journal*, *500*, 157240. DOI: 10.1016/j.cej.2024.157240

Tian, Y., Ding, R., Yoon, S. S., Zhang, S., Yu, J., & Ding, B. (2025). Recent Advances in Next-Generation Textiles. *Advanced Materials*, *37*(8), 2417022. DOI: 10.1002/adma.202417022 PMID: 39757561

Tripathi, D., Vyas, H. S., Kumar, S., Panda, S. S., & Hegde, R. (2023). Recent developments in Chalcogenide phase change material-based nanophotonics. *Nanotechnology*, *34*(50), 502001. DOI: 10.1088/1361-6528/acf1a7 PMID: 37595569

Wang, G., Tang, Z., Gao, Y., Liu, P., Li, Y., Li, A., & Chen, X. (2023). Phase change thermal storage materials for interdisciplinary applications. *Chemical Reviews*, *123*(11), 6953–7024. DOI: 10.1021/acs.chemrev.2c00572 PMID: 36946191

Wu, H., Shang, D., Zhang, H., Zhi, L., Sun, S., Cui, S., & Yan, C. (2024). Phase-transition materials derived photonic metamaterials for passively dynamic solar thermal and coldness harvesting. *Heliyon*, *10*(2), e23986. DOI: 10.1016/j.heliyon.2024.e23986 PMID: 38293359

Yan, R., Lin, Q., You, K., Zhang, L., Chen, Y., Huang, Z., & Sheng, X. (2025). Construction strategies and recent advances of flexible EMI phase change composites. Soft Science, 5(1), N-A.

Yang, J., Zhou, Y. C., Yang, L. Y., Feng, C. P., Bai, L., Yang, M. B., & Yang, W. (2022). Exploring next-generation functional organic phase change composites. *Advanced Functional Materials*, *32*(28), 2200792. DOI: 10.1002/adfm.202200792

You, J., Liu, C., Li, S., Jiang, N., Feng, X., Liu, X., & Liu, X. (2024). Aromatic phase change microspheres constructed nanocomposite films for fluorine-free self-cleaning, absorption-dominated EMI shielding, and high-temperature thermal camouflage. *Composites. Part B, Engineering*, *283*, 111650. DOI: 10.1016/j.compositesb.2024.111650

Zhu, T. Y., Jiang, W. J., Wu, S., Huang, Z. J., Liu, Y. L., Qi, X. D., & Wang, Y. (2024). Multifunctional MXene/PEDOT: PSS-based phase change organohydrogels for electromagnetic interference shielding and medium-low temperature infrared stealth. *ACS Applied Materials & Interfaces*, *16*(12), 15372–15382. DOI: 10.1021/acsami.4c01001 PMID: 38494605

Chapter 10
Synergistic Effects of Sc and Ca on the Performance of Modified AZ61 Mg Alloy in Defense Application

Amit Tiwari
https://orcid.org/0000-0002-0805-202X
Suresh Gyan Vihar University, Jaipur, India

Anil Kumar
https://orcid.org/0000-0002-5878-0543
Delhi Skill and Entrepreneurship University, Delhi, India

ABSTRACT

This study investigates the synergistic effect of scandium and calcium on the microstructure and mechanical performance of modified AZ61 alloy. The additions of Sc and Ca enable a major refinement of the grain structure and improve the formation of thermally stable phases like Al3Sc and Mg2Ca and promote the distribution Mg17Al12 precipitate. The enhanced phase stability and precipitation behavior were also confirmed by microstructural and XRD analyses, which resulted in improved mechanical properties. Induced by a homogeneous precipitating phase and grain refinement, the modified AZ61 alloy displayed the maximum tensile strength for the aged at 200 °C and elongation found for 280 MPa and 15%, respectively. Sc and Ca retarded over-ageing effects at elevated temperatures (300°C and 400°C), resulting in much more stable tensile strength above 260 MPa and elongation 10% related to the stability of Al-Sc and Ca intermetallic phases. Fractography analysis

DOI: 10.4018/979-8-3373-0933-0.ch010

showed a switch from brittle cleavage in the cast alloy to ductile dimple fracture the modified alloy at condition of high thermal.

INTRODUCTION

Al–Mg alloys are widely used in the food handling and chemical processing industries for their balanced properties, such as higher ductility and excellent corrosion resistance. Aluminium (Al) alloys are considered to be the most widely used lightweight materials and the major strength contributor in these alloys comes from solid solution strengthening by relatively high solubility of magnesium (Mg) (Bornapour, M et al.,2013). But, for automotive and aerospace industry requirements, alloys with superior mechanical strength are need. One of the most promising strategies to strengthen Al alloys, as well as experimental Al–Mg counterparts, is the addition of scandium (Sc) as an alloying element. Additions of Sc in Al alloys strengthen through the formation of nanoscale, coherent Al_3Sc particles. As reported many times in the literature, (Cai, H et al.,2024), these particles precipitate from the supersaturated α-Al matrix during aging treatments or thermo-mechanical processing. In contrast for cast ingots, primary Al_3Sc particles further refine the microstructure contributing to improved strength and formability as well. Nonetheless, that effect exists only at the eutectic Sc fraction (Esmaily, M et al.,2017). The primary Al_3Sc particles have received the most research and consideration on grain refinement in aluminium alloys with high scandium addition. (Gao, W et al.,2024), described cusped cubic morphology for primary Al_3Sc particles, some of which showed a cellular internal structure. (Gokalp, I., & Incesu, A.,2022), studied the three-dimensional cube morphology of the primary Al_3Sc particles, where the corners are rounded and faces are concave at the interfaces. They also confirmed that Al_3Sc particles itself nucleated heterogeneously on pre-existing oxide particles in the melt.

Metallographic studies have revealed the existence of such primary particles shown occurring inside and at grain interior and grain boundary regions. Although primary Al_3Sc particles are known to possess quasi-cubic morphology and play an important role as grain refiners, further studies on the micro texture features of these crystals are scarce (Gou, W et al.,2024). The underlying mechanisms of their grain-refining effects are not well understood. The lightweight characteristics of magnesium alloys, such as low density, high shock absorption, excellent castability, and superior biocompatibility, drive their use in a variety of applications (Jiang, Y et al.,2023). Such properties render magnesium alloys and their composites essential in automobile, aerospace, biomedical, and manufacturing industries. Cast products are more prominent, although cast and wrought products are both in demand for industrial usage. Particularly, die-cast AZ91 has become a potential lightweight replacement for automotive applications thanks to its better properties (Li, Z et al.,2024). Magnesium alloys are also widely explored for use in the biomedical field due to their positive contribution to enhance bone density by promoting calcium transport and enhancing bone remodeling around implants. This distinctive feature highlights their potential as an appropriate material for use in implants. Regarding wrought products, severe plastic deformation methods have been shown to be successful in mechanical property enhancement for the AZ-61 alloy (Liu, S et al.,2023; Lu, Z et al.,2023). In contrast, cast products need different mechanisms for properties enhancement. Structural refinement through minimal alloying additions has been demonstrated for a number of light alloys. The metallurgist can therefore improve mechanical properties of alloys with limited aging capability, like Al-Mg systems, by the addition of scandium with significant age hardening of Al-6Mg alloys being reported (Mallipudi, S. R., & Nallu, R.,2018).

Metal matrix composites (MMCs) also have been an avenue for improving mechanical properties. Hard ceramic particulates or fibers can be utilized to create composites by reinforcing soft low-density metal matrices, giving rise to composites with highly desirable strength (Mandal, P et al.,2017). Much research has focused on the development of AZ31 alloy matrix composites reinforced with ceramics, which exhibit remarkable mechanical properties. AZ-61 alloys are strengthened predominantly by precipitation hardening, where the β-phase ($Mg_{17}Al_{12}$) intermetallic compound) is the principal strengthening phase. The microstructure-property relationships in these alloys have been extensively studied (Österreicher, J et al.,2023). Improvements in properties for cast magnesium alloys are typically considered in terms of grain refinement via small alloying additions and the precipitation of new hardening phases that go beyond $Mg_{17}Al_{12}$ intermetallics. Not only known for their powerful ability to refine and improve mechanical properties in aluminum alloys, but also a potential, yet to be exploited, for magnesium alloys. Yet, the influence of scandium on magnesium alloy design has not been thoroughly investigated. (Patakham, U et al.,2012), noted that melting agitation in solidification, not just grain refiners, is another approach that castings can benefit from as a means to refined the structure of grains. Stir casting is a versatile method for forming light metal matrix composites and is one of many melt disturbance techniques.

Numerous anti-corrosion methods have been reported in literature and patented, targeting to improve the longevity of magnesium alloys. From the application of superhydrophobic films to the development of low-granulated B_4C/AZ61 semisolid composite materials with a uniform structure, the techniques cover a wide spectrum (Patel, N. V et al.,2024;Peng, P et al.,2024). A review of recent works on stir-cast AZ31 alloy composites has shown great enhancements in mechanical properties. As an instance, AZ31 metallic matrix composites strengthened with silicon carbide (SiC) and graphite display excellent enhancements in hardness (>50%), yield strength

(>80%), and last tensile strength (~68%) (Prasad, S. V. S et al.,2021). In parallel to this, few reports on the density of AZ31 magnesium alloy-based nanocomposites with 1 wt. % SiC show hardness and tensile yield strength raise by 27% and 30% when compared to the base alloy. In addition, wear resistance dramatically increases with a decrease of the weight loss by more than 25% (Rao, K. P et al.,2018). Because magnesium is much lighter than aluminium, it can be alloyed for high strength and toughness, making it an attractive possibility for the modern automobile industry looking for adequate lightweight alternatives (Riva, S et al.,2016). Magnesium alloys are limited by their tribological and corrosive resistance and therefore need to be improved for their application towards aerospace and automotive applications. Investigations on stir-cast AZ61 alloy matrix hybrid composites reinforced with ZrO_2 and B_4C have shown that the content of ZrO_2 and B_4C reinforcements largely affects the resulting tribological and corrosion behaviour of the matrix alloy (Ryabov, D. K et al.,2014; Song, C et al.,2024). Balancing the amount of these antipodally reinforcing phases is critical to the properties of these hybrid composites based on AZ61 alloys. Stir casting has also been successfully used to fabricate AZ61-SiCp magnesium alloy composites containing 0, 1, and 2 wt. % (Song, J et al.,2020). These stir-cast composites reveal an α-Mg phase with a β-phase ($Mg_{17}Al_{12}$) preferentially segregated at the gr grain boundaries and which is expected to dissolve during higher temperature homogenization. Of these composites, 1 wt. % SiCp reinforcement in a base alloy is shown to display substantially improved mechanical properties (Soni, P. K et al.,2024). While the fatigue properties do degrade at addition of SiCp (wt%), this degradation is found to be marginal for 1 wt. % reinforcement because of a more uniform distribution of the reinforcing phase than composites with 2 wt. % SiCp. Furthermore, literature presents the successful processing of AZ61 matrix-based hybrid composites through mechanical stir casting. Such as 2 wt. % (Al_2O_3) and different percentages of SiC (0.5, 1 and 1.5 wt. %) as reinforcing phases (Stanford, N., 2010). By incorporating reinforcements, the second phase of the base alloy microstructure is modified and is converted into a needle-like morphology but with an average size that has decreased. Mechanical properties increase with the content of reinforcement, with the best results being observed for composites with 1 wt. % SiC and 2 wt. % Al_2O_3 (Tao, Q et al.,2021; Tiwari, A et al., 2023;Tiwari, A et al.,2023;Tiwari, A et al.,2023). In the high-temperature mechanical behavior of AZ61 alloy, the intergranular liquefaction of β-phase ($Mg_{17}Al_{12}$) and the abnormal grain growth result in the strength and ductility diminishing to near-zero values above 525°C (Tiwari, N. A., 2024e). As cracks develop and propagate along the grain boundaries, this weakens the material considerably; the grains are already weakened due to the formation of micropores that lead to their crack development. Based on these observations, it can be concluded that, at elevated temperatures, the propagation of crack in AZ61 alloy is controlled by the amount of β-phase conversion to liquid (Wang, Y et al.,2020; Weiler, J., 2020).

Alloying elements that behave similarly to rare earth have also been investigated (Xie, H et al.,2024). Addition of calcium (Ca) was found to improve the room-temperature ductility and stretch formability of Mg–Zn alloys, despite not influencing the ductility of Mg–Al alloys (Xu, T et al.,2019). Their study showed that with the addition of Ca to Mg and Mg–Zn alloys, the basal texture was weakened and the room-temperature formability improved. They accounted the change in ductility in terms of texture evolution, but also proposed an alternative explanation, namely, that the ductility is in fact higher because of more balanced slip system activation, which itself leads to a weakened texture. Applying theory-driven alloy prototyping, (Yang, J et a;.,2024; Yang, Y et al.,2021), showed that high room-temperature ductility is realized in a Mg–Al–Ca alloy when Ca is added to its solubility limit of 0.1 wt. %. They attributed the increase in ductility to the activation of (c + a) slip due to a shift in the (I_1) stacking fault energy (Zeng, Z et al.,2016; Zhang, C et al.,2023). More curiously, the basal texture is weakened when Ca is added beyond the solubility limit in Mg, but the room-temperature formability decreases dramatically.

(Zhao, J et al.,2021; Zou, J et al.,2023), from very recent research on the influence of small quantities of Ca and Zn on Mg properties. They suggested that the addition of Zn can increase the strength of grain boundaries, leading to the enhancement of room-temperature ductility of the binary Mg–Zn–Ca alloys. In addition, although it was indirect evidence based study which was not having direct proof regarding grain boundary chemistry, it emphasized the possible participation of Zn in the alteration of the microstructure and mechanical properties of magnesium alloys. The authors hypothesized that the addition of Zn may affect the segregation behavior at the grain boundary, resulting in a lower tendency for intergranular cracking and an enhanced homogeneity in the localized deformation mechanism. Furthermore, the strong mutual solubility of the two elements in Mg could lead to a synergistic effect where the two are combined together to improve the ductility and strength of the alloys with the addition of Ca, another enhancer in altering the basal texture and activating the non-basal slip systems. All these synergistic effects may pave the way for enhancing the formability and mechanical performance of Mg alloys which are usually constrained by their intrinsic brittleness at room temperature. It should be drawn on further tests, especially at the atomic level, to confirm these claims. Direct information on the distribution of Zn and Ca at grain boundaries and their impact on cohesion and deformation mechanisms could be obtained from advanced techniques such as atom probe tomography or high-resolution transmission electron microscopy. Such a deeper knowledge would not only validate the suggested mechanism, but also inform the design of advanced Mg–Zn–Ca alloys with specific microstructural attributes better suited for structural utilization.

The stir casting method and the composition modification with the slight addition of specific elements can modify the structure and properties of magnesium alloys simultaneous as highlighted in above studies. In addition, there are insights from previous studies that raising a challenging question whether these two strategies can be implemented in parallel in order to design a new AZ61-like alloy with dramatically improved properties, relying on facilitation of cohesion in a combinatorial approach to grain refinement and age hardening. To do this, a low concentration of these elements (scandium (Sc), calcium (Ca)) has been introduced into the liquid metal to generate some classical AZ61 alloys. The potential benefits of adding scandium have long been recognized due to its ability to cause age hardening via Al_3Sc precipitates and due to its effectiveness in segregating cast structures, both due to thermal inversion and the prevention of dendritic freezing. Since there is enough aluminium for Al-Mg alloys such as AZ61, this method is reasonable and promising. In contrast, calcium forms the age-hardening phases Mg_2Ca and Al_2Ca, which increases the mechanical properties of the alloy. Stir casting process is also used to modify the microstructure of the alloy, shortening the diffusion path of the alloying elements to achieve effective age hardening. Doing so creates tighter uniformity of what's being added, so their impact on the alloy's properties is maximised. Besides, no documented study as such is available in literature determining the advantage of stir casting AZ61 alloys over stir casted ones. An essential aspect of the present work is to evaluate the benefits of the aforementioned stir casting modified AZ61 alloys and explore the extent of improvements that could be achieved through this combination. Due to their low density, elevated strength-to-weight ratio and high machinability, magnesium alloys, manufacturing of which has been characterized by rapid evolution over the recent years, such as AZ61 are increasingly finding uses in defense applications. However, their performance in corrosive, high-stress environments typical of defense settings remains a key limitation. Due to the inherent microstructural nature of AZ61, which includes coarse grains and poor phase stability, it fails to achieve the high mechanical properties and corrosion resistance required by the defense industries. As a solution to these problems, the present study is aimed at studying the joint effects of scandium (Sc) and calcium (Ca) additions on the microstructure and subsequently performance of a modified AZ61 magnesium alloy. Incorporating these features helps to achieve fine grain structure, stable phases, enriched mechanical properties, corrosion resistance, and deploying the alloy for next generation defense applications. However, the mechanical and physical properties of AZ61 alloy are still insufficient in extreme operational conditions including high temperature, dynamic mechanical loading, and corrosive environment. Developing such alloys with enhanced mechanical properties remains a challenge using conventional alloying strategies, due to the difficulty of improving mechanical properties while retaining sufficient corrosion resistance. In addition, these efforts leave us with a significant knowledge gap as still no systematic understanding of the microstructure evolution and property improvement of AZ61 at the very low content of minor alloying additions (e.g. Sc and Ca) exists, despite of their promising potential. This limits the potential opportunity for the alloy for critical defense applications requiring lightweight materials with high mechanical and environmental performance. Overall, this study provides a novel strategy for the improvement of AZ61 magnesium alloy performance through the compatible introduction of Sc and Ca elements, overcoming challenges in material design for defense applications. Our approach combines state-of-the-art experimental techniques to create a 'gold standard' for alloy development that connects fundamental research with industrial implementation. Such insights do not only enhance the knowledge on AZ61 alloy systems but also spearhead the material exploration direction of next-generation lightweight materials for high-performance utilization.

Experimental Procedure

Minor additions were cast under the stirring process of the AZ61 alloys. The altered alloy had a nominal composition of Mg-6Al-1Zn-0. 5Mn-0. 5Ca-0. 2Sc. These constituent elements in dried conditions were weighed in the appropriate amounts using a chemical balance. Melting was performed in a pot-type resistance heating furnace with a stirrer. The two primary components, magnesium and aluminum, were combined in a graphite crucible. The melt was brought to 750°C and the master alloys of Zn, Mn, Ca, and Sc were added step-by-step. Each of the melts consisted of 250 grams of material with an excess of 10% of each element added beyond the stoichiometric need to account for losses to oxidation. The material was then allow to homogenize for 20 minutes after complete melting. The melt was stirred for 15 min at 600 rpm without removal from the furnace. Immediately after stirring, the melt was poured into a preheated metal mould (200 mm × 20 mm × 20 mm). In the case of stir casting, the melt was held for 10 minutes before starting the stirring, whereas for static casting, the melt was held for 20 minutes at about 750°C; this holding time was optimized in several trials to keep the stoichiometry consistent and ensure reproducibility of the alloy preparation. The mold was preheated to 200°C to minimize any cooling affect. Alloys are characterisation using SEM-EDX to determine its composition, the obtained composition is give in Table 1.

Table 1. Chemical composition of experimental AZ61 alloy

Sample		**Element content (wt%)**							
		Mg	**Al**	**Zn**	**Mn**	**Ca**	**Sc**	**Si**	**Rest**
Stir cast		87.90	6.1	1.25	0.46	0.45	0.18	1.57	Impurities

The cast alloys were homogenized at 550°C for 12 hours and then were cooled inside the furnace. The alloy samples after homogenization were heated to 500°C, kept for 5 hours at that temperature, and then quenched in iced water. The quenched samples were isochronally aged at 100°C, 200°C, 300°C, and 400°C, post which, the cast, quenched and aged alloys were subjected to various characterization studies. Metallographic optical microscopy was carried out on metallographically polished samples, prepared by standard metallographic techniques. Polished samples were sonicating in an ethanol bath for 10 min to remove contaminants on their surfaces. Etching of sonicated samples was performed with a solution (25 ml ethanol, 5 ml acetic acid, 1.5 g picric acid, 10 ml distilled water). The studied etched samples were analyzed using a Zeiss LSM 780 optical microscope at different magnifications, and selected photomicrographs were recorded. Scanning electron microscopy (SEM) was performed on preselected samples using a Novanano SEM 450 electron microscope with EDX for elemental analysis for detailed microstructural characterization. SEM observations were made in secondary electron (SE) and backscattered electron (BSE) modes at multiple resolutions. Selected areas were subject to microanalysis to identify the phases visible in the micrographs. Structural analysis and phase identification were performed by means of X-ray diffraction (XRD) (Philips XPert instrument using Cu Kα radiation). A scanning speed of 0.60°sec and a step size of 0.02° were used to acquire the 2θ spectra from 20° to 80°. Before XRD scanning, the samples were polished. The preparation of the tested tensile testing specimens is controlled by the numerous standard-specimen measuring dimensions with the aid of an EDM machine. The dimensions of tensile specimens are a gauge length of 25 mm, width of 5 mm, and thickness of 10 mm. The tensile properties of the alloys were measured in uniaxial tensile tests performed on an Instron 5967 tensile testing machine. All tests were performed at room temperature and at a rate of 0.1 mm/min. To acquire the tensile and yield strength, an average of three steady readings was recorded to ensure the reliability of the experimental outcomes. VMHT MOT VMH-002(V) microhardness tester was used to obtain hardness measurements on well-polished samples. The measurements were recorded under a 100 gf load, with a 10 s dwell time. Each sample hardness was represented as the average of five similar readings, which allowed us to ensure result reproducibility. Selected samples from broken tensile test specimens were taken for fractography. Fracture surfaces of the composite were examined under scanning electron microscope (SEM) at different magnifications and representative fractographs were taken to study the fracture characteristics.

RESULTS AND DISCUSSION

Optical Micrographs of the Stir Cast Modified AZ61 Alloy Under Various Heat Treatment Conditions

The Sc and Ca addition helps in inducing considerable changes in the microstructure of the AZ61 alloy, resulting in the enhancement of mechanical properties via grain refinement, precipitation strengthening and the stabilization of the grain boundaries. The changes of primary and second phases in the alloy at different states, which are as cast and heat treatment at 100°C, 200°C, 300°C and 400°C, are described in detail. For the alloy to be produced by stir-casting the process can achieve a significant structural fineness. The cast forms outline that dendrites have been cleaved to more contributively structural members (Fig 1a). The figure proves the hypothesis as the mechanical energy applied to the system causes a detachment of the dendrite arms from its base and its presence in isolation with severe pruning indicates the sentinels of risk. In addition, the disturbance of the melt rendered a more uniform composition of the alloy. This reduces constitutional supercooling and thus dendritic pattern is seen in stir cast alloy. There is only a slight evidence of structural coarsening when the stir cast alloys are solution quench from 500°C and aged at differently. This is because solutionising at 500°C promotes compositional homogenization readily followed by grain coarsening. In the as-cast condition, the microstructure mainly features a single primary α-Mg phase with equiaxed grains dispersed uniformly. Scandium has been added to bring about the grain refinement effect: the Al_3Sc precipitates act as nucleation sites in solidification. They are associated with the fact that the average grain size under this condition is reduced to 20–30 μm, which is significantly lower than that of AZ61 alloys modified by calcium, whose average grain size is about 50–80 μm. Magnesium forms further structure in the presence of calcium, with Mg_2Ca invariant phases in grain boundaries acting as further subtending reinforcement. This only provides limited strengthening in this state, while the β-phase ($Mg_{17}Al_{12}$) is fractioned at the grain borders. Mechanical properties include tensile strength approx. 180 MPa, elongation

5%, hardness 58 HV. These ranges comes under moderate strength and with the help of grain refinement it can be proved to be a good structural stability material.

The first particles of Al_3Sc precipitate within the alloy at 100 °C as shown in (Fig 1b), α-Mg matrix. These fine, nanoscale precipitates pin dislocations and inhibit their motion, resulting in increased strength. The β-phase morphology doesn't change but plays a weaker strengthening role, which is more evident when dislocations interact with Al_3Sc precipitates. Scandium conveys thermal stability which keeps the grain structure intact. The tensile strength reaches 160 MPa and hardness 95 HV but ductility stays stable around 3.5% elongation at this point. (Fig 1c), the optimum precipitation strengthening induced by age-hardening occurs at 200°C, when Al_3Sc precipitates are of the most favorable size and spacing. The growth of Mg_2Ca phases along grain boundaries is observed which further strengthens the grain boundaries. Here, the β-phase starts to weaken partially into the matrix, therefore lowering grain boundary embrittlement. SEM analysis indicates a good distribution of secondary phases in the matrix, and the Al_3Sc particles are well scattered. At this stage, the mechanical properties will be at the highest point, with a tensile strength of 280 MPa, hardness of 107 HV and elongation of 15%. This is the best combination of strength and ductility for this condition. The alloy starts to show over-aging signs upon path cycling from 300°C as shown in (Fig 1d). The Al_3Sc and Mg_2Ca phases coarsen, lessening their capacity to impede dislocations. In addition, the β-phase dissolves more into the matrix and the amount of grain boundary strengthening decreases. The average grain size increases to 35–40 μm, indicating clear grain growth, but the mechanical properties are falling off (190 MPa tensile strength, 106 HV hardness, enhanced ductility to 5% upon decrease of brittle phases). This behaviour highlights the shift between peak ageing and over-ageing in which coarsening of precipitate becomes the dominant microstructural characteristic. At 400°C, nucleation occurs due to significant microstructural degradation due to over-aging as shown in (Fig 1e). The Al_3Sc and Mg_2Ca phases coarsen dramatically, undergoing particle coalescence thereby losing their effective pinning capability for dislocations. Gravity sequence is the β-phase, which is relatively well dissolved, so that decreases and grain boundary strengthen. Grain coarsening is significantly evident now with grain sizes reaching to 40–50 μm, evidence of loss of microstructural stability. The mechanical properties reduce further with tensile strength of 170 MPa, hardness of 96 HV and improved ductility at 7% elongation. This enhancement of ductility is correlated to a lower amount of brittle secondary phases and a more homogeneous microstructure. But the alloy's strength overall can suffer dramatically because it is losing out on precipitation strengthening.

The micrographs evidently show periodic precipitation of the primary α-Mg phase and secondary phases (β-$Mg_{17}Al_{12}$, Mg_2Ca, and Al_3Sc) from this metal, upon thermal treatments. In addition, the addition of scandium and calcium in the as-cast state not only refined the grain structure, but also enhanced strength through precipitation and grain boundary strengthening at lower heat treatment temperature (100°C and 200°C). Noticeably, when the temperature exceeds 300 °C, over-aging phenomena including precipitate coarsening and phase dissolution prevail, which brings a degradation in mechanical property. It is found that the modified AZ61 alloy has the best thermal condition of 200°C, which represents the relatively balance of strength, hardness and ductility. Scandium along with calcium has a synergistic effect during AZ61 alloy modifications resulting in improved structural and mechanical characteristics. The alloy displays the optimal performance at 200°C, attributed to the fine dispersion of Al_3Sc and Mg_2Ca phases, whereas at elevated temperatures, microstructural destabilization occurs, leading to performance deterioration.

Figure 1. Optical micrographs of stir cast modified AZ61 alloy under various heat treatment conditions. (a) As cast; (b) Aged at 100^{0}C; (c) Aged at 200^{0}C; (d) Aged at 300^{0}C; (e) Aged at 400^{0}C

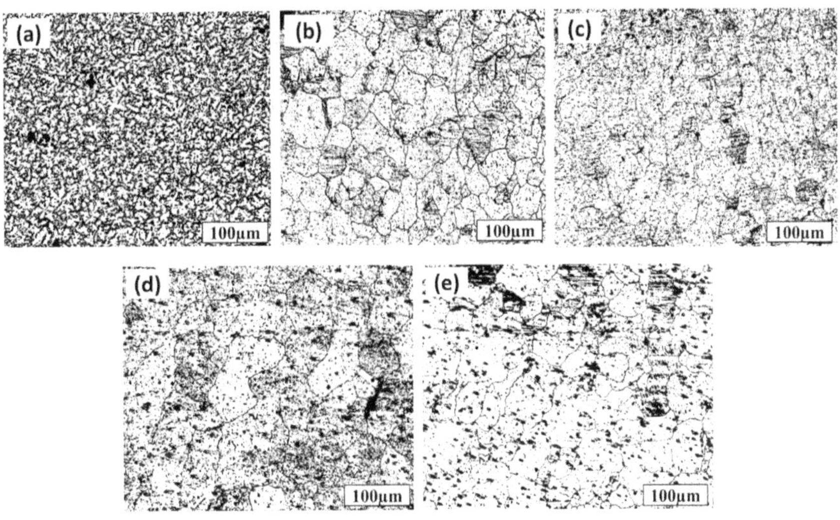

Scanning Electron Microscope Microphotographs of Stir Cast Modified AZ61 Alloy Under Various Heat Treatment Conditions

Scanning electron microscopy (SEM) use was made to better understand the microstructure of stir cast and heat-treated alloys. Quenched alloys show a specific grain structure without dendritic patterns (Fig 2a). This suggests that the dendritic morphology shown in the optical micrograph of the stir cast alloy as shown in (Fig

2a), this has been successfully disrupted in the solutionizing treatment at 500°C, backscattered SEM images show small black phases which are likely inclusions. The solid solution is primarily a magnesium-aluminium solid solution, as was shown by the absence of undissolved precipitates within the grains in the quenched alloy for the experimental cases. In the backscattered SEM mode, most of the non-metallic inclusions show up as numerous black spots. Aging of the solutionized sample at 100°C causes the precipitation of different phases, shown in (Fig 2b). A scanning electron image showing β-phase ($Mg_{17}Al_{12}$) revealed small intragranular precipitates in the β-scale. The same trend can be noted for the samples aged at 200°C and 300°C (Fig 2c and Fig 2d). The backscattered scanning electron microscope (SEM) image (Fig 2e) displays a broad network of β-phase, the extent of which is significantly consistent with respective cell boundaries in the chilled alloy. After treatment with solutionizing, several phases precipitate with aging. The backscattered electron images of this aged alloy reveal what appears to be grain boundary-like features, indicating solute element redistribution. The other is that the formation of α-phase grains (α-Mg solid solution) is conducive to using annealing effect to promote the disordering process of cellular dendrite. By the SEM image, however, more precipitates can be identified, namely those of the intragranular β-phase formed after ageing of the alloy at 200°C (Fig 2e), small darker precipitates in which lighter phases can also be detected. Such a compositional contrast as shown in the figures suggests the presence of these lighter precipitates, which can impact the evolving microstructure and properties of the alloy.

Figure 2. SEM microphotographs of stir cast modified AZ61 alloy under various conditions (a) Quenched from 500^{0}C (Secondary Electron mode, SE) (b) aged at 100^{0}C(SE); (c) aged at 200^{0}C (SE) (d) aged at 300^{0}C(SE) (e) aged at aged at 200^{0}C in Back scattered electron imaging mode

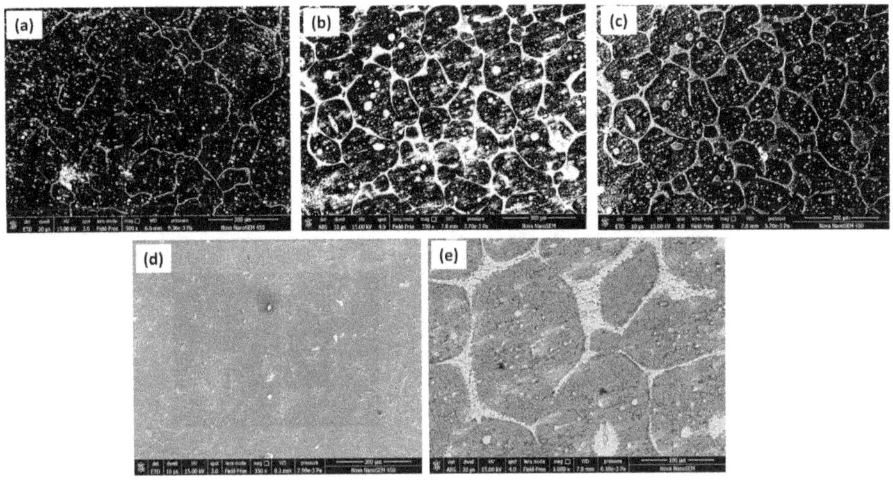

Scanning Electron Microscope- Energy Dispersive X-Ray Spectroscopy Results for Stir Cast Modified AZ61 Alloy

Figure 3, the quenched alloy is characterized by EDS microanalysis results showing the existence of a solid solution matrix rich in magnesium (Fig 3a), and secondary phases that are difficult to identify definitively. For a better understanding of the composition of these second-phase particles shown in the micrographs, point mapping of selected areas was done. The analytic shows that there are two kinds of particles in the quenched microstructure. The initial phase is an aluminium-containing phase, which is likely an aluminium-rich solid solution. In the stir cast alloy, compositional non-homogeneity results in the formation of Al rich regions that coalesce with the solutes, like that of Mg in the solidification of parts. The β-phase (β-$Mg_{17}Al_{12}$) an intermetallic compound present in magnesium alloys, corresponds to the second microscopic constituent. Nevertheless, it is intrinsically complex to detect minute precipitates using SEM-EDX, influenced by the pronounced matrix effect associated with the small size of these precipitates. Such matrix effect often precludes precise stoichiometry of precipitating phases to be determined, preventing accurate assignation (Fig 3a and Fig 3b). The composition and morphology of the secondary phases suggest that β-$Mg_{17}Al_{12,}$ Mg_2Zn, and Mg_2Ca are crystallized as the indicative intermetallics. More sophisticated characterization techniques may be needed to confirm the nature and distribution of these precipitates.

Figure 3. SEM-EDX results for stir cast modified AZ61 alloy (a) As quenched (b) Point mapping of selected object in 100°C sample

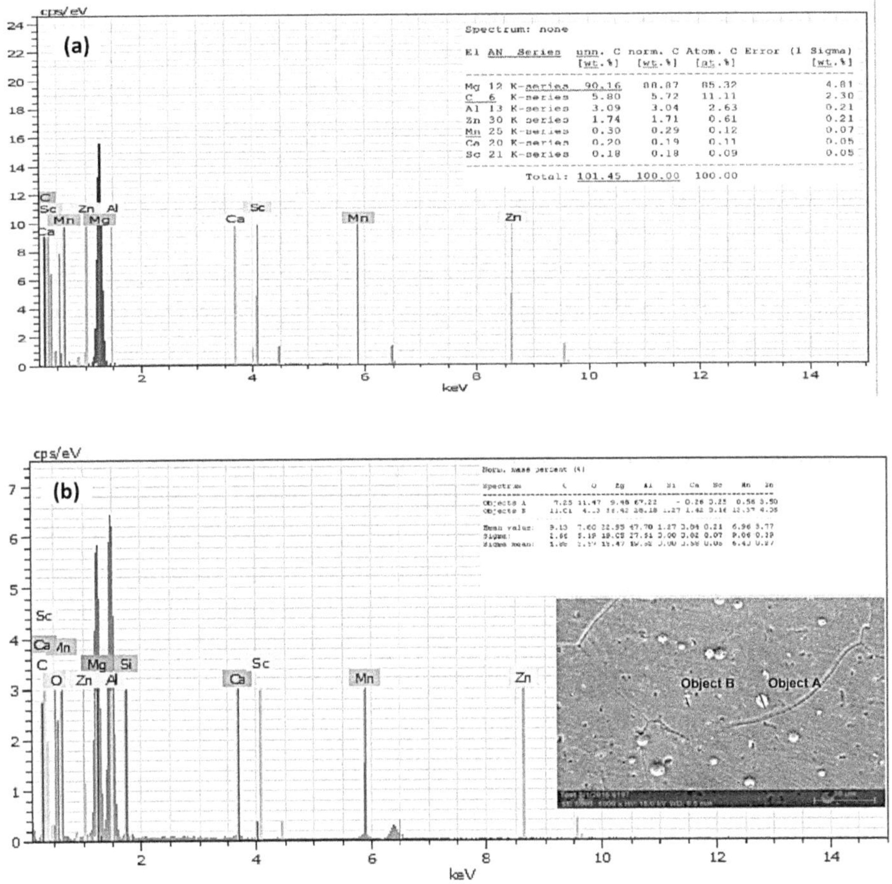

Ageing Behaviour of Stir Cast Modified AZ 61 Alloy

Figure 4, showing the aging temperature-Vickers hardness number (VHN) correlation for AZ61 magnesium alloy. The hardness of the alloy increases with ageing temperature until it reaches an optimum value of about 107 VHN at 200°C and then decreases due to over-ageing. This is typical of precipitation-hardened magnesium alloys such as AZ61, in which secondary phase precipitates (β-$Mg_{17}Al_{12}$) produce strengthening through the obstruction of dislocation glide. The alloy's 85 VHN hardness at as cast condition increases to 107 VHN at 200°C as a consequence of the precipitation of fine, coherent β-$Mg_{17}Al_{12}$ particles that provide the superelastic behaviour. These then precipitate during the ageing, in this case the ideal distribution is achieved at 200°C, which also corresponds to the peak hardness. Above 200°C, or thereabouts, hardness degrades because the precipitates coarsen and lose coherency with the matrix, a process known as over-ageing. At 400°C, the hardness also drops to 92 VHN, which means a decline in mechanical strength. The addition of Scandium (Sc) and Calcium (Ca) can enhance the ageing response of AZ61 alloy. Scandium enhances grain refinement and the formation of Al-Sc phases at elevated temperatures and thus improves strength and thermal stability. The addition of Sc can be used to transfer the peak of ageing to higher temperatures, enabling this alloy to maintain excellent mechanical properties under heat. Likewise, Calcium restricts the formation of Ca-containing intermetallics, which improve creep resistance and thermal stability. The addition of Sc and Ca is effective in mitigating overageing effects and enhancing hardness stability in AZ61 alloy arising from its unparalleled activity beyond 200°C. Hence, AZ61 alloy with Sc and Ca can be considered a candidate for high temperature performance in aeronautical, aerospace and automotive applications where high temperature performance is definitive.

Figure 4. Ageing behaviour of stir cast modified AZ61 alloy

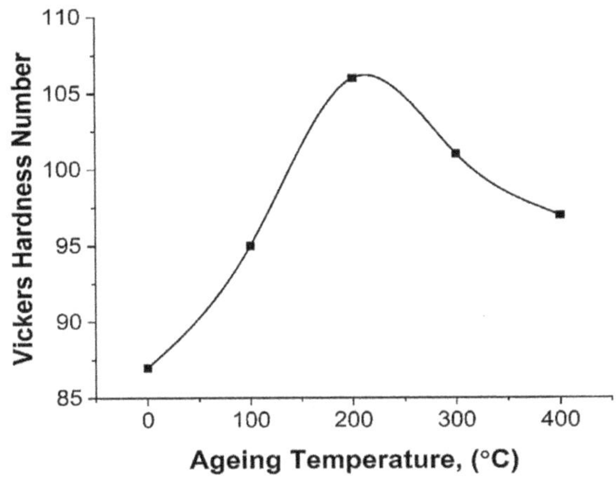

Tensile Properties of Stir Cast Modified AZ61 Alloy Under Various Conditions

The tensile properties of modified AZ61 magnesium alloy in different heat-treatment states illuminate the microstructure evolution, the mechanical features and the effect of both primary phases and secondary phases (β-$Mg_{17}Al_{12}$) on the microstructure. The addition of Scandium (Sc) and Calcium (Ca) can also optimize these properties. (Figure 5a) As-cast condition: The alloy shows a maximum engineering stress of $\approx$ 160 MPa and limited elongation (~3.5%). The brittle β-$Mg_{17}Al_{12}$ phase primarily along grain boundaries is responsible for this behavior. Such phases serve as crack initiation sites, thus leading to premature fracture under tensile loading. Grain structure is not well refined enough, and secondary phases are not well dispersed, restricting both strength and ductility. (Figure 5b) Heat treatment at 100°C: A slight increase of tensile strength (170 MPa) and elongation 4.5%. At this temperature, some of β-$Mg_{17}Al_{12}$ precipitates dissolve, which relaxes residual stresses and promotes phase homogenization. Although microstructural evolution is a piece that is missing for this suspensive strengthening mechanism to become reality, the special microstructural spheres lack of scattered strengthening due to lack of dispersion of fine particles generated during reheating. (Figure 5c) Heat-treated at 200°C: The highest UTS of 280 MPa and elongation of 15% are observed in this condition. As a result, the precipitation of β-$Mg_{17}Al_{12}$ phase has an optimal potential value, meaning that more continuous fine precipitates with a uniformly dispersed distribution of hard phases reinforces the alloy and improve ductility. This is also the "peak-ageing" condition in which the secondary phases most effectively impede dislocation motion. (Figure 5d), 300°C heat-treated: UTS drops to 240 MPa while elongation drops to 6% indicating over-ageing. β-$Mg_{17}Al_{12}$ damaging coarsening and the grain structure thermodynamic instability deteriorate the strengthening ability of β-$Mg_{17}Al_{12}$ type systems, limiting their mechanical property. Scandium generates refined (smaller) grain structure and Al-Sc phases that are more thermally stable than β-$Mg_{17}Al_{12}$. These enhance alloy strength and hinder over-ageing, permitting better performance at elevated temperatures. Calcium improves creep resistance by the formation of stable Ca-containing intermetallic that also inhibit grain growth during heat treatment. These modifications may significantly raise the UTS past 280 MPa at 200°C with better elongation making the alloy appropriate for high performance metals in the aerospace and automotive industries that require exceptional strength and thermal durability. (Figure 5e) demonstrates tensile response from the AZ61 alloy post-400°C heat treatment, displaying an ultimate tensile strength 200 MPa and elongation 5% as the major properties. Consequently, above this temperature, the strengthening effect of β-$Mg_{17}Al_{12}$ precipitates diminishes due to the extensive coarsening of precipitates and the loss of fine distribution as well as coherency with the matrix at high temperature. This coarsening further reduces the effectiveness of the alloy to hinder the motion of dislocations, ultimately leading to lower mechanical properties than the optimal aged condition at 200°C, which exhibits the highest tensile properties overall; the observed ductility, although slightly higher than both the as-cast and over-aged conditions is inadequate for applications where strength and elongation must be optimised. (Fig 5f) also further demonstrates the connection between ageing temperatures yield strength and elongation. Yield strength reaches a maximum value (120 MPa) at a temperature of 200 °C showing an optimal precipitation hardening when the elongation is also at its maximum value (15%). Sc–Ca doped AZ61 magnesium alloys exhibit a promising balance of high strength, thermal stability, and corrosion resistance, making them well-suited for defense applications. Their elevated yield strength (up to 120 MPa) and usable temperature beyond 200 °C support deployment in **aircraft skin panels** and **missile casings**, where lightweight thermal resistance is critical. Grain refinement and fatigue resilience also qualify these alloys for **light armor plating**, offering improved survivability without the weight burden of traditional steels or titanium.

. At 400°C, the yield strength is reduced to 90 MPa, indicative of over-ageing, but the elongation is still reasonably high (7%). Aluminium is often alloyed with Scandium and Calcium, which can both promote fine grain structure and precipitation of long-lived phases (e.g., Al-Sc and Ca-containing intermetallics) that increase yield strength retention and over-ageing resistance, enhancing strength and ductility at elevated temperatures. As such AZ61 alloys with Sc and Ca addition are promising candidates for use where high thermal stability and mechanical performance are vital.

Figure 5. Variation of tensile properties of stir cast modified AZ61 alloy under various heat treatment conditions (a) As Cast; (b) Aged at 100°C; (c) Aged at 200°C; (d) Aged at 300°C; (e) Aged at 400°C; (f) Error bar of yield strength and Elongation of composite

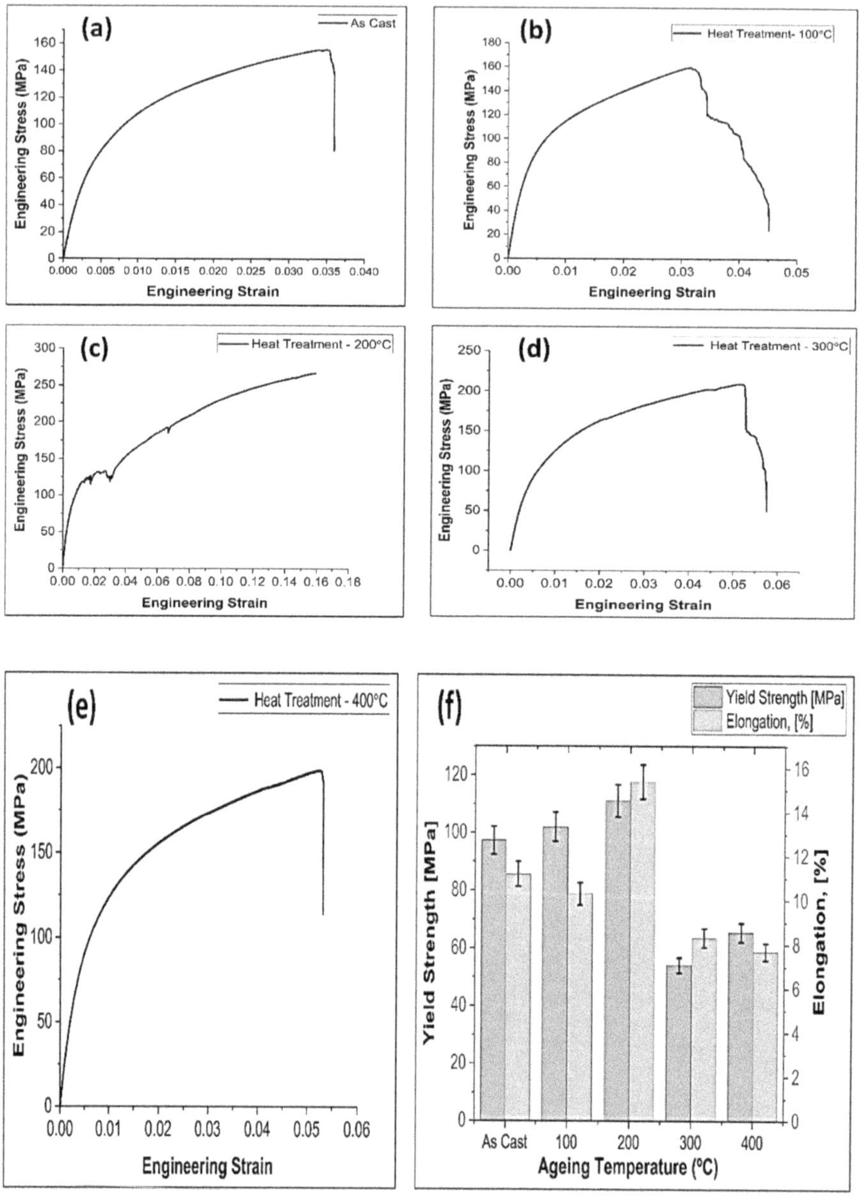

Fractography Analysis of Modified AZ61 Magnesium Alloy Under Various Conditions

A fractography analysis of AZ61 magnesium alloy, which shows the difference fracture mechanism in different heat-treatment conditions, reveals the microstructure evolution and its correlation with the mechanical performance. Discussion surrounding primary and secondary phases (β-$Mg_{17}Al_{12}$) and the effects of adding Scandium (Sc) and Calcium (Ca) are also explored further. (Figure 6a) As-cast condition: The fracture surface mainly shows cleavage fractures, characteristic of brittle failure. The crude β-$Mg_{17}Al_{12}$ phase is not well distributed along the grain boundaries and is a stress concentration area. These attributes suppress the plastic deformation process and manifest as a brittle fracture mode under tensile loading with limited elongation (~3.5%). (Figure 6b) Heat treatment at 100°C: At this temperature, the dimple features from the fracture surface with some cleavage facets further indicates mixed-mode failure. Brittleness is decreased via partial dissolution and redistribution of β-$Mg_{17}Al_{12}$, which enhances ductility (4.5%) and strength (170 MPa) to a small extent. Plastic deformation permits some movement of dislocation before failure in its secondary phase refinement. (Figure 6c) 200°C heat treatment: A fully ductile fracture surface displaying fine and evenly distributed dimples is represented here, denoting large plastic deformation prior to fracture. It further enhances the matrix strength via the optimized precipitation of fine β-$Mg_{17}Al_{12}$ particles while retaining ductility. This condition displays maximum mechanical properties where ultimate elongation (15%) and the ultimate tensile strength (280 MPa) optimized the trade-off between strength and toughness. (Fig 6d) Fracture at 300°C: The fracture surface shows the mode mix, with the Shape of shallow dimples and some cleavage. Over-ageing results in rapid coarsening that makes them less effective at pinning dislocations. At 240 MPa, the tensile strength is lowered and elongation is reduced to 6%, indicating the beginning of degradation of mechanical properties. (Fig 6e) Fracture at 400°C: The fractured surface displays micro pores and mixed mode of ductile and brittle characteristics. Elongated coarse β-$Mg_{17}Al_{12}$ precipitates and grain growth alleviate the contribution to strengthening. Cracks nucleate in the catalysis sites (micro-pores), restricting the elongation (5%) and tensile strength (200 MPa). Scandium can modify grain structure and also lead to the formation of stable Al-Sc precipitates, resulting in enhancement in elevated temperature strength and fracture toughness. These phases are thermally stable which forestalls the effects of over-ageing. Calcium forms stable Ca-containing intermetallics that support creep resistance and inhibit micro-void formation. Sc and Ca working together yields reduced cleavage fracture and micro-pore initiation while retaining ductility at high temperatures for better mechanical performance. AZ61 enables enhancing fracture resistance and maintaining tensile strength within a wider range of heat-treatment conditions (including temperatures above 300°C), by incorporating these elements. This proves AZ61 useful in high-stress and high-temperature applications.

Figure 6. Fractography of tensile samples of stir cast modified AZ61 alloy under various heat treatment condition (a) As Cast; (b) Aged at 100°C; (c) Aged at 200°C; (d) Aged at 300°C; (e) Aged at 400°C

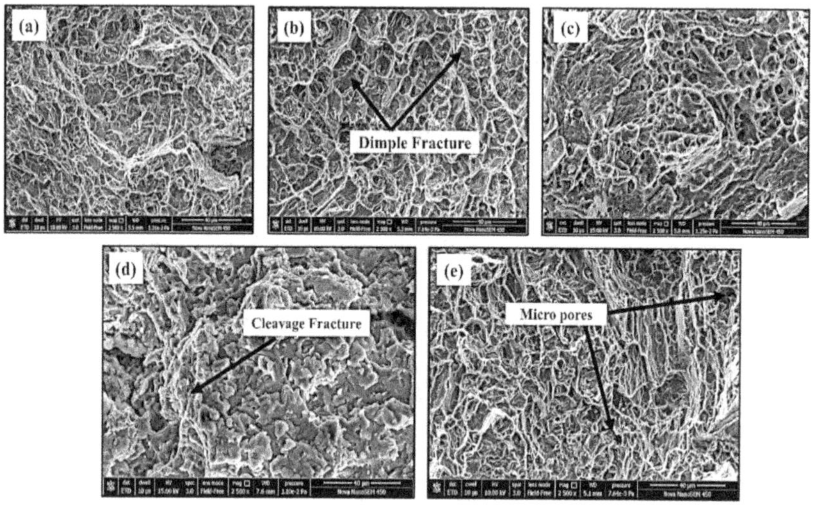

XRD Analysis of Stir Cast Modified AZ61 Alloy Under Various Conditions

The XRD patterns (Figure 7), indicate the crystallographic nature of the modified AZ61 magnesium alloy in the as-cast and heat treated conditions (heat treatment at 200°C and heat treatment at 400°C) showing note worthy peaks at different diffraction angles for the specific phase including α-Mg (hexagonal close-packed magnesium matrix), β-$Mg_{17}Al_{12}$ (precipitates), Mg_2Ca (intermetallic phase) and Al_3Sc (thermally stable phase), which contributes to the overall alloy properties. The peaks in the as-cast form are predominantly characteristic of α-Mg with some peaks arising from β-$Mg_{17}Al_{12}$ trace in the grain boundaries. Such a microstructure is detrimental to both ductility and toughness, as β-$Mg_{17}Al_{12}$ forms thick and continuous networks that serve as the seed of cracks, which in turn, decrease the mechanical performance significantly. Upon heat treatment at 200°C, XRD patterns in figure 4 show the peaks for α-Mg and β-$Mg_{17}Al_{12}$ are sharper and more intense, suggesting that crystallinity was enhanced, and precipitation was optimized. Moreover, the appearance of the peaks Al_3Sc and Mg_2Ca is observed, which means the formation of these thermally stable intermetallics. The addition of Al_3Sc both helps refine the grains and reinforces the matrix by impeding dislocation motion, while the Mg_2Ca complex improves creep resistance and stability at elevated temperatures. These phases balance strength and ductility and improve the tensile strength (280 MPa) and elongation (18%). At 400°C, the contribution of α-Mg peaks becomes markedly dominant, yet fine β-$Mg_{17}Al_{12}$ has clearly coarsened, impairing its role as a strengthening phase. In contrast, Al_3Sc and Mg_2Ca peaks do not broaden, indicating that their high-temperature stability is consistent. This stability reduces the over-ageing effect, allowing the alloy to maintain a high tensile strength (260 MPa) and moderate elongation (10%) of the as cooled die castings. For defense applications, the synergistic effects of these various phases are critical. The matrix α-Mg characterizes lightweight properties, the β-$Mg_{17}Al_{12}$ phase improves the room-temperature strength, Al_3Sc stabilizes microstructure at thermal loadings, and Mg_2Ca provides long-term thermal stability. The resultant AZ61 alloy exhibits these crystallographic characteristics, rendering excellent mechanical properties requisite for defence and aerospace applications, withstanding the most extreme conditions of high-stress, high-temperature.

Figure 7. XRD analysis of stir cast modified AZ61 alloy under cast and heat treatment conditions

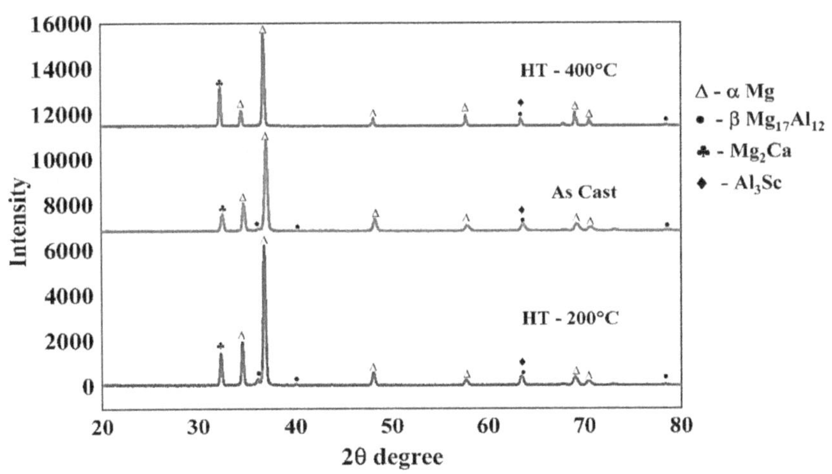

CALPHAD-Inspired Phase Stability Map for the Mg–Al–Sc System

Figure 8, illustrates the relative phase stability index as a function of Scandium content (0–1 wt%) and temperature (100–600 °C). Peak stability is observed near 0.2 wt% Sc and ~350–375 °C, suggesting optimal conditions for Al_3Sc precipitation during heat treatment. This guides alloy design and aging strategies for enhanced high-temperature strength and microstructural stability.

Figure 8. CALPHAD-inspired phase stability map for the Mg–Al–Sc system

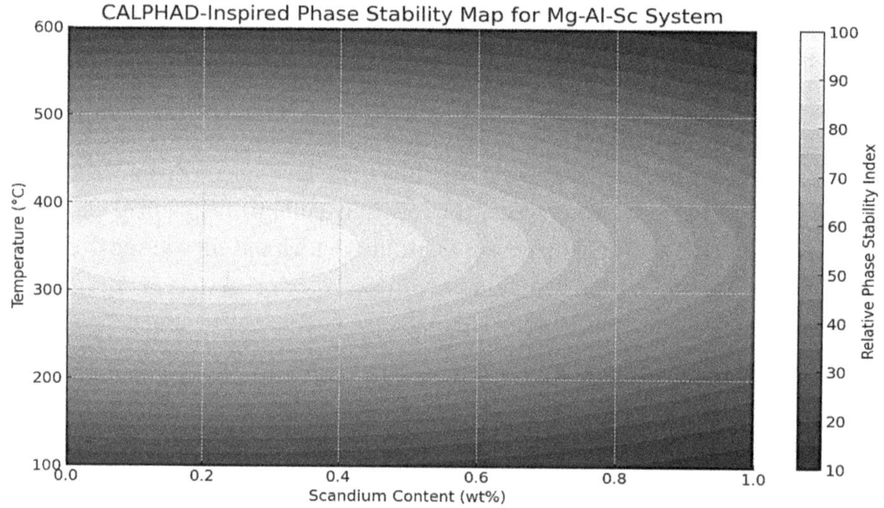

Maximum phase stability is observed near 0.2 wt% Scandium at ~350 °C, which aligns with the optimal precipitation window for Al_3Sc. At lower Sc levels (<0.1 wt%) or temperatures beyond 500 °C, the stability of Al_3Sc-type precipitates diminishes significantly. This map supports alloy design strategies that target narrow Sc additions and controlled aging treatments to exploit high-strength precipitation while maintaining cost-efficiency.

CALPHAD-Inspired Phase Stability Map for the Mg–Ca–Sc System

Figure 9, presents the relative phase stability index as a function of Calcium content (0–1 wt%) and temperature (100–600 °C). Maximum phase stability is observed near 0.4 wt% Ca and ~275 °C, indicating optimal conditions for Mg_2Ca

formation. A secondary stability region around 0.2 wt% Ca at ~350 °C suggests potential synergistic phase behavior with Sc additions, aiding in thermal resistance and grain boundary stabilization in Sc–Ca co-doped magnesium alloys.

Figure 9. CALPHAD-inspired phase stability map for the Mg–Ca–Sc system

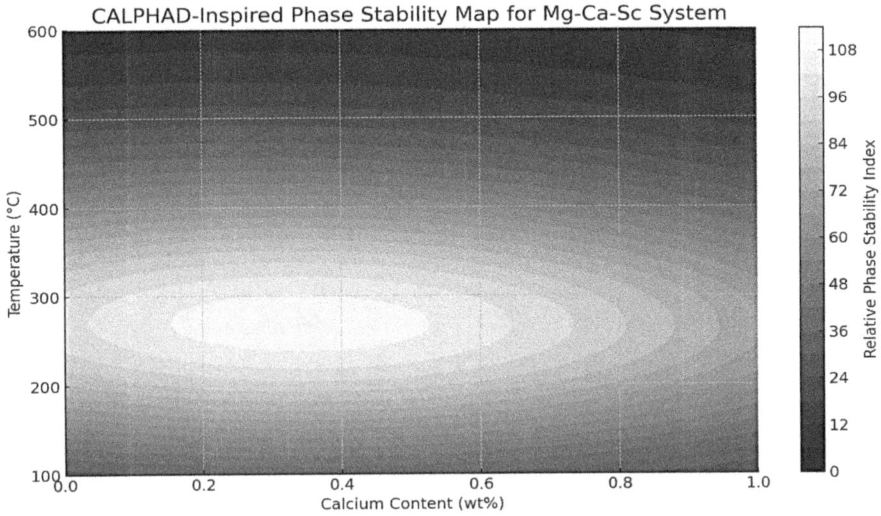

Mg_2Ca phase shows peak stability around 0.4 wt% Ca and ~250 °C, which is consistent with reported experimental observations. The secondary peak near 0.2 wt% Ca and 350 °C suggests a synergistic stability field when Sc is co-doped—highlighting the potential of dual-phase precipitation ($Mg_2Ca + Al_3Sc$) in enhancing mechanical strength and thermal resistance. The broader stability region offers flexibility in heat treatment design and suggests composition windows where co-precipitation may be thermodynamically feasible.

Ashby Plot of Yield Strength vs. Elongation to Failure for Various Mg-Based Alloys

Figure 10, illustrates the trade-off between yield strength and ductility in magnesium alloys containing different strengthening phases such as Al_3Sc, Mg_2Ca, β-$Mg_{17}Al_{12}$, and $MgZn_2$. Alloys with Al_3Sc and rare-earth additions (e.g., Mg-Sc, Mg-Gd) exhibit high strength but reduced ductility, suitable for high-performance structural applications. In contrast, systems like Mg-Zn and Mg-Zn-Ca provide better elongation with moderate strength, favorable for formability-critical applications. This Ashby plot guides alloy selection based on mechanical performance requirements.

Figure 10. Illustrates the mechanical trade-off between yield strength and ductility among various Mg-based alloys. Alloys with Al_3Sc or Mg_5Gd show high strength at the expense of ductility, while Mg-Zn-Ca offers a better balance

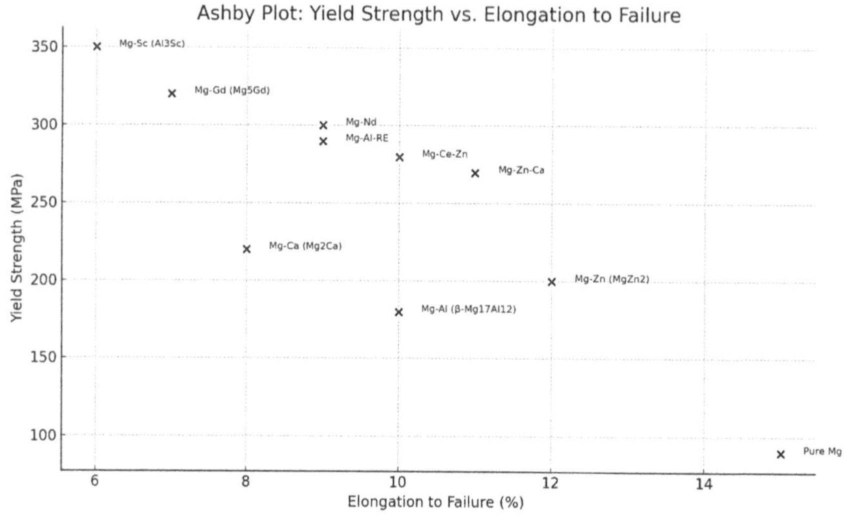

Ashby Plot of Yield Strength vs. Maximum Usable Temperature for Mg-Based Alloys

Figure 11, illustrates the relationship between yield strength and thermal stability of various magnesium alloys. Alloys containing Al_3Sc and rare-earth elements (e.g., Mg-Sc, Mg-Gd, Mg-Nd) show superior strength retention at elevated temperatures (up to 500 °C), making them suitable for aerospace and defense applications. In contrast, traditional systems such as Mg-Al (β-$Mg_{17}Al_{12}$) and Mg-Zn exhibit lower thermal tolerance, limiting their use in high-temperature environments. This chart aids in selecting magnesium alloys based on strength requirements and service temperature constraints.

Figure 11. Demonstrates the service temperature limitations of Mg alloys. Al_3Sc and rare-earth reinforced systems (Mg-Gd, Mg-Nd) retain mechanical strength up to 400–500 °C, while β-$Mg_{17}Al_{12}$ and Mg_2Ca lose effectiveness above 200–250 °C

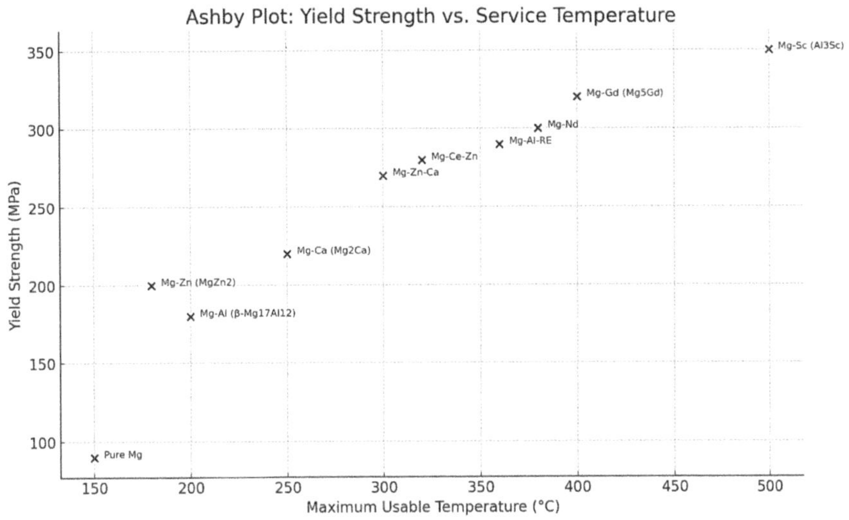

Nyquist Plot (EIS) for AZ61 and Its Scandium- and Calcium-Doped Variants in 3.5% NaCl Solution

The Nyquist Figure 12, illustrates the electrochemical behavior of AZ61 and its Sc- and Ca-doped variants in 3.5% NaCl solution. The base AZ61 alloy shows the smallest semicircle with a charge transfer resistance (Rct) of approximately 1.2 Ω·cm^2, indicating lower corrosion resistance. Upon Sc addition, Rct increases to 2.4 Ω·cm^2, while Ca doping raises it further to 2.8 Ω·cm^2. The AZ61 + Sc + Ca alloy exhibits the largest semicircle, with an Rct of 3.6 Ω·cm^2, confirming the superior corrosion protection due to synergistic grain boundary stabilization and passive film integrity. This validates the effectiveness of dual-element doping in corrosive environments.

Figure 12. Nyquist plots for AZ61 and alloyed variants in 3.5 wt% NaCl solution. Co-doped AZ61 + Sc + Ca shows the highest charge transfer resistance (R_{ct}), indicating superior corrosion protection

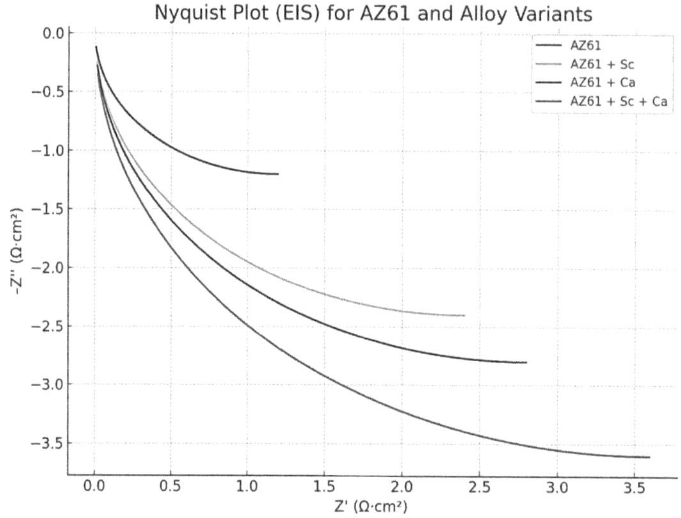

Corrosion Current Density (I_{corr}) of AZ61 and Its Scandium- and Calcium-Doped Variants in 3.5% NaCl Solution

Figure 13, presents corrosion current density (Icorr) values for AZ61 and its Sc- and Ca-modified variants in a 3.5% NaCl environment. The base AZ61 alloy shows the highest corrosion rate with an Icorr of 45.3 μA/cm^2, indicating poor resistance. Sc addition significantly reduces Icorr to 28.7 μA/cm^2, while Ca alone lowers it further to 25.4 μA/cm^2. The dual-doped AZ61 + Sc + Ca alloy exhibits the best corrosion performance with an Icorr of just 17.9 μA/cm^2. This ~60% reduction from the baseline highlights the enhanced passive film formation and electrochemical stability due to synergistic Sc–Ca doping. To capitalize on current findings, a parametric design map can be developed using data on phase stability, mechanical performance, and corrosion resistance of Sc–Ca doped AZ61 alloys. This map would guide alloy selection across Technology Readiness Levels (TRLs)—from lab-scale innovation (TRL 1–3) to field-deployable systems (TRL 7–9). For example, compositions with 0.2 wt% Sc and 0.15 wt% Ca, optimized for high strength (280MPa) and corrosion resistance (Icorr = 17.9 μA/cm^2), can be scaled for UAVs and armor systems. This approach ensures systematic material tailoring, bridging research outputs to practical defense-grade applications efficiently and strategically.

Figure 13. Bar plot of corrosion current density (I_{corr}) values. The lowest I_{corr} of 17.9 μA/cm² is observed for AZ61 + Sc + Ca, representing over 60% improvement compared to base AZ61.

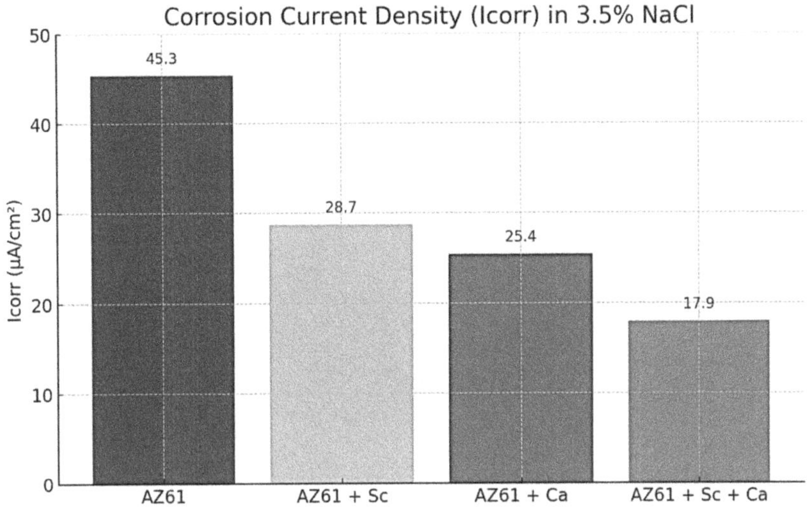

Simulated Phase Stability Map for AZ61 Magnesium Alloy with Scandium and Calcium Additions

Fgure 14, illustrates the predicted phase stability regions as a function of combined Sc + Ca content (0–1 wt%) and temperature (200–600 °C). The diagram shows the transition from a single α-Mg phase (blue) to multiphase zones featuring Al_3Sc (orange) and Mg_2Ca (red) precipitates. Al_3Sc stabilizes in the range of ~0.2–0.6 wt% alloying at intermediate temperatures (~350–450 °C), while Mg_2Ca forms at higher dopant levels and lower temperatures. This map guides compositional tailoring and heat treatment strategies for achieving targeted mechanical and thermal properties in Sc–Ca-doped AZ61 alloys.

Figure 14. Simulated phase diagram for AZ61 alloy with increasing combined Sc + Ca content

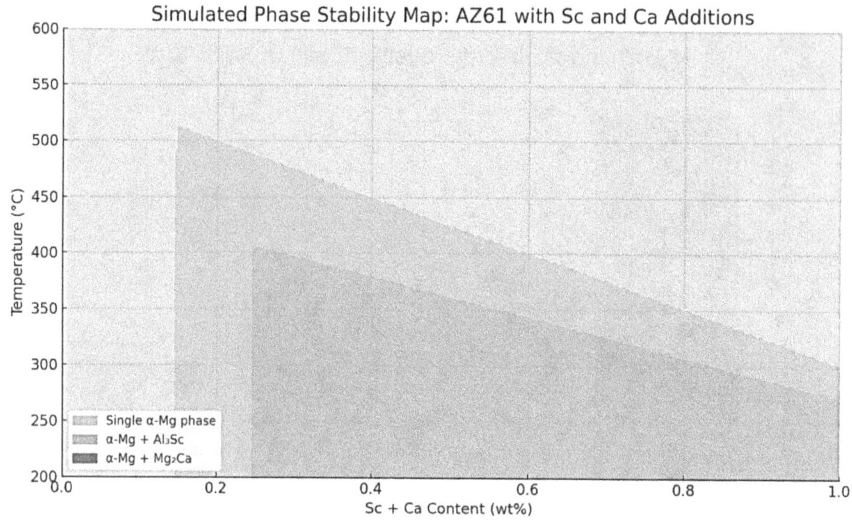

Table 2, includes a wide range of intermetallic phases and precipitation types across various Mg-based systems (e.g., Mg-Al, Mg-Zn, Mg-RE, Mg-Ca, Mg-Sc), with their mechanisms, formation conditions, mechanical performance, and unique attributes.

Table 2. Comparative table of strengthening mechanisms in Mg alloys

Strengthening Phase and Ref.	Composition	Mechanism	Formation Temp (°C)	Yield Strength (MPa)	Thermal Stability	Key Attributes
β-$Mg_{17}Al_{12}$ (Gao, W et al., 2024)	Mg-Al	Precipitation Hardening	150–200	170–230	Moderate (unstable >200°C)	Forms along GBs; coarse precipitates
Al_3Sc (Jiang, Y et al., 2023)	Mg-Al-Sc	Precipitation + Grain Refinement	300–350	300–400	Excellent (up to 500°C)	Coherent nanoscale precipitates; cost-sensitive
Mg_2Ca (Liu, S et al., 2023)	Mg-Ca	Laves Phase Strengthening	150–200	200–270	Moderate	Improves corrosion resistance; Laves phase

continued on following page

Table 2. Continued

Strengthening Phase and Ref.	Composition	Mechanism	Formation Temp (°C)	Yield Strength (MPa)	Thermal Stability	Key Attributes
$MgZn_2$ (Österreicher, J et al.,2023)	Mg-Zn	Laves Phase Strengthening	120–180	180–250	Low	Used in Mg-Zn-based alloys; limited high-T stability
Al_2Y (Peng, P et al., 2024)	Mg-Al-Y	Grain Boundary Pinning	300–400	220–280	High	Rare-earth addition; improves GB cohesion
Mg_5Gd (Song, J et al.,2020)	Mg-Gd	Age Hardening	250–350	280–350	Excellent	Used in high-performance Mg-RE alloys
$Mg_{12}Nd$ (Yang, Y et al.,2021)	Mg-Nd	Age Hardening	250–350	250–320	Good	Stable rare-earth phase; refined grains
$Mg_{14}Ce_2Zn$ (Peng, P et al.,2024)	Mg-Zn	Multiphase Precipitation	200–300	270–330	Moderate	Multimodal strengthening effect; improves ductility
Mg_4Zn_7 (Zou, J et al.,2023)	Mg-Zn	Dislocation Blocking	120–180	200–260	Low	Zinc-rich precipitates; effective at low temp
Al_4RE (e.g., Al_4Y, Al_4Nd) (Zhao, J et a., 2021)	Mg-Al-RE	Stable Intermetallic Formation	300–400	260–310	High	RE intermetallics with high thermal resistance

CONCLUSION

In this presented study we investigate the synergetic effects of Scandium (Sc) and Calcium (Ca) on the structure and mechanical properties of modified AZ61 magnesium alloys, revealing insights for further improvement in defense applications. Micro structural refinement, improved phase stability and the best possible mechanical performance at different heat-treatment conditions are the major improvements found in the study of materials. Microstructural characterization showed that coarse grains and continuous $Mg_{17}Al_{12}$ phases distributed along the grain boundaries in the as-cast AZ61 alloy, serving as potential crack initiation sites and severely restricting both the tensile strength (160 MPa) and the elongation (3.5%). Both XRD analysis confirmed that those additions indeed generated Al-Sc and Ca-containing intermetallic phases that embrittled the alloy and suppressed grain growth. Thermally stable precipitation phases, specifically Al_3Sc and Mg_2Ca, were identified as beneficial to precipitation behaviour and high-temperature strength. The modified AZ61 alloy reached the peak-aged condition with a maximal tensile strength of 280 MPa and 15% elongation after aging (heat treatment) at 200°C. The XRD patterns revealed a relatively homogenous distribution of fine $Mg_{17}Al_{12}$ precipitates, leading to pronounced strengthening. The coarse $Mg_{17}Al_{12}$ in the alloy at elevated ageing temperature (300 °C, 400 °C) indicated overageing. Meanwhile, the tensile strength and elongation of the alloy maintains above 260 MPa and 10%, respectively. The fracture evaluation demonstrated that additions of Sc and Ca reduced the cleavage fractures and micro-pores thereby facilitating a ductile mode of fracture even at increased temperature. These characteristics are important in defence applications such as armor systems, aerospace, and high-performance structures where lightweight materials with a combination of high strength, toughness, and thermal stability are desired. This work is the first to report the cooperative effect of Sc and Ca in refining grain size, stabilizing intermetallic phases, and optimizing the microstructure of AZ61 alloys. These results open the door to engineering the next generation of magnesium-alloys specifically designed for extreme conditions in defense, performing better, weighing less, and operating more reliably in mission-critical scenarios.

ACKNOWLEDGMENT

The authors wish to acknowledge the support received from the Suresh Gyan Vihar University while conducting the research.

REFERENCES

Bornapour, M., Celikin, M., Cerruti, M., & Pekguleryuz, M. (2013). Magnesium implant alloy with low levels of strontium and calcium: The third element effect and phase selection improve bio-corrosion resistance and mechanical performance. *Materials Science and Engineering C*, *35*, 267–282. DOI: 10.1016/j.msec.2013.11.011 PMID: 24411378

Cai, H., Zhang, N., Liu, L., Su, J., Li, Y., Kang, Y., & Guo, F. (2024). Effects of cooling rate on the microstructure and properties of magnesium alloy-a review. *Journal of Magnesium and Alloys*, *12*(8), 3094–3114. DOI: 10.1016/j.jma.2024.06.021

Esmaily, M., Svensson, J., Fajardo, S., Birbilis, N., Frankel, G., Virtanen, S., Arrabal, R., Thomas, S., & Johansson, L. (2017b). Fundamentals and advances in magnesium alloy corrosion. *Progress in Materials Science*, *89*, 92–193. DOI: 10.1016/j.pmatsci.2017.04.011

Gao, W., Wang, X., Lin, Y., Wang, X., Liu, D., & Sun, X. (2024). Achieving ultra-high strength and ductility in a rare-earth-free magnesium alloy via precisely controlled secondary hot extrusion process with an extremely low extrusion speed. *Journal of Magnesium and Alloys*. DOI: 10.1016/j.jma.2024.07.015

Gokalp, I., & Incesu, A. (2022). Effect of CA addition to the elevated temperature mechanical properties of AZ series magnesium alloys. *International Journal of Metalcasting*, *17*(2), 1402–1412. DOI: 10.1007/s40962-022-00872-z

Gou, W., Shi, Z., Zhu, Y., Gu, X., Dai, F., Gao, X., & Wang, L. (2024). Multi-objective optimization of three mechanical properties of Mg alloys through machine learning. *Materials Genome Engineering Advances*, *2*(3), e54. Advance online publication. DOI: 10.1002/mgea.54

Jiang, Y., Le, Q., Zhu, Y., Liao, Q., Wang, T., Bao, L., & Wang, P. (2023). Review on forming process of magnesium alloy characteristic forgings. *Journal of Alloys and Compounds*, *970*, 172666. DOI: 10.1016/j.jallcom.2023.172666

Li, Z., Ning, F., Ma, L., Jia, W., & Shi, Q. (2024). Influence of heat treatment and forming cycle on the precise forming of AZ31 magnesium alloy sheet. *Materials Today Sustainability*, *27*, 100940. DOI: 10.1016/j.mtsust.2024.100940

Liu, S., Wang, C., Ning, H., Meng, Z., Guan, K., & Wang, H. (2023). Achieving high ductility and low in-plane anisotropy in magnesium alloy through a novel texture design strategy. *Journal of Magnesium and Alloys*, *12*(7), 2863–2873. DOI: 10.1016/j.jma.2023.01.017

Lu, Z., Jiang, Y., Yu, L., Xu, J., Peng, J., & Xiang, K. (2023). The effect of minor scandium addition on microstructure evolution and mechanical properties of spray formed Al-Zn-Mg-Cu alloy. *Journal of Alloys and Compounds*, *948*, 169710. DOI: 10.1016/j.jallcom.2023.169710

Mallipudi, S. R., & Nallu, R. (2018). Effect of scandium and zirconium additions on mechanical properties of AL–MG–MN alloy. *Transactions of the Indian Institute of Metals*, *72*(1), 227–238. DOI: 10.1007/s12666-018-1476-8

Mandal, P., Anant, R., Kumar, R., & Muthaiah, V. R. (2017). Effect of scandium on ageing kinetics in cast Al-Zn-Mg alloys. *Materials Science and Engineering A*, *696*, 257–266. DOI: 10.1016/j.msea.2017.04.067

Österreicher, J., Nebeling, D., Grabner, F., Cerny, A., Zickler, G., Eriksson, J., Wikström, G., Suppan, W., & Schlögl, C. (2023). Secondary ageing and formability of an Al-Cu-Mg alloy (2024) in W and under-aged tempers. *Materials & Design*, *226*, 111634. DOI: 10.1016/j.matdes.2023.111634

Patakham, U., Kajornchaiyakul, J., & Limmaneevichitr, C. (2012). Grain refinement mechanism in an Al–Si–Mg alloy with scandium. *Journal of Alloys and Compounds*, *542*, 177–186. DOI: 10.1016/j.jallcom.2012.07.018

Patel, N. V., Sharma, N. P., Tiwari, N. A., & Vasnani, N. H. (2024b). Magnesium alloys in the automotive industry: Alloying elements and their impact on material performance. *Global Journal of Engineering and Technology Advances*, *21*(1), 130–145. DOI: 10.30574/gjeta.2024.21.1.0194

Peng, P., Xue, H., She, J., Zhang, J., Tang, A., Long, S., Zhang, C., Yang, Q., & Pan, F. (2024). Ultrafine-grained Mg alloy: Preparation, properties, design strategy. *Journal of Materials Research and Technology*, *29*, 4480–4504. DOI: 10.1016/j.jmrt.2024.02.131

Prasad, S. V. S., Prasad, S. B., Verma, K., Mishra, R. K., Kumar, V., & Singh, S. (2021c). The role and significance of Magnesium in modern day research-A review. *Journal of Magnesium and Alloys*, *10*(1), 1–61. DOI: 10.1016/j.jma.2021.05.012

Rao, K. P., Prasad, Y. V. R. K., Dharmendra, C., Suresh, K., Hort, N., & Dieringa, H. (2018). Review on hot working behavior and strength of Calcium-Containing Magnesium Alloys. *Advanced Engineering Materials, 20*(9), 1701102. Advance online publication. DOI: 10.1002/adem.201701102

Riva, S., Yusenko, K. V., Lavery, N. P., Jarvis, D. J., & Brown, S. G. R. (2016). The scandium effect in multicomponent alloys. *International Materials Reviews, 61*(3), 203–228. DOI: 10.1080/09506608.2015.1137692

Ryabov, D. K., Kolobnev, N. I., & Samohvalov, S. V. (2014). Effect of Scandium addition on mechanical properties and corrosion resistance of medium strength AL-ZN-MG-CU) alloy. *Materials Science Forum, 794–796*, 241–246. . DOI: 10.4028/www.scientific.net/MSF.794-796.241

Song, C., Dong, B., Zhang, S., Yang, H., Liu, L., Kang, J., Meng, J., Luo, C., Wang, C., Cao, K., Qiao, J., Shu, S., Zhu, M., Qiu, F., & Jiang, Q. (2024). Recent progress of Al–Mg alloys: Forming and preparation process, microstructure manipulation and application. *Journal of Materials Research and Technology, 31*, 3255–3286. DOI: 10.1016/j.jmrt.2024.07.051

Song, J., She, J., Chen, D., & Pan, F. (2020d). Latest research advances on magnesium and magnesium alloys worldwide. *Journal of Magnesium and Alloys, 8*(1), 1–41. DOI: 10.1016/j.jma.2020.02.003

Soni, P. K., Tiwari, A., Nayak, S., Vasnani, H., & Jha, S. K. (2024c). Ultrafine grain refinement and improved mechanical strength of compositionally modified AZ 61 alloy composite. In *Advances in chemical and materials engineering book series* (pp. 327–362). DOI: 10.4018/979-8-3373-0669-8.ch009

Stanford, N. (2010). The effect of calcium on the texture, microstructure and mechanical properties of extruded Mg–Mn–Ca alloys. *Materials Science and Engineering A, 528*(1), 314–322. DOI: 10.1016/j.msea.2010.08.097

Tao, Q., Xu, P., Li, M., & Lu, W. (2021). Machine learning for perovskite materials design and discovery. *npj Computational Materials, 7*(1), 23. Advance online publication. DOI: 10.1038/s41524-021-00495-8

Tiwari, A., Kumar, N., & Banerjee, M. (2023c). Ageing characteristics of stir cast AZ 61 alloy with minor additions. *Recent Patents on Engineering, 18*(6), e090623217838. Advance online publication. DOI: 10.2174/1872212118666230609122741

Tiwari, A., Kumar, N., & Banerjee, M. (2023i). Applications of genetic algorithm in prediction of the best achievable combination of hardness and tensile strength for graphene reinforced magnesium alloy (AZ61) matrix composite. *Results in Control and Optimization*, *14*, 100334. DOI: 10.1016/j.rico.2023.100334

Tiwari, A., Kumar, N., & Banerjee, M. K. (2023m). Effect of Hot Rolling on Microstructure and Mechanical Properties of Stir Cast AZ 61 Alloy with Minor Additions. *Indian Journal of Science and Technology*, *16*(24), 1810–1822. DOI: 10.17485/IJST/v16i24.302

Tiwari, N. A. (2024e). Prediction model for hardness and tensile strength of graphene reinforced AZ 61 alloy based composite using Metaheuristic Algorithm. *Global Journal of Engineering and Technology Advances, 20*(3), 042–052. DOI: 10.30574/gjeta.2024.20.3.0167

Wang, Y., Liu, G., Wei, Y., & Qiao, Y. (2020). Effects of Magnesium-Calcium Alloys with Different Calcium Content on Their Mechanical Properties. *IOP Conference Series. Materials Science and Engineering*, *735*(1), 012010. DOI: 10.1088/1757-899X/735/1/012010

Weiler, J. (2020). Exploring the concept of castability in magnesium die-casting alloys. *Journal of Magnesium and Alloys*, *9*(1), 102–111. DOI: 10.1016/j.jma.2020.05.008

Xie, H., Zhao, H., Guo, X., Li, Y., Hu, H., Song, J., Jiang, B., & Pan, F. (2024). Recent progress on cast magnesium alloy and components. *Journal of Materials Science*, *59*(22), 9969–10002. DOI: 10.1007/s10853-024-09459-x

Xu, T., Yang, Y., Peng, X., Song, J., & Pan, F. (2019d). Overview of advancement and development trend on magnesium alloy. *Journal of Magnesium and Alloys*, *7*(3), 536–544. DOI: 10.1016/j.jma.2019.08.001

Yang, J., Zhu, Z., Han, S., Gu, Y., Zhu, Z., & Zhang, H. (2024). Evolution, limitations, advantages, and future challenges of magnesium alloys as materials for aerospace applications. *Journal of Alloys and Compounds*, *176707*, 176707. Advance online publication. DOI: 10.1016/j.jallcom.2024.176707

Yang, Y., Xiong, X., Chen, J., Peng, X., Chen, D., & Pan, F. (2021b). Research advances in magnesium and magnesium alloys worldwide in 2020. *Journal of Magnesium and Alloys*, *9*(3), 705–747. DOI: 10.1016/j.jma.2021.04.001

Zeng, Z., Bian, M., Xu, S., Davies, C., Birbilis, N., & Nie, J. (2016). Effects of dilute additions of Zn and Ca on ductility of magnesium alloy sheet. *Materials Science and Engineering A*, *674*, 459–471. DOI: 10.1016/j.msea.2016.07.049

Zhang, C., Li, Z., Zhang, J., Tang, H., & Wang, H. (2023b). Additive manufacturing of magnesium matrix composites: Comprehensive review of recent progress and research perspectives. *Journal of Magnesium and Alloys*, *11*(2), 425–461. DOI: 10.1016/j.jma.2023.02.005

Zhao, J., You, C., Chen, M., Lyu, S., Tie, D., & Liu, H. (2021). Effect of calcium oxide particle size on microstructure and properties of AZ91 Mg alloy. *Journal of Alloys and Compounds*, *886*, 160970. DOI: 10.1016/j.jallcom.2021.160970

Zou, J., Ma, L., Zhu, Y., Qin, L., & Yuan, Y. (2023). Gradient microstructure and superior strength–ductility synergy of AZ61 magnesium alloy bars processed by radial forging with different deformation temperatures. *Journal of Materials Science and Technology*, *170*, 65–77. DOI: 10.1016/j.jmst.2023.07.003

Chapter 11
Ultra-Wide Bandgap ß-gallium Oxide (Ga2O3) Advancing Next-Generation Defense Technologies

Pramod Mandal
https://orcid.org/0000-0001-9797-0377
University of Pardubice, Czech Republic

Sudesna Roy
https://orcid.org/0000-0003-3129-0891
KIIT University, India

ABSTRACT

Ultra-wide bandgap β-Gallium oxide (Ga_2O_3) has gained great attention for next-generation defense applications due to its outstanding properties, including a wide energy bandgap (Eg) (~4.6-4.8 eV), high breakdown electric field (~8 MV/cm), high melting temperature (1795 ºC), and excellent chemical and thermal stability in harsh environments. These characteristics make it ideal for next-generation defense technologies such as solar-blind UV photodetectors (for missile and rocket plume detection), sensors (for surveillance), and power electronics (for energy storage and propulsion systems). This review explores the background, properties, and potential of β-Ga_2O_3 in next generation defense applications, also discussing various thin-film growth techniques, etc. Here, it also covers challenges related to growth, thermal management, and dopant material to enhance the optical and electrical properties. The review explored by outlining opportunities and future scope for β-Ga_2O^3-based

DOI: 10.4018/979-8-3373-0933-0.ch011

next-generation devices for compact, and resilient defense systems.

INTRODUCTIONS

Wide-bandgap (WBG) semiconductors provide an excellent material characteristic that could allow the functioning of prospective power devices at higher temperatures, voltages, and switching rates than what is now possible with Silicon technology (D. Liu et al., 2024). It provides the foundation for the development of several essential technologies, including electronics, optoelectronics, communications, and sensing (Wong et al., 2021). WBG semiconductors are those whose bandgap is greater than 2.2 eV (H. Wu et al., 2024)(Wong et al., 2021). Using WBG semiconductors-based devices can increase the ability to convert electrical energy and improve power converter performance for next-generation defense technology (Capan, 2024). In recent years, WBG materials such as Ga_2O_3(Gallium oxide), SiC (Silicon carbide), Carbon(diamond), AlGaN(aluminum gallium nitride), ZnO(zinc oxide), ZnMgO(zinc magnesium oxide), etc, have been extensively used in several applications in the field of optoelectronics, and electronics industries for solar-blind UV detectors, LEDs, sensors, etc. (Capan, 2024)(Rogalski et al., 2023). Comparison to above mentioned WBG materials, the performance of the Ga_2O_3-based device is exceptional(Higashi-waki & Wong, 2024), as some of the important characteristics have been discussed in **Table 1**. As per performance, devices such as photodetectors fabricated through these above materials showed poor performance and lower efficiency compared to Ga_2O_3. In the case of AlGaN-based solar-blind photodetectors, results showed poor results as the Al-doped concentration was increased (Shen et al., 2018). To achieve an energy bandgap at an optimal level, an Al 40% doped concentration (atom%), which causes structural defects (Sang et al., 2013). Similarly, other wide-bandgap semiconductor, ZnMgo-based solar-blind photodetectors, showed a limited ability to achieve energy bandgap (up to 4.5 eV), thus this tends to degrade the performance of the device. For natural diamond-based solar-blind photodetectors, the band energy gap achieved was too wide, 5.5 eV, and there also exists a limitation, such as sensitive wavelengths less than 225 nm, high cost, and difficulty in doping(Z. Liu et al., 2019). Therefore, due to the lower performance and limitations of wide-bandgap semiconductors such as AlGaN, ZnMgo, and diamond-based solar-blind photodetectors, it is critical to find alternative semiconductors with a suitable energy bandgap. Hence, Ga_2O_3 can be considered

Table 1. Comparison of Ga_2O_3 with other wide bandgap materials

Properties	β-Ga_2O_3	SiC	Diamond	AlGaN	ZnO	ZnMgO
Energy bandgap (eV)	~4.6-4.9	3.2(4H-SiC)	5.5	3.4-6.2	3.3	3.4-7.8
Breakdown field (MV/cm)	8	3	10	3-5	3	3-5
Electron mobility (cm^2/V-s)	300	950	2200	1000 (for GaN)	200	100
Melting point (°C)	1795	2830	Sublimations at ~3550	~1700	~1975	~1975
BFOM	3444	340	2000	870	500	700
Thermal stability	Yes	yes	No	limited	yes	low

Ga_2O_3 is extensively used in solar-blind ultraviolet (UV) photodetector, sensors, and power electronic devices that are used for next-generation applications in defense, such as missile and rocket plume detection using solar-blind UV photodetector. While in the case of a sensor, it can be used for x-ray radiation detection and surveillance imaging, as well as in power electronics for power storage and advanced propulsion systems. In this chapter, the β-Ga_2O_3-based device for advancing the next-generation defense technologies are explored. There are few similar studies that explore related to β-Ga_2O_3, such as (Higashiwaki & Wong, 2024) who explore the β-Ga_2O_3 based material and device technologies where they cover the several important characteristics which is relevant for the electronic device applications, considering the bulk growth and epitaxial growth methods. Similarly, (Hu et al., 2023) had explored the next generation β-Ga_2O_3 nanomaterial for electronics devices, where they reported recent development of 1D β-Ga_2O_3 nanostructure for electronic devices, including growth technique and mechanism for gas sensor, photocatalyst, and photodetector, etc. But, related to the β-Ga_2O_3 based device for advancing the next-generation in defense technologies has been not explored till date, so in this chapter we have reviewed the background, properties, and potential of β-Ga_2O_3 based device in next generation defense applications, with including various thin-film growth techniques, and challenges associated with the growth, thermal management, and dopant material to enhance the optical and electrical properties.

Background of Ga_2O_3

Ga_2O_3 has a long history of research and more than 70 years of development, but it has received attention as a semiconductor material in the last decade (Higashiwaki, 2021)(Busch, 1989). Silicon-based technology had initially dominated the semiconductor industry for applications in power devices, electronics, and optoelectronic devices. But it has fallen short of meeting the demand over time. The void has been filled by the wide bandgap semiconductors like GaN, InGaN, and SiC. These wide-band semiconductors have been successfully industrialized, with an impact on our daily lives. But production of these semiconductors is hampered by technological difficulties in obtaining a large volume of high-quality crystals. Among all UWBG semiconductors, only Ga_2O_3 could be grown in large quantities. Gallium oxide is a transparent semiconducting oxide, that belongs to a group of ultra-wide band gap semiconductors discovered in 1875 by **Lecoq et al**. (de Boisbaudran, 1875). Ga_2O_3 consists of five polymorphs i.e. α, β, γ, δ, and Ɛ-gallium oxide, which was discovered by **Roy et al.**(Roy et al., 1952) in 1952. Each polymorph has its own set of space groups and coordination. The physical structure of all the polymorphs is listed in **Table 2.** Amongst all the polymorphs, β-Ga_2O_3 is the most stable one, as shown in **Figure 1**(Mandal et al., 2024).

Table 2. Properties of the different polymorphs of Ga_2O_3

Phases	System	Space group	Lattice parameters	References
α	Hexagonal	Rc	a = 4.9825, b = 13.433	(Marezio & Remeika, 1967)
β	Monoclinic	C2/m	a = 12.214, b = 3.0371, c = 5.7981, β = 103.83	(Mandal et al., 2025)(Mandal et al., 2024)
γ	Cubic	Fdm	a = 8.22	(Pohl, 1968)
K	Orthorhombic	Pna21	a = 5.0463, b = 8.7020, c = 9.2833	(Cora et al., 2017)
δ	Cubic	Ia3	a = 9.491	(Yoshioka et al., 2007)
Ɛ	Hexagonal	P63mc	a = 2.9036, a = 9.2554	(Playford et al., 2013)

Figure 1. The unit cell structure of β-Ga_2O_3

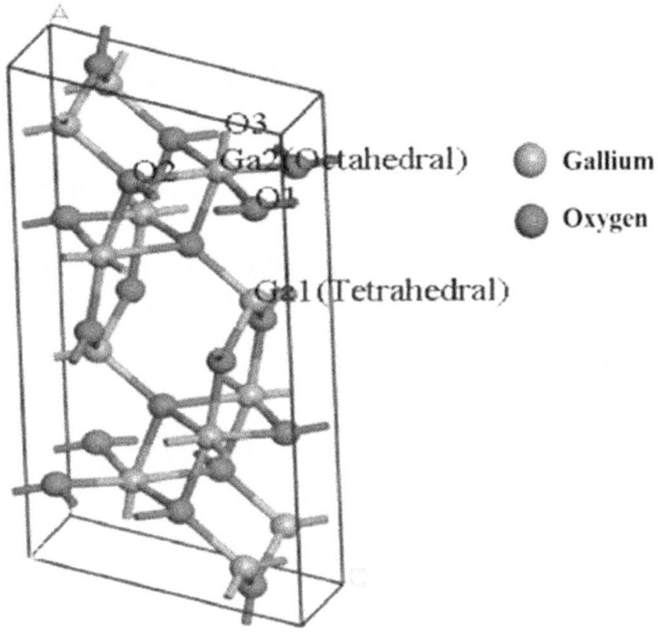

α, β, γ, δ, k and Ɛ-Ga_2O_3 polymorphs are meta-stable and can be converted to β-phase by thermal treatment from 300°C to 870°C (Tak et al., 2021)(Mandal et al., 2025). **Figure 2** shows a schematic depiction of the phase changes in gallium oxide, showing the regions of stability (Zhu et al., 2022)(Tak et al., 2021). β-Ga_2O_3 consists of ultra-wideband energy gap of 4.6-4.9 eV, a melting point of 1795°C, a high Baliga figure of merits (BFOM) of 3214.1, a high electron mobility (μ) ($cm^{-2}V^{-1}S^{-1}$) of 300, a dielectric constant (ε) of ~10, a breakdown electric field (MV/cm) of 8, and a thermal conductivity (λ) ($Wcm^{-1}K^{-1}$) of 0.11(X. Liu et al., 2020)(J. Xu et al., 2019)(Higashiwaki et al., 2014)(J. Zhang et al., 2020)(Mandal et al., 2024).

Figure 2. Phase diagram of Ga_2O_3

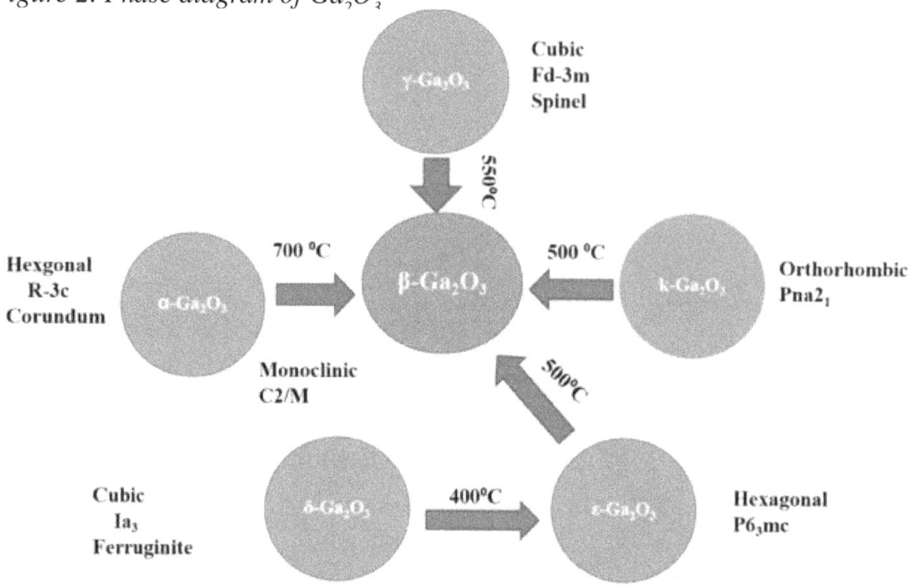

Properties of β-Ga_2O_3

Structural Property

β-Ga_2O_3 has a monoclinic structure with a lattice parameter of a=12.23 Å, b=3.04 Å, c=5.8 Å, β= 103.82° and space group of no.12, C2/m in the monoclinic system(Mandal et al., 2025). This is the most chemically and thermally stable polymorph of gallium oxide. It is amphoteric in nature, with stability from 750 °C to 900 °C and have a high melting temperature of 1795 °C and a relative density of 5.95 g/cm^3(Stepanov et al., 2016)(Galazka, 2022). It has an oxidation state of +3 which is highly stable.

Electronic Properties

The bonding β-Ga_2O_3 is predominantly ionic with the complete exchange of electrons between gallium and oxygen. The electronic structure of gallium is [Ar] $3d^{10}$ $4s^2$ $4p^1$. This shows fully filled d and s orbitals. The electronic band structure is further explored using the density functional theory (DFT). DFT is a computational quantum mechanical modelling method used to investigate the electronic structure of materials in physics, chemistry, and materials science(Kohn et al., 1996). In this, the local density approximation, or the generalized gradient approximation (GGA) has been used, yielding accurate results for many structural and energetic properties of bulk materials as well as surfaces, interfaces, and point defects. Based on ground-state theory, this calculates the electronic band structure. Its goal is to derive a quantitative understanding of material properties from quantum mechanics' fundamental laws. **Dong et al.**(Dong et al., 2017), and **Mandal et al.**(Mandal, Singh, et al., 2023)(Mandal et al., 2025) have studied Ga_2O_3 electronic properties theoretically using DFT and scissor operator approximation for correcting bandgap energy. The band structure of β-Ga_2O_3 using DFT is to be found ~1.91-1.98 eV and that using the scissor operator to be ~4.6-4.9 eV as shown in **Figure 3.**

Figure 3. Band structure of β-Ga_2O_3 without and with scissors operator

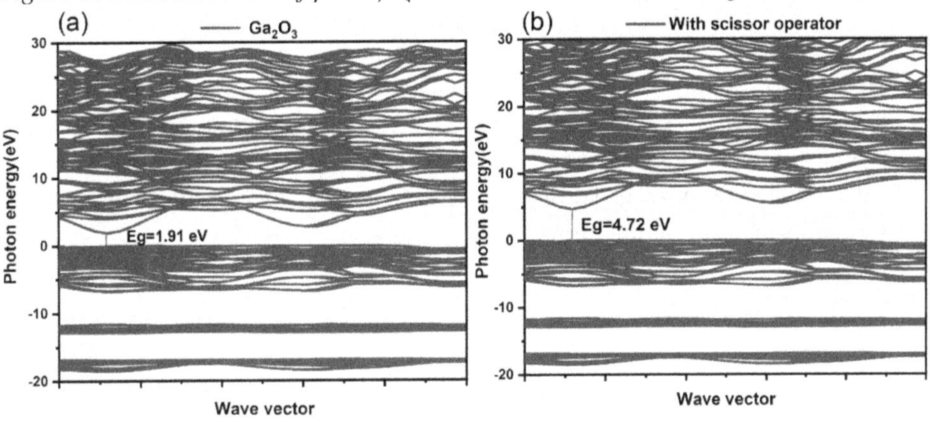

They concluded that although DFT modelling has provided an adequate qualitative description of the electronic structure, but it largely underestimates the band gap values during simulation. This is because the DFT simulation is based on the ground state theory, which results in an understated exchange-correlation potential (U_{ax}) among excited electrons.

Thermal Properties

β-Ga_2O_3 consists of very low thermal conductivity compared to other the wide band gap semiconductors i.e., GaN (210 W/mK), and SiC (270 W/mK). It has a different thermal conductivity in different directions as thermal conductivity is 10.9 and 13 W/mK in the [100] direction using a time-domain thermo-reflectance (TDTR) and laser flash method, while 23 and 27 W/mK in the [010] direction using laser flash and the TDTR method (Galazka, 2022)(Guo et al., 2015).

Optical Properties

β-Ga_2O_3 crystal has optical properties with high transmittance values. However, the addition of impurities or dopants causes coloration in the crystal. Due to free carrier absorption in the red or near-infrared regions (NIR) of the spectrum, conductive n-type crystals of Ga_2O_3 have a bluish coloration, while insulating crystals have colourless or light yellowish colour (maybe due to minor absorption in the blue part of the visible spectrum). The transmittance observed in the UV region is nearly the same for all crystals and is independent of carrier density. However, in the visible and near-infrared wavelength regions, transmittance tends to decrease as carrier density increases.

Luminescent Properties

The β-Ga_2O_3 emits three different types of light i.e., UV (ultraviolet), blue, and green. UV luminescence has no intrinsic properties (Onuma et al., 2013). The concentration of oxygen vacancies and electrical conductivity has a great impact on the UV and blue luminescence properties of Ga_2O_3. Also, the resistivity of a crystal has a great influence on blue luminescence.

Electrical Properties

β-Ga_2O_3 is usually classified as an insulator due to its wide band gap. However, Ga_2O_3 synthesized under reducing conditions showed high n-type conductivity, because of the lack of trace oxygen in the crystal lattice(Hajnal et al., 1999). The conductivity of Ga_2O_3 crystals can be controlled from 10^{-9} to 38 $\Omega^{-1}.cm^{-1}$ by changing the oxygen content of the growth atmosphere(Ueda et al., 1997). There have always been questions about electrical conductivity, **Varley et al**. (Varley et al., 2010) investigated oxygen vacancies in Ga_2O_3 using the first-principles study and concluded that oxygen vacancies, as deep donors with ionization energies greater than 1 eV, cannot contribute to n-type conductivity (W. Zhou et al., 2017). Therefore, the presence of hydrogen may be responsible for gallium oxide's n-type conductivity.

Fabrications Method of Gallium Oxide (β-Ga_2O_3)

β-Ga_2O_3 thin film has been fabricate via different deposition techniques and processes. Several deposition method has been applied such as; pulse laser deposition technique(PLD)(An et al., 2019), molecular beam epitaxy (MBE)(Singh Pratiyush et al., 2017), metal organic chemical vapor deposition (MOCVD)(Park et al., 2019), physical vapor deposition (PVD) RF sputtering(Mandal, Roy, et al., 2023)(Mandal et al., 2025), mist chemical vapor deposition(CVD)(Y. Xu et al., 2018), and Sol gel method(Shen et al., 2018) etc., has been used for fabrication of Ga_2O_3 as thin films. These methods are illustrated in **Figure** 4.

Figure 4. Different types of deposition methods for β-Ga_2O_3

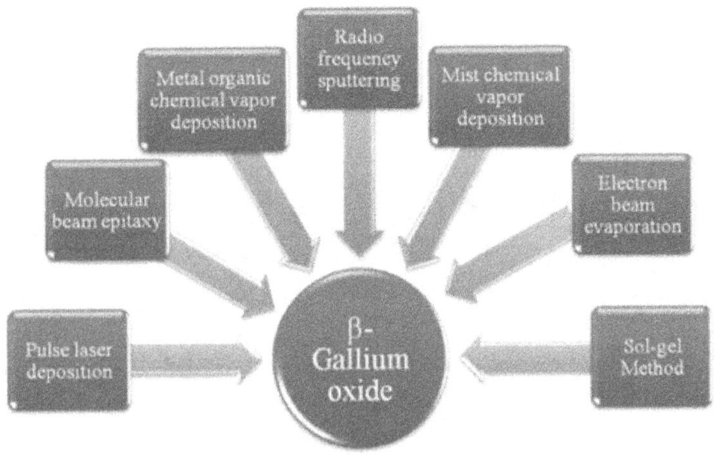

Pulsed Laser Deposition (PLD)

Figure 5 shows the schematic diagram of PLD process, this technique consists of three primary steps: ablation of the target material, production of a high-energy plume, and film formation on the substrate (Haider et al., 2022). This process entails vaporizing the surface of a solid object inside a vacuum chamber and condensing the vapor on a substrate to form a thin layer. By adjusting the deposition parameters, it is possible to achieve the required coating thickness, shape, and composition.

Figure 5. Schematic diagram of PLD for depositing thin films

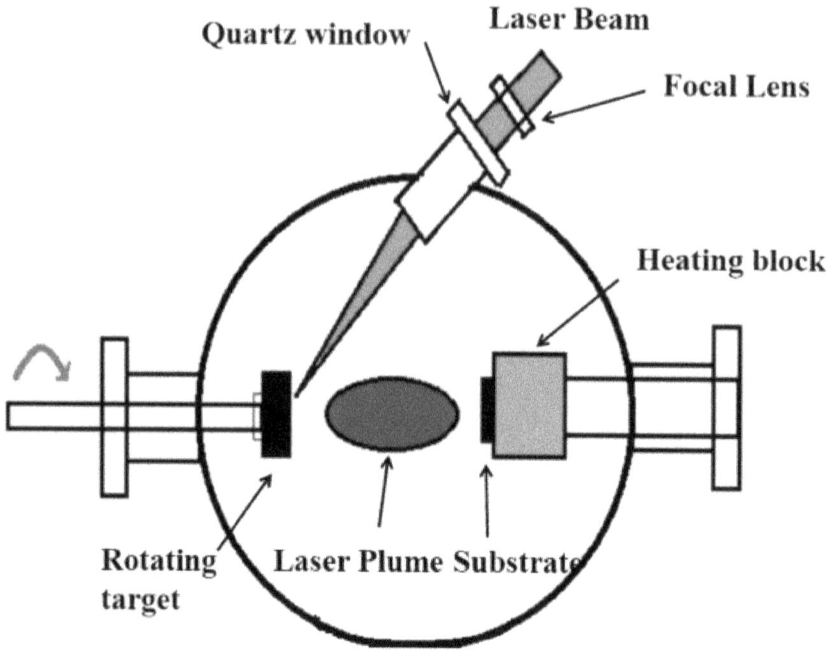

Molecular Beam Epitaxy (MBE)

Figure 6 shows the schematic diagram of the MBE process, this method is a complex, precisely regulated process for creating single-layer crystal epitaxial films under high vacuum. This procedure is slow, more expensive, and produces high-quality thin films. This process involves arranging solid source materials in evaporation cells around a heated substrate in the center. The molecular beam condition occurs when the mean free path of the vapor exceeds the practical vacuum chamber dimensions (1 m) at pressures less than 10^{-5}Torr. To avoid contamination from residual gases, ultrahigh vacuum (UHV) conditions are required (Martín-Palma & Lakhtakia, 2013).

Figure 6. Schematic diagram of MBE for thin film deposition

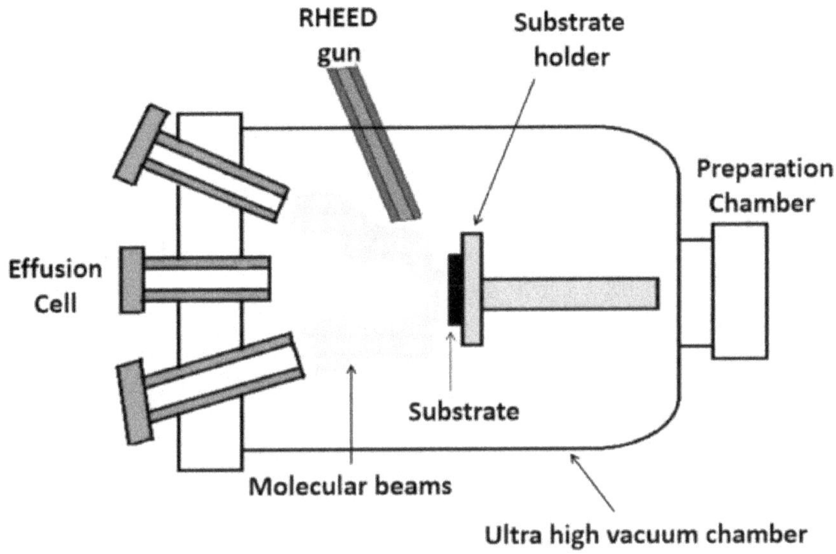

Metal Organic Chemical Vapor Deposition (MOCVD)

MOCVD is an advanced type of chemical vapor deposition (CVD) that uses organometallic precursors. It is also referred to as MOVPE or OMVPE, the schematic diagram of this process is explained in **Figure 7**. The principle of MOCVD deposition is to make contact between the volatile chemical material to be deposited and a substrate heated under a vacuum. Following the carrier gas argon (Ar), nitrogen (N2) delivers the precursors to the substrate, where they become adsorbed on the surface. This technology is used to produce high-purity crystalline semiconducting thin films and micrometer/nanometer structures (Buckner et al., 2016).

Figure 7. Schematic diagram of MOCVD method for thin film deposition

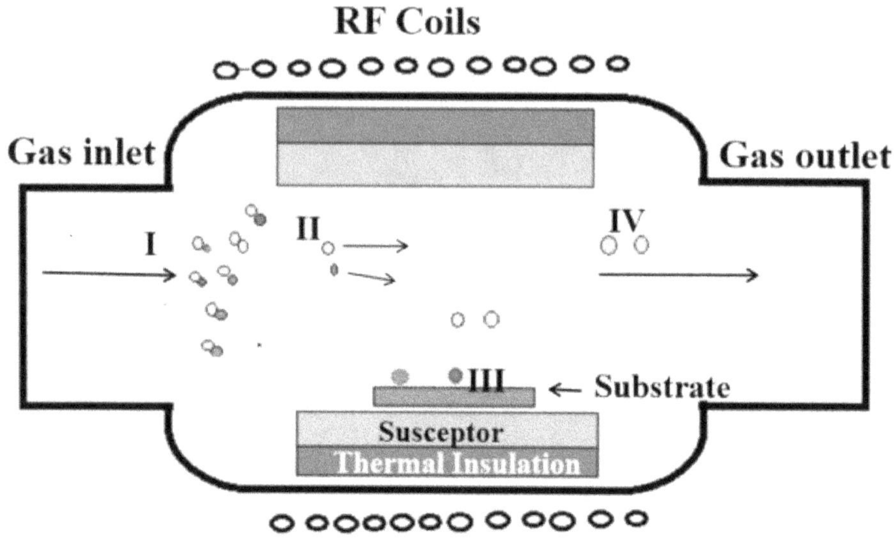

Radio Frequency Sputtering

Radiofrequency (RF) sputtering is a high-rate vacuum coating method that deposits metals, alloys, and compounds on a substrate with thicknesses ranging from nanometers to millimeters. In this procedure, the target materials and cleaned substrate are initially linked to the chamber, and a low vacuum pressure (to evacuate) of ~10^{-6}-10^{-8} mbar is maintained. As argon gas is introduced into a vacuum chamber, charge electrons accelerate neutral argon (Ar), forming Ar^+, which is then accelerated towards the target (cathode) and strikes the surface. As a result, atoms become loosened from the target material, and they begin to eject. Atoms are deposited on the substrate atom by atom (Martín-Palma & Lakhtakia, 2013)(Buckner et al., 2016). The schematic diagram of this RF sputtering process is explained in **Figure 8.**

Figure 8. Schematic diagram of working of RF sputtering method

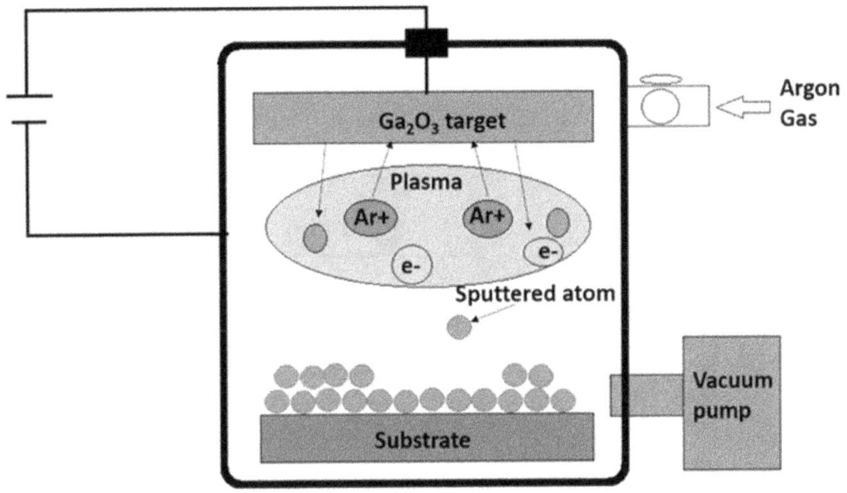

Mist Chemical Vapor Deposition

The schematic diagram of Mist or low-pressure chemical vapor deposition is shown in Figure **9**, this process is a non-vacuum solution-based technique in which ultrasonically produced aerosols are conveyed using gas. This technology is effective for creating thin films over wide regions at a high rate and low cost, while maintaining excellent control over contaminants(Ha et al., 2019). The disadvantage of developing thin films with this technology is that the films' growth rates are rather low.

Figure 9. Schematic diagram of Mist-CVD method

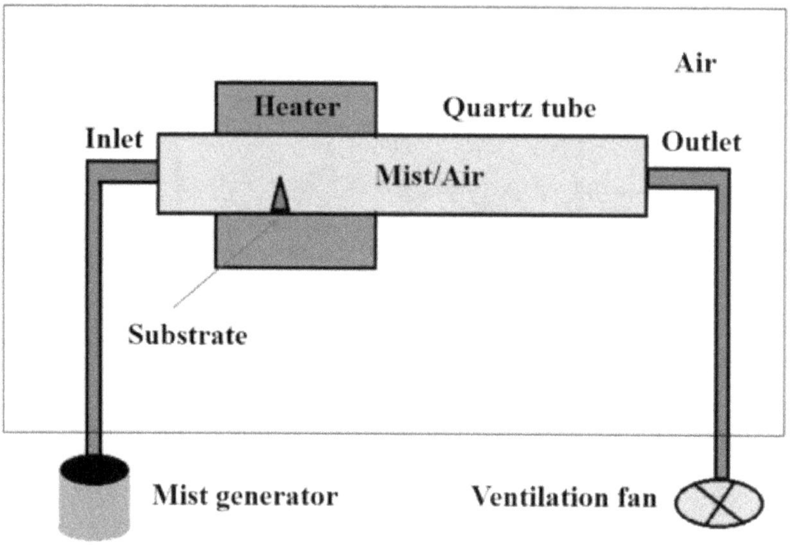

Low Pressure Chemical Vapor Deposition

Low-Pressure Chemical Vapor Deposition (LPCVD) is a thin-film deposition process that deposits a thin film on a substrate by use of a chemical reaction between precursor gases in the vapor phase at low pressure. The schematic diagram of the process of LPCVD is shown in **Figure 10** (J. Zhou et al., 2024). This method is widely used in the semiconductor and microelectronics industries to produce thin films with excellent quality and consistency. It requires several components for the successful deposition of the thin film i.e., reactor chamber, substrate, heating system, vacuum system, etc.

Figure 10. Schematic diagram of LPCVD method

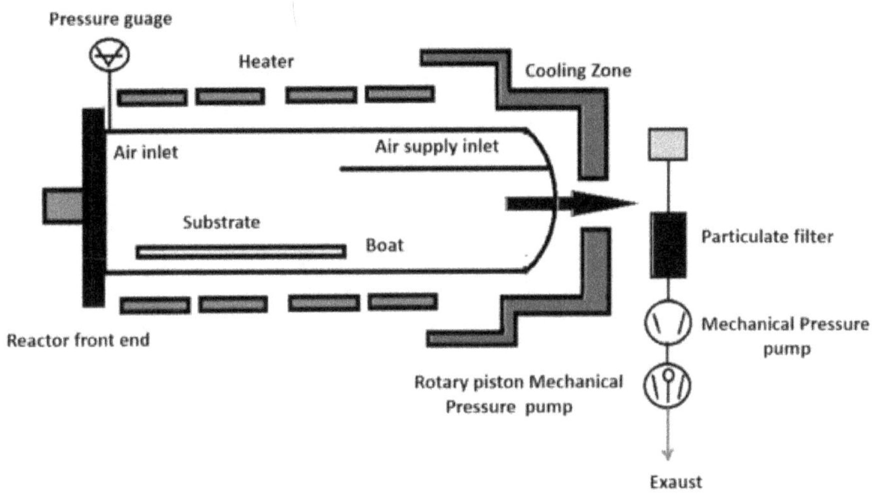

The key components of the LPCVD method are described below:

I. *Reactor Chamber:* A sealed, low-pressure area used for deposition and chemical reactions.
II. *Substrate:* The material such as sapphire/silicon wafers on which the thin film is deposited.
III. *Heating System:* The substrate is maintained at a high temperature (typically 300–1050°C) to encourage chemical reactions.
IV. *Vacuum System:* Maintains the low pressure (0.1-1 Torr) in the chamber, ensuring uniform deposition and reducing contamination.

Sol-Gel Method

The sol-gel process is widely utilized for the synthesis of inorganic and organic-inorganic hybrid materials, including nanoparticles, nanorods, thin films, and monoliths. In this technique, sol is first obtained through hydrolysis/polymerization reactions, which involve the addition of appropriate chemicals to the precursor solution. Two processes, spin coating, and dip coating, are employed to deposit the solution as thin films onto desired substrates. The resultant solution is then condensed or polymerized to form a gel. Depending on the future processing processes, this gel can be used to produce nanoparticles, xerogel, glass, or ceramic materials. The nanoparticles and xerogels can also be obtained by simple evaporation of the solvent (Bokov et al., 2021). The schematic diagram is shown in **Figure 11**.

Figure 11. Sol-gel method for synthesis of thin film

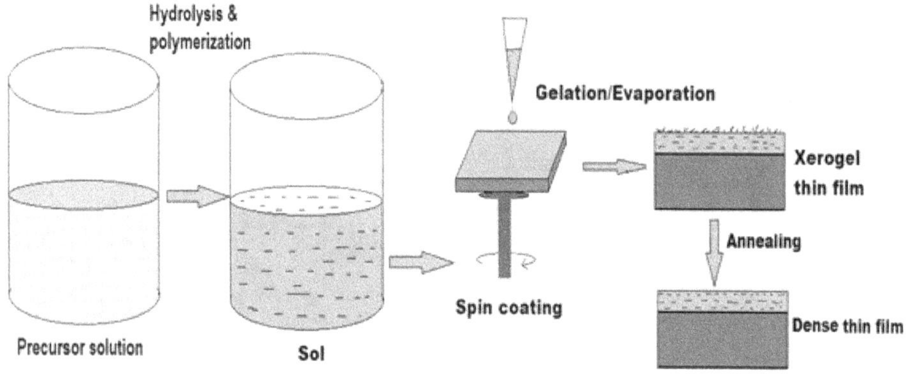

However, among all the methods the most accepted methods of deposition are PLD, PVD sputtering, and vapor phase epitaxy for depositing Ga_2O_3 thin films with control over thickness and adhesiveness. The difficulty with PLD and vapor phase epitaxy is they are restricted to lower deposition rates and smaller deposition areas. On another hand, PVD RF sputtering has the capability to produce thin films at a high deposition rate as well as a large deposition area.

Dopant Material Used in β-Ga_2O_3

In thin-film technology, dopant materials are essential because they improve the optical characteristics, structural stability, and electrical conductivity of thin films. Dopant material maximizes the mechanical strength of the material and allows for exact control over semiconductor behavior. In context to β-Ga_2O_3, material like rare earth Eu, metal (Al and Sn) has been successfully doped with β-Ga_2O_3 for various device applications. Rare earth europium (Eu), aluminum (Al), and tin (Sn) have altered the material properties of β-Ga_2O_3 i.e., morphological properties (crystallinity and grain size), electrical properties (conductivity), and optical properties (energy bandgap)(Mandal et al., 2022; Mandal, Singh, et al., 2023; Mandal et al., 2024). Regarding the rare earth Eu, it belongs to lanthanide groups, which have symbols of Eu, atomic number 63, and atomic molar mass 167.26 g mol^{-1}. It has two types of oxidation states ie; Eu^{2+} and Eu^{3+}, which pose divalent in the metallic state and can be converted into a trivalent state with the help of the application of pressure(Song et al., 2012). However, among, Eu^{2+} and Eu^{3+}, Eu^{3+} are the most stable oxidation states of europium. Since it consists of narrow emission lines in the visible range, hence, can be used as phosphor materials for display and imagining applications. For next-generation β-Ga_2O_3-based devices like; Solar-blind UV photodetectors, IR sensors, and power devices (SBD and MOSFET), it would be beneficial to use a dopant atom that could increase the electrical properties as well as optical properties of β-Ga_2O_3. As we know β-Ga_2O_3 acts as an insulator in room temperature conditions, in this context, metal (Al and Sn) can be a beneficial dopant to enhance the material properties (Y. Zhang et al., 2010; Mandal et al., 2024). This metal has been extensively used as a dopant material to modulate the conductivity as well as energy bandgap Ga_2O_3(Mandal et al., 2024; Mandal et al., 2022) In the case of Al, it consist of a band energy gap ranging from 6.4 eV for amorphous films to 8.8 eV for bulk material(Mandal et al., 2022). It, therefore, is capable of tuning the band energy gap of β-Ga_2O_3 as well as altering its electrical properties(Ma et al., 2017; Saikumar et al., 2021). Whereas Sn is the most promising dopant material in β-Ga_2O_3 in recent due to successfully enhancing the electrical as well as physical properties. According to our published findings, when Sn is doped with β-Ga_2O_3, the free electrons of the Sn-5s orbital state will fill the conduction band, improving the conductivity and physical characteristics of intrinsic β-Ga_2O_3 (Mandal et al., 2024).

Thus, we can state that dopant materials such as Eu, Al, and Sn can play a significant role in improving the physical, optical, and electrical characteristics of β-Ga_2O_3. This makes it a very promising material for next-generation defence applications, which will ultimately help the military and aerospace industries advance technologically.

ULTRA-WIDE BANDGAP Β-GALLIUM OXIDE (GA_2O_3) FOR ADVANCING NEXT-GENERATION DEFENSE TECHNOLOGIES: A REVIEW

WBG material β-Ga_2O_3 has attracted significant attention as a material alternative for next-generation defense technologies, owing to its excellent material properties. It stands out due to its ultrawide energy bandgap, high breakdown electric field strength, excellent thermal stability, and scalability as already discussed in the introduction section. These characteristics distinguish it as a versatile material capable of meeting a variety of important needs in defense systems, particularly in high-performance electronics working in severe circumstances.

One of the most promising applications of β-Ga_2O_3 in defense is in solar-blind UV photodetectors, particularly for missile and rocket plume detection. In addition to photodetectors, β-Ga_2O_3 shows immense potential for advanced radiation sensors and imaging systems. Its strong resistance to radiation damage makes it an ideal candidate for use in harsh environments, such as space or nuclear applications. These sensors can be integrated into surveillance and reconnaissance systems, offering improved sensitivity and reliability in detecting radiation signatures or imaging in hostile conditions. Furthermore, it is revolutionizing the field of power electronics, which plays a pivotal role in modern defense technologies. The high breakdown electric field of β-Ga_2O_3 allows for the development of compact, high-efficiency power devices capable of operating at high voltages and temperatures. These devices are particularly valuable in advanced propulsion systems, such as those used in electric vehicles, drones, and aircraft, where weight, efficiency, and thermal management are critical factors.

This review explored the potential of β-Ga_2O_3 in next-generation defense technologies in categories field as discussed below,

2.1 Solar-blind UV photodetector
2.2 Sensor
2.3 Power-electronics

Solar-Blind UV Photodetector (PDs)

Photodetectors are devices that convert light into electrical signals, helping to detect and measure various types of electromagnetic radiation. **Figure 12** shows the electromagnetic radiation spectrum showing ultraviolet rays i.e., UVA, UVB, UVC & EVC, visible region, and infrared region (Suchikova et al., 2024). The UV spectrum between 200 to 280 nm wavelength is referred to as the "solar-blind region"(H. Liu et al., 2022). This region has a low background level on the planet's surface because the atmosphere absorbs almost all of it. These can be detected without the influence of the Sun's radiation. Solar-blind UV photodetectors in this region can reduce the possibility of false radiation by detecting a very weak signal against a strong earthly background, and before any disastrous occurrences occur, they can quickly identify what is frequently called early warning signals(Arora & Kumar, 2020). In recent, wide bandgap β-Ga_2O_3 has emerged as the best choice for solar-blind UV photodetector in comparison to other wide band gap materials such as SiC, diamond, AlGaN, Zno, ZnMgO, etc. based solar-blind photodetector. Since it consists of ultrawide energy bandgap and makes it highly sensitive to UV-C radiation (wavelength less than <280 nm), while remaining effectively "blind" to visible and infrared radiation. This characteristic eliminates interference from solar background radiation, enabling accurate detection of missile launches even in broad daylight.

Figure 12. Electromagnetic spectrum showing ultraviolet rays i.e., UVA, UVB, UVC & EVC, visible region and infrared

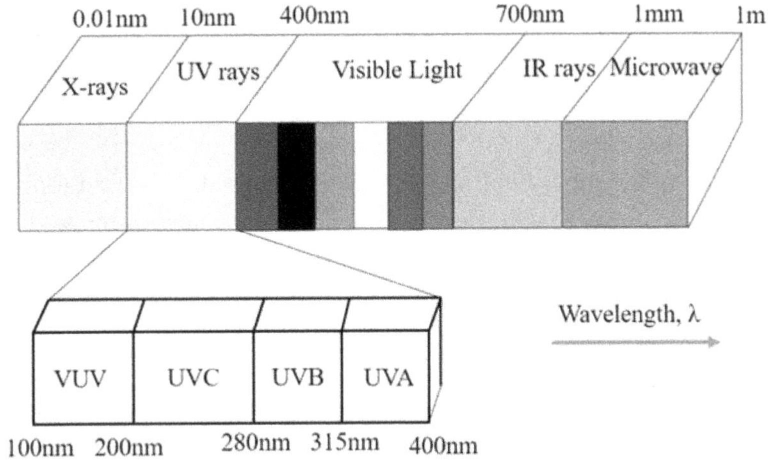

Solar-blind UV photodetector is a device that is used to convert the intensity of solar-blind light into a measurable electrical signal(C. Wu et al., 2022). It involved mainly three processes i.e., (Kaur & Kumar, 2021).

I. Creation of photo-generated carrier in the device by incident light,
II. Photo-generated carrier transport with any multiplication by the current-gain mechanism,
III. Charge carrier extraction at the end electrode to produce the output signal.

Performance Characteristics of Solar-Blind UV Photodetector (PDs)

The working principle of PDs can be understood based on the performance characteristics i.e., dark current (I_{dark}), photo current (I_{photo}), photo to dark current ratio (PDCR), quantum efficiency, responsivity (R), response time, signal to noise ratio, and specific detectivity. The performance of a photodetector is being valued through listed performance characteristics.

Dark Current (I_{dark}). Dark current is the residual current (tiny current) that flows in the photodetector even when no photon is present [20]. It consists of a semiconductor junction saturation current and radiation current. It is generally associated with specific defects of the depletion layer, which reflects the selectivity of the device to some extent (A)(J. Xu et al., 2019). It is expressed as Amperes.

Photo Current (I_{photo}). When an energetic photon strikes the photodetector, an electron is in excited mode and forming an electron-hole pair. If the photon absorption occurs in the depletion layer, the internal electric field in this region eliminates the barrier between electrons and holes, allowing holes to move toward the anode and electrons to move toward the cathode; thus, photocurrent is generated (J. Xu et al., 2019).

Quantum Efficiency. Quantum efficiency is the capacity of the photodetector to transform the input light into an electrical signal. It is used to measure a device's sensitivity to optical radiation. These are categorized as:

I. Internal quantum efficiency (IQE)
II. External quantum efficiency (EQE).

The IQE is defined as the ratio of the number of electron-hole pairs produced in the semiconductor to the number of incident photons per second responsivity. While the EQE is defined as the ratio of the number of electron-hole pairs gathered to the number of input photons per second. The EQE has typically been considered in the case of gallium oxide-based solar-blind photodetector (Kaur & Kumar, 2021). It is expressed as in equation (1);

$$EQE = \frac{Rhc}{e\lambda} \quad (1)$$

Where, R is the responsivity, c is the speed of light, h is planks constant, e is electronic charge and λ is the wavelength of photons.

Responsivity. The responsivity, R is defined as the ratio of generated photocurrent to incident optical power. It is expressed as the following equation (2).

$$R = \frac{I_{photo} - I_{dark}}{\text{Indicent optical Power}} \quad (2)$$

Response Time. The response time is the amount of time it takes for the detector output to alter in response to changes in the input light intensity. The rising time and the decay time are the two components of response time (Kaur & Kumar, 2021).

I. Rise time: The rise time is the time it takes for the photodetector output level to change from 10% to 90% of its maximum level.
II. Decay time is the time it takes for the output level to change from 90% to 10% of its maximum level.

Signal to Noise Ratio. The signal-to-noise ratio (SNR) describes the relative intensity of the required signal in comparison to the background noise. It is the ratio of useful signal power to noise power as expressed in following equation (3) (Kaur & Kumar, 2021);

$$A = 10log\frac{P_{signal}}{P_{noise}} \quad (3)$$

Specific Detectivity (D*). Specific detectivity (D*) describes the degree to which a weak signal can be recognized in relation to detector noise. It is expressed as in equations (4)(Kaur & Kumar, 2021);

$$\text{Specific detectivity} = R.\sqrt{\frac{A_{OP}}{2.e.I_{dark}}} \quad (4)$$

Here, A_{op} is the effective area under illumination, I_{dark} is the dark current, e is the electronic charge, and R is the responsivity in A.W-1.

Moreover, PDs are a key component for the functionality of any optoelectronic system, that transforms the photon energy of light into an electrical signal (Suchikova et al., 2024)(Kumar & Kaur, 2024). It has numerous uses in military and civilian domains, including missile guidance as well as location navigation for next-generation defense applications. These photodetectors are remarkably excellent at detecting missile plumes because they can identify specific UV emissions from missile exhaust (Suchikova et al., 2024). These detectors function in the UV-C region, making use of the Earth's atmosphere's ability to absorb nearly all solar energy below 280 nm. This distinctive characteristic minimizes interference from solar background radiation, allowing reliable detection of missile launches even in the daytime. Solar-blind UV photodetectors are perfect for continuous monitoring and early warning systems since they do not suffer from solar influence. When a missile is fired, the burning of its rocket engines leaves distinct UV traces. These photodetectors are extremely sensitive to such emissions, allowing them to detect missiles shortly after launch. This early detection capability is crucial for defense methods, allowing enough time to respond to any threats. **Figure 13** shows the β-Ga_2O_3 based Schottky contact metal semiconductor metal (MSM) structures photodetector.

Figure 13. Schematic diagram of Sn-doped β-Ga_2O_3 based solar-blind UV photodetector of M-S-M structure (Mondal et al., 2020)

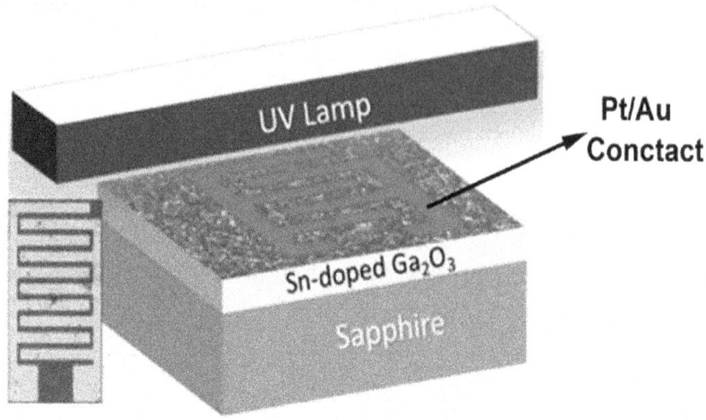

Recently, many investigations on Ga_2O_3 thin-based solar-blind UV photodetectors deposited by various methods have been performed and reported for social, defense, and industrial purposes.

Here, we will explore the performance of the few published articles based on β-Ga_2O_3 based solar-blind UV photodetector for next-generation defense technology through the different deposition methods, as discussed below.

PLD Method

(Yu et al., 2015) had fabricated the β-Ga_2O_3 thin film on (0001) sapphire substrate by varying the deposition temperature from 400 °C to 1000 °C through the PLD method for solar-blind UV photodetector. The samples deposited at temperatures 600 °C and 800 °C were chosen for the fabrication of photodetector. The device's electrical properties were further examined under 254 nm ultraviolet (UV) light illumination with a biasing voltage of 5V. With the reported dark current, photocurrent, PDCR, and responsivity of 1.2×10^{-11} A, 1.8×10^{-6} A, 10^5, and 0.903 A/W (at a wavelength of 254 nm), the solar-blind UV photodetector made with 800 °C grown film performs better than that at 600 °C. The response time was further examined, and it was found that the rise time (from 10% to 90%) and fall time (from 90% to 10%) were less than 1 s and 3 s, respectively. Also, (Blumenschein et al., 2019) have deposited β-Ga_2O_3 thin films on a c-plane sapphire substrate by different growth pressures (1×10^{-4} to 1×10^{-2} torr) for solar-blind UV photodetector using the PLD method. The device was fabricated by depositing Ni/Au contacts on the surface of all Ga_2O_3 deposited samples for MSM structured device. The I-V properties of the photodetectors developed at various pressures were investigated using a 6V bias voltage and UV light with a wavelength of 250 nm. The best electrical properties were reported for samples prepared at growth pressure of 1x 10^{-3} torr and 896 °C growth temperature, with an ON/OFF ratio of 90 and a responsivity of 30.45 AW^{-1}.

MOCVD Method

(Zhi et al., 2021) used the MOCVD method to fabricate Si-doped β-Ga_2O_3 thin films on a sapphire substrate (0001) for a solar-blind photodetector. Further Ti/Au contacts comprising of 54 pairs of fingers was deposited on the Si-doped β-Ga_2O_3 thin film's surface using photolithography and magnetron sputtering method simultaneously for the device. The electrical I-V characteristics of the device were further examined under the illumination of a 254 nm UV lamp with a biasing voltage of 20 V. The reported result produces a high photo-responsivity of 788 $A{\cdot}W^{-1}$ under 200 μWcm^{-2} illumination and a photocurrent range of 11.2 mA to 21.7 mA. Furthermore, the rise and fall time of the device were reported to be 0.63 and 0.44 seconds, respectively, suggesting a reasonable response to the 254 nm UV light. Also (Sheoran et al., 2022) deposited a UID-doped β-Ga_2O_3 thin film on a c-plane sapphire substrate for solar-blind photodetector. For the MSM structured photodetector, Ni and Au finger electrode were deposited using the e-beam evaporation process with a mask aligner system. The device's electrical properties were examined; at higher biasing voltages >5V, the dark current, photo-to-dark current ratio, responsivity, and detectivity were reported to be 254 pA, 10^5, 6.62 A/W, and 10^{14} jones, respectively. The device demonstrated the capacity to detect a very weak signal.

RF Sputtering Method

Researcher, **Liu *et al.***(Z. Liu et al., 2019) had fabricated a β-Ga_2O_3 based solar-blind UV photodetector on a c-plane sapphire substrate at optimal condition i.e. sputtering power of 120W. The electrical properties of the device were analysed and reported, it showed a high photo-to-dark current ratio (PDCR) of 2.83×10^5, with a dark current of 1.32×10^{-11} A, a responsivity of 144.46 A/W at the illumination of 250 nm wavelength UV light. Moreover, the external quantum efficiency of 64711% and a detectivity of 7.29×10^{14}Jones respectively were analysed. **Kaur *et al.***(Kaur et al., 2023) had also developed a polycrystalline β-Ga_2O_3-based solar-blind UV photodetector of MSM structures. The electrical properties of the devicc were reported as a PDCR of about 54 at a bias voltage of 5 V, a responsivity of 0.6 A/W, and rise and decay periods of 0.003 and 0.012 seconds, respectively. **Chen *et al.***(X. Chen et al., 2020) had fabricated Ga_2O_3 thin for fabricating β-Ga_2O_3 based solar blind UV photodetector device on MgO substrate. The thin film was deposited using a sputtering power of 100 W with changing growth temperatures of 600 °C,

650 °C, and 700 °C respectively. In comparison to the thin films prepared at growth temperatures of 600 °C and 650 °C, 700 °C showed the best result. Therefore, the thin film deposited at a growth temperature of 700 °C was observed to be the β-phase of Ga_2O_3 and was selected for fabricating solar-blind UV photodetector. Furthermore, an MSM structured device was fabricated using a platinum (Pt) electrode of 100 nm thickness using a mask. The I-V characteristic of the device was achieved at the luminous intensity of 70 μW/cm^2 and a bias voltage of 10V. The dark current was observed to be 3.4 pA and photocurrent at 254 nm wavelength was 42 nA. The photocurrent to dark current ratio was achieved to be over 104 with short rise time (0.07 s/0.53 s) and decay time (0.06 s/0.16 s). Further, photoresponsivity was measured to be 30 mA/W. The photocurrent was observed to be higher when the biasing voltage was increased from 1V to 10V. Moreover, **Wang *et al.*** (C. Wang et al., 2022) deposited tin (Sn)-doped β-Ga_2O_3 thin film on a sapphire substrate for fabricating solar-blind UV photodetector. The parameters they used for the deposition of the thin film were sputtering power of 500 W and variation of oxygen flow ratio of O_2/(O_2+Ar) as 0%, 0.5%, 1%,1.5%, 2.0%, and 2.5%, respectively. All samples deposited on sapphire substrate were observed to be amorphous. Furthermore, to investigating the effect of oxygen gas flow on the electrical characteristics of devices, an MSM (metal semiconductor metal) structure UV photodetector was fabricated. The I-V characteristic of the device was investigated using an illumination light of 254 nm wavelength and bias voltage of -5 V to 5V. The dark current of the photodetector prepared at 0% oxygen gas flow was reported as 4.89 pA, while at 1% oxygen gas flow dark current was somewhat decreased, however, above 1%, the dark current was reported to increase to 2.5%. The reason might be a higher oxygen flow ratio and SnO phase generations. The photo-to-dark current ratio (PDCR=812) of the oxygen flow ratio of 1% was more, and the responsivity(R=27.8) was less amongst all fabricated devices. The photodetector fabricated at an oxygen flow ratio of 1% was reported optimal with a rise and decay time of 0.05s/0.12s.

Low Pressure CVD Method

Mondal *et al.*(Mondal et al., 2022) had fabricated Sn-doped β-Ga_2O_3 based solar-blind photodetectors on c-plane sapphire at a growth temperature of 1050 °C while maintaining the vacuum pressure and growth pressure of 4.9 x 10^{-2} mbar and 1 mbar respectively. The electrode Pt/Au material by DC sputtering method of five fingers was deposited. The electrical characteristics of the device were analyzed in the illumination of a 254 nm wavelength UV light source using a biasing voltage of 50 V. The performance characteristics such as PDCR, responsivity, detectivity, external quantum efficiency, and decay time (response time) of SBPD were reported as 9 x 10^4, 1532 A/W, 1.7 x 10^{15} jones, 7.4 x $10^{5\%}$, and ~0.2 respectively. Also, **Chen**

et al.(S. Zhang et al., 2024) deposited β-Ga_2O_3 thin film on the c-plane sapphire substrate for solar-blind UV photodetector. For the deposition of Ga_2O_3 thin films different weight of Ga atoms was chosen i.e., 7 mg, 8 mg, 10 mg, and 12 mg and named samples as W7, W8, W9, W10, and W12. These thin films were characterized using different methods and upon analysis of the results, W10 showed better crystallinity amongst all. Furthermore, the performance properties of all devices were studied using the ultraviolet photodetector test system (Zolix DSR-3110-UV-F) with illumination of light of 245 nm wavelength. The photocurrent obtained for the W10 sample was reported to be 210 μA under 20 bias voltage, which was higher than other devices (W8 and W12). Similarly, the responsivity of the W8, W10, and W12 samples was reported as 11.3 mA/W, 75.4 A/W, and 57.3 A/W respectively, while response time (rising time/decay time) of W10 was reported as 0.29 s/0.31 s, which was almost like W8 samples, however, W12 samples response time (2.94 s/2.15 s) was slightly higher than other. Moreover, detectivity of samples (W8, W10, and W12) was reported as 9.7×10^6, 5.9×10^{11}jones and 3.0×10^{11} jones respectively. Corresponding to the performance of all photodetectors working in the illumination of 245 nm wavelength, the sample W10 showed a good result.

Mist CVD Method

Zhang *et al.*(S. Zhang et al., 2024) had fabricated a β-Ga_2O_3-based solar-blind UV photodetector fabrication. Before fabricating the device, the thin film was deposited by varying the temperature from 400 °C to 800 °C as well as oxygen gas flow rate from 0 to 2 L/min respectively. From the analysis of solar-blind UV photodetector I-V characteristics of all samples, oxygen gas flow rate played an important role to obtaining high PDCR of 3.0×10^7, larger detectivity of 3.06×10^{11} Jones and fast response time of 0.528 s/ 0.158 s of sample deposited at parameters i.e. growth temperature- 600 °C, oxygen gas flow rate - 0.5 L/min (S62).

ALD Method

Lee *et al.* (S. H. Lee et al., 2017) constructed β-Ga_2O_3/polyimide-based photodetector and investigated their electrical and responsivity. The device (Ga_2O_3 thin film of 30 nm thick) has a reported dark current of 200 pA, a rise time of 1 to 6 μ.sec, and a high responsivity of 45.11 A/W at illumination of 253 nm wavelength UV light and biasing voltage of 10 V. The photocurrent was reported to be enhanced by ~10^4 times more than the dark current. Despite being based on an atomically thin (3 nm) Ga_2O_3 thin film, the gadget maintained its fast detection speed. The Ga_2O_3/

polyimide-based device demonstrated its feasibility in flexible DUV photodetector applications.

These above results of solar-blind UV photodetector suggesting that these can be used for applications for next generation defense technology.

Sensor

A sensor is a device that detects physical inputs like light, heat, or motion and converts them into signals for processing or display. The SBD-based sensors are shown in **Figure 14** (Zhu et al., 2022). A sensor can be useful in the applications of sensing missiles of long-range distance, and imaging systems for surveillance. In the case of missile defense, it has become increasingly urgent, requiring individuals and decision-makers to access accurate and reliable information (Suchikova et al., 2024)(Fontana & Di Lauro, 2022). A defense sensor system's features are directly tied to the dangers it must handle (Fontana & Di Lauro, 2022).

Figure 14. Cross sectional view of β-Ga_2O_3 based SBD sensor

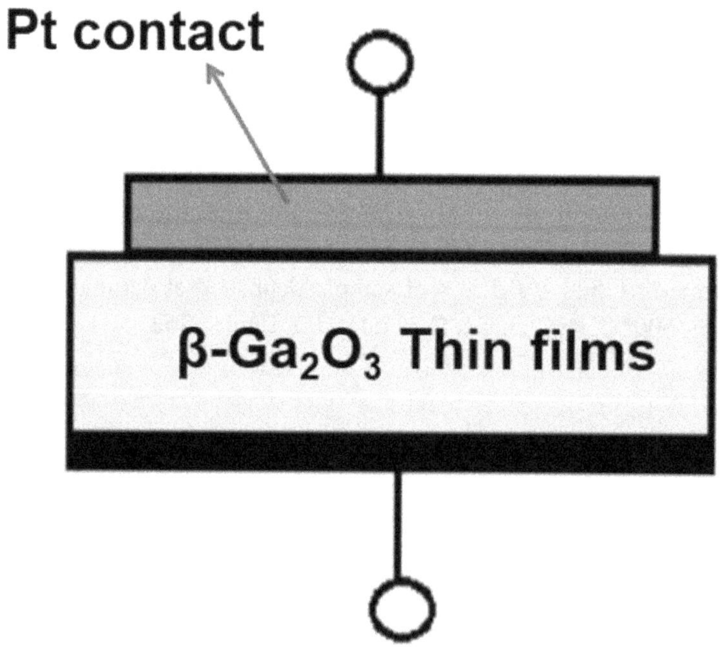

The characteristics of sensors play a crucial part in the functioning of a sensor and can be explained as input signal, output signal, accuracy, calibration error, hysteresis, nonlinearity, saturation, repeatability, dead band, resolution, output impedance, excitation, dynamic range, and reliability (Vetelino & Reghu, 2017) (Panda et al., 2024)(Han et al., 2011). The brief details of these function parameters are explained below.

I. Input Signal:

An input signal is a physical phenomenon, such as temperature, pressure, light, or motion, that a sensor detects. The sensor then transforms this input into a measurable output, usually an electrical signal, which it can subsequently analyze to monitor and respond to environmental changes or circumstances(Vetelino & Reghu, 2017)(Panda et al., 2024).

II. Output Signal:

The output signal is the sensor's response to the observed input signal, which is usually in the form of electrical, optical, or mechanical energy. This output is proportional to the input stimulus and allows the sensor to send measured data to external systems for analysis, control, or display(Panda et al., 2024).

III. Accuracy:

Accuracy is the degree of agreement between a sensor's output and the true value of the measured input, which is frequently expressed as a percentage or in absolute units(Vetelino & Reghu, 2017).

IV. Calibration Error:

Calibration error is the difference between a sensor's measured output and the reference value, even after calibration.

V. Hysteresis:

Hysteresis in sensors signifies the difference in output when the input signal follows an increasing or decreasing route, even if the input value remains constant. This develops because the sensor is unable to return to the same state for identical inputs in different situations(Vetelino & Reghu, 2017).

VI. Nonlinearity:

Nonlinearity in sensors can be characterized as the deviation of the sensor's real output from a straight-line (linear) relationship with the input, showing that the sensor's response is not completely proportionate throughout the measurement range (Vetelino & Reghu, 2017).

VII. Saturation:

Sensor saturation occurs when the output reaches its maximum and no longer changes, even if the input signal increases beyond that threshold(Vetelino & Reghu, 2017).

VIII. Repeatability:

Repeatability in sensors refers to the ability to consistently generate the same output while measuring the same input over and over under same conditions(Vetelino & Reghu, 2017)(Panda et al., 2024).

IX. Dead Band:

Dead band in sensors refers to a location of input signals where the sensor does not create a quantifiable output, even if the input has changed(Vetelino & Reghu, 2017).

X. *Resolution:*

The sensor can detect and output even the tiniest change in the input.

XI. Output Impedance:

Output impedance in sensors relates to the resistance the sensor provides for the output signal, which influences how well the signal can be conveyed to the associated system without distortion or loss.

XII. Excitation:

Excitation in sensors refers to the external energy, such as voltage or current, that active sensors required to function. This energy is required for the sensor to convert the detected physical stimulus into a useful output signal, hence sustaining its sensing function and assuring precise performance.

XIII. Reliability:

Sensor reliability correlates to their capacity to perform reliably over time, without failure, under specified environmental conditions, and during the intended lifespan. Reliable sensors preserve accuracy, stability, and performance, reducing the chance of malfunction or performance degradation as they age or are used repeatedly(Panda et al., 2024).

Selected publications demonstrating the performance of β-Ga_2O_3 based sensor for next generation device applications have been discussed below.

Kalygina *et al.*(Kalygina et al., 2023) had used RF sputtering to fabricate a Ga_2O_3/$ZnGeP_2$ heterostructure-based sensor. The device's I-V characteristics were further investigated under illumination of 254 nm ultraviolet (UV) light. The unannealed sample displayed a low sensitivity to UV and IR (808 nm to 1064 nm wavelengths), whereas the annealed sample at 400 °C had a high sensitivity in both. These characteristics favor the application of sensors in radiation detectors. Also, **Almaev *et al.***(Almaev et al., 2020) had used the RF magnetron sputtering method to fabricate a chromium doped Ga_2O_3-based sensor on a sapphire substrate. The performance of the sensor was investigated at temperatures ranging from 300 to 700 °C. Starting around 300 °C, sensors show sensitivity to oxygen; the addition of chromium greatly enhances sensor sensitivity to oxygen. As the oxygen content rises from 9 to 100 vol%, sensor resistance rises in accordance with the power law, with an index of around ~0.5 in the 500-700 °C temperature range. In a nitrogen-containing atmosphere with a 21% O_2 exposure, the sensor had a reaction time of 20 s at 700 °C and a recovery time of 52 s. Ga_2O_3 thin films with Cr additions improve oxygen detection performance in the 500-700 °C range. **Chen *et al.***(M. Chen et al., 2021) deposited crystalline β-Ga_2O_3 thin film on quartz substrate through electron beam evaporation method, further Au 10 finger electrodes was deposited by DC magnetron sputtering method for fabrication of MSM structured Au/Ga_2O_3/Au sensor for x-ray detection. Furthermore, their electrical characteristics and sensitivities were investigated at a 10 V bias. The x-ray detection sensitivity was reported to be 138.80 $\mu CmGy_{air}^{-1}$ cm^{-3}, with a dark current of 50 pA, rise and fall times of 35 m.s. The x-ray response current and the sensitivity of the crystalline Ga_2O_3 based x-ray detector was reported to improves with increasing bias voltage.

Power Electronics

Power electronic devices such as Schottky Barrier Diodes (SBDs) and metal oxide semiconductor (MOS) field effect transistor (FET) have played a crucial role for the defense applications, where high power, efficiency, and reliability are necessary.

Figure 15(a) and **(b)** shows the β-Ga_2O_3 based Schottky barrier diode (SBD) and a MOSFET device(Ji et al., 2024)(Y. B. Wang et al., 2020).

Figure 15. Cross sectional view of (a) β-Ga_2O_3 based SBD, and (b) β-Ga_2O_3 based MOSFET.

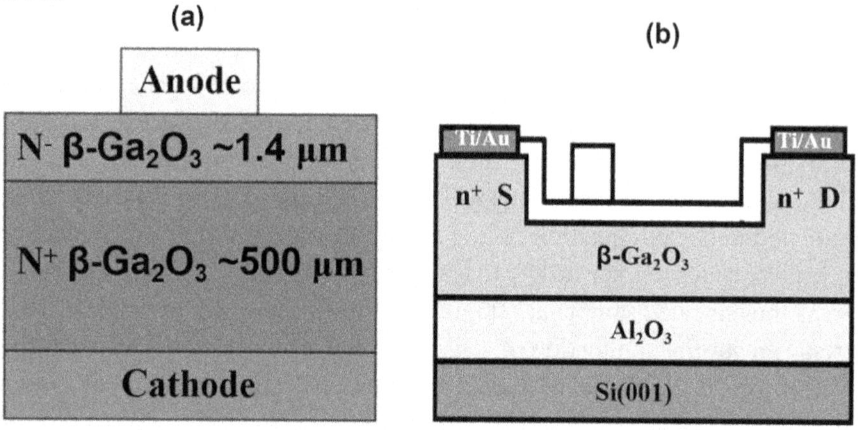

These devices are particularly valuable in advanced propulsion systems, such as those used in electric vehicles, drones, and aircraft, where weight, efficiency, and thermal management are critical factors. β-Ga_2O_3's adaptability and ability to function in severe settings make it a revolutionary material for next-generation defense applications.

Some important basic parameters of power devices Schottky Barrier Diodes (SBDs) and transistors e.g., MOSFETs in switching circuits for power control and conversions are On-resistance (R_{on}), breakdown voltage (V_{br}) and barrier height (φ_B), which is explained below.

I. On-resistance (R_{on})

The on resistance is one of the most essential factors for power devices. It can be expressed as the following formula in the equations 5 (Shivani et al., 2022),

$$R_{on} = \rho . \frac{W}{A} = \frac{W}{q\mu . Nb} \quad (5)$$

Where, ρ *is* the resistivity of semiconductor, and W is the optimum drift thickness, also mobility of majority carrier in drift and channel region, q is the charge of electron and N_b is the doping level of drift region.

In the case of Schottky diode, The total specific R_{on} of a unipolar vertical Schottky diode depends on contact resistance ($R_{contacts}$), substrate resistance ($R_{substrate}$) and drift region resistance ($R_{drift\ layer}$), which can be expressed as the following equations 6 (Shivani et al., 2022)

$$R_{on\ total} = R_{Contacts} + R_{Substrate} + R_{drift\ layer} \tag{6}$$

And in case of MOSFET, on-resistance is the sum of source-to-drain contact resistance (R_c), intrinsic channel resistance (R_i), gate-to-drain/gate-to-source resistance (R_{GD}/R_{GS}) and drift resistance (R_{drift}), which is expressed as in the following equations 7 (Shivani et al., 2022)

$$R_{on} = 2.R_C + R_i + R_{GS} + R_{GD} + R_{drift} \tag{7}$$

II. Breakdown voltage (V_{br})

Breakdown voltage is a power device-specific parameters that is impacted by its setup and geometry, including the thickness of the drift layer. It refers to the maximum voltage that a device can resist before completely failure. A material's breakdown voltage is determined by the critical electric field required to induce carrier multiplication at the junction. This critical electric field is also influenced by the material's bandgap energy (Eg). When a material's bandgap expands, more energy is required to break its bonds, therefore ultra-wide bandgap materials required in power device that can withstand much greater breakdown fields (Shivani et al., 2022).

III. Barrier height (φ_B)

The barrier height of SBD device can be defined as the energy difference between the metal's work function and the semiconductor's electron affinity. It plays an important role in SBDs, this can be adjusted by altering the top metal thin film or the bottom semiconductor (S. W. Lee et al., 2021).

The performance of β-Ga_2O_3 based power electronics devices like; Schottky barrier diode and MOSFET have been discussed below.

Schottky Barrier Diodes (SBDs)

Heinselman *et al.* (Heinselman et al., 2021) fabricated a vertical β-Ga_2O_3 SBD with platinum (Pt) Schottky & titanium (Ti) Ohmic contacts and gold (Au) contact pads on n-type single crystal UID β-Ga_2O_3 substrates. Furthermore, the device's *J-V* properties were analyzed up to 400 °C and cycled numerous times to investigate degradation. At room temperature, the devices had a turn-on voltage near 1 V, an ideality factor of 1.02, a series resistance of 0.123 Ω.cm^2, and an on/off ratio of > 10^9 at ±2 V was reported, heating up to 400 °C reduced the turn-on voltage to 0.1 V, raised the ideality factor, and dropped the on/off ratio at ±2 V to 17. The Pt/Ga_2O_3 Schottky barrier height was determined to be close to 1.2 eV, with a Richardson constant of 20 A/cm^2.K^2. Moreover, the contacts were observed to degrade during repeated cycles at temperatures as high as 400 °C, resulting in 1000 times increase in the diodes' series resistance when measured at ambient temperature following cycling. STEM EDS data revealed that this degradation was due in part to Ti migration and oxidation on the top surface of the Au contact pads. Also, **Zhang *et al.***(S. Zhang et al., 2021) had reported Ni/Ga_2O_3/Ti SBDs, where photolithography, lift-off, and electron beam vapor methods were utilized to fabricate Ni/Au Schottky and Ti/Au Ohmic connections on a single crystal Ga_2O_3 substrate. Electrical parameters, including on-state resistance (R_{on}), Schottky barrier height (φ_B), and ideal factor (n), were determined by examining the current density-voltage (J-V) and Current-voltage (C-V) curves of the Ga_2O_3-based SBDs. The current density-voltage (J-V) curves of the SBD were investigated further. The output current density was reported as 3.04 × 10^{-2} A/cm^2 and 1.25 × 10^{-7} A/cm^2 at 2 and -2 V, respectively. Extrapolating the log-scaled J-V curve's linear area yielded an estimated R_{on} of 51.5 Ω.cm^2. The n, φ_B, and turn-on voltage (V_{on}) were calculated to be 1.34, 0.93 eV, and 0.52, respectively. The Ni/Ga_2O_3/Ti Schottky barrier diode has reported excellent electrical properties and shown potential for next-generation devices. **Zhang *et al.***(Q. Zhang et al., 2023) had fabricated a β-Ga_2O_3-based SBDs, then deposited a Ti/Au bilayer using Rf magnetron sputtering and annealed at 500 °C to provide ohmic connections to form Au/Ti/β-Ga_2O_3. The effect of annealing temperature (200-400 °C) on electrical characteristics was discussed further. The as-fabricated SBD achieved a maximum current density (J_{max}) of 905.2 A/cm^2 and an ON/OFF ratio of 10^8. Increasing the annealing temperature to 200 °C resulted in improved values of 950.2 A/cm^2 and 1.5 × 10^{11}. When the annealing temperature was increased to 400°C, the J_{max} and ON/OFF ratio decreased to 717.1 A/cm^2 and 10^8 levels, respectively. The SBD had n and φ_B values ranging from 1.5 to 2.5 and 0.8 to 1.0 eV, respectively.

MOSFET

Park *et al.* (Razeghi et al., 2019) used the MOCVD approach to fabricate an n-type Ga_2O_3:Si-based MOSFET on a c-plane sapphire substrate, and Ti/Au metal contacts were deposited further for the source and drain electrodes. The e-beam evaporation method was employed to deposit thin films of the SiO2 dielectric layer. Furthermore, the electrical properties of fabricated Ga_2O_3:Si-based MOSFET were analyzed and reported. The circular and square MOSFET devices had maximum IDs of 26.7 mA mm^{-1} and 9.12 mA mm^{-1}, with excellent breakdown voltages of 360 V and 390 V, respectively. Both devices had on/off ratios of more than 10^7 to 10^8. Furthermore, this device was performed effectively up to 250 °C without substantial degradation. The device had shown excellent results and showed great potential for power electronic devices. Also, **Wang *et al.*** (Y. B. Wang et al., 2020) employed the ion-cutting deposition method to fabricate a β-Ga_2O_3-based MOSFET device on a heterogeneous Ga_2O_3-Al_2O_3-Si (GaOISi) substrate. Further, the device was characterized, as the enhancement (E)-mode Ga_2O_3 transistor, with a 15 nm channel thickness (Tch), has a high threshold voltage (VTH) of around 8 V as per the report. The thermal stability of MOSFETs was reported to be outstanding for both ON and OFF current (I_{ON}/I_{OFF}) modes. The breakdown voltages (V_{br}) of 522 and 391 V at 25°C and 200°C, respectively were reported for the device.

Thus, according to the results above discussed few of published report, which are based on Ga_2O_3-based solar blind UV photodetector, sensors, and power electronic (SBDs and MOSFET), indicating excellent performance, that is, their outstanding sensitivity, quick reaction, and durability. These characteristics could improve danger detection, secure communication, imagining system and missile tracking, and guaranteeing strong and dependable defence systems for upcoming security and surveillance missions.

Challenges Associated with the β-Ga_2O_3 Based Next Generation Devices. Ga_2O_3 has been attracted and acts as a beneficial material for next generation defense applications, however there are few challenges to be faced to improving the technology. Some of challenge are explained here.

I. Growth technology:

Growth of β-Ga_2O_3 is still a big challenge for the researcher, scientist as well as industries, achieving high quality epitaxial thin films growth related to control over the grain boundaries, unform doping of materials, phase purity and defects. So, to overcome these challenges, there is a need to develop high advance growth technologies, as well as optimized deposition process for fabrication of any device.

II. Dopant material selection

The second most challengingly job is the selection of the dopant material which can suitably affect the electrical as well as optical properties of the device. As we know that Ga_2O_3 is acts an insulator in room temperature conditions, therefore it needs a dopant material for enhanced these properties(Mandal et al., 2024). In case of device application, a proper dopant material is needed, otherwise there might be chances of lattice mismatch of thin films with the substrate that poses additional growth and/or device characterization.

III. Device contacts fabrication

Device contact is a very key element in the performance of any device. Contacts like Schottky and ohmic might vary with the deposition method as well as the way of preparation. Sometimes without cleaning properly, contamination or dust creates issues in the contacts, therefore, proper cleaning is important. Also, the fabrication of stable and low resistance ohmic contact at the same time fabrication of Schottky contact is still a big barrier for the researcher as well as the scientist.

IV. Long duration of performance of a device.

For any device, the main task is long durable performance in extremely harsh conditions. Here, due to the excellent chemical and thermal stability of Ga_2O_3 is still an emerging material for next generation devices, therefore, device based on Ga_2O_3 require optimized parametric condition to be used in any next generation devices.

V. Thermal management

As we know, that Ga_2O_3 have a very low thermal conductivity of 0.11 $Wcm^{-1}K^{-1}$, which creating issue as self-heating in any power device, therefore Ga_2O_3 thin properties should be enhanced with proper doping and deposition condition before integrating for any device applications.

VI. Compatibility with existing devices

Ga_2O_3-based devices should be compatible to integrate with existing technology i.e. SiC-based devices for testing and commercialization.

CONCLUSION

This review presents a comprehensive overview of the background, physical properties, and potential applications of ultra-wide bandgap β-Ga_2O_3 for next-generation defence technologies. Key thin-film growth techniques, including PLD, MBE, MOCVD, PVD RF sputtering, mist-CVD, and Sol-gel method, are briefly discussed. Additionally, critical challenges such as growth-induced defects, thermal management, and effective doping strategies for optimizing optical and electrical properties are covered. Despite these challenges, ongoing research and technological advancements in fabrication methods, material engineering, and device integration are expected to unlock the full potential of β-Ga_2O_3. Progress in these areas could lead to the realization of high-performance, compact, and resilient defence systems, marking significant strides in next-generation electronic and optoelectronic applications.

CURRENT & FUTURE DEVELOPMENTS

β-Ga_2O_3 has been a much-researched next generation material to scientists, researchers and semiconductor industry, particularly in Défense applications, due to its exceptional material properties compared to other wide bandgap (WBG) semiconductors like SiC, diamond, AlGaN, ZnO, and ZnMgO, as discussed earlier (J. Xu et al., 2019) (Chi et al., 2022). Due to the remarkable material properties, including a wide energy bandgap (~4.6-4.9 eV), a high breakdown voltage (~8 MV/cm), thermal stability at harsh conditions, and cost-effective fabrication process, making it suitable for next-generation defines applications (J. Xu et al., 2019). In terms of performance for next-generation defence applications, devices produced using alternative wide bandgap semiconductors such as SiC, diamond, AlGaN, ZnO, ZnMgO, etc. fared badly when compared to Ga_2O_3-based next-generation devices (J. Xu et al., 2019). In the case of SiC-based devices, it has been showing considerable potential for optoelectronic devices; nevertheless, they are becoming increasingly limited due to their higher breakdown voltage and manufacturing costs when compared to Ga_2O_3-based next-generation devices. Moreover, AlGaN-based devices have shown poor performance due to a lack of power handling capacity and versatility. Similarly, ZnMgO-based next-generation devices exhibited a narrow achieving band energy gap (up to 4.5 eV), thus tend to reduce performance and natural diamond-based devices are difficult to process because of their extremely high thermal conductivity, which limits their ability to expand.

As a result, with improvement in epitaxial growth, thermal management etc as discussed above in challenge associated sections, wide bandgap β-Ga_2O_3-based devices have the potential to entirely alter the next generations defence technology by providing systems that are more durable, portable, and more efficient.

ACKNOWLEDGEMENT

I would like to thanks to the KIIT Deemed to be University, India and University of Pardubice, Czech Republic for their kind support for writing this book chapter, also the AI tool to improve the language of the manuscript.

REFERENCES

Almaev, A. V., Chernikov, E. V., Davletkildeev, N. A., & Sokolov, D. V. (2020). Oxygen sensors based on gallium oxide thin films with addition of chromium. *Superlattices and Microstructures*, *139*(January), 106392. DOI: 10.1016/j.spmi.2020.106392

An, Y., Dai, L., Wu, Y., Wu, B., Zhao, Y., Liu, T., Hao, H., Li, Z., Niu, G., Zhang, J., Quan, Z., & Ding, S. (2019). Epitaxial growth of β -Ga2O3 thin films on Ga2O3 and Al2O3 substrates by using pulsed laser deposition. *Journal of Advanced Dielectrics*, *9*(4), 3–9. DOI: 10.1142/S2010135X19500322

Arora, K., & Kumar, M. (2020). Sputtered-Growth of High-Temperature Seed-Layer Assisted β-Ga 2 O 3 Thin Film on Silicon-Substrate for Cost-Effective Solar-Blind Photodetector Application. *ECS Journal of Solid State Science and Technology : JSS*, *9*(6), 065013. DOI: 10.1149/2162-8777/aba7fd

Blumenschein, N., Paskova, T., & Muth, J. F. (2019). Effect of Growth Pressure on PLD-Deposited Gallium Oxide Thin Films for Deep-UV Photodetectors. *Physica Status Solidi (A). Physica Status Solidi. A, Applications and Materials Science*, *216*(20), 1900098. Advance online publication. DOI: 10.1002/pssa.201900098

Bokov, D., Turki Jalil, A., Chupradit, S., Suksatan, W., Javed Ansari, M., Shewael, I. H., Valiev, G. H., & Kianfar, E. (2021). Nanomaterial by Sol-Gel Method: Synthesis and Application. *Advances in Materials Science and Engineering*, *2021*(December), 5102014. Advance online publication. DOI: 10.1155/2021/5102014

Buckner, C. A., Lafrenie, R. M., Dénommée, J. A., Caswell, J. M., Want, D. A., Gan, G. G., Leong, Y. C., Bee, P. C., Chin, E., Teh, A. K. H., Picco, S., Villegas, L., Tonelli, F., Merlo, M., Rigau, J., Diaz, D., Masuelli, M., Korrapati, S., Kurra, P., . . . Mathijssen, R. H. J. (2016). We are IntechOpen, the world ' s leading publisher of Open Access books Built by scientists, for scientists TOP 1%. *Intech, 11*(tourism), 13. https://www.intechopen.com/books/advanced-biometric-technologies/liveness-detection-in-biometrics

Busch, G. (1989). Early history of the physics and chemistry of semiconductors-from doubts to fact in a hundred years. *European Journal of Physics, 10*(4), 254–264. DOI: 10.1088/0143-0807/10/4/002

Capan, I. (2024). Wide-Bandgap Semiconductors for Radiation Detection: A Review. *Materials (Basel), 17*(5), 1147. Advance online publication. DOI: 10.3390/ma17051147 PMID: 38473617

Chen, M., Zhang, Z., Zhan, R., She, J., Deng, S., Xu, N., & Chen, J. (2021). Fast-response X-ray detector based on nanocrystalline Ga2O3 thin film prepared at room temperature. *Applied Surface Science, 554*(December 2020), 149619. DOI: 10.1016/j.apsusc.2021.149619

Chen, X., Mi, W., Wu, J., Yang, Z., Zhang, K., Zhao, J., Luan, C., & Wei, Y. L. (2020). A solar-blind photodetector based on β-Ga2O3 film deposited on MgO (100) substrates by RF magnetron sputtering. *Vacuum, 180*(May), 109632. DOI: 10.1016/j.vacuum.2020.109632

Cora, I., Mezzadri, F., Boschi, F., Bosi, M., Čaplovičová, M., Calestani, G., Dódony, I., Pécz, B., & Fornari, R. (2017). The real structure of ε-Ga2O3 and its relation to κ-phase. *CrystEngComm, 19*(11), 1509–1516. DOI: 10.1039/C7CE00123A

de Boisbaudran, L. (1875). On the chemical and spectroscopic characters of a new metal (gallium). *The London, Edinburgh and Dublin Philosophical Magazine and Journal of Science, 50*(332), 414–416. DOI: 10.1080/14786447508641312

Dong, L., Jia, R., Xin, B., Peng, B., & Zhang, Y. (2017). Effects of oxygen vacancies on the structural and optical properties of β-Ga2O3. *Scientific Reports, 7*(September 2016), 1–12. DOI: 10.1038/srep40160

Du, X., Mei, Z., Liu, Z., Guo, Y., Zhang, T., Hou, Y., Zhang, Z., Xue, Q., & Kuznetsov, A. Y. (2009). Controlled growth of high-quality ZnO-based films and fabrication of visible-blind and solar-blind ultra-violet detectors. *Advanced Materials, 21*(45), 4625–4630. DOI: 10.1002/adma.200901108

Fontana, S., & Di Lauro, F. (2022). An Overview of Sensors for Long Range Missile Defense. *Sensors (Basel), 22*(24), 1–19. DOI: 10.3390/s22249871 PMID: 36560236

Galazka, Z. (2022). Growth of bulk β-Ga2O3single crystals by the Czochralski method. *Journal of Applied Physics, 131*(3), 031103. Advance online publication. DOI: 10.1063/5.0076962

Gorokhov, E. V., Magunov, A. N., Feshchenko, V. S., & Altukhov, A. A. (2008). Solar-blind UV flame detector based on natural diamond. *Instruments and Experimental Techniques, 51*(2), 280–283. DOI: 10.1134/S002044120802022X

Guo, Z., Verma, A., Wu, X., Sun, F., Hickman, A., Masui, T., Kuramata, A., Higashiwaki, M., Jena, D., & Luo, T. (2015). Anisotropic thermal conductivity in single crystal β-gallium oxide. *Applied Physics Letters*, *106*(11), 111909. Advance online publication. DOI: 10.1063/1.4916078

Ha, M.-T., Kim, K.-H., Kwon, Y.-J., Kim, C.-J., Jeong, S.-M., & Bae, S.-Y. (2019). Understanding Thickness Uniformity of Ga 2 O 3 Thin Films Grown by Mist Chemical Vapor Deposition. *ECS Journal of Solid State Science and Technology : JSS*, *8*(7), Q3206–Q3212. DOI: 10.1149/2.0381907jss

Haider, A. J., Alawsi, T., Haider, M. J., Taha, B. A., & Marhoon, H. A. (2022). A comprehensive review on pulsed laser deposition technique to effective nanostructure production: Trends and challenges. *Optical and Quantum Electronics*, *54*(8), 1–25. DOI: 10.1007/s11082-022-03786-6

Hajnal, Z., Miró, J., Kiss, G., Réti, F., Deák, P., Herndon, R. C., & Kuperberg, J. M. (1999). Role of oxygen vacancy defect states in the n-type conduction of β-Ga2O3. *Journal of Applied Physics*, *86*(7), 3792–3796. DOI: 10.1063/1.371289

Han, B., Yu, X., Zhang, K., Kwon, E., & Ou, J. (2011). Sensing properties of CNT-filled cement-based stress sensors. *Journal of Civil Structural Health Monitoring*, *1*(1–2), 17–24. DOI: 10.1007/s13349-010-0001-5

Heinselman, K., Walker, P., Norman, A., Parilla, P., Ginley, D., & Zakutayev, A. (2021). Performance and reliability of β-Ga2O3 Schottky barrier diodes at high temperature. *Journal of Vacuum Science & Technology. A, Vacuum, Surfaces, and Films*, *39*(4), 1–7. DOI: 10.1116/6.0001003

Higashiwaki, M. (2021). β-Gallium Oxide Devices: Progress and Outlook. *Physica Status Solidi. Rapid Research Letters*, *15*(11), 2100357. Advance online publication. DOI: 10.1002/pssr.202100357

Higashiwaki, M., Sasaki, K., Kuramata, A., Masui, T., & Yamakoshi, S. (2014). Development of gallium oxide power devices. *Physica Status Solidi (A). Physica Status Solidi. A, Applications and Materials Science*, *211*(1), 21–26. DOI: 10.1002/pssa.201330197

Higashiwaki, M., & Wong, M. H. (2024). Beta-Gallium Oxide Material and Device Technologies. *Annual Review of Materials Research*, *54*(1), 175–198. DOI: 10.1146/annurev-matsci-080921-104058

Hu, J., Yu, B., & Zhou, J. (2023). New Generation Beta-Gallium Oxide Nanomaterials: Growth and Performances in Electronic Devices. *Advanced Engineering Materials*, *25*(19), 1–18. DOI: 10.1002/adem.202300688

Ji, X., Wang, J., Qi, S., Liang, Y., Hu, S., Zheng, H., Zhang, S., Yue, J., Qi, X., Li, S., Liu, Z., Shu, L., Tang, W., & Li, P. (2024). Improvement of Ga2O3 vertical Schottky barrier diode by constructing NiO/Ga2O3 heterojunction. *Journal of Semiconductors, 45*(4), 042503. Advance online publication. DOI: 10.1088/1674-4926/45/4/042503

Kalygina, V., Podzyvalov, S., Yudin, N., Slyunko, E., Zinoviev, M., Kuznetsov, V., Lysenko, A., Kalsin, A., Kopiev, V., Kushnarev, B., Oleinik, V., Baalbaki, H., & Yunin, P. (2023). Effect of UV and IR Radiation on the Electrical Characteristics of Ga2O3/ZnGeP2 Hetero-Structures. *Crystals, 13*(8), 1203. Advance online publication. DOI: 10.3390/cryst13081203

Kaur, D., Debata, S., Pratap Singh, D., & Kumar, M. (2023). Strain effects on the optoelectronic performance of ultra-wide band gap polycrystalline β-Ga2O3 thin film grown on differently-oriented Silicon substrates for solar blind photodetector. *Applied Surface Science, 616*(November 2022), 156446. DOI: 10.1016/j.apsusc.2023.156446

Kaur, D., & Kumar, M. (2021). A Strategic Review on Gallium Oxide Based Deep-Ultraviolet Photodetectors: Recent Progress and Future Prospects. *Advanced Optical Materials, 9*(9), 1–34. DOI: 10.1002/adom.202002160

Kohn, W., Becke, A. D., & Parr, R. G. (1996). Density functional theory of electronic structure. *Journal of Physical Chemistry, 100*(31), 12974–12980. DOI: 10.1021/jp9606691

Kumar, K., & Kaur, D. (2024). A review on recent advancements in the growth of MoS2 based flexible photodetectors. *Solar Energy Materials and Solar Cells, 268*(August 2023), 112736. DOI: 10.1016/j.solmat.2024.112736

Lee, S. H., Kim, S., Moon, Y.-J., Kim, S. M., Jung, H. J., Seo, M. S., Lee, K. M., Kim, S.-K., & Lee, S. W. (2017). High-Responsivity Deep-Ultraviolet-Selective Photodetectors Using Ultrathin Gallium Oxide Films. *ACS Photonics, 4*(11), 2937–2943. DOI: 10.1021/acsphotonics.7b01054

Lee, S. W., Lee, H., Park, Y., Kim, H., Somorjai, G. A., & Park, J. Y. (2021). Surface chemistry of hot electron and metal-oxide interfaces. *Surface Science Reports, 76*(3), 100532. DOI: 10.1016/j.surfrep.2021.100532

Liu, D., Huang, Y., Zhang, Z., Li, Z., Yan, Y., Chen, D., Zhao, S., Feng, Q., Zhang, J., Zhang, C., & Hao, Y. (2024). Strategy to solve the thermal issue of ultra-wide bandgap semiconductor gallium oxide field effect transistor. *Journal of Alloys and Compounds, 986*(March), 174143. DOI: 10.1016/j.jallcom.2024.174143

Liu, H., Zhou, S., Zhang, H., Ye, L., Xiong, Y., & Yu, P. (2022). *Ultrasensitive fully transparent amorphous Ga 2 O 3 solar-blind deep-ultraviolet photodetector for.*

Liu, X., Liu, Q., Zhao, B., Ren, Y., Tao, B. W., & Zhang, W. L. (2020). Comparison of β-Ga2O3 thin films grown on r-plane and c-plane sapphire substrates. *Vacuum, 178*(April), 109435. DOI: 10.1016/j.vacuum.2020.109435

Liu, Z., Wang, X., Liu, Y., Guo, D., Li, S., Yan, Z., Tan, C. K., Li, W., Li, P., & Tang, W. (2019). A high-performance ultraviolet solar-blind photodetector based on a β-Ga2O3 Schottky photodiode. *Journal of Materials Chemistry. C, Materials for Optical and Electronic Devices*, *7*(44), 13920–13929. DOI: 10.1039/C9TC04912F

Ma, X., Zhang, Y., Dong, L., & Jia, R. (2017). First-principles calculations of electronic and optical properties of aluminum-doped β-Ga2O3 with intrinsic defects. *Results in Physics*, *7*, 1582–1589. DOI: 10.1016/j.rinp.2017.04.023

Mandal, P., Kumar, S., Pandey, A., Katariya, L., Mondal, A., & Bag, A. (2024). Unveiling structural and optical properties of Sn-doped β-Ga2O3: A correlation of experimental and theoretical observations. *Materials Science and Engineering: B, 302*(December 2023). DOI: 10.1016/j.mseb.2024.117266

Mandal, P., Mondal, A., Pandey, A., Romeo, S., & Bag, A. (2025). Investigating role of annealing in shaping morphological, structural, tribological, and optical characteristics of gallium oxide (Ga 2 O 3): Insights from DFT analysis. *Journal of Alloys and Compounds, 1011*(December 2024), 178367. DOI: 10.1016/j.jallcom.2024.178367

Mandal, P., Roy, S., & Singh, U. P. (2022). Investigation on the optical and electrical performance of aluminium doped gallium oxide thin films. *Optical and Quantum Electronics*, *54*(8), 1–15. DOI: 10.1007/s11082-022-03851-0

Mandal, P., Roy, S., & Singh, U. P. (2023). Optimization of RF sputtering parameters for improving the optical and electrical properties of β-gallium oxide thin films. *Bulletin of Materials Science*, *46*(2), 64. Advance online publication. DOI: 10.1007/s12034-023-02888-5

Mandal, P., Singh, U. P., & Roy, S. (2023). Structure-property correlation of europium doped β-Ga2O3: An experimental and first-principle study. *Journal of Materials Science Materials in Electronics*, *34*(30), 1–16. DOI: 10.1007/s10854-023-11455-x

Marezio, M., & Remeika, J. P. (1967). Bond lengths in the α-Ga2O3 structure and the high-pressure phase of Ga2-xFexO3. *The Journal of Chemical Physics*, *46*(5), 1862–1865. DOI: 10.1063/1.1840945

Martín-Palma, R. J., & Lakhtakia, A. (2013). Vapor-Deposition Techniques. In *Engineered Biomimicry*. Elsevier Inc., DOI: 10.1016/B978-0-12-415995-2.00015-5

Mondal, A., Nandi, S., Yadav, M. K., Nandi, A., & Bag, A. (2022). Broad Range (254-302 nm) and High Performance Ga2O3:SnO2Based Deep UV Photodetector. *IEEE Transactions on Nanotechnology*, *21*, 320–327. DOI: 10.1109/TNANO.2022.3186472

Mondal, A., Yadav, M. K., Shringi, S., & Bag, A. (2020). Extremely low dark current and detection range extension of Ga2O3 UV photodetector using Sn alloyed nanostructures. *Nanotechnology*, *31*(29), 294002. Advance online publication. DOI: 10.1088/1361-6528/ab82d4 PMID: 32208373

Onuma, T., Fujioka, S., Yamaguchi, T., Higashiwaki, M., Sasaki, K., Masui, T., & Honda, T. (2013). Correlation between blue luminescence intensity and resistivity in β-Ga2O3 single crystals. *Applied Physics Letters*, *103*(4), 22–25. DOI: 10.1063/1.4816759

Panda, S., Mehlawat, S., Dhariwal, N., Sanger, A., & Kumar, A. (2024). Comprehensive review on gas sensors: Unveiling recent developments and addressing challenges. *Materials Science and Engineering B*, *308*(August), 117616. DOI: 10.1016/j.mseb.2024.117616

Park, J. H., McClintock, R., Jaud, A., Dehzangi, A., & Razeghi, M. (2019). MOCVD grown β-Ga2O3 metal-oxide-semiconductor field effect transistors on sapphire. *Applied Physics Express*, *12*(9), 0–5. DOI: 10.7567/1882-0786/ab3b2a

Playford, H. Y., Hannon, A. C., Barney, E. R., & Walton, R. I. (2013). Structures of uncharacterised polymorphs of gallium oxide from total neutron diffraction. *Chemistry (Weinheim an der Bergstrasse, Germany)*, *19*(8), 2803–2813. DOI: 10.1002/chem.201203359 PMID: 23307528

Pohl, K. (1968). Hydrothermale Bildung von γ-Ga2O3. *Naturwissenschaften*, *55*(2), 82. DOI: 10.1007/BF00599490

Razeghi, M., Park, J.-H., & McClintock, R. (2019). (Invited) MOCVD-Grown Ga 2 O 3 Field Effect Transistors on Sapphire. *ECS Meeting Abstracts*, *MA2019-02*(31), 1356–1356. DOI: 10.1149/MA2019-02/31/1356

Rogalski, A., Bielecki, Z., Mikołajczyk, J., & Wojtas, J. (2023). Ultraviolet Photodetectors: From Photocathodes to Low-Dimensional Solids. *Sensors (Basel)*, *23*(9), 280–315. DOI: 10.3390/s23094452 PMID: 37177656

Roy, R., Hill, V. G., & Osborn, E. F. (1952). Polymorphism of Ga2O3and the System Ga2O3−H2O. *Journal of the American Chemical Society*, *74*(3), 719–722. DOI: 10.1021/ja01123a039

Saikumar, A. K., Nehate, S. D., & Sundaram, K. B. (2021). A review of recent developments in aluminum gallium oxide thin films and devices. *Critical Reviews in Solid State and Material Sciences*, *0*(0), 1–32. DOI: 10.1080/10408436.2021.1922357

Sang, L., Liao, M., & Sumiya, M. (2013). A comprehensive review of semiconductor ultraviolet photodetectors: From thin film to one-dimensional nanostructures. *Sensors (Basel)*, *13*(8), 10482–10518. DOI: 10.3390/s130810482 PMID: 23945739

Shen, H., Yin, Y., Tian, K., Baskaran, K., Duan, L., Zhao, X., & Tiwari, A. (2018). Growth and characterization of β-Ga2O3 thin films by sol-gel method for fast-response solar-blind ultraviolet photodetectors. *Journal of Alloys and Compounds*, *766*, 601–608. DOI: 10.1016/j.jallcom.2018.06.313

Sheoran, H., Fang, S., Liang, F., Huang, Z., Kaushik, S., Manikanthababu, N., Zhao, X., Sun, H., Singh, R., & Long, S. (2022). High Performance of Zero-Power-Consumption MOCVD-Grown β-Ga2O3-Based Solar-Blind Photodetectors with Ultralow Dark Current and High-Temperature Functionalities. *ACS Applied Materials & Interfaces*, *14*(46), 52096–52107. DOI: 10.1021/acsami.2c08511 PMID: 36346904

Shivani, K., Kaur, D., Ghosh, A., & Kumar, M. (2022). A strategic review on gallium oxide based power electronics: Recent progress and future prospects. *Materials Today. Communications*, *33*(August), 104244. DOI: 10.1016/j.mtcomm.2022.104244

Singh Pratiyush, A., Krishnamoorthy, S., Vishnu Solanke, S., Xia, Z., Muralidharan, R., Rajan, S., & Nath, D. N. (2017). High responsivity in molecular beam epitaxy grown β-Ga2O3 metal semiconductor metal solar blind deep-UV photodetector. *Applied Physics Letters*, *110*(22), 221107. Advance online publication. DOI: 10.1063/1.4984904

Song, Z., Liao, J., Ding, X., Zhou, T., & Liu, Q. L. (2012). Stability of divalent/trivalent oxidation state of europium in some Sr-based inorganic compounds. *Journal of Luminescence*, *132*(7), 1768–1773. DOI: 10.1016/j.jlumin.2012.02.013

Stepanov, S. I., Nikolaev, V. I., Bougrov, V. E., & Romanov, A. E. (2016). Gallium oxide: Properties and applications - A review. *Reviews on Advanced Materials Science*, *44*(1), 63–86.

Suchikova, Y., Nazarovets, S., & Popov, A. I. (2024). Ga_2O_3 Solar-Blind Photodetectors: From Civilian Applications to Missile Detection and Research Agenda. *Optical Materials*, *157*(P3), 116397. DOI: 10.1016/j.optmat.2024.116397

Tak, B. R., Kumar, S., Kapoor, A. K., Wang, D., Li, X., Sun, H., & Singh, R. (2021). Recent advances in the growth of gallium oxide thin films employing various growth techniques - A review. *Journal of Physics. D, Applied Physics*, *54*(45), 453002. Advance online publication. DOI: 10.1088/1361-6463/ac1af2

Ueda, N., Hosono, H., Waseda, R., & Kawazoe, H. (1997). Synthesis and control of conductivity of ultraviolet transmitting β-Ga2O3 single crystals. *Applied Physics Letters*, *70*(26), 3561–3563. DOI: 10.1063/1.119233

Varley, J. B., Weber, J. R., Janotti, A., & Van De Walle, C. G. (2010). Oxygen vacancies and donor impurities in β-Ga2 O 3. *Applied Physics Letters*, *97*(14), 10–13. DOI: 10.1063/1.3499306

Vetelino, J. F., & Reghu, A. (2017). Introduction to sensors. *Introduction to Sensors*, (November), 1–180. DOI: 10.1201/9781315218274

Wang, C., Fan, W. H., Zhang, Y. C., Kang, P. C., Wu, W. Y., Wuu, D. S., Lien, S. Y., & Zhu, W. Z. (2022). Effect of oxygen flow ratio on the performance of RF magnetron sputtered Sn-doped Ga2O3 films and ultraviolet photodetector. *Ceramics International*, (August). Advance online publication. DOI: 10.1016/j.ceramint.2022.11.251

Wang, Y. B., Xu, W. H., You, T. G., Mu, F. W., Hu, H. D., Liu, Y., Huang, H., Suga, T., Han, G. Q., Ou, X., & Hao, Y. (2020). β-Ga2O3 MOSFETs on the Si substrate fabricated by the ion-cutting process. *Science China. Physics, Mechanics & Astronomy*, *63*(7), 1–4. DOI: 10.1007/s11433-020-1533-0

Wong, M. H., Bierwagen, O., Kaplar, R. J., & Umezawa, H. (2021). Ultrawide-bandgap semiconductors: An overview. *Journal of Materials Research*, *36*(23), 4601–4615. DOI: 10.1557/s43578-021-00458-1

Wu, C., Wu, F., Hu, H., Wang, S., Liu, A., & Guo, D. (2022). Review of self-powered solar-blind photodetectors based on Ga2O3. *Materials Today Physics*, *28*(July), 100883. DOI: 10.1016/j.mtphys.2022.100883

Wu, H., Meng, W., Zhu, C., Tian, Z., Ma, F., & Jiao, Y. (2024). Two-dimensional monolayer BiSnO3: A novel wide-band-gap semiconductor with high stability and strong ultraviolet absorption. *Journal of Physics Condensed Matter*, *36*(30), 305501. Advance online publication. DOI: 10.1088/1361-648X/ad3da5 PMID: 38604165

Xu, J., Zheng, W., & Huang, F. (2019). Gallium oxide solar-blind ultraviolet photodetectors: A review. *Journal of Materials Chemistry. C, Materials for Optical and Electronic Devices*, *7*(29), 8753–8770. DOI: 10.1039/C9TC02055A

Xu, Y., An, Z., Zhang, L., Feng, Q., Zhang, J., Zhang, C., & Hao, Y. (2018). Solar blind deep ultraviolet β-Ga 2 O 3 photodetectors grown on sapphire by the Mist-CVD method. *Optical Materials Express*, *8*(9), 2941. DOI: 10.1364/OME.8.002941

Yoshioka, S., Hayashi, H., Kuwabara, A., Oba, F., Matsunaga, K., & Tanaka, I. (2007). Structures and energetics of Ga2O3 polymorphs. *Journal of Physics Condensed Matter*, *19*(34), 346211. Advance online publication. DOI: 10.1088/0953-8984/19/34/346211

Yu, F.-P., Ou, S.-L., & Wuu, D.-S. (2015). Pulsed laser deposition of gallium oxide films for high performance solar-blind photodetectors. *Optical Materials Express*, *5*(5), 1240. DOI: 10.1364/OME.5.001240

Zhang, J., Shi, J., Qi, D. C., Chen, L., & Zhang, K. H. L. (2020). Recent progress on the electronic structure, defect, and doping properties of Ga2O3. *APL Materials*, *8*(2), 020906. Advance online publication. DOI: 10.1063/1.5142999

Zhang, Q., Liu, J., Tu, C., Zhai, D., He, M., & Lu, J. (2023). High-performance β-Ga2O3 Schottky barrier diodes and metal-semiconductor field-effect transistors on a high doping level epitaxial layer. *Journal of Alloys and Compounds*, *939*, 168732. DOI: 10.1016/j.jallcom.2023.168732

Zhang, S., Liu, Z., Liu, Y., Zhi, Y., Li, P., Wu, Z., & Tang, W. (2021). Electrical characterizations of planar ga2o3 schottky barrier diodes. *Micromachines*, *12*(3), 1–8. DOI: 10.3390/mi12030259 PMID: 33802423

Zhang, S., Wang, J., Ji, X., Yan, Z., Ye, L., Zheng, H., Liu, Y., Chen, X., & Li, P. (2024). Oxygen vacancies modulating performance for Ga2O3 solar-blind photodetectors via low-cost mist chemical vapor deposition. *Materials Today. Communications*, *39*(February), 108717. DOI: 10.1016/j.mtcomm.2024.108717

Zhang, Y., Yan, J., Zhao, G., & Xie, W. (2010). First-principles study on electronic structure and optical properties of Sn-doped β-Ga2O3. *Physica B, Condensed Matter*, *405*(18), 3899–3903. DOI: 10.1016/j.physb.2010.06.024

Zhi, Y. S., Jiang, W. Y., Liu, Z., Liu, Y. Y., Chu, X. L., Liu, J. H., Li, S., Yan, Z. Y., Wang, Y. H., Li, P. G., Wu, Z. P., & Tang, W. H. (2021). High-responsivity solar-blind photodetector based on MOCVD-grown Si-doped β-Ga2O3thin film. *Chinese Physics B*, *30*(5), 057301. Advance online publication. DOI: 10.1088/1674-1056/abe37a

Zhou, J., Zhan, J., Lv, B., Guo, Y., & Jiang, B. (2024). Deposition Contribution Rates and Simulation Model Refinement for Polysilicon Films Deposited by Large-Sized Tubular Low-Pressure Chemical Vapor Deposition Reactors. *Materials (Basel), 17*(23), 1–13. DOI: 10.3390/ma17235952 PMID: 39685388

Zhou, W., Xia, C., Sai, Q., & Zhang, H. (2017). Controlling n-type conductivity of β-Ga2O3 by Nb doping. *Applied Physics Letters, 111*(24), 1–5. DOI: 10.1063/1.4994263

Zhu, J., Xu, Z., Ha, S., Li, D., Zhang, K., Zhang, H., & Feng, J. (2022). Gallium Oxide for Gas Sensor Applications: A Comprehensive Review. *Materials (Basel), 15*(20), 7339. Advance online publication. DOI: 10.3390/ma15207339 PMID: 36295403

Compilation of References

Abbas, M., Akhai, S., Abbas, U., Jafri, R., & Arif, S. M. (2025). AI-enabled sustainable urban planning and management. In *Real-World Applications of AI Innovation* (pp. 233–260). IGI Global Scientific Publishing.

Abbass, M., Abbas, U., Jafri, R., Arif, S. M., & Akhai, S. (2025). AI and Machine Learning Applications in Sustainable Smart Cities. In *Sustainable Smart Cities and the Future of Urban Development* (pp. 1–32). IGI Global Scientific Publishing.

Abdel Hamid, Z. (2014). Review article: Composite and nanocomposite coatings. *J Metall Eng*, *3*, 29–42.

Abdelgawad, A., & Yelamarthi, K. (2017). Internet of things (IoT) platform for structure health monitoring. *Wireless Communications and Mobile Computing*, *2017*(1), 6560797. DOI: 10.1155/2017/6560797

Abed, M. S., & Jawad, Z. A. (2022). Nanotechnology for defence applications. In *Nanotechnology for electronic applications* (pp. 187–205). Springer Nature Singapore. DOI: 10.1007/978-981-16-6022-1_10

Ablakatov, I. K., Baiserikov, B. M., Ismailov, M. B., & Nurgozhin, M. R. (2022). Aluminum-lithium alloys: types, properties, application, and production technologies. Overview. *Kompleksnoe Ispolzovanie Mineralnogo Syra= Complex use of mineral resources, 323*(4), 5-14.

Abo-Neima, S. E., Motaweh, H. A., & Elsehly, E. M. (2020). Antimicrobial activity of functionalised carbon nanotubes against pathogenic microorganisms. *IET Nanobiotechnology / IET*, *14*(6), 457–464. DOI: 10.1049/iet-nbt.2019.0342 PMID: 32755954

Adeleke, A. K., Montero, D. J. P., Lottu, O. A., Ninduwezuor-Ehiobu, N., & Ani, E. C. (2024). 3d printing in aerospace and defense: A review of technological breakthroughs and applications. *World Journal of Advanced Research and Reviews*, *21*(2), 1149–1160. DOI: 10.30574/wjarr.2024.21.2.0558

Adir, O., Poley, M., Chen, G., Froim, S., Krinsky, N., Shklover, J., Shainsky-Roitman, J., Lammers, T., & Schroeder, A. (2020). Integrating artificial intelligence and nanotechnology for precision cancer medicine. *Advanced Materials*, *32*(13), 1901989. DOI: 10.1002/adma.201901989 PMID: 31286573

Agarwal, N., Rangamani, A., Bhavsar, K., Virnodkar, S. S., Fernandes, A. A. A., Chadha, U., Srivastava, D., Patterson, A. E., & Rajasekharan, V. (2024). An overview of carbon-carbon composite materials and their applications. *Frontiers in Materials*, *11*, 1374034. DOI: 10.3389/fmats.2024.1374034

Aglawe, K., Giri, S., Dhande, M., & Shelare, S. (2023, September). Application of aluminum alloys in aviation industry: A review. In *AIP Conference Proceedings* (Vol. 2800, No. 1). AIP Publishing. DOI: 10.1063/5.0163002

Agrawal, R., & Mukhopadhyay, A. (2023). The use of machine learning and meta-heuristic algorithm for wear performance optimization of AISI 1040 steel and investigation of corrosion resistance. *Proceedings of the Institution of Mechanical Engineers. Part J, Journal of Engineering Tribology*, *237*(4), 701–717. DOI: 10.1177/13506501221094416

Aizenberg, J., Weaver, J. C., Thanawala, M. S., Sundar, V. C., Morse, D. E., & Fratzl, P. (2005). Skeleton of Euplectella sp.: Structural hierarchy from the nanoscale to the macroscale. *Science*, *309*(5732), 275–278. DOI: 10.1126/science.1112255 PMID: 16002612

Ajirotutu, R. O., Adeyemi, A. B., Ifechukwu, G.-O., Iwuanyanwu, O., Ohakawa, T. C., & Garba, B. M. P. (2024). Future cities and sustainable development: Integrating renewable energy, advanced materials, and civil engineering for urban resilience. *International Journal of Sustainable Urban Development*.

Akafuah, N. K., Poozesh, S., Salaimeh, A., Patrick, G., Lawler, K., & Saito, K. (2016). Evolution of the automotive body coating process—A review. *Coatings*, *6*(2), 24. DOI: 10.3390/coatings6020024

Akhai, S. (2024). A Review on Optimizations in μ-EDM Machining of the Biomedical Material Ti6Al4V Using the Taguchi Method: Recent Advances Since 2020. *Latest Trends in Engineering and Technology*, 395-402.

Akhai, S. (2024). Innovating Healthcare Delivery: Harnessing Drones for Medical Services. In *Modelling of Virtual Worlds Using the Internet of Things* (Vol. 1, pp. 232-250). CRC Press.

Akhai, S. (2024). Smart Home Environment for Society: An Overview of IoT-Enabled Air Conditioners. In *Modelling of Virtual Worlds Using the Internet of Things* (Vol. 1, pp. 74-87). CRC Press.

Akhai, S., & Abbass, M. (2025). Toward Resilient Futures: The Role of AI-Driven Strategies in Climate Adaptation. In *Nexus of AI, Climatology, and Urbanism for Smart Cities* (pp. 325-340). IGI Global Scientific Publishing.

Akhai, S., & Khang, A. (2024). Energy Efficiency and Human Comfort: AI and IoT Integration in Hospital HVAC Systems. *Medical Robotics and AI-Assisted Diagnostics for a High-Tech Healthcare Industry*, 93-108.

Akhai, S., & Khang, A. (2024). Innovations in Medical Manufacturing: A Review of 3D Printing, Robotics, and Internet of Things (IoT). In *The Quantum Evolution* (Vol. 1, pp. 226-241). CRC Press.

Akhai, S., & Kumar, V. (2024). Quantum resilience and distributed trust: The promise of blockchain and quantum computing in defense. In *Sustainable Security Practices Using Blockchain, Quantum and Post-Quantum Technologies for Real Time Applications* (pp. 125-153). Springer Nature Singapore.

Akhai, S., Abbass, M., Kaur, P., & Kaur, T. (2025). Digital Transformation Across Generations: Robotics and AI in Action. In *Impacts of Digital Technologies Across Generations* (pp. 23-40). IGI Global Scientific Publishing.

Akhai, S., Singh, S., Kaur, S., & Kaur, K. (2024). Advancing Aerospace Inspection With Terahertz Technology. In *New Innovations in AI, Aviation, and Air Traffic Technology* (pp. 195-211). IGI Global. DOI: 10.4018/979-8-3693-1954-3.ch010

Akhai, S., Walia, J., Bhullar, B. S., & Wadhwa, A. S. (2024). Advances in Bandgap and Conductivity Engineering of Graphene-Based Semiconductors. In Carbon-Based Materials and Environmental Remediation: Graphene, Biochar, and More (pp. 211-228). IGI Global Scientific Publishing. DOI: 10.4018/979-8-3693-8257-8.ch007

Akhai, S. (2023). Navigating the Potential Applications and Challenges of Intelligent and Sustainable Manufacturing for a Greener Future. *Evergreen*, *10*(4), 2237–2243. DOI: 10.5109/7160899

Akhai, S., Srivastava, P., Sharma, V., & Bhatia, A. (2021). Investigating weld strength of AA8011-6062 alloys joined via friction-stir welding using the RSM approach. [). IOP Publishing.]. *Journal of Physics: Conference Series*, *1950*(1), 012016. DOI: 10.1088/1742-6596/1950/1/012016

Akhai, S., & Wadhwa, A. S. (2024). Recent Advances in Bio-Tribology From Joint Lubrication to Medical Implants: A Review. *Journal of Materials*, *2*(2), 125–135.

Akhai, S., & Wadhwa, A. S. (2024). Wireless networking solutions for climate control and comfort in modern air conditioning systems. In *Intelligent Networks* (pp. 165–181). CRC Press. DOI: 10.1201/9781003541363-11

Akhtar, N., Ahmad, T., Husain, D., Majdi, A., Alam, M. T., Husain, N., & Wayal, A. K. S. (2022). Ecological footprint and economic assessment of conventional and geopolymer concrete for sustainable construction. *Journal of Cleaner Production*, *380*, 134910. DOI: 10.1016/j.jclepro.2022.134910

Alajami, A. A., Moreno, G., & Pous, R. (2022). Design of a uav for autonomous rfid-based dynamic inventories using stigmergy for mapless indoor environments. *Drones (Basel)*, *6*(8), 208. DOI: 10.3390/drones6080208

Alamdari, H. (2017). Aluminium production process: Challenges and opportunities. *Metals*, *7*(4), 133. DOI: 10.3390/met7040133

Al-Atroush, M. E., Marouf, A., Aloufi, M., & Marouf, M. (2023). A novel application of the geothermal asphalt pavement: A feasible E-fuel source. *Transportation Engineering*, *12*, 100183. DOI: 10.1016/j.treng.2023.100183

Ali, A. H., Rajesh, G., Kumar, M. K., Lakhanpal, S., Parashar, A. K., & Kaushik, A. (2024). Advancing aluminum-based composite *manufacturing: leveraging tio2 reinforcement through stir casting technique. E3S Web of Conferences, 507, 01042. DOI: 10.1051/e3sconf/202450701042

Ali, A., & Hafez, W. A. (2023). Overcoming challenges in the powder metallurgy process of metal matrix composites. *SVU-International Journal of Engineering Sciences and Applications*, *5*(1), 13–18. DOI: 10.21608/svusrc.2023.228672.1144

Aljasmi, N., Almaeeni, K., Alyammahi, M., Panda, A. S., Almheiri, A., Joy, A., Susantyoko, R. A., & Subramanian, V. (2024). Productivity Analysis of Automated 3D Printing System. *ASME International Mechanical Engineering Congress and Exposition, 88605*, V002T03A017. DOI: 10.1115/IMECE2024-145736

Aljibori, H. S., Alamiery, A., & Kadhum, A. A. H. (2023). Advances in corrosion protection coatings: A comprehensive review. *Int. J. Corros. Scale Inhib*, *12*(4), 1476–1520.

Almaev, A. V., Chernikov, E. V., Davletkildeev, N. A., & Sokolov, D. V. (2020). Oxygen sensors based on gallium oxide thin films with addition of chromium. *Superlattices and Microstructures*, *139*(January), 106392. DOI: 10.1016/j.spmi.2020.106392

Alovsat, T. Y.Tanriverdiyev Yusif Alovsat. (2024). Classification and properties of composites based on matrix material. *World Journal of Advanced Research and Reviews*, *24*(2), 2278–2282. DOI: 10.30574/wjarr.2024.24.2.3608

Althoey, F., Zaid, O., Alsulamy, S., Martínez-García, R., de Prado-Gil, J., & Arbili, M. M. (2023). Experimental study on the properties of ultra-high-strength geopolymer concrete with polypropylene fibers and nano-silica. *PLoS One*, *18*(4), e0282435. DOI: 10.1371/journal.pone.0282435 PMID: 37079561

Aluminium for modern marine applications. (2025). *Hydro*. https://www.hydro.com/en-SE/aluminium/industries/marine-offshore

Amraei, J., & Katunin, A. (2022). Recent advances in limiting fatigue damage accumulation induced by self-heating in polymer–matrix composites. *Polymers*, *14*(24), 5384. DOI: 10.3390/polym14245384 PMID: 36559751

Anirudh, S., Krishnamurthy, S., Kandasubramanian, B., & Kumar, P. (2023). Probing into atomically thin layered nano-materials protective coating for aerospace and strategic defence application–A review. *Journal of Alloys and Compounds*, *968*, 172203. DOI: 10.1016/j.jallcom.2023.172203

An, Y., Dai, L., Wu, Y., Wu, B., Zhao, Y., Liu, T., Hao, H., Li, Z., Niu, G., Zhang, J., Quan, Z., & Ding, S. (2019). Epitaxial growth of β -Ga2O3 thin films on Ga2O3 and Al2O3 substrates by using pulsed laser deposition. *Journal of Advanced Dielectrics*, *9*(4), 3–9. DOI: 10.1142/S2010135X19500322

Appleton, B. R., Sartwell, B., Peercy, P. S., Schaefer, R., & Osgood, R. (1985). Ion-beam-and laser-induced surface modifications. *Materials Science and Engineering*, *70*, 23–51. DOI: 10.1016/0025-5416(85)90267-8

Arango, S., Peláez-Vargas, A., & García, C. (2012). Coating and surface treatments on orthodontic metallic materials. *Coatings*, *3*(1), 1–15. DOI: 10.3390/coatings3010001

Aravinthan, T., & Manalo, A. (2012). Field applications and case studies of FRP in civil infrastructure: The Australian experience. *Proceedings of the 6th International Conference on Frp Composites in Civil Engineering (CICE 2012).*

Armstrong, G. (2010). Engineered coatings for composites and polymers used in defence and aerospace: now and the future. *Transactions of the IMF, 88*(4), 179-184. DOI: 10.1179/002029610X12734927543102

Arora, K., & Kumar, M. (2020). Sputtered-Growth of High-Temperature Seed-Layer Assisted β-Ga 2 O 3 Thin Film on Silicon-Substrate for Cost-Effective Solar-Blind Photodetector Application. *ECS Journal of Solid State Science and Technology : JSS*, *9*(6), 065013. DOI: 10.1149/2162-8777/aba7fd

Arora, N., & Akhai, S. (2015). Reclaiming EN-14b steel grade implements by hardfacing. *International Journal of Scientific Research*, *4*(10), 14–16.

Artesani, A., Di Turo, F., Zucchelli, M., & Traviglia, A. (2020). Recent advances in protective coatings for cultural heritage–an overview. *Coatings*, *10*(3), 217. DOI: 10.3390/coatings10030217

Arunkumar, K., Raghu, R., Jule, L. T., & Krishnaraj, R. (2022). Effect of graphene filler on mechanical properties of cotton/viscose hybrid composite. *Materialwissenschaft und Werkstofftechnik*, *53*(3), 298–307. DOI: 10.1002/mawe.202100138

Asghar Yaseen, A., Waqar, T., Khan, M. A. A., Asad, M., & Djavanroodi, F. (2021). Fish Scales and Their Biomimetic Applications. *Frontiers in Materials.*

Ashby, M., Cope, E., & Cebon, D. (2013). Materials selection for engineering design. In *Informatics for materials science and engineering* (pp. 219–244). Butterworth-Heinemann. DOI: 10.1016/B978-0-12-394399-6.00010-2

Asim, N., Badiei, M., Samsudin, N. A., Mohammad, M., Razali, H., & Hui, D. (2024). Clean technology option development for smart and multifunctional construction materials: Sustainable geopolymer composites. *Journal of Building Engineering*, *94*, 109932. DOI: 10.1016/j.jobe.2024.109932

Atkins, C., Chahid, Y., Wells, J. T., Westsik, M., Morris, K. C., Breen, C., Macleod, A., Bissell, L., Cochrane, W., Kotlewski, R., McPhee, S., Snell, R. M., Todd, I., Beardsley, M., Harris, M., Pimentel, M., McKegney, S., Orr, J., Alcock, S. G., & Cottarelli, S. (2023). Understanding the role of additive manufacturing in the development of astronomical hardware. *Astronomical Optics: Design, Manufacture, and Test of Space and Ground Systems*, *IV*, 36. DOI: 10.1117/12.2677452

Awaad, S., Mansour, D. M., Mahdi, I., & Abdelrasheed, I. (2024). Impact of material supply chain on the productivity optimization for the construction of roads projects. *Scientific Reports*, *14*(1), 3294. DOI: 10.1038/s41598-024-53660-6 PMID: 38332188

Azizi-Lalabadi, M., Hashemi, H., Feng, J., & Jafari, S. M. (2020). Carbon nanomaterials against pathogens; the antimicrobial activity of carbon nanotubes, graphene/graphene oxide, fullerenes, and their nanocomposites. *Advances in Colloid and Interface Science*, *284*, 102250–102267. DOI: 10.1016/j.cis.2020.102250 PMID: 32966964

Bai, J. (2022). *Advanced fibre-reinforced polymer (FRP) composites for structural applications.* Woodhead Publishing.

Bai, Y., Jantunen, H., & Juuti, J. (2018). Hybrid, multi-source, and integrated energy harvesters. *Frontiers in Materials*, *5*, 65. DOI: 10.3389/fmats.2018.00065

Bakar, M. S. A., Asyraf, M. R. M., Syamsir, A., Ma, Q., Hazrati, K. Z., Azlin, M. N. M., & Thiagamani, S. M. K. (2024). Kenaf/glass fiber-reinforced polymer composites: Pioneering sustainable materials with enhanced mechanical and tribological properties. *Polymer Composites*, *45*(16), 14421–14447. DOI: 10.1002/pc.28785

Baker, D. (2024). *US Air Power 1945-1990: Volume 1-US Fighters and Fighter-Bombers, 1945-1949* (Vol. 2). Helion and Company.

Banerjee, R., Pal, A., Ghosh, D., Ghosh, A. B., Nandi, M., & Biswas, P. (2021). Improved photocurrent response, photostability and photocatalytic hydrogen generation ability of CdS nanoparticles in presence of mesoporous carbon. *Materials Research Bulletin*, *134*, 111085–111098. DOI: 10.1016/j.materresbull.2020.111085

Barthelat, F., & Espinosa, H. D. (2007). An experimental investigation of deformation and fracture of nacre–mother of pearl. *Experimental Mechanics*, *47*(3), 311–324. DOI: 10.1007/s11340-007-9040-1

Barthelat, F., Li, C. M., Comi, C., & Espinosa, H. D. (2006). Mechanical properties of nacre constituents and their impact on mechanical performance. *Journal of Materials Research*, *21*(8), 1977–1986. DOI: 10.1557/jmr.2006.0239

Baskaran, J. (2024a). Advancements in Metal Additive Manufacturing Technology for Biomedical Applications. In *Trends and Applications in Mechanical Engineering, Composite Materials and Smart Manufacturing* (pp. 1-34): IGI Global. DOI: 10.4018/979-8-3693-1966-6.ch001

Baskaran, J. (2024b). Roadmap to Achieve Sustainable and Eco-Friendly Metal Additive Manufacturing Processes. In *Futuristic Technology for Sustainable Manufacturing* (pp. 185–215). IGI Global. DOI: 10.4018/979-8-3693-1862-1.ch011

Baskaran, J., Muthukannan, D., Shukla, R., & Gokuldoss, P. K. (2024). Manufacturability and deformation studies on a novel metallic lattice structure fabricated by Selective Laser Melting. *Vacuum*, *222*, 113065. DOI: 10.1016/j.vacuum.2024.113065

Becker, L. (2006, February). Influence of IR sensor technology on the military and civil defense. In *Quantum Sensing and Nanophotonic Devices III* (Vol. 6127, pp. 180–194). SPIE. DOI: 10.1117/12.640529

Belloni, E., Celli, I., Nakamura, S., Pisello, A., Villacci, D., & Cotana, F. (2023). A POWER-EFFICIENT SMART LASER-PHOTOLUMINESCENT-LIGHT (LPL) WITH PV-SYSTEM INTEGRATION: EXPERIMENTAL ANALYSIS AND OPTIMIZATION FOR PEDESTRIAN ROADS. *Conference Proceeding by Commission Internationale de L'Eclairage CIE 2023*, 1686–1695. DOI: 10.25039/x50.2023.PO004

Benmohamed, M., Benmounah, A., Haddad, A., & Yahi, S. (2022). The effect of inhibiting molybdate used in anodizing-conversion treatment to improve corrosion protection of AA2030 aluminum alloy in different steps. *Journal of Engineering and Applied Sciences (Asian Research Publishing Network)*, *69*(1), 40. Advance online publication. DOI: 10.1186/s44147-022-00090-8

Bhatnagar, A., & Lang, D. (2006). Military and law enforcement applications of lightweight ballistic materials. In *Lightweight Ballistic Composites* (pp. 364–397). Woodhead Publishing. DOI: 10.1533/9781845691554.2.364

Bhoominathan, R., Boopathy, G., Jacob, S., Pakiyaraj, A., & Karthik, M. K. (2024). Analysis of failure modes and energy absorption in glass fiber and chopped basalt sandwich panels under low-velocity impact. *Materials Today: Proceedings*, *85*, 123–129. DOI: 10.1016/j.matpr.2024.05.054

Bianchi, G., Fanelli, C., Freddi, F., Giuliani, F., & La Placa, A. (2025). Systematic review railway infrastructure monitoring: From classic techniques to predictive maintenance. *Advances in Mechanical Engineering*, *17*(1), 16878132241285632. DOI: 10.1177/16878132241285631

Bian, X., Yao, W., Liao, G., Liang, G., She, A., & Wei, Y. (2024). Hydrophobic cement-based materials with micro-nano hierarchical structures: Preparation, characterization, and stability assessment. *Frontiers in Materials*, *11*, 1512374. DOI: 10.3389/fmats.2024.1512374

BinSabt, M. H., Azeez, F. A., & Suleiman, N.BinSabt. (2023). Eco-friendly silane-based coating for mitigation of carbon steel corrosion in marine environments. *ACS Omega*, *8*(14), 12886–12898. DOI: 10.1021/acsomega.3c00013 PMID: 37065042

Blumenschein, N., Paskova, T., & Muth, J. F. (2019). Effect of Growth Pressure on PLD-Deposited Gallium Oxide Thin Films for Deep-UV Photodetectors. *Physica Status Solidi (A). Physica Status Solidi. A, Applications and Materials Science*, *216*(20), 1900098. Advance online publication. DOI: 10.1002/pssa.201900098

Bokov, D., Turki Jalil, A., Chupradit, S., Suksatan, W., Javed Ansari, M., Shewael, I. H., Valiev, G. H., & Kianfar, E. (2021). Nanomaterial by Sol-Gel Method: Synthesis and Application. *Advances in Materials Science and Engineering*, *2021*(December), 5102014. Advance online publication. DOI: 10.1155/2021/5102014

Boopathy, G., Gurusami, K., Chinnapandian, M., & Vijayakumar, K. R. (2022). Optimization of process parameters for injection moulding of Nylon6/SiC and Nylon6/B_4C polymer matrix composites. *Fluid Dynamics & Materials Processing*, *18*(2), 223–232. DOI: 10.32604/fdmp.2022.018225

Boopathy, G., Vijayakumar, K. R., Chinnapandian, M., & Gurusami, K. (2019). Development and experimental characterization of fibre metal laminates to predict the fatigue life. *International Journal of Innovative Technology and Exploring Engineering*, *8*(10), 2815–2819. DOI: 10.35940/ijitee.J9597.0881019

Bornapour, M., Celikin, M., Cerruti, M., & Pekguleryuz, M. (2013). Magnesium implant alloy with low levels of strontium and calcium: The third element effect and phase selection improve bio-corrosion resistance and mechanical performance. *Materials Science and Engineering C*, *35*, 267–282. DOI: 10.1016/j.msec.2013.11.011 PMID: 24411378

Bouhamed, A., Missaoui, S., Ben Ayed, A., Attaoui, A., Missaoui, D., Jeder, K., Guesmi, N., Njeh, A., Khemakhem, H., & Kanoun, O. (2024). A comprehensive review of strategies toward efficient flexible piezoelectric polymer composites based on BaTiO3 for next-generation energy harvesting. *Energies*, *17*(16), 4066. DOI: 10.3390/en17164066

Boylu, E. E., & Ekinci, S. (2024). Examination of Material Selection's Effect on Lighweightness on Structures. *Journal of Architectural Sciences and Applications*, *9*(2), 798–814. DOI: 10.30785/mbud.1475316

Brummer, M., Raddatz, K. J., Schmitt, M., Schlick, G., Tobie, T., Daub, R., & Stahl, K. (2023). Static load-carrying behavior and material properties of additively manufactured gears (pbf-lb/m, 16mncr5). *Rapid Prototyping Journal*, *29*(11), 117–130. DOI: 10.1108/RPJ-02-2023-0035

Buckner, C. A., Lafrenie, R. M., Dénommée, J. A., Caswell, J. M., Want, D. A., Gan, G. G., Leong, Y. C., Bee, P. C., Chin, E., Teh, A. K. H., Picco, S., Villegas, L., Tonelli, F., Merlo, M., Rigau, J., Diaz, D., Masuelli, M., Korrapati, S., Kurra, P., . . . Mathijssen, R. H. J. (2016). We are IntechOpen, the world ' s leading publisher of Open Access books Built by scientists, for scientists TOP 1%. *Intech, 11*(tourism), 13. https://www.intechopen.com/books/advanced-biometric-technologies/liveness-detection-in-biometrics

Busch, G. (1989). Early history of the physics and chemistry of semiconductors-from doubts to fact in a hundred years. *European Journal of Physics*, *10*(4), 254–264. DOI: 10.1088/0143-0807/10/4/002

Butenegro, J. A., Bahrami, M., Abenojar, J., & Martínez, M. Á. (2021). Recent progress in carbon fiber reinforced polymers recycling: A review of recycling methods and reuse of carbon fibers. *Materials (Basel)*, *14*(21), 6401. DOI: 10.3390/ma14216401 PMID: 34771926

Butt, M. A. (2022). Thin-film coating methods: A successful marriage of high-quality and cost-effectiveness—a brief exploration. *Coatings*, *12*(8), 1115. DOI: 10.3390/coatings12081115

Cai, H., Zhang, N., Liu, L., Su, J., Li, Y., Kang, Y., & Guo, F. (2024). Effects of cooling rate on the microstructure and properties of magnesium alloy-a review. *Journal of Magnesium and Alloys*, *12*(8), 3094–3114. DOI: 10.1016/j.jma.2024.06.021

Calabrese, F., Cataldo, M., Corallo, A., De Pascalis, A., Mancarella, L., Ostuni, L., & Zizzari, A. A. (2012). Damage Control System: an application for ship safety and security. *IFAC Proceedings Volumes, 45*(27), 103-108.

Calle, L., Hintze, P., Li, W., & Buhrow, J. (2010). Smart coatings for autonomous corrosion detection and control. In *AIAA SPACE 2010 Conference & Exposition* (p. 8877). DOI: 10.2514/6.2010-8877

Campilho, R. D. S. G., & Silva, F. J. G. (2023). Industrial process improvement by automation and robotics. [). MDPI.]. *Machines (Basel)*, *11*(11), 1011. DOI: 10.3390/machines11111011

Cao, S., & Yu, J. (2014). g-C_3N_4-Based Photocatalysts for Hydrogen Generation. *The Journal of Physical Chemistry Letters*, *5*(12), 2101–2107. DOI: 10.1021/jz500546b PMID: 26270499

Capan, I. (2024). Wide-Bandgap Semiconductors for Radiation Detection: A Review. *Materials (Basel)*, *17*(5), 1147. Advance online publication. DOI: 10.3390/ma17051147 PMID: 38473617

Caruso, M. C., Pascale, C., Camacho, E., Scalari, S., Animato, F., Alonso, M. C., Gimenez, M., & Ferrara, L. (2020). Life cycle assessment on the use of ultra high performance fibre reinforced concretes with enhanced durability for structures in extremely aggressive environments: case study analyses. *Concrete Durability and Service Life Planning: Proceedings of ConcreteLife'20 3*, 121–125.

Çava, K., İpek, H., Uşun, A., & Aslan, M. (2024). Examine the mechanical properties of woven glass fiber fabric reinforced composite plate manufactured with vat-photopolymerization. *Polymer Composites*, *45*(18), 17105–17120. DOI: 10.1002/pc.28955

Çelik, M., Noble, T. A., Jorge, F., Jian, R., Brádaigh, C. M. Ó., & Robert, C. (2022). Influence of line processing parameters on properties of carbon fibre epoxy towpreg. *Journal of Composites Science*, *6*(3), 75. DOI: 10.3390/jcs6030075

Cerchier, P., Pezzato, L., Gennari, C., Moschin, E., Moro, I., & Dabalà, M. (2020). PEO coating containing copper: A promising anticorrosive and antifouling coating for seawater application of AA 7075. *Surface and Coatings Technology*, *393*, 125774. DOI: 10.1016/j.surfcoat.2020.125774

Chandler, K. A., & Bayliss, D. A. (1985). Corrosion protection of steel structures.

Chauhan, H. R., Variya, S., Solanki, A., & Saladi, S. (2024). Development of graphene nanoplatelets reinforced hvof sprayed cr3c2-nicr coatings. *Journal of Thermal Spray and Engineering*, *4*, 81–86. DOI: 10.52687/2582-1474/401

Chen, M., Zhang, Z., Zhan, R., She, J., Deng, S., Xu, N., & Chen, J. (2021). Fast-response X-ray detector based on nanocrystalline Ga2O3 thin film prepared at room temperature. *Applied Surface Science, 554*(December 2020), 149619. DOI: 10.1016/j.apsusc.2021.149619

Chen, E., Berrocal, C. G., Löfgren, I., & Lundgren, K. (2022). Comparison of the service life, life-cycle costs and assessment of hybrid and traditional reinforced concrete through a case study of bridge edge beams in Sweden. *Structure and Infrastructure Engineering*, *19*(1), 39–57. DOI: 10.1080/15732479.2021.1919720

Cheng, K., Guo, Z., Zhang, P., Feng, L., Zhou, Y., Li, L., Song, H., Wang, T., Zhao, Y., & Zhao, L. (2025). A Universal Strategy for Multicolor Mechanoluminescence via Radiative Energy Transfer Based on Ultraviolet Mechanoluminescent Material Ca9Al (PO4) 7: Ce3+. *Laser & Photonics Reviews*, *19*(7), 2401524. DOI: 10.1002/lpor.202401524

Chen, P. Y., & Ritchie, R. O. (2016). Bioinspired materials: Advancements, limitations, and future directions. *Materials Today*, *19*(8), 414–421.

Chen, W., Patel, D., & Radhakrishnan, R. (2019). Biomimetic infrared sensors for defense applications. *Sensors (Basel)*, *19*(7), 1206. DOI: 10.3390/s19051206 PMID: 30857309

Chen, X., Mi, W., Wu, J., Yang, Z., Zhang, K., Zhao, J., Luan, C., & Wei, Y. L. (2020). A solar-blind photodetector based on β-Ga2O3 film deposited on MgO (100) substrates by RF magnetron sputtering. *Vacuum*, *180*(May), 109632. DOI: 10.1016/j.vacuum.2020.109632

Chen, Y., Cheng, Y., Liu, Y., Wang, Y., Qu, Y., Jiang, D., Qin, Z., & Yuan, Y. (2024). Achieving long-lived shallow trapping states in carbon nitride through the n-π*electronic transition for enhanced photocatalytic hydrogen generation. *Applied Catalysis B: Environmental*, *342*, 123453–123467. DOI: 10.1016/j.apcatb.2023.123453

Chen, Y., Dang, B., Fu, J., Zhang, J., Liang, H., Sun, Q., Zhai, T., & Li, H. (2022). Bioinspired construction of micronano lignocellulose into an impact resistance "wooden armor" with bouligand structure. *ACS Nano*, *16*(5), 7525–7534. DOI: 10.1021/acsnano.1c10725 PMID: 35499235

Cherif, N., Jaafar, W., Vinogradov, E., Yanıkömeroğlu, H., Pollin, S., & Yongaçoğlu, A. (2023). Ituavs: Intermittently tethered uavs for future wireless networks. *IEEE Wireless Communications*, *30*(4), 124–130. DOI: 10.1109/MWC.018.2100720

Chi, J., Wang, C., Zhou, G., Fu, X., Chen, X., Yin, X., Zhang, Z., & Wang, Y. (2023). A critical review on separation and extraction of scandium from industrial wastes: Methods, difficulties, and mechanism. *Journal of Environmental Chemical Engineering*, *11*(5), 111068. DOI: 10.1016/j.jece.2023.111068

Choi, M., Park, J., Jie, M., & Choi, W. (2023). Development of a chaff dispense program for target tracking radar deception. *Proceedings of Engineering and Technology Innovation*, *24*, 22–28. DOI: 10.46604/peti.2023.11150

Choy, K. L. (2000). Innovative and cost effective deposition of coatings using ES-AVD method. *Surface Engineering*, *16*(6), 465–468.

Choy, S. Y., Sun, C.-N., Leong, K. F., & Wei, J. (2017). Compressive properties of functionally graded lattice structures manufactured by selective laser melting. *Materials & Design*, *131*, 112–120. DOI: 10.1016/j.matdes.2017.06.006

Civera, M., Naseem, A., & Chiaia, B. (2024). Recent advances in embedded technologies and self-sensing concrete for structural health monitoring. *Structural Concrete*, suco.202400714. DOI: 10.1002/suco.202400714

Clark, D. E. (1993). Environmental stability of materials—Issues and opportunities. *MRS Bulletin*, *18*(9), 25–31. DOI: 10.1557/S0883769400037994

Continental Steel & Tube Company. (2020, March 26). *A guide to understanding marine grade aluminum*. https://continentalsteel.com/blog/marine-grade-aluminum/

Cooper, M. (2024). *Navigating Complexity: Insights into Strategic Procurement Practices in Global Supply Chains.*

Cora, I., Mezzadri, F., Boschi, F., Bosi, M., Čaplovičová, M., Calestani, G., Dódony, I., Pécz, B., & Fornari, R. (2017). The real structure of ε-Ga2O3 and its relation to κ-phase. *CrystEngComm*, *19*(11), 1509–1516. DOI: 10.1039/C7CE00123A

Croll, S. G. (2020). Surface roughness profile and its effect on coating adhesion and corrosion protection: A review. *Progress in Organic Coatings*, *148*, 105847. DOI: 10.1016/j.porgcoat.2020.105847

Cui, P., Chen, J., Fu, K., Deng, J., Sun, T., Chen, K., & Yin, P. (2024). Bioinspired bouligand-structured cellulose nanocrystals/poly(vinyl alcohol) composite hydrogel for enhanced impact resistance. ACS Applied Materials &Amp. *Interfaces*, *16*(39), 53022–53032. DOI: 10.1021/acsami.4c13264 PMID: 39306751

Czél, G., Sycheva, A., & Janovszky, D. (2023). Effect of different fillers on thermal conductivity, tribological properties of polyamide 6. *Scientific Reports*, *13*(1), 845. Advance online publication. DOI: 10.1038/s41598-023-27740-y PMID: 36646774

Czerwinski, F. (2020). Cerium in aluminum alloys. *Journal of Materials Science*, *55*(1), 24–72. DOI: 10.1007/s10853-019-03892-z PMID: 33066314

Das, J. P., Nardekar, S. S., Ravichandran, V., & Kim, S. (2024). From Friction to Function: A High-Voltage Sliding Triboelectric Nanogenerator for Highly Efficient Energy Autonomous IoTs and Self-Powered Actuation. *Small*, *20*(48), 2405792. DOI: 10.1002/smll.202405792 PMID: 39221685

Daş, O., Aybarç, U., Öztürk, Y., Kıral, B. G., Öztürk, H., Kıral, Z., & Uyulgan, B. (2023). A preliminary study on the manufacturing of aluminum–potassium feldspar metal matrix composite materials. *Materialwissenschaft und Werkstofftechnik*, *54*(1), 109–119. DOI: 10.1002/mawe.202200100

Das, S., Pillai, A. U., Chemenghat, A., & Rahaman, M. M. (2023). Determining the role of microstructural topology on the mechanical performance of nacre-inspired composites using a phase-field model. *Materials Today. Communications*, *37*, 107453. DOI: 10.1016/j.mtcomm.2023.107453

Davis, J. (2003). Coating technology: The future: New and emerging technologies. *Materials Performance*, *42*(11), 24–27.

De Belie, N., Gruyaert, E., Al-Tabbaa, A., Antonaci, P., Baera, C., Bajare, D., Darquennes, A., Davies, R., Ferrara, L., Jefferson, T., Litina, C., Miljevic, B., Otlewska, A., Ranogajec, J., Roig-Flores, M., Paine, K., Lukowski, P., Serna, P., Tulliani, J.-M., & Jonkers, H. M. (2018). A review of self-healing concrete for damage management of structures. *Advanced Materials Interfaces*, *5*(17), 1800074. DOI: 10.1002/admi.201800074

de Boisbaudran, L. (1875). On the chemical and spectroscopic characters of a new metal (gallium). *The London, Edinburgh and Dublin Philosophical Magazine and Journal of Science*, *50*(332), 414–416. DOI: 10.1080/14786447508641312

De Jong, A., Jansen, M., Van Dijk, J., & Meyer, J. (2021). Analysis of Innovative Practices in Advanced Materials and Structural Engineering. *Fusion of Multidisciplinary Research. International Journal (Toronto, Ont.)*, *2*(1), 178–188.

Deng, H., Huang, J., Liu, Q., Zhao, T., Zhou, C., & Gao, J. (2023). A distributed collaborative allocation method of reconnaissance and strike tasks for heterogeneous uavs. *Drones (Basel)*, *7*(2), 138. DOI: 10.3390/drones7020138

Dennis, R. V., Patil, V., Andrews, J. L., Aldinger, J. P., Yadav, G. D., & Banerjee, S. (2015). Hybrid nanostructured coatings for corrosion protection of base metals: A sustainability perspective. *Materials Research Express*, *2*(3), 032001. DOI: 10.1088/2053-1591/2/3/032001

Dezdarani, V. H., & Mohammadi, M. H. P. (n.d.). *Tu Designing Smart Roads with Climate-Responsive Materials for Enhanced Durability and Maintenance.*

Dhakal, H. N., & Ismail, S. O. (2020). *Sustainable composites for lightweight applications.*

Dhinakaran, V., Stalin, B., Ravichandran, M., Balasubramanian, M., Chairman, C. A., & Pritima, D. (2020). Wire Arc Additive Manufacturing Perspectives and Recent Developments. *IOP Conference Series. Materials Science and Engineering*, *988*(1), 12102. DOI: 10.1088/1757-899X/988/1/012102

Dickie, R. A. (1994). Paint adhesion, corrosion protection, and interfacial chemistry. *Progress in Organic Coatings*, *25*(1), 3–22. DOI: 10.1016/0300-9440(94)00500-1

Dighade, R., Gomase, V., Peshattiwar, R., Selokar, A., Sangidwar, N., Peshattiwar, S., & Malve, S. (2024). Emission of carbon footprint from building construction materials: A review. *IOP Conference Series. Earth and Environmental Science*, *1409*(1), 12010. DOI: 10.1088/1755-1315/1409/1/012010

Disegi, J. A. (1997, April). Anodizing treatments for titanium implants. In *Proceedings of the 1997 16 Southern Biomedical Engineering Conference* (pp. 129-132). IEEE. DOI: 10.1109/SBEC.1997.583231

Djokić, S. S. (2002). Electroless deposition of metals and alloys. In *Modern aspects of electrochemistry* (pp. 51–133). Springer US. DOI: 10.1007/0-306-47604-5_2

Djokić, S. S., & Cavallotti, P. L. (2010). Electroless deposition: Theory and applications. *Electrodeposition. Theory into Practice*, •••, 251–289.

Dong, L., Jia, R., Xin, B., Peng, B., & Zhang, Y. (2017). Effects of oxygen vacancies on the structural and optical properties of β-Ga2O3. *Scientific Reports, 7*(September 2016), 1–12. DOI: 10.1038/srep40160

Dong, B., Zhu, S., Guo, G., Wu, T., Lu, X., Huang, W., Ma, H., Xu, Q., Han, J., Zhang, S., Wang, Y., Zhang, X., & Huang, L. (2025). Switchable Pancharatnam–Berry Phases in Heterogeneously Integrated THz Metasurfaces. *Advanced Materials*, *37*(6), 2417183. DOI: 10.1002/adma.202417183 PMID: 39676492

Dong, G., Li, X., & Ji, Y. (2023). An experimental and numerical study on the ballistic performance of a lightweight ceramic composite armor. *Journal of Physics: Conference Series*, *2478*(7), 072030. DOI: 10.1088/1742-6596/2478/7/072030

Dong, L. (2021). Mechanical responses of Ti-6Al-4V truss lattices having a combined simple-cubic and body-centered-cubic (SC-BCC) topology. *Aerospace Science and Technology*, *116*, 106852. DOI: 10.1016/j.ast.2021.106852

Dong, Y., Fu, Y., He, C., & Fang, D. (2023). A hyper-pseudoelastic model of cyclic stress-softening effect for rubber composites. *Polymers*, *15*(14), 3033. DOI: 10.3390/polym15143033 PMID: 37514422

Dougherty, N. (2010). Simulated Altitude Liquid Rocket Testing at the USAF AEDC: Atlas/Mercury, Titan/Gemini, Saturn/Apollo, Space Shuttle, ISS. In *46th AIAA/ASME/SAE/ASEE Joint Propulsion Conference & Exhibit* (p. 6982). DOI: 10.2514/6.2010-6982

Du Plessis, A., Broeckhoven, C., Yadroitsava, I., Yadroitsev, I., Hands, C. H., Kunju, R., & Bhate, D. (2019). Beautiful and functional: A review of biomimetic design in additive manufacturing. *Additive Manufacturing*, *27*, 408–427. DOI: 10.1016/j.addma.2019.03.033

Duclos, A., Bernardi, E., Robbiola, L., Deshayes, C., de Souza Machado, L., Martini, C., Chiavari, C., Balbo, A., Escobar, C., Guilminot, E., Malard, B., & Brunet, M. (2024). Duralumin Alloys in World War II Heritage Aircraft: Correlations Between Manufacturing Parameters and Alloys' Properties. *Heritage*, *7*(12), 6772–6790. DOI: 10.3390/heritage7120313

Duigou, J. L., Gulbrandsen-Dahl, S., Vallet, F., Söderberg, R., Eynard, B., & Perry, N. (2016). Optimization and lifecycle engineering for design and manufacture of recycled aluminium parts. *CIRP Annals*, *65*(1), 149–152. DOI: 10.1016/j.cirp.2016.04.111

Dutta, A. K. (2023). Metal Chalcogenides-Based Multifunctional Nanomaterials for Solar-Light-Driven Photo-Catalysis and Electro-Chemical Sensing. In Kumar, A. (Ed.), *Innovative Multifunctional Nanomaterial for Photocatalysis, Sensing, and Imaging* (pp. 1–16). IGI Global Scientific Publishing., DOI: 10.4018/978-1-6684-8743-3.ch001

Dutta, A. K., Maji, S. K., & Adhikary, B. (2014). γ–Fe2O3 nanoparticles: An easily recoverable effective photo-catalyst for the degradation of rose bengal and methylene blue dyes in the waste-water treatment plant. *Materials Research Bulletin*, *49*, 28–34. DOI: 10.1016/j.materresbull.2013.08.024

Dutta, A. K., Maji, S. K., Srivastava, D. N., Mondal, A., Biswas, P., Paul, P., & Adhikary, B. (2012). Synthesis of FeS and FeSe Nanoparticles from a Single Source Precursor: A Study of Their Photocatalytic Activity, Peroxidase-Like Behavior, and Electrochemical Sensing of H2O2. *ACS Applied Materials & Interfaces*, *4*(4), 1919–1927. DOI: 10.1021/am300408r PMID: 22448782

Du, X., Mei, Z., Liu, Z., Guo, Y., Zhang, T., Hou, Y., Zhang, Z., Xue, Q., & Kuznetsov, A. Y. (2009). Controlled growth of high-quality ZnO-based films and fabrication of visible-blind and solar-blind ultra-violet detectors. *Advanced Materials*, *21*(45), 4625–4630. DOI: 10.1002/adma.200901108

Edinburgh, D. S. T. (2018). The Future of Advanced Materials and Manufacturing for Defence.

Elgendy, I. M., Elkaliny, N. E., Saleh, H. M., Darwish, G. O., Almostafa, M. M., Metwally, K., Yahya, G., & Mahmoud, Y. A. G. (2024). Bacteria-powered self-healing concrete: Breakthroughs, challenges, and future prospects. *Journal of Industrial Microbiology & Biotechnology*, *52*, kuae051. DOI: 10.1093/jimb/kuae051 PMID: 39673695

El-hadad, A. S., Al Hariri, M. O. R., Gamal, A., El-Gammal, M. A., Badawy, A. H., & El-Feky, M. S. (2024). Unlocking the potential of carbon nanotubes and nano clay in concrete: Assessing the bond strength and failure modes. *Engineering Research Journal (Shoubra)*, *53*(4), 89–97. DOI: 10.21608/erjsh.2024.279620.1287

Elsawy, E. E., Seleem, M. H., Esa, M., & Osman, A. (2024). Numerical evaluation of the performance of different composite bridge decks. *IOP Conference Series. Earth and Environmental Science*, *1396*(1), 012009. DOI: 10.1088/1755-1315/1396/1/012009

Emmanuel, A. O., Fayomi, O. S. I., & Akande, I. G. (2021). Aluminium Alloys as Advanced Materials: A short communication. *IOP Conference Series. Materials Science and Engineering*, *1107*(1), 12024. DOI: 10.1088/1757-899X/1107/1/012024

Endre, R., & Popoviç, M. (2006). Problems and prospect of Al-Mg alloys application in marine constructions. *Metalurgija*, *12*(4), 297. https://scindeks.ceon.rs/article.aspx?artid=0354-63060604297R

Esmaily, M., Svensson, J., Fajardo, S., Birbilis, N., Frankel, G., Virtanen, S., Arrabal, R., Thomas, S., & Johansson, L. (2017b). Fundamentals and advances in magnesium alloy corrosion. *Progress in Materials Science, 89*, 92–193. DOI: 10.1016/j.pmatsci.2017.04.011

Espejo-Antúnez, L., Fernández-Morales, C., Moreno-Vázquez, J. M., Tabla-Hinojosa, F. B., Cardero-Durán, M. Á., & Albornoz-Cabello, M. (2022). Assessment from a biopsychosocial approach of flight-related neck pain in fighter pilots of spanish air force. an observational study. *Diagnostics (Basel), 12*(2), 233. DOI: 10.3390/diagnostics12020233 PMID: 35204324

Faisal, T., & Park, C. (2016). *Spider-Web Inspired Mechanical Metamaterials for Defense Applications.ArXiv Preprint,* 1607.01407. Retrieved from https://arxiv.org/abs/1607.01407

Fauchais, P., & Vardelle, A. (2012). Thermal sprayed coatings used against corrosion and corrosive wear. *Advanced plasma spray applications, 10*, 34448.

Fayomi, O. S. I., Agboola, O., Akande, I. G., & Emmanuel, A. O. (2020, December). Challenges of coatings in aerospace, automobile and marine industries. In *AIP Conference Proceedings* (Vol. 2307, No. 1). AIP Publishing. DOI: 10.1063/5.0033579

Feli, S., Asgari Marnani, J., Ghasemi, A., & Monajemi Nejad, S. (2024). Seismic performance assessment of reinforced concrete coupled walls reinforced with different types of shape memory alloys. *Structural Design of Tall and Special Buildings, 33*(18), e2185. DOI: 10.1002/tal.2185

Fernando, R. H. (2009). Nanocomposite and nanostructured coatings: Recent advancements.

Ferreira, M. G. S., Zheludkevich, M. L., & Tedim, J. (2011). Advanced protective coatings for aeronautical applications. In *Nanocoatings and ultra-thin films* (pp. 235–279). Woodhead Publishing. DOI: 10.1533/9780857094902.2.235

Finkenstadt, D. J., & Eapen, T. T. (2024). Prominence Strategies. In *Bioinspired Strategic Design* (pp. 101–154). Productivity Press., DOI: 10.4324/9781032715315-4

Florea, M., Constantin, V. S., Bucur, A. C., Faur, R., Predu, D., & Cazacu, A. (2024). Materials and Structures Used in Aeronautics: Present and Future Perspectives. *INCAS Bulletin, 16*(4).

Fontana, S., & Di Lauro, F. (2022). An Overview of Sensors for Long Range Missile Defense. *Sensors (Basel), 22*(24), 1–19. DOI: 10.3390/s22249871 PMID: 36560236

Fratzl, P., & Weinkamer, R. (2007). Nature's hierarchical materials. *Progress in Materials Science*, *52*(8), 1263–1334. DOI: 10.1016/j.pmatsci.2007.06.001

Freitag, S., Goetz, S., & Wartzack, S. (2023). Continuous simulation of variations during the design of endless fiber reinforced composite structure assemblies. DS 125: *Proceedings of the 34th Symposium Design for X (DFX2023)*. DOI: 10.35199/dfx2023.02

Friedl, K. E. (2018). Military applications of soldier physiological monitoring. *Journal of Science and Medicine in Sport*, *21*(11), 1147–1153. DOI: 10.1016/j.jsams.2018.06.004 PMID: 29960798

Funke, W. (1985). The role of adhesion in corrosion protection by organic coatings. *Journal of the oil and colour chemists' association, 68*(9), 229-232.

Furtado, C., Kopp, R., Ni, X., Sarrado, C., Kalfon-Cohen, E., Wardle, B. L., & Camanho, P. (2024). J-integral experimental reduction reveals fracture toughness improvements in thin-ply carbon fiber laminates with aligned carbon nanotube interlaminar reinforcement. ACS Applied Materials &Amp. *Interfaces*, acsami.3c17333. Advance online publication. DOI: 10.1021/acsami.3c17333 PMID: 38624137

Galazka, Z. (2022). Growth of bulk β-Ga2O3single crystals by the Czochralski method. *Journal of Applied Physics*, *131*(3), 031103. Advance online publication. DOI: 10.1063/5.0076962

Gambardella, A., Esperto, V., Tucci, F., & Carlone, P. (2022). Defects reduction in the robotic layup process. *Key Engineering Materials*, *926*, 1437–1444. DOI: 10.4028/p-7v9349

GangaRao, H. (2017). Infrastructure applications of fiber-reinforced polymer composites. In *Applied Plastics Engineering Handbook* (pp. 675–695). Elsevier. DOI: 10.1016/B978-0-323-39040-8.00032-8

Gao, W., Wang, X., Lin, Y., Wang, X., Liu, D., & Sun, X. (2024). Achieving ultra-high strength and ductility in a rare-earth-free magnesium alloy via precisely controlled secondary hot extrusion process with an extremely low extrusion speed. *Journal of Magnesium and Alloys*. DOI: 10.1016/j.jma.2024.07.015

Gao, H., Ji, B., Jäger, I. L., Arzt, E., & Fratzl, P. (2003). Materials science: Biological materials and bioinspired structures. *Materials Science and Engineering A*, *86*(3), 213–221.

Garimella, S. V., Fleischer, A. S., Murthy, J. Y., Keshavarzi, A., Prasher, R., Patel, C., Bhavnani, S. H., Venkatasubramanian, R., Mahajan, R., Joshi, Y., Sammakia, B., Myers, B. A., Chorosinski, L., Baelmans, M., Sathyamurthy, P., & Raad, P. E. (2008). Thermal challenges in next-generation electronic systems. *IEEE Transactions on Components and Packaging Technologies, 31*(4), 801–815. DOI: 10.1109/TCAPT.2008.2001197

Gaur, U. P., & Kamari, E. (2024). Applications of thermal spray coatings: A review. *J. Therm. Spray Eng, 4*, 106–114. DOI: 10.52687/2582-1474/405

Ge, G., Hu, J., Hu, Z., Li, H., Huo, Y., Tang, S., Liu, Y., Ding, J., Hou, S., & Gao, Y. (2024). The influence of alloy composition on microstructure and performance of mixed-smelting alloy and weld metal. *Materials (Basel), 17*(19), 4708. DOI: 10.3390/ma17194708 PMID: 39410282

Georgantzinos, S. K., Giannopoulos, G. I., Stamoulis, K., & Markolefas, S. (2023). Composites in aerospace and mechanical engineering. *Materials (Basel), 16*(22), 7230. DOI: 10.3390/ma16227230 PMID: 38005159

Georgilas, I. P., & Tourassis, V. D. (2007, December). Quality issues in enameling of ceramic industry products. In *2007 IEEE International Conference on Industrial Engineering and Engineering Management* (pp. 1226-1230). IEEE. DOI: 10.1109/IEEM.2007.4419387

Ghazlan, A., Ngo, T., Tan, P., Xie, Y. M., & Tran, P. (2021). Inspiration from Nature's body armours – A review of biological and bioinspired composites. *Composites Part B: Engineering* Healing Platform with Water-Controlled Adhesiveness for Highly Effective Antibacterial Application and Data Encryption–Decryption'. *ACS Applied Materials & Interfaces, 12*(52), 57686–57694. DOI: 10.1021/acsami.0c15364

Ghosh, A., Dhiman, S., Gupta, A., & Jain, R. (2022). Process evaluation of scandium production and its environmental impact. *Environments (Basel, Switzerland), 10*(1), 8. DOI: 10.3390/environments10010008

Gialanella, S., Malandruccolo, A., Gialanella, S., & Malandruccolo, A. (2020). Alloys for aircraft structures. *Aerospace alloys*, 41-127.

Gibson, I., Rosen, D. W., & Stucker, B. (2015). *Additive manufacturing technologies* (2nd ed.). Springer. DOI: 10.1007/978-1-4939-2113-3

Gisario, A., Kazarian, M., Martina, F., & Mehrpouya, M. (2019). Metal additive manufacturing in the commercial aviation industry: A review. *Journal of Manufacturing Systems, 53*, 124–149. DOI: 10.1016/j.jmsy.2019.08.005

Glaser, K. F., & Shockley, G. R. (1970, April 21). Aluminum In Marine Construction. *Offshore Technology Conference*. https://doi.org/DOI: 10.4043/1197-MS

Gokalp, I., & Incesu, A. (2022). Effect of CA addition to the elevated temperature mechanical properties of AZ series magnesium alloys. *International Journal of Metalcasting*, *17*(2), 1402–1412. DOI: 10.1007/s40962-022-00872-z

Gopal, M., Lurdhumary, J., Bathrinath, S., Priya, A. P., Sarojwal, A., & Boopathi, S. (2024). Energy Harvesting and Smart Highways for Sustainable Transportation Infrastructure: Revolutionizing Roads Using Nanotechnology. In *Principles and Applications in Speed Sensing and Energy Harvesting for Smart Roads* (pp. 136–165). IGI Global.

Gorokhov, E. V., Magunov, A. N., Feshchenko, V. S., & Altukhov, A. A. (2008). Solar-blind UV flame detector based on natural diamond. *Instruments and Experimental Techniques*, *51*(2), 280–283. DOI: 10.1134/S002044120802022X

Gou, W., Shi, Z., Zhu, Y., Gu, X., Dai, F., Gao, X., & Wang, L. (2024). Multi-objective optimization of three mechanical properties of Mg alloys through machine learning. *Materials Genome Engineering Advances*, *2*(3), e54. Advance online publication. DOI: 10.1002/mgea.54

Govindarajan, B., Jaganraj, R., Srinivasan, V., & Kumaran, T. (2025). Evolution of UAV technology from early innovations to future horizons. In *Innovations and developments in unmanned aerial vehicles* (pp. 53–94). IGI Global Scientific Publishing., DOI: 10.4018/978-1-6684-9999-9.ch003

Green, J. L., Cooke, D., Beckman, A. B. W., Ramos, K., & Donahue, B. B. (2024). Maximizing the Potential of NASA's Super Heavy Launch Vehicles: Lessons From Saturn V. In *AIAA AVIATION FORUM AND ASCEND 2024* (p. 4879).

Greiling, P., Kirkpatrick, C., & Valentine, G. (1995). Military applications for heterostructure microelectronics technology. *Solid-State Electronics*, *38*(9), 1559–1567. DOI: 10.1016/0038-1101(95)00069-6

Groetsch, T., Maghe, M., Creighton, C., & Varley, R. J. (2023). Economic and environmental effects of precursor variation in a continuous carbon fibre manufacturing process. *Journal of Industrial and Engineering Chemistry*, *127*, 554–566. DOI: 10.1016/j.jiec.2023.07.041

Guimarães, J. S., Peixoto, M. S., & Sales, A. S. B. (2024). Biological concrete, a viable alternative in modern civil construction. *International Journal of Scientific Management and Tourism*, *10*(6), e1233–e1233. DOI: 10.55905/ijsmtv10n6-030

Gunawan, H., Khaerudin, K., Deksino, G. R., & Gultom, R. A.George Royke DeksinoRudy AG Gultom. (2024). Optimising offset defense implementation as fulfilment efforts in fighters aircraft materials and technology. [IJHESS]. *International Journal of Humanities Education and Social Sciences*, *3*(4). Advance online publication. DOI: 10.55227/ijhess.v3i4.724

Guo, Q., Huang, B., Lu, C., Zhou, T., Su, G., Jia, L., & Zhang, X. (2019). A cephalopod-inspired mechanoluminescence material with skin-like self-healing and sensing properties. *Materials Horizons*, *6*(5), 996–1004. DOI: 10.1039/C8MH01624K

Guo, Q., & Peng, J. (2024). Persistent monitoring for points of interest with different data update deadlines. *Sensors (Basel)*, *24*(4), 1224. DOI: 10.3390/s24041224 PMID: 38400382

Guo, Z., Verma, A., Wu, X., Sun, F., Hickman, A., Masui, T., Kuramata, A., Higashiwaki, M., Jena, D., & Luo, T. (2015). Anisotropic thermal conductivity in single crystal β-gallium oxide. *Applied Physics Letters*, *106*(11), 111909. Advance online publication. DOI: 10.1063/1.4916078

Gupta, B. D., Pathak, A., & Semwal, V. (2019). Carbon-Based Nanomaterials for Plasmonic Sensors: A Review. *Sensors (Basel)*, *19*(16), 3536. DOI: 10.3390/s19163536 PMID: 31412590

Gupta, S., Modgil, S., Choi, T.-M., Kumar, A., & Antony, J. (2023). Influences of artificial intelligence and blockchain technology on financial resilience of supply chains. *International Journal of Production Economics*, *261*, 108868. DOI: 10.1016/j.ijpe.2023.108868

Hadi, S. P., Prabawani, B., & Qomariah, A. (2021). Circular Initiatives for Industrial Sustainability. *E3S Web of Conferences, 317*, 1078. https://doi.org/DOI: 10.1051/e3sconf/202131701078

Haider, A. J., Alawsi, T., Haider, M. J., Taha, B. A., & Marhoon, H. A. (2022). A comprehensive review on pulsed laser deposition technique to effective nanostructure production: Trends and challenges. *Optical and Quantum Electronics*, *54*(8), 1–25. DOI: 10.1007/s11082-022-03786-6

Hajjioui, E. A., Bouchaâla, K., Faqir, M., & Essadiqi, E. (2023). A review of manufacturing processes, mechanical properties and precipitations for aluminum lithium alloys used in aeronautic applications. *Heliyon*, *9*(3), e12565. DOI: 10.1016/j.heliyon.2022.e12565 PMID: 36895401

Hajnal, Z., Miró, J., Kiss, G., Réti, F., Deák, P., Herndon, R. C., & Kuperberg, J. M. (1999). Role of oxygen vacancy defect states in the n-type conduction of β-Ga2O3. *Journal of Applied Physics*, *86*(7), 3792–3796. DOI: 10.1063/1.371289

Ha, M.-T., Kim, K.-H., Kwon, Y.-J., Kim, C.-J., Jeong, S.-M., & Bae, S.-Y. (2019). Understanding Thickness Uniformity of Ga 2 O 3 Thin Films Grown by Mist Chemical Vapor Deposition. *ECS Journal of Solid State Science and Technology : JSS*, *8*(7), Q3206–Q3212. DOI: 10.1149/2.0381907jss

Han, B., Yu, X., Zhang, K., Kwon, E., & Ou, J. (2011). Sensing properties of CNT-filled cement-based stress sensors. *Journal of Civil Structural Health Monitoring*, *1*(1–2), 17–24. DOI: 10.1007/s13349-010-0001-5

Han, C., Kundu, B. K., Liang, Y., & Sun, Y. (2024). Near-Infrared Light-Driven Photocatalysis with an Emphasis on Two-Photon Excitation: Concepts, Materials, and Applications. *Advanced Materials*, *36*(5), 2307759–2307772. DOI: 10.1002/adma.202307759 PMID: 37703435

Hao, C., Zhao, B., Shao, L., Cao, Y., Fei, M., Liu, W., Liu, T., Chang, Y.-C., Simmons, K. L., & Zhang, J. (2024). Mild chemical recycling of carbon fiber-reinforced epoxy composites in aqueous buffers and development of hydrothermally recyclable vitrimer composites from recyclates. *Resources, Conservation and Recycling*, *207*, 107668. DOI: 10.1016/j.resconrec.2024.107668

Harrison, W. N., Moore, D. G., & Richmond, J. C. (1947). *Review of an investigation of ceramic coatings for metallic turbine parts and other high-temperature applications* (No. NACA-TN-1186).

Hasan, H. R., Musamih, A., Salah, K., Jayaraman, R., Omar, M., Arshad, J., & Boscovic, D. (2024). Smart agriculture assurance: IoT and blockchain for trusted sustainable produce. *Computers and Electronics in Agriculture*, *224*, 109184. DOI: 10.1016/j.compag.2024.109184

Hasanov, R. I. (2023). Towards a sustainable future in aluminium production: Environmental and economic benefits of revolutionary inert anode technology. *Journal of Sustainability Science and Management*, *18*(10), 176–186. DOI: 10.46754/jssm.2023.10.012

Hassan, H. Z., & Saeed, N. M. (2024). Advancements and applications of light-weight structures: A comprehensive review. *Discover Civil Engineering*, *1*(1), 47. DOI: 10.1007/s44290-024-00049-z

Hassan, H., Ismail, A., Ahmad, S., & Soon, C. F. (2017). Super-hydrophobic green corrosion inhibitor on carbon steel. *IOP Conference Series. Materials Science and Engineering*, *215*(1), 12023. DOI: 10.1088/1757-899X/215/1/012023

He, C., Zu, Q., Chen, J., & Noori, M. N. (2015). A review of the mechanical properties of beetle elytra and development of the biomimetic honeycomb plates. *The Journal of Sandwich Structures & Materials*, *17*(4), 399–416. DOI: 10.1177/1099636215576881

Heinselman, K., Walker, P., Norman, A., Parilla, P., Ginley, D., & Zakutayev, A. (2021). Performance and reliability of β-Ga2O3 Schottky barrier diodes at high temperature. *Journal of Vacuum Science & Technology. A, Vacuum, Surfaces, and Films*, *39*(4), 1–7. DOI: 10.1116/6.0001003

Helbig, W. A. (1946). Activated carbon. *Journal of Chemical Education*, *23*(2), 98–104. DOI: 10.1021/ed023p98

Hernández Rodríguez, G., Gastaldo, G., Stendardo, L., Rafik, Y., Pothin, J., Budinger, M., Antonini, C., Pommier-Budinger, V., & Coclite, A. M. (2024). Icephobic Gradient Polymer Coatings Coupled with Electromechanical De-icing Systems: A Promising Ice Repellent Hybrid System. *Advanced Engineering Materials*, •••, 2401532. DOI: 10.1002/adem.202401532

Hernández-Sánchez, D., Navarro-Rosales, M., Sala-Arceo, A., Miramontes-Ortega, A., & Priego-Mondragón, E. (2024). Construction Innovation-Graphene Oxide as the Differentiating Nanoadditive for Concrete and Coatings. *Materials Science Forum*, *1142*, 97–116. DOI: 10.4028/p-l9NF7J

Higashiwaki, M. (2021). β-Gallium Oxide Devices: Progress and Outlook. *Physica Status Solidi. Rapid Research Letters*, *15*(11), 2100357. Advance online publication. DOI: 10.1002/pssr.202100357

Higashiwaki, M., Sasaki, K., Kuramata, A., Masui, T., & Yamakoshi, S. (2014). Development of gallium oxide power devices. *Physica Status Solidi (A). Physica Status Solidi. A, Applications and Materials Science*, *211*(1), 21–26. DOI: 10.1002/pssa.201330197

Higashiwaki, M., & Wong, M. H. (2024). Beta-Gallium Oxide Material and Device Technologies. *Annual Review of Materials Research*, *54*(1), 175–198. DOI: 10.1146/annurev-matsci-080921-104058

Hirsch, P., & Theumer, T. (2022). Comparative study on polyamide 11 and polyamide 10.10 as matrix polymers for biogenic wood-plastic composites. *Macromolecular Symposia*, *403*(1), 2100492. Advance online publication. DOI: 10.1002/masy.202100492

Hopper, D. G., & Desjardins, D. D. (1995, June). Requirements for AMLCDs in US military applications. In *Cockpit Displays II* (Vol. 2462, pp. 130–141). SPIE. DOI: 10.1117/12.210949

Hossain, M. D., Deb, A. K., Hossain, M. M., Azam, F. A. B., & Chowdhury, M. (2023). Modification of footwear pvc outsoles into graphene oxide-based pvc (pvc/go) soling composites for the enhancement of better physical properties. Textile &Amp. *Leather Review*, *6*, 271–283. DOI: 10.31881/TLR.2023.035

Hossain, M. R., Sultana, R., Patwary, M. M., Khunga, N., Sharma, P., & Shaker, S. J. (2022). Self-healing concrete for sustainable buildings. A review. *Environmental Chemistry Letters*, *20*(2), 1–9. DOI: 10.1007/s10311-021-01375-9

Hosseinabadi, O. F., & Khedmati, M. R. (2021). A review on ultimate strength of aluminium structural elements and systems for marine applications. *Ocean Engineering*, *232*, 109153. DOI: 10.1016/j.oceaneng.2021.109153

Hsiung, B. K., He, H., & Dunlop, J. W. (2020). Trends in bioinspired material research for structural and defense applications. *Advanced Functional Materials*, *30*(5), 1909854.

Hu, C., Pan, F., Ye, W., & Chen, J. (2024). P-7.2: study on the cone-jet mode spraying mechanism of electrohydrodynamic atomization printing for oled organic thin-film encapsulation. SID Symposium Digest of Technical Papers, 55(S1), 1054-1058. DOI: 10.1002/sdtp.17275

Huang, Z., Zhan, Y., Li, W., Li, X., & Amirfazli, A. (2024). Superhydrophobic Coatings with Synergistically Enhanced Anti/Deicing Performance by Optically/Electrically Assisted Heating. *Advanced Engineering Materials*, *26*(24), 2401627. DOI: 10.1002/adem.202401627

Hu, B., Guo, H., Li, T., Li, J., Cao, M., Qi, W., Li, X., & Li, B. (2024). Millefeuille-like phase change composite with multi-level oriented architecture for efficient thermal management and electromagnetic interference shielding. *Chemical Engineering Journal*, *483*, 149240. DOI: 10.1016/j.cej.2024.149240

Huber, T., Grasl, P., Kleiser, M., Kromoser, B., & Preinstorfer, P. (2024). Holistic life cycle cost analysis of road bridges with non-metallic reinforcement. *Developments in the Built Environment*, *20*, 100533. DOI: 10.1016/j.dibe.2024.100533

Hu, J., Yu, B., & Zhou, J. (2023). New Generation Beta-Gallium Oxide Nanomaterials: Growth and Performances in Electronic Devices. *Advanced Engineering Materials*, *25*(19), 1–18. DOI: 10.1002/adem.202300688

Huynh, V., Ngo, N. K., & Golden, T. D. (2019). Surface activation and pretreatments for biocompatible metals and alloys used in biomedical applications. *International Journal of Biomaterials*, *2019*(1), 3806504. DOI: 10.1155/2019/3806504 PMID: 31275394

Ibhadode, A. O. (2023). *Engine lightweighting: use of green materials as reinforcement in aluminum metal matrix composites*. Renewable Energy - Recent Advances., DOI: 10.5772/intechopen.108273

Iijima, S., & Ichihashi, T. (1993). Single-shell carbon nanotubes of 1-nm diameter. *Nature*, *363*(6430), 603–605. DOI: 10.1038/363603a0

Im, H., Kim, W., Noh, Y., Hong, I., & Yook, J. (2022). Rcs estimation of singly curved dielectric shell structure with PMCHWT method and experimental verification. *Sensors (Basel)*, *22*(3), 734. DOI: 10.3390/s22030734 PMID: 35161480

Islam, M. H., Afroj, S., & Karim, N. (2023). Toward sustainable composites: Graphene-modified jute fiber composites with bio-based epoxy resin. *Global Challenges (Hoboken, NJ)*, *7*(9), 2300111. Advance online publication. DOI: 10.1002/gch2.202300111 PMID: 37745826

Izanker, S. V., Dhole, A., & Kumar, P. (2023, November). Navigating the Nexus: Exploring the Fusion of AI and Nanotechnology for Cutting-Edge Advances. In *2023 1st DMIHER International Conference on Artificial Intelligence in Education and Industry 4.0 (IDICAIEI)* (Vol. 1, pp. 1-5). IEEE.

Jagadeesh, B., Duraiselvam, M., & Prashanth, K. (2023). Deformation behavior of metallic lattice structures with symmetrical gradients of porosity manufactured by metal additive manufacturing. *Vacuum*, *211*, 111955. DOI: 10.1016/j.vacuum.2023.111955

Jagadeesh, B., Duraiselvam, M., & Prashanth, K. (2024). Deformation and energy absorption studies on FBCC and FBCCz lattice structures with symmetrical density gradients produced by L-PBF of Ti-6Al-4V alloy. *Materials Today: Proceedings*. Advance online publication. DOI: 10.1016/j.matpr.2024.02.008

Jiang, B., Chen, Q., & Yang, J. (2020). Advances in joining technology of carbon fiber-reinforced thermoplastic composite materials and aluminum alloys. *International Journal of Advanced Manufacturing Technology*, *110*(9), 2631–2649. DOI: 10.1007/s00170-020-06021-2

Jiang, Y., Le, Q., Zhu, Y., Liao, Q., Wang, T., Bao, L., & Wang, P. (2023). Review on forming process of magnesium alloy characteristic forgings. *Journal of Alloys and Compounds*, *970*, 172666. DOI: 10.1016/j.jallcom.2023.172666

Jirků, P., Urban, J., Müller, M., Kolář, V., Chandan, V., Svobodová, J., Mishra, R. K., & Jamshaid, H. (2023). Evaluation of mechanical properties and filler interaction in the field of sla polymeric additive manufacturing. *Materials (Basel)*, *16*(14), 4955. DOI: 10.3390/ma16144955 PMID: 37512230

Ji, X., Wang, J., Qi, S., Liang, Y., Hu, S., Zheng, H., Zhang, S., Yue, J., Qi, X., Li, S., Liu, Z., Shu, L., Tang, W., & Li, P. (2024). Improvement of Ga2O3 vertical Schottky barrier diode by constructing NiO/Ga2O3 heterojunction. *Journal of Semiconductors*, *45*(4), 042503. Advance online publication. DOI: 10.1088/1674-4926/45/4/042503

Jongejan, M., Mackintosh, A., & de Kluijver, D. (2015). Cost and Weight Reduction by Choosing Aluminium for Offshore Structures. *Abu Dhabi International Petroleum Exhibition and Conference*. https://doi.org/DOI: 10.2118/177449-MS

Joseph, O., & Aluko, M. O. (2022). Effect of synthetic materials in reinforcement of aluminium matrix composites. *Materials Science Forum*, *1076*, 3–11. DOI: 10.4028/p-o2816k

Joseph, W. A., Loto, R. T., Babalola, P., & Okeniyi, J. O. (2023). Analysis of the effect of zea mays husk particulate reinforcement on 1170 aluminium corrosion in a simulated industrial environment. *Key Engineering Materials*, *943*, 97–107. DOI: 10.4028/p-6b99x5

Jung, S., Kara, L. B., Nie, Z., Simpson, T. W., & Whitefoot, K. S. (2023). Is additive manufacturing an environmentally and economically preferred alternative for mass production? *Environmental Science & Technology*, *57*(16), 6373–6386. DOI: 10.1021/acs.est.2c04927 PMID: 37066969

Junior, A. B., Espinosa, D. C. R., Vaughan, J., & Tenório, J. A. S. (2021). Recovery of scandium from various sources: A critical review of the state of the art and future prospects. *Minerals Engineering*, *172*, 107148. DOI: 10.1016/j.mineng.2021.107148

Ju, X., Xiao, J., Wang, D., Zhao, C., Gao, T., & Wang, X. (2022). Effect of gaps/overlaps induced waviness on the mechanical properties of automated fiber placement (afp)-manufactured composite laminate. *Materials Research Express*, *9*(4), 045305. DOI: 10.1088/2053-1591/ac6858

Kalygina, V., Podzyvalov, S., Yudin, N., Slyunko, E., Zinoviev, M., Kuznetsov, V., Lysenko, A., Kalsin, A., Kopiev, V., Kushnarev, B., Oleinik, V., Baalbaki, H., & Yunin, P. (2023). Effect of UV and IR Radiation on the Electrical Characteristics of Ga2O3/ZnGeP2 Hetero-Structures. *Crystals*, *13*(8), 1203. Advance online publication. DOI: 10.3390/cryst13081203

Kanani, M., Danaee, I., & Maddahy, M. H. (2014). Microstructural characteristics and corrosion behavior of cerium oxide conversion coatings on AA6063. *Materials and Corrosion*, *65*(11), 1073–1079. DOI: 10.1002/maco.201307539

Kang, S. H., Lee, D., & Kim, H. (2018). Adaptive camouflage systems: The next generation of stealth materials. *Advanced Materials*, *30*(25), 1706698. PMID: 29709083

Kang, S., Pinault, M., Pfefferle, L. D., & Elimelech, M. (2007). Single-walled carbon nanotubes exhibit strong antimicrobial activity. *Langmuir*, *23*(17), 8670–8673. DOI: 10.1021/la701067r PMID: 17658863

Kannan, G. R., Sivakumar, M., Jagadeesh, B., & Balaji, N. (2024). Powder Bed Fusion Process–State of Art. In *Advances in Additive Manufacturing* (pp. 89-107). DOI: 10.1002/9781394238316.ch5

Kannan, G., Sujatha, E. R., Almajed, A., & Moghal, A. A. B. (2025). Microbial-Derived Biopolymers: A Pathway to Sustainable Civil Engineering. *Polymers*, *17*(2), 172. DOI: 10.3390/polym17020172 PMID: 39861245

Kannappan, R. P., & Nachimuthu, M. (2024). *4D Printing-3D Printing of Smart Materials*.

Kant, R., Chauhan, P. S., Bhatt, G., & Bhattacharya, S. (2019). Corrosion monitoring and control in aircraft: A review. *Sensors for Automotive and Aerospace Applications*, 39-53.

Kappenthuler, S., & Seeger, S. (2020). Holistic evaluation of the suitability of metal alloys for sustainable marine construction from a technical, economic and availability perspective. *Ocean Engineering*, *219*, 108378. DOI: 10.1016/j.oceaneng.2020.108378

Karadimas, G., & Salonitis, K. (2023). Ceramic matrix composites for aero engine applications—A review. *Applied Sciences (Basel, Switzerland)*, *13*(5), 3017. DOI: 10.3390/app13053017

Karbhari, V. M. (2007). Introduction: The use of composites in civil structural applications. In *Durability of composites for civil structural applications* (pp. 1–10). Elsevier.

Karthik, S., & Mohan, K. S. R. (2021). A taguchi approach for optimizing design mixture of geopolymer concrete incorporating fly ash, ground granulated blast furnace slag and silica fume. *Crystals*, *11*(11), 1279. DOI: 10.3390/cryst11111279

Kartsonakis, I. A., Kontiza, A., & Kanellopoulou, I. A. (2024). Advanced Micro/ Nanocapsules for Self-Healing Coatings. *Applied Sciences (Basel, Switzerland), 14*(18), 8396. DOI: 10.3390/app14188396

Kasten, M. (2025). Aluminum Strength vs Steel Strength. http://www.kastenmarine.com/alumVSsteel.htm

Katti, P. P. (2023). Corrosion-resistant metal nanoparticle composite coatings for industrial applications: Effect of niso4, nicl2, mo, nh4cl, and h3bo3 concentrations on nano particle concentration. *European Chemical Bulletin, 11*(12). Advance online publication. DOI: 10.48047/ecb/2022.11.12.015

Kaufman, J. G., & Rooy, E. L. (2004). *Aluminum Alloy Castings: Properties, Processes And Applications*. http://ci.nii.ac.jp/ncid/BA7486842X

Kaur, D., Debata, S., Pratap Singh, D., & Kumar, M. (2023). Strain effects on the optoelectronic performance of ultra-wide band gap polycrystalline β-Ga2O3 thin film grown on differently-oriented Silicon substrates for solar blind photodetector. *Applied Surface Science, 616*(November 2022), 156446. DOI: 10.1016/j.apsusc.2023.156446

Kaur, D., & Kumar, M. (2021). A Strategic Review on Gallium Oxide Based Deep-Ultraviolet Photodetectors: Recent Progress and Future Prospects. *Advanced Optical Materials*, *9*(9), 1–34. DOI: 10.1002/adom.202002160

Kaushik, V., Sharma, P., Priyanka, P., & Mali, H. S. (2022). Numerical modeling of fiber reinforced polymer textile composites for characterizing the mechanical behavior – a review. *Materialwissenschaft und Werkstofftechnik, 53*(10), 1263–1289. DOI: 10.1002/mawe.202200032

Khalaf, H. I., Al-Sabur, R., Abdullah, M. E., Kubit, A., & Derazkola, H. A. (2022). Effects of Underwater Friction Stir Welding Heat Generation on Residual Stress of AA6068-T6 Aluminum Alloy. *Materials (Basel)*, *15*(6), 2223. DOI: 10.3390/ma15062223 PMID: 35329680

Khang, A., & Akhai, S. (2024). Green intelligent and sustainable manufacturing: key advancements, benefits, challenges, and applications for transforming industry. *Machine Vision and Industrial Robotics in Manufacturing*, 405-417.

Khang, A., & Akhai, S. (2025). E-Waste and Lithium-ion Battery Recycling Insights for Sustainable Transportation. In *Driving Green Transportation System Through Artificial Intelligence and Automation: Approaches, Technologies and Applications* (pp. 203-230). Springer Nature Switzerland. DOI: 10.1007/978-3-031-72617-0_11

Khatri, W. A., Al Mehthel, M., Salazar, O., Baig, M., & Al Wohaibi, S. (2024). Use of Polymeric Materials in Construction to Improve Durability & Sustainability. *World Journal of Engineering and Technology*, *13*(1), 12–38. DOI: 10.4236/wjet.2025.131002

Khovalyg, D., Mudry, A., & Keller, T. (2023). Prefabricated thermally-activated fiber-polymer composite building slab p-tacs: Toward the multifunctional and pre-fabricated structural elements in buildings. *Journal of Building Physics*, *46*(6), 686–707. DOI: 10.1177/17442591221150257

Khurshid, M. F., Hasan, M. M. B., Abdkader, A., & Cherif, C. (2022). Processing of waste carbon and polyamide fibers for high performance thermoplastic composites: A novel manufacturing technology for unidirectional tapes structure. *Journal of Industrial Textiles*, *51*(4, suppl), 7256S–7276S. DOI: 10.1177/15280837221077705

Kiani, A., Waraich, P. S., Venkatakrishnan, K., & Tan, B. (2022). Synthesis of 3d nanostructured metal alloy of immiscible materials induced by megahertz-repetition femtosecond laser pulses.. DOI: 10.32920/21740780

Kim, J., Kwon, Y., & Kang, N. (2024). Deep Generative Design for Mass Production. *ArXiv Preprint ArXiv:2403.12098.*

Kim, C. K., Cho, D., Kim, S., Song, S. W., Seo, K. M., & Cho, Y. T. (2023). High-Throughput Metal 3D Printing Pen Enabled by a Continuous Molten Droplet Transfer. *Advancement of Science*, *10*(6), 2205085. Advance online publication. DOI: 10.1002/advs.202205085 PMID: 36526589

Kim, K. J., & Lee, J. (2023). Computational evaluation of automotive camber component for light-weight design. *Materialwissenschaft und Werkstofftechnik*, *54*(4), 423–428. DOI: 10.1002/mawe.202200279

Koga, G. Y., Ferreira, T., Guo, Y., Coimbrão, D. D., Jorge, A. M.Jr, Kiminami, C. S., Bolfarini, C., & Botta, W. J. (2021). Challenges in optimizing the resistance to corrosion and wear of amorphous Fe-Cr-Nb-B alloy containing crystalline phases. *Journal of Non-Crystalline Solids*, *555*, 120537. DOI: 10.1016/j.jnoncrysol.2020.120537

Kohn, W., Becke, A. D., & Parr, R. G. (1996). Density functional theory of electronic structure. *Journal of Physical Chemistry*, *100*(31), 12974–12980. DOI: 10.1021/jp9606691

Kotari, S., Punna, E., Reddy, S., & Venukumar, S. (2020). Mechanical and micro structural behaviour of flux coated GTAW and FSW joined AA6061 aluminium alloy. *Materials Today: Proceedings*, *27*, 1660–1667. DOI: 10.1016/j.matpr.2020.03.562

Kretzer, M. (2016). *Information materials: Smart materials for adaptive architecture*. Springer.

Kulkarni, P., Mali, K. D., & Singh, S. (2022). Experimental investigations on flexural performance of polymer composites with non-uniform out-of-plane waviness defect. *Journal of Thermoplastic Composite Materials*, *36*(7), 2703–2717. DOI: 10.1177/08927057221108393

Kumar, K., & Kaur, D. (2024). A review on recent advancements in the growth of MoS2 based flexible photodetectors. *Solar Energy Materials and Solar Cells, 268*(August 2023), 112736. DOI: 10.1016/j.solmat.2024.112736

Kumar, D., & Thakur, L. (2022). A study of processing and parametric optimization of wear-resistant az91-tib2 composite fabricated by ultrasonic-assisted stir casting process. *Surface Topography : Metrology and Properties*, *10*(2), 025024. DOI: 10.1088/2051-672X/ac7065

Kumar, H., Wadhwa, A. S., Akhai, S., & Kaushik, A. (2024). Parametric analysis, modeling and optimization of the process parameters in electric discharge machining of aluminium metal matrix composite. *Engineering Research Express*, *6*(2), 025542. DOI: 10.1088/2631-8695/ad4ba9

Kumar, H., Wadhwa, A. S., Akhai, S., & Kaushik, A. (2024). Parametric optimization of the machining performance of Al-SiCp composite using combination of response surface methodology and desirability function. *Engineering Research Express*, *6*(2), 025505. DOI: 10.1088/2631-8695/ad38ff

Kumar, K. R. V., Prabhu, L., Subin, B. S., Satheen, S., & Vaishnav, K. (2020). Development of Hybrid Aluminium Metal Matrix Composites for Marine Applications. *IOP Conference Series. Materials Science and Engineering*, *993*(1), 12016. DOI: 10.1088/1757-899X/993/1/012016

Kumar, M. (2023). Design and analysis of smart materials and structures for aerospace applications. *NeuroQuantology : An Interdisciplinary Journal of Neuroscience and Quantum Physics*. Advance online publication. DOI: 10.48047/nq.2022.20.10.NQ551262

Kumar, P., Kumar, D., Akhai, S., Singh, J., Bansal, J., & Chauhan, A. (2024). Investigating the Microstructure and Mechanical Performance of Ceria and Graphene-Reinforced Aluminum Hybrid Composites. *Scientia Iranica*, *0*(0), 0. DOI: 10.24200/sci.2024.61986.7593

Kumar, P., Sharma, V., Kumar, D., & Akhai, S. (2023). Morphology and mechanical behavior of friction stirred aluminum surface composite reinforced with graphene. *EVERGREEN Joint Journal of Novel Carbon Resource Sciences & Green Asia Strategy*, *10*(1), 105–110. DOI: 10.5109/6781056

Kumar, V., Angra, S., & Singh, S. (2023). Influence of rare earth elements on aluminium metal matrix composites: A review. *Mater. Phys. Mech*, *51*(2), 1–20.

Kusuma, C., Himam, F., & Kusdiana, W. (2023). Experimental study of sandwich strength materials made from banana fiber, as advance materials of the unmanned mine sweeper (ums) hull. *Key Engineering Materials*, *941*, 249–255. DOI: 10.4028/p-7byi0f

Lakshmi, P. V., Sumalatha, A., Rani, K. S., Kalyan, K., Sekhar, O. C., & Aasritha, K. S. (2024). Thermoelectric Energy Harvesting From Non-Biodegradable Dry Waste. *2024 5th International Conference on Recent Trends in Computer Science and Technology (ICRTCST)*, 472–475.

Landry, S. J., & Bihlman, W. (2023). Aerospace systems automation. In *Springer Handbook of Automation* (pp. 1115–1137). Springer International Publishing. DOI: 10.1007/978-3-030-96729-1_51

Langan, T. J., & Dorin, T. (2023). Scandium in Commercial Wrought Aluminum Alloys. *Rare Earth Metals and Minerals Industries: Status and Prospects*, 359-389.

Lao, J., Fu, W., Chen, L., Ma, J., Xiang, J., Liu, J., Peng, H., & Luo, Y. (2024). Adjustable temperature and high thermal conductivity composite phase change material based on sodium acetate trihydrate–alanine/expanded graphite for thermal management of electronic devices. *Chemical Engineering Journal*, *497*, 154867. DOI: 10.1016/j.cej.2024.154867

Latha, K. A., Navya, A., Prakash Raju, R., Poorna Chandar, V., & Bharath Raj, K. (2024). Multipurpose Smart Bridge Health Monitoring System Using Ardunio Based Sensors. *Applied Mechanics and Materials*, *924*, 91–100. DOI: 10.4028/p-H2r4sq

Lau, V. W., Klose, D., Kasap, H., Podjaski, F., Pignié, M. C., Reisner, E., Jeschke, G., & Lotsch, B. V. (2017). Dark Photocatalysis: Storage of Solar Energy in Carbon Nitride for Time-Delayed Hydrogen Generation. *Angewandte Chemie International Edition in English*, *56*(2), 510–514. DOI: 10.1002/anie.201608553 PMID: 27930846

Lee, H., Scherer, N. F., & Messersmith, P. B. (2007). Single-molecule mechanics of mussel adhesion. *Proceedings of the National Academy of Sciences of the United States of America*, *104*(43), 17345–17350. PMID: 16920796

Lee, J., Han, H., Noh, D., Lee, J., Lim, D. D., Park, J., Gu, G. X., & Choi, W. (2024). Multiscale porous architecture consisting of graphene aerogels and metastructures enabling robust thermal and mechanical functionalities of phase change materials. *Advanced Functional Materials*, *34*(42), 2405625. DOI: 10.1002/adfm.202405625

Lee, S. H., Kim, S., Moon, Y.-J., Kim, S. M., Jung, H. J., Seo, M. S., Lee, K. M., Kim, S.-K., & Lee, S. W. (2017). High-Responsivity Deep-Ultraviolet-Selective Photodetectors Using Ultrathin Gallium Oxide Films. *ACS Photonics*, *4*(11), 2937–2943. DOI: 10.1021/acsphotonics.7b01054

Lee, S. W., Lee, H., Park, Y., Kim, H., Somorjai, G. A., & Park, J. Y. (2021). Surface chemistry of hot electron and metal-oxide interfaces. *Surface Science Reports*, *76*(3), 100532. DOI: 10.1016/j.surfrep.2021.100532

Lee, T., Kim, D., Cho, S., & Kim, M. O. (2025). Advancements in Surface Coatings and Inspection Technologies for Extending the Service Life of Concrete Structures in Marine Environments: A Critical Review. *Buildings (Basel, Switzerland)*, *15*(3), 304. DOI: 10.3390/buildings15030304

Liang, L., Sheng, P., Yao, G., Huang, Z., Lin, Y., & Jiang, B. (2025). Integration of Flexible Thermoelectric Energy Harvesting System for Self-Powered Sensor Applications. *ACS Applied Materials & Interfaces*, *17*(2), 3656–3664. DOI: 10.1021/acsami.4c20424 PMID: 39757409

Li, M., Wang, X., Liu, H., Sun, X., Binwen, S., Fu, T., & Yao, T. (2024). Multi-objective optimization of the design of a protective-components-to-chassis protection module for unmanned armored vehicles. DOI: 10.21203/rs.3.rs-5350741/v1

Li, S. S., Yue, X., Li, Q. Y., Peng, H. L., Dong, B. X., Liu, T. S., Yang, H.-Y., Fan, J., Shu, S.-L., Qiu, F., & Jiang, Q. C. (2023). Development and applications of aluminum alloys for aerospace industry. *Journal of Materials Research and Technology*, *27*, 944–983. DOI: 10.1016/j.jmrt.2023.09.274

Lisiecki, A., & Kurc-Lisiecka, A. (2023). Laser cladding of nicrbsi/wc + w2c composite coatings. *Coatings*, *13*(3), 576. DOI: 10.3390/coatings13030576

Li, T., Yu, G., Zhou, S., Zheng, Z., Gao, X., & Song, Y. (2024). In-plane shear failure behavior and mechanical properties of sic/sic composites. *Journal of Physics: Conference Series*, *2730*(1), 012026. DOI: 10.1088/1742-6596/2730/1/012026

Liu, H., Zhou, S., Zhang, H., Ye, L., Xiong, Y., & Yu, P. (2022). *Ultrasensitive fully transparent amorphous Ga 2 O 3 solar-blind deep-ultraviolet photodetector for.*

Liu, C., Ji, Y., Tang, J., Otsuka, K., Wang, Y., Hou, M., Hao, Y., Ren, S., Luo, P., Ma, T., Wang, D., & Ren, X. (2022). A lightweight strain glass alloy showing nearly temperature-independent low modulus and high strength. *Nature Materials*, *21*(9), 1003–1007. DOI: 10.1038/s41563-022-01298-y PMID: 35788570

Liu, D., Huang, Y., Zhang, Z., Li, Z., Yan, Y., Chen, D., Zhao, S., Feng, Q., Zhang, J., Zhang, C., & Hao, Y. (2024). Strategy to solve the thermal issue of ultra-wide bandgap semiconductor gallium oxide field effect transistor. *Journal of Alloys and Compounds*, *986*(March), 174143. DOI: 10.1016/j.jallcom.2024.174143

Liu, J., Wang, S., Yang, H., Zhu, D., & Yue, G. (2024). Processing and characterization of high-performance thermoplastic composites manufactured by laser-assisted automated fiber placement in-situ consolidation and hot-press. *Journal of Engineered Fibers and Fabrics*, *19*, 15589250241254440. Advance online publication. DOI: 10.1177/15589250241254440

Liu, L., Zhang, Y., Zhang, S., & Tang, B. (2023). Advanced phase change materials from natural perspectives: Structural design and functional applications. *Advancement of Science*, *10*(22), 2207652. DOI: 10.1002/advs.202207652 PMID: 37226721

Liu, M., Chen, Z., Chen, M., Shao, Y., Liu, B., & Tang, Y. (2020). Scale-up strategy study of coating furnace for TRISO particle fabrication based on numerical simulations. *Nuclear Engineering and Design*, *357*, 110413. DOI: 10.1016/j.nucengdes.2019.110413

Liu, S., Wang, C., Ning, H., Meng, Z., Guan, K., & Wang, H. (2023). Achieving high ductility and low in-plane anisotropy in magnesium alloy through a novel texture design strategy. *Journal of Magnesium and Alloys*, *12*(7), 2863–2873. DOI: 10.1016/j.jma.2023.01.017

Liu, X. (2024). Graphene: in-depth research, applications, and challenges in advanced technology. *International Conference on Electrical Engineering and Intelligent Control (EEIC 2024)*, 2024, 580–583.

Liu, X., Liu, Q., Zhao, B., Ren, Y., Tao, B. W., & Zhang, W. L. (2020). Comparison of β-Ga2O3 thin films grown on r-plane and c-plane sapphire substrates. *Vacuum*, *178*(April), 109435. DOI: 10.1016/j.vacuum.2020.109435

Liu, Z., Wang, X., Liu, Y., Guo, D., Li, S., Yan, Z., Tan, C. K., Li, W., Li, P., & Tang, W. (2019). A high-performance ultraviolet solar-blind photodetector based on a β-Ga2O3 Schottky photodiode. *Journal of Materials Chemistry. C, Materials for Optical and Electronic Devices*, *7*(44), 13920–13929. DOI: 10.1039/C9TC04912F

Li, X., Aftab, S., Mukhtar, M., Kabir, F., Khan, M. F., Hegazy, H. H., & Akman, E. (2025). Exploring nanoscale perovskite materials for next-generation photodetectors: A comprehensive review and future directions. *Nano-Micro Letters, 17*(1), 28. DOI: 10.1007/s40820-024-01501-6 PMID: 39343866

Li, X., Wang, Y., & Zhang, X. (2024). Strong, healable materials with bio-like ordered architectures and versatile functionality. *SusMat, 4*(6), e248. DOI: 10.1002/sus2.248

Li, Y., Yu, C., Wang, W., Li, H., & Xu, J. (2022). A review on structural failure of composite pressure hulls in deep sea. *Journal of Marine Science and Engineering, 10*(10), 1456. DOI: 10.3390/jmse10101456

Li, Z., Ning, F., Ma, L., Jia, W., & Shi, Q. (2024). Influence of heat treatment and forming cycle on the precise forming of AZ31 magnesium alloy sheet. *Materials Today Sustainability, 27*, 100940. DOI: 10.1016/j.mtsust.2024.100940

Lowe, T. C., Valiev, R. Z., Li, X., & Ewing, B. R. (2021). Commercialization of bulk nanostructured metals and alloys. *MRS Bulletin, 46*(3), 265–272. DOI: 10.1557/s43577-021-00060-0

Lu, Z., Jiang, Y., Yu, L., Xu, J., Peng, J., & Xiang, K. (2023). The effect of minor scandium addition on microstructure evolution and mechanical properties of spray formed Al-Zn-Mg-Cu alloy. *Journal of Alloys and Compounds, 948*, 169710. DOI: 10.1016/j.jallcom.2023.169710

Lu, Z., Wang, P., & Zhang, D. (2014). Super-hydrophobic film fabricated on aluminium surface as a barrier to atmospheric corrosion in a marine environment. *Corrosion Science, 91*, 287–296. DOI: 10.1016/j.corsci.2014.11.029

Madhankumar, A., & Xavior, M. A. (2024). Graphene reinforced ceramic matrix composite (grcmc) – state of the art. *Engineering Research Express, 6*(2), 022503. DOI: 10.1088/2631-8695/ad476a

Magotra, S., & Jee, A. A. (2024). A Review on durability and microstructure of Fly-Ash based geopolymer concrete (FA-GPC). *Materials Today: Proceedings*. Advance online publication. DOI: 10.1016/j.matpr.2024.06.011

Mahamood, R. M., Akinlabi, E. T., Mahamood, R. M., & Akinlabi, E. T. (2017). Additive manufacturing of funtionally graded materials. *Functionally graded materials*, 47-68.

Mahesh, V., Mahesh, V., & Harursampath, D. (2024). On low velocity impact response of sandwich composite with jute/epoxy facesheet and cenosphere reinforced functionally graded core: Experimental and finite element approach. *Polymer Composites, 45*(14), 13151–13163. DOI: 10.1002/pc.28693

Majumdar, S., Sinha, A., Das, A., Datta, P., & Nag, D. (2024). An Insight View of Evolution of Advanced Aluminum Alloy for Aerospace and Automotive Industry: Current Status and Future Prospects. *Journal of The Institution of Engineers (India): Series D*, 1-18.

Makhlouf, A. S. H. (2014). Protective coatings for automotive, aerospace and military applications: Current prospects and future trends. In *Handbook of Smart Coatings for Materials Protection* (pp. 121–131). Woodhead Publishing. DOI: 10.1533/9780857096883.1.121

Mala, S., Deep, U., Mala, E., & Akhai, S. (2025). Digital Health Technologies for Preventive Cardiology. In *Impact of Digital Solutions for Improved Healthcare Delivery* (pp. 379–400). IGI Global.

Malik, B., Masud, J., & Akhtar, S. (2022). Objective comparison of numerical spin study with aircraft model free-flight spin tests. *Proceedings of the Institution of Mechanical Engineers. Part G, Journal of Aerospace Engineering*, *237*(2), 419–432. DOI: 10.1177/09544100221103011

Malik, M., & Amor, B. (2023). Optimizing Bridge Rehabilitation: A Life Cycle Assessment and Cost Analysis of Conventional and UHPC Overlays. *Interdisciplinary Symposium on Smart & Sustainable Infrastructures*, 971–984.

Mallipudi, S. R., & Nallu, R. (2018). Effect of scandium and zirconium additions on mechanical properties of AL–MG–MN alloy. *Transactions of the Indian Institute of Metals*, *72*(1), 227–238. DOI: 10.1007/s12666-018-1476-8

MAMULAT. (2024). *S., BYCHKOV, A., & KHAYDAROV, B.* Geopolymer Composite Materials for Low-Carbon Hydraulic Engineering and Infrastructure Construction in the Arctic Regions.

Mandal, P., Kumar, S., Pandey, A., Katariya, L., Mondal, A., & Bag, A. (2024). Unveiling structural and optical properties of Sn-doped β-Ga2O3: A correlation of experimental and theoretical observations. *Materials Science and Engineering: B*, *302*(December 2023). DOI: 10.1016/j.mseb.2024.117266

Mandal, P., Mondal, A., Pandey, A., Romeo, S., & Bag, A. (2025). Investigating role of annealing in shaping morphological, structural, tribological, and optical characteristics of gallium oxide (Ga 2 O 3): Insights from DFT analysis. *Journal of Alloys and Compounds*, *1011*(December 2024), 178367. DOI: 10.1016/j.jallcom.2024.178367

Mandal, P., Anant, R., Kumar, R., & Muthaiah, V. R. (2017). Effect of scandium on ageing kinetics in cast Al-Zn-Mg alloys. *Materials Science and Engineering A*, *696*, 257–266. DOI: 10.1016/j.msea.2017.04.067

Mandal, P., Roy, S., & Singh, U. P. (2022). Investigation on the optical and electrical performance of aluminium doped gallium oxide thin films. *Optical and Quantum Electronics*, *54*(8), 1–15. DOI: 10.1007/s11082-022-03851-0

Mandal, P., Roy, S., & Singh, U. P. (2023). Optimization of RF sputtering parameters for improving the optical and electrical properties of β-gallium oxide thin films. *Bulletin of Materials Science*, *46*(2), 64. Advance online publication. DOI: 10.1007/s12034-023-02888-5

Mandal, P., Singh, U. P., & Roy, S. (2023). Structure-property correlation of europium doped β-Ga2O3: An experimental and first-principle study. *Journal of Materials Science Materials in Electronics*, *34*(30), 1–16. DOI: 10.1007/s10854-023-11455-x

Marezio, M., & Remeika, J. P. (1967). Bond lengths in the α-Ga2O3 structure and the high-pressure phase of Ga2-xFexO3. *The Journal of Chemical Physics*, *46*(5), 1862–1865. DOI: 10.1063/1.1840945

Maris Njørdal, V. I. A. (2024). *Speira Newsroom*. https://www.speira.com/newsroom/press/water-revolution-foundation-certifies-via-maris-njoerdal/

Maritimer. (1967). The use of aluminium-rich alloys in shipbuilding. *Anti-Corrosion Methods and Materials, 14*(3), 12. https://doi.org/DOI: 10.1108/eb005203

Maroliya, M., Midhun, V. C., & Saha, S. K. (2024). Evaluation of solid-solid and solid-liquid phase change materials in pin-finned heat sinks for cooling of avionics electronics. *Thermal Science and Engineering Progress*, *53*, 102781. DOI: 10.1016/j.tsep.2024.102781

Marshall, J. S., Young, J. W., & Hallam, J. (2008). Bio-Inspired Approaches to Sensing for Defence and Security Applications. *Journal of the Royal Society, Interface*, *5*(21), 467–477. DOI: 10.1098/rsif.2008.0109.focus

Martín-Palma, R. J., & Lakhtakia, A. (2013). Vapor-Deposition Techniques. In *Engineered Biomimicry*. Elsevier Inc., DOI: 10.1016/B978-0-12-415995-2.00015-5

Ma, X., Zhang, Y., Dong, L., & Jia, R. (2017). First-principles calculations of electronic and optical properties of aluminum-doped β-Ga2O3 with intrinsic defects. *Results in Physics*, *7*, 1582–1589. DOI: 10.1016/j.rinp.2017.04.023

Mbam, S. O., Nwonu, S. E., Orelaja, O. A., Nwigwe, U. S., & Gou, X. F. (2019). Thin-film coating; historical evolution, conventional deposition technologies, stress-state micro/nano-level measurement/models and prospects projection: A critical review. *Materials Research Express*, *6*(12), 122001. DOI: 10.1088/2053-1591/ab52cd

McCafferty, E. E., Hubler, G. K., Natishan, P. M., Moore, P. G., Kant, R. A., & Sartwell, B. D. (1987). Naval research laboratory surface modification program: Ion beam and laser processing of metal surfaces for improved corrosion resistance. *Materials Science and Engineering*, *86*, 1–17. DOI: 10.1016/0025-5416(87)90438-1

McIntyre, T. B. (2024). *The Heroes We Needed: The B-29ers Who Ended World War II and My Fight to Save the Forgotten Stories of the Greatest Generation*. Helion and Company.

Mei, L., Xiang, D., & Huang, Y. (2024). Research on the mechanical properties of eps lightweight soil mixed with slag. *PLoS One*, *19*(1), e0297372. DOI: 10.1371/journal.pone.0297372 PMID: 38265991

Melhem, G. N., Munroe, P. R., & Vithal, A. (2024). *Supersonic Effects on Rivets*. CRC Press. DOI: 10.1201/9781003516903

Menzie, W. D., Barry, J. J., Bleiwas, D. I., Bray, E. L., Goonan, T. G., & Matos, G. (2010). *The global flow of aluminum from 2006 through 2025*. US Department of the Interior, US Geological Survey. DOI: 10.3133/ofr20101256

Meyers, M. A., Lin, A. Y. M., Olevsky, E. A., & Chen, P. Y. (2008). Biological materials: Structure and mechanical properties. *Progress in Materials Science*, *53*(1), 1–206. DOI: 10.1016/j.pmatsci.2007.05.002

Mirzendehdel, A. M., & Suresh, K. (2016). Support structure constrained topology optimization for additive manufacturing. *Computer Aided Design*, *81*, 1–13. DOI: 10.1016/j.cad.2016.08.006

Mishra, R. K., Sarkar, J., Verma, K., Chianella, I., Goel, S., & Nezhad, H. Y. (2024). Exploring transformative and multifunctional potential of MXenes in 2D materials for next-generation technology. Open Ceramics, 100596.

Mohamed, H. M., Ibrahim, B., & ElZefzafy, H. (2024). Sustainable Innovative New Advanced Materials for Underground Infrastructure Projects. *2024 International Conference on Decision Aid Sciences and Applications (DASA)*, 1–4. DOI: 10.1109/DASA63652.2024.10836421

Mohammad, Q., Kaliyaperumal, G., Poornima, E., Singh, N., Sethi, V. A., & Kumari, V. (2024). Aluminum-alumina composite manufacturing: unlocking potential with friction stir processing. E3S Web of Conferences, 507, 01034. DOI: 10.1051/e3sconf/202450701034

Mohammadi, H., Ahmad, Z., Mazlan, S. A., Johari, M. A. F., Siebert, G., Petrů, M., & Koloor, S. S. R. (2022). Lightweight glass fiber-reinforced polymer composite for automotive bumper applications: A review. *Polymers*, *15*(1), 193. DOI: 10.3390/polym15010193 PMID: 36616541

Moharekpour, M., Shokri, M., Wellerdick, E., Traverso, M., Oeser, M., & Liu, P. (2024). An evaluation of the economic viability and accessibility of CRCP and JPCP: A comparative analysis. *Sustainability (Basel)*, *16*(3), 1108. DOI: 10.3390/su16031108

Mondal, A., Nandi, S., Yadav, M. K., Nandi, A., & Bag, A. (2022). Broad Range (254-302 nm) and High Performance Ga2O3:SnO2Based Deep UV Photodetector. *IEEE Transactions on Nanotechnology*, *21*, 320–327. DOI: 10.1109/TNANO.2022.3186472

Mondal, A., Yadav, M. K., Shringi, S., & Bag, A. (2020). Extremely low dark current and detection range extension of Ga2O3 UV photodetector using Sn alloyed nanostructures. *Nanotechnology*, *31*(29), 294002. Advance online publication. DOI: 10.1088/1361-6528/ab82d4 PMID: 32208373

Mondal, P., Satra, J., Srivastava, D. N., Bhadu, G. R., & Adhikary, B. (2021). $Pd^{\delta+}$-Mediated Surface Engineering of AgMnO4 Nanorods as Advanced Bifunctional Electrocatalysts for Highly Efficient Water Electrolysis. *ACS Catalysis*, *11*(6), 3687–3703. DOI: 10.1021/acscatal.0c05638

Monteiro, O., & Davies, B. (1998). Cost Effective Plasma-Based Process for Depositing a Wide Variety of High Temperature Protective Coatings. *Materials Technology*, *13*(1), 3–4. DOI: 10.1080/10667857.1998.11752756

Montemor, M. F. (2016). Corrosion issues in joining lightweight materials: A review of the latest achievements. *Physical Sciences Reviews*, *1*(2), 20150011. DOI: 10.1515/psr-2015-0011

Moore, B., Asadi, E., & Lewis, G. (2017). Deposition methods for microstructured and nanostructured coatings on metallic bone implants: A review. *Advances in Materials Science and Engineering*, *2017*(1), 5812907. DOI: 10.1155/2017/5812907

Moore, R., Torres, H., DeLuca, R., Gentile, M., Zadrovicz, K. C., Albrecht, M., & Kohli, S. (2023, June). Salt spray resistant silver coatings for aerospace and defense applications. In *Advanced Optics for Imaging Applications: UV through LWIR VIII* (Vol. 12530, pp. 46–55). SPIE. DOI: 10.1117/12.2663465

Morganti, L., Vandi, L., Larraz, J. A., García-Jaca, J., Muedra, A. N., & Pracucci, A. (2024). A1–a5 embodied carbon assessment to evaluate bio-based components in façade system modules. *Sustainability (Basel)*, *16*(3), 1190. DOI: 10.3390/su16031190

Morse, L., & Mallardo, V. (2023). Bottom-up manufacturing cost optimisation of composite aircraft structures: Manual layup vs. automated layup. *AIP Conference Proceedings*, 020027. Advance online publication. DOI: 10.1063/5.0145177

Morstein, C. E., & Dienwiebel, M. (2021). Graphite lubrication mechanisms under high mechanical load. *Wear*, *477*, 203794. DOI: 10.1016/j.wear.2021.203794

Motia, K., Kumar, R., & Akhai, S. (2024). AI and Smart Manufacturing: Building Industry 4.0. In *Modern Management Science Practices in the Age of AI* (pp. 1-28). IGI Global.

Mouritz, A. P. (2012). *Introduction to aerospace materials*. Elsevier.

Murcia, D. H., & Taha, M. R. (2023). Emerging Materials and Technologies for Next-Generation Sustainable and Resilient Polymer Concrete. *International Congress on Polymers in Concrete*, 47–58.

Muribwathoho, O., Msomi, V., & Mabuwa, S. (2022). Metal Matrix Composite Fabricated with 5000 Series Marine Grades of Aluminium Using FSP Technique: State of the Art Review. *Applied Sciences (Basel, Switzerland)*, *12*(24), 12832. DOI: 10.3390/app122412832

Mushtaq, A., Ali, S., Chaudhry, A. H., Sial, N., Aslam, M., & Batool, H. (2025). Geopolymers as Supplementary Cementitious Materials to Reduce Carbon Dioxide Emissions. *Nature Environment & Pollution Technology, 24*.

Naik, G. G., & Jagtap, V. A. (2024). Two heads are better than one: Unravelling the potential Impact of Artificial Intelligence in nanotechnology. *Nano TransMed*, 100041.

Najem, O., Benbrahim, M., Kabbaj, M. N., & Boumhidi, J. (2025). Synergy between Embedded and Non-Embedded Sensors in IoT Networks for Structural Health Monitoring of Concrete Structures: A Literature Review. *E3S Web of Conferences, 601*, 28.

Naleway, S. E., Porter, M. M., McKittrick, J., & Meyers, M. A. (2015). Structural design elements in biological materials: Application to bioinspiration. *Advanced Materials*, *27*(37), 5455–5476. DOI: 10.1002/adma.201502403 PMID: 26305858

Nan, H., Wang, P., Zhang, Q., Meng, D., & Lei, Q. (2023). Study on the mechanical properties of continuous composite beams under coupled slip and creep. *Materials (Basel)*, *16*(13), 4741. DOI: 10.3390/ma16134741 PMID: 37445055

Narkhede, G., Samuel, C., Mahajan, S., Verma, D., Sakhare, N., & Chaudhari, T. (2024). Beyond traditional supply chain management: Addressing sociopolitical challenges in increasingly turbulent global trade landscape. *Business Strategy & Development*, *7*(2), e397. DOI: 10.1002/bsd2.397

Nartita, R., Ionita, D., & Demetrescu, I. (2021). Sustainable coatings on metallic alloys as a nowadays challenge. *Sustainability (Basel)*, *13*(18), 10217. DOI: 10.3390/su131810217

Nasiri, H. (n.d.). *Innovations in Construction: Exploring the Future with New Materials*.

Nathanael, A. J., & Oh, T. H. (2020). Biopolymer coatings for biomedical applications. *Polymers*, *12*(12), 3061. DOI: 10.3390/polym12123061 PMID: 33371349

National Materials and Manufacturing Board (NMMB). (2021). Bio-Inspired Manufacturing for Future Defense Applications. Retrieved from www.nas.edu

National Research Council. (2012). *Opportunities and challenges in biomimetics*. The National Academies Press.

Nawaz, F., Ali, M., Ahmad, S., Yong, Y., Rahman, S., Naseem, M., Hussain, S., Razzaq, A., Khan, A., Ali, F., Balushi, R., Al-Hinaai, M., & Ali, N. (2024). Carbon based nanocomposites, surface functionalization as a promising material for VOCs (volatile organic compounds) treatment. *Chemosphere*, *364*, 143014–143025. DOI: 10.1016/j.chemosphere.2024.143014 PMID: 39121955

Nelson, R. L., Ramsay, J. D. F., Woodhead, J. L., Cairns, J. A., & Crossley, J. A. A. (1981). The coating of metals with ceramic oxides via colloidal intermediates. *Thin Solid Films*, *81*(4), 329–337. DOI: 10.1016/0040-6090(81)90517-4

Nguyen, K. T., Ahn, N., Le, T. A., & Lee, K. (2016). Theoretical and experimental study on mechanical properties and flexural strength of fly ash-geopolymer concrete. *Construction & Building Materials*, *106*, 65–77. DOI: 10.1016/j.conbuildmat.2015.12.033

Nguyen, V. G. (2024). Applying FRP Composites Materials for Repairing and Reinforcing Dien Bien Phu Bridge, Binh Thanh District, HCMC: Effective and Economic Solution. *International Conference Series on Geotechnics, Civil Engineering and Structures*, 141–148. DOI: 10.1007/978-981-97-1972-3_12

Nikolov, S., Fabritius, H., Petrov, M., Friák, M., Lymperakis, L., Sachs, C., Raabe, D., & Neugebauer, J. (2011). Robustness and optimal use of design principles of arthropod exoskeletons studied by ab initio-based multiscale simulations. *Journal of the Mechanical Behavior of Biomedical Materials*, *4*(2), 129–145. DOI: 10.1016/j.jmbbm.2010.09.015 PMID: 21262491

Nilimaa, J. (2023). Smart materials and technologies for sustainable concrete construction. *Developments in the Built Environment*, *15*, 100177. DOI: 10.1016/j.dibe.2023.100177

Nine, M. J., Cole, M. A., Tran, D. N., & Losic, D. (2015). Graphene: A multipurpose material for protective coatings. *Journal of Materials Chemistry. A, Materials for Energy and Sustainability*, *3*(24), 12580–12602. DOI: 10.1039/C5TA01010A

Ni, X., Gao, Y., Zhang, X., Lei, Y., Sun, G., & You, B. (2021). An eco-friendly smart self-healing coating with NIR and pH dual-responsive superhydrophobic properties based on biomimetic stimuli-responsive mesoporous polydopamine microspheres. *Chemical Engineering Journal*, *406*, 126725. DOI: 10.1016/j.cej.2020.126725

Nunes, H., Emadinia, O., Vieira, M. F., & Reis, A. (2023). Low- and High-Pressure Casting Aluminum Alloys: A Review. *IntechOpen eBooks*. https://doi.org/DOI: 10.5772/intechopen.109869

Nwokediegwu, Z. Q. S., Ilojianya, V. I., Ibekwe, K. I., Adefemi, A., Etukudoh, E. A., & Umoh, A. A. (2024). Advanced materials for sustainable construction: A review of innovations and environmental benefits. *Engineering Science & Technology Journal*, *5*(1), 201–218. DOI: 10.51594/estj.v5i1.744

Ogata, K. A., Lazarevic, S., & Miller, S. F. (2014). Dissimilar Material Joint Strength and Structure for Friction Stir Forming Process. https://doi.org/DOI: 10.1115/MSEC2014-4044

Ogunsuji, Y. M., Amosu, O. R., Choubey, D., Abikoye, B. E., Kumar, P., & Umeorah, S. C. (2024). Sourcing renewable energy components: building resilient supply chains, reducing dependence on foreign suppliers, and enhancing energy security. *World Journal of Advanced Research and Reviews*, *23*(2).

Ojha, A., Kumar, A., Das, D., & Bandyopadhyay, T. K. (2025). Bio-Concrete and Its Self-Healing Capacity to Next-generation Sustainable Architecture: A Comprehensive Review. *Exon*, *2*(1), 1–15. DOI: 10.69936/en01y0025

Ojo, O. O., & Taban, E. (2023). Post-processing treatments–microstructure–performance interrelationship of metal additive manufactured aerospace alloys: A review. *Materials Science and Technology*, *39*(1), 1–41. DOI: 10.1080/02670836.2022.2130530

Omrany, H., Al-Obaidi, K. M., Husain, A., & Ghaffarianhoseini, A. (2023). Digital twins in the construction industry: A comprehensive review of current implementations, enabling technologies, and future directions. *Sustainability (Basel)*, *15*(14), 10908. DOI: 10.3390/su151410908

Önal, M., Neşer, G., & Gürsel, K. T. (2020). Environmental impacts of steel ship hulls building and recycling by life cycle assessment (LCA). *Ships and Offshore Structures*, *16*(10), 1061–1066. DOI: 10.1080/17445302.2020.1816706

Onuma, T., Fujioka, S., Yamaguchi, T., Higashiwaki, M., Sasaki, K., Masui, T., & Honda, T. (2013). Correlation between blue luminescence intensity and resistivity in β-Ga2O3 single crystals. *Applied Physics Letters*, *103*(4), 22–25. DOI: 10.1063/1.4816759

Österreicher, J., Nebeling, D., Grabner, F., Cerny, A., Zickler, G., Eriksson, J., Wikström, G., Suppan, W., & Schlögl, C. (2023). Secondary ageing and formability of an Al-Cu-Mg alloy (2024) in W and under-aged tempers. *Materials & Design*, *226*, 111634. DOI: 10.1016/j.matdes.2023.111634

Pakhomov, A., Pisano, D., Sicignano, A., & Goldburt, T. (2005, May). High performance seismic sensor requirements for military and security applications. In *Unattended Ground Sensor Technologies and Applications VII* (Vol. 5796, pp. 117–124). SPIE. DOI: 10.1117/12.604358

Panda, B. (2021). Corrosion Resistant Superhydrophobic Aluminum Alloy: A Review. *IOP Conference Series. Materials Science and Engineering*, *1017*(1), 12008. DOI: 10.1088/1757-899X/1017/1/012008

Panda, S., Mehlawat, S., Dhariwal, N., Sanger, A., & Kumar, A. (2024). Comprehensive review on gas sensors: Unveiling recent developments and addressing challenges. *Materials Science and Engineering B*, *308*(August), 117616. DOI: 10.1016/j.mseb.2024.117616

Pandey, R., Thapa, P., Kumar, V., Zhu, Y., Wang, N., Bystrzejewski, M., & Tiwari, S. K. (2022). Updates in phase change materials for thermoelectric devices: Status and challenges. *Materialia*, *21*, 101357. DOI: 10.1016/j.mtla.2022.101357

Paranjape, A. A., & Ananthkrishnan, N. (2023). Combat aircraft agility metrics - a review. *Journal of Aerospace Sciences and Technologies*, •••, 55–66. DOI: 10.61653/joast.v58i2.2006.706

Paraye, P., & Sarviya, R. (2024). Advances in polymer composites, manufacturing, recycling, and sustainable practices. *Polymer-Plastics Technology Materials*, 1-24.

Park, J. H., McClintock, R., Jaud, A., Dehzangi, A., & Razeghi, M. (2019). MOCVD grown β-Ga2O3 metal-oxide-semiconductor field effect transistors on sapphire. *Applied Physics Express*, *12*(9), 0–5. DOI: 10.7567/1882-0786/ab3b2a

Paruthi, S., Rahman, I., Khan, A. H., Sharma, N., & Alyaseen, A. (2024). Strength, durability, and economic analysis of GGBS-based geopolymer concrete with silica fume under harsh conditions. *Scientific Reports*, *14*(1), 31572. DOI: 10.1038/s41598-024-77801-z PMID: 39738079

Parveez, B., Kittur, M. I., Badruddin, I. A., Kamangar, S., Hussien, M., & Umarfarooq, M. A. (2022). Scientific advancements in composite materials for aircraft applications: A review. *Polymers*, *14*(22), 5007. DOI: 10.3390/polym14225007 PMID: 36433134

Patakham, U., Kajornchaiyakul, J., & Limmaneevichitr, C. (2012). Grain refinement mechanism in an Al–Si–Mg alloy with scandium. *Journal of Alloys and Compounds*, *542*, 177–186. DOI: 10.1016/j.jallcom.2012.07.018

Patchen, A., Young, S., Goodbred, L., Puplampu, S., Chawla, V., & Penumadu, D. (2023). Lower carbon footprint concrete using recycled carbon fiber for targeted strength and insulation. *Materials (Basel)*, *16*(15), 5451. DOI: 10.3390/ma16155451 PMID: 37570158

Patel, N. V., Sharma, N. P., Tiwari, N. A., & Vasnani, N. H. (2024b). Magnesium alloys in the automotive industry: Alloying elements and their impact on material performance. *Global Journal of Engineering and Technology Advances*, *21*(1), 130–145. DOI: 10.30574/gjeta.2024.21.1.0194

Patil, A. Y., Hegde, C., Savanur, G., Kanakmood, S. M., Contractor, A. M., Shirashyad, V. B., Chivate, R. M., Kotturshettar, B. B., Mathad, S. N., Patil, M. B., Soudagar, M. E. M., & Fattah, I. M. R. (2022). Biomimicking nature-inspired design structures—An experimental and simulation approach using additive manufacturing. *Biomimetics*, *7*(4), 186. DOI: 10.3390/biomimetics7040186 PMID: 36412714

Patscheider, J. (2003). Nanocomposite hard coatings for wear protection. *MRS Bulletin*, *28*(3), 180–183. DOI: 10.1557/mrs2003.59

Paz Martínez-Viademonte, M., Abrahami, S. T., Hack, T., Burchardt, M., & Terryn, H. (2020). A review on anodizing of aerospace aluminum alloys for corrosion protection. *Coatings*, *10*(11), 1106. DOI: 10.3390/coatings10111106

Peckerar, M. C., & Neidert, R. E. (2005). High-speed microelectronics for military applications. *Proceedings of the IEEE*, *71*(5), 657–666. DOI: 10.1109/PROC.1983.12647

Peng, P., Xue, H., She, J., Zhang, J., Tang, A., Long, S., Zhang, C., Yang, Q., & Pan, F. (2024). Ultrafine-grained Mg alloy: Preparation, properties, design strategy. *Journal of Materials Research and Technology*, *29*, 4480–4504. DOI: 10.1016/j.jmrt.2024.02.131

Picraux, S. T., & Pope, L. E. (1984). Tailored surface modification by ion implantation and laser treatment. *Science*, *226*(4675), 615–622. DOI: 10.1126/science.226.4675.615 PMID: 17774924

Playford, H. Y., Hannon, A. C., Barney, E. R., & Walton, R. I. (2013). Structures of uncharacterised polymorphs of gallium oxide from total neutron diffraction. *Chemistry (Weinheim an der Bergstrasse, Germany)*, *19*(8), 2803–2813. DOI: 10.1002/chem.201203359 PMID: 23307528

Pohl, K. (1968). Hydrothermale Bildung von γ-Ga2O3. *Naturwissenschaften*, *55*(2), 82. DOI: 10.1007/BF00599490

Pokorný, P., & Kouřil, M. (2024). Predicted corrosion performance of organofunctional silane coated steel reinforcement for concrete structures: An overview. *Buildings (Basel, Switzerland)*, *14*(6), 1756. DOI: 10.3390/buildings14061756

Prajwal, L., Balagi, G. M., & Kumar, C. (n.d.). Study of FRP on Preloaded Beam as A Method of Retrofitting-A Review. *International Research Journal on Advanced Engineering and Management (IRJAEM)*, *2*(10).

Prasad, S. V. S., Prasad, S. B., Verma, K., Mishra, R. K., Kumar, V., & Singh, S. (2021c). The role and significance of Magnesium in modern day research-A review. *Journal of Magnesium and Alloys*, *10*(1), 1–61. DOI: 10.1016/j.jma.2021.05.012

Quek, J., & Rajeev, R. (2023). Composite System for Structural Enhancement and Efficient Heritage Conservation: A Case Study. *The International Conference on Sustainable Civil Engineering and Architecture*, 419–428.

Qureshi, A. H., Ahmad, N., Rana, M. A. A., Manzoor, B., & Zayed, T. (2024). Construction Sector Transition towards Smart Applications of Graphene Oxide in Cement-Based Composites: A Scientometric Review and Bibliometric Analysis. *Buildings (Basel, Switzerland)*, *14*(10), 3042. DOI: 10.3390/buildings14103042

Raabe, D., Ponge, D., Uggowitzer, P. J., Roscher, M., Paolantonio, M., Liu, C., Antrekowitsch, H., Kozeschnik, E., Seidmann, D., Gault, B., De Geuser, F., Deschamps, A., Hutchinson, C., Liu, C., Li, Z., Prangnell, P., Robson, J., Shanthraj, P., Vakili, S., & Pogatscher, S. (2022). Making sustainable aluminum by recycling scrap: The science of "dirty" alloys. *Progress in Materials Science*, *128*, 100947. DOI: 10.1016/j.pmatsci.2022.100947

Radić, A., Fuentes, J. M. A., Ariza-Montes, A., Han, H., & Law, R. (2020). Job demands–job resources (JD-R) model, work engagement, and well-being of cruise ship employees. *International Journal of Hospitality Management*, *88*, 102518. DOI: 10.1016/j.ijhm.2020.102518

Radushkevich, L. V. (1952). On the Structure of Carbon Formed by Thermal Decomposition of Carbon Oxide at an Iron Contact. *J. Phys. Chem. Russ.*, *26*, 88–95.

Raj, S. H. K., Boopathy, G., Vijayananth, S., Balaji, G., & Ramanan, N. (2023). Feasibility and insights into the optimization and characterization of friction welded aluminum–steel dissimilar joints. In *International Conference on Smart Sustainable Materials and Technologies* (pp. 121–129). Springer Nature Switzerland. https://doi.org/DOI: 10.1007/978-3-031-35362-4_12

Rajabi, H., Gorb, S. N., & Jafarpour, M. (2016). Challenges in bio-inspired material design: Scaling up from laboratory to industrial production. *Bioinspiration & Biomimetics*, *11*(2), 026005. PMID: 27033298

Raja, T., & Devarajan, Y. (2024). Study on the mechanical and thermal properties of aramid fibre reinforced with sawdust particulates in an epoxy matrix composite: A novel material for structural applications. *Polymer International*, *73*(12), 1063–1070. DOI: 10.1002/pi.6685

Raja, V., Gnanasekaran, R. K., Kaladgi, A. R., Rajendran, P., Khan, S. A., & Asif, M. (2022). Multi-disciplinary computational investigations on asymmetrical failure factors of disc brakes for various cfrp materials: A validated approach. *Symmetry*, *14*(8), 1616. DOI: 10.3390/sym14081616

Rajesh, P. V., Gupta, K. K., Čep, R., Ramachandran, M., Kouřil, K., & Kalita, K. (2022). Optimizing friction stir welding of dissimilar grades of aluminum alloy using waspas. *Materials (Basel)*, *15*(5), 1715. DOI: 10.3390/ma15051715 PMID: 35268941

Raj, R. J., Selvam, P., & Pughalendi, M. (2021). A review of aluminum alloys in aircraft and aerospace industry. *J. Huazhong Univ. Sci. Technol*, *1671*, 4512.

Ramesha, K., Sudersanan, P. D., Santosh, N., & Jangam, S. (2021). Sustainable Processing of Dissimilar Aluminium Alloy Joints by Friction Stir Welding. *IOP Conference Series. Materials Science and Engineering*, *1132*(1), 12029. DOI: 10.1088/1757-899X/1132/1/012029

Rammohan, S., Kumaran, S. T., Uthayakumar, M., Korniejenko, K., Nykiel, M., & Velayutham, A. (2022). Prediction of abrasive waterjet machining parameters of military-grade armor steel by semi-empirical and regression models. *Materials (Basel)*, *15*(12), 4368. DOI: 10.3390/ma15124368 PMID: 35744427

Ramraji, K., Rajkumar, K., Rajesh, M., & Bhaskar, A. (2023). Mechanical and vibration behavior of surface-modified titanium sheet interleaved with woven basalt/flax fiber metal laminates. *Polymer Composites*, *44*(12), 8442–8453. DOI: 10.1002/pc.27709

Rana, A., Rana, M., Rashmi, R., & Das, P. (2024). The Future of Smart Textiles: Combining Wearable Technology With 4D Printing. *Available atSSRN* 5088887.

Rangapuram, M., Dasari, S., Abutunis, A., Chandrashekhara, K., Lua, J., & Li, R. (2024). Experimental and numerical investigation of heat evolution inside the autoclave for composite manufacturing. *Composites and Advanced Materials*, *33*, 26349833241273506. Advance online publication. DOI: 10.1177/26349833241273506

Ranjan, M., Kumar, S., & Sinha, S. (2024). Nanosilica's Influence on Concrete Hydration, Microstructure, and Durability: A Review. *Journal of Applied Engineering Sciences*, *14*(2), 322–335. DOI: 10.2478/jaes-2024-0040

Rao, K. P., Prasad, Y. V. R. K., Dharmendra, C., Suresh, K., Hort, N., & Dieringa, H. (2018). Review on hot working behavior and strength of Calcium-Containing Magnesium Alloys. *Advanced Engineering Materials*, *20*(9), 1701102. Advance online publication. DOI: 10.1002/adem.201701102

Rao, M. S., & Padal, K. B. (2022, November). New developments of aluminium alloys for future generation applications-A review. In. AIP Conference Proceedings: Vol. 2648. *No. 1* (p. 030011). AIP Publishing LLC. DOI: 10.1063/5.0114421

Raval, S. K., & Judal, K. B. (2020). Recent Advances in Dissimilar Friction Stir Welding of Aluminum to Magnesium Alloys. *Materials Today: Proceedings*, *22*, 2665–2675. DOI: 10.1016/j.matpr.2020.03.398

Razeghi, M., Park, J.-H., & McClintock, R. (2019). (Invited) MOCVD-Grown Ga 2 O 3 Field Effect Transistors on Sapphire. *ECS Meeting Abstracts, MA2019-02*(31), 1356–1356. DOI: 10.1149/MA2019-02/31/1356

Reddy, S. (2018). Recent trends in biomimetic materials research and applications. *Materials Today*, *21*(1), 65–78.

Rhamdani, A. R., Rhamdhani, M. A., Brooks, G., Pownceby, M. I., Nugraha Tha-ha, Y., Abbott, T. B., Grandfield, J., & Hartley, C. (2025). The production of rare earth based magnesium and aluminium alloys–a review. *Mineral Processing and Extractive Metallurgy Review*, *46*(1), 44–67. DOI: 10.1080/08827508.2023.2243009

Richards, B. S., Hudry, D., Busko, D., Turshatov, A., & Howard, I. A. (2021). Photon Upconversion for Photovoltaics and Photocatalysis: A Critical Review. *Chemical Reviews*, *121*(15), 9165–9195. DOI: 10.1021/acs.chemrev.1c00034 PMID: 34327987

Ricks, T. M., Pineda, E. J., Bednarcyk, B. A., McCorkle, L., Miller, S. G., Murthy, P. L. N., & Segal, K. N. (2022). Multiscale progressive failure analysis of 3d woven composites. *Polymers*, *14*(20), 4340. DOI: 10.3390/polym14204340 PMID: 36297918

Ritchie, R. O. (2011). The conflicts between strength and toughness. *Nature Materials*, *10*(11), 817–822. DOI: 10.1038/nmat3115 PMID: 22020005

Riva, S., Yusenko, K. V., Lavery, N. P., Jarvis, D. J., & Brown, S. G. R. (2016). The scandium effect in multicomponent alloys. *International Materials Reviews*, *61*(3), 203–228. DOI: 10.1080/09506608.2015.1137692

Rogalski, A., Bielecki, Z., Mikołajczyk, J., & Wojtas, J. (2023). Ultraviolet Photodetectors: From Photocathodes to Low-Dimensional Solids. *Sensors (Basel)*, *23*(9), 280–315. DOI: 10.3390/s23094452 PMID: 37177656

Roy, M. (2023). Materials and Coatings for Defence Application. *Defence Science Journal*, *73*(2), 119–120. DOI: 10.14429/dsj.73.18849

Roy, R., Hill, V. G., & Osborn, E. F. (1952). Polymorphism of Ga2O3and the System Ga2O3–H2O. *Journal of the American Chemical Society*, *74*(3), 719–722. DOI: 10.1021/ja01123a039

Roy, S., & Sahoo, P. (2014). Parametric optimization of corrosion and wear of electroless Ni-P-Cu coating using grey relational coefficient coupled with weighted principal component analysis. *International Journal of Mechanical and Materials Engineering*, *9*(1), 1–15. DOI: 10.1186/s40712-014-0010-y

Russo, A., Russo, F., Palumbo, C., & Riccio, A. (2022). Numerical assessment on the fatigue behavior of composite open-hole tensile specimens. *Macromolecular Symposia, 404*(1), 2100454. Advance online publication. DOI: 10.1002/masy.202100454

Rusu, I., Judele, L., & Lepadatu, D. (2024). POLYMERIC COATING FOR THE INSULATION, PROTECTION AND SURFACES FINISHING OF THE REINFORCED BUILDINGS CONCRETE EXPOSED TO CORROSIVE ENVIRONMENTS. *International Multidisciplinary Scientific GeoConference: SGEM, 2024*(6.1), 279–285.

Ryabov, D. K., Kolobnev, N. I., & Samohvalov, S. V. (2014). Effect of Scandium addition on mechanical properties and corrosion resistance of medium strength AL-ZN-MG-CU) alloy. *Materials Science Forum, 794–796*, 241–246. . DOI: 10.4028/www.scientific.net/MSF.794-796.241

Sadiku, E. R., Agboola, O., Mochane, M. J., Fasiku, V. O., Owonubi, S. J., Ibrahim, I. D., & Ndamase, A. S. (2019). The use of polymer nanocomposites in the aerospace and the military/defence industries. In *Polymer nanocomposites for advanced engineering and military applications* (pp. 316–349). IGI Global. DOI: 10.4018/978-1-5225-7838-3.ch011

Saevarsdottir, G., Kvande, H., & Welch, B. J. (2020). Reducing the carbon footprint: aluminium smelting with changing energy systems and the risk of carbon leakage. In *Light metals 2020* (pp. 726–734). Springer International Publishing. DOI: 10.1007/978-3-030-36408-3_98

Saikumar, A. K., Nehate, S. D., & Sundaram, K. B. (2021). A review of recent developments in aluminum gallium oxide thin films and devices. *Critical Reviews in Solid State and Material Sciences, 0*(0), 1–32. DOI: 10.1080/10408436.2021.1922357

Sales, A. S. B., Nahime, B. O., Lima, E. C., & Reis, I. C. (2024). Study of tire rubber residue in cementitious composite compared to expanded polystyrene, to make a lightweight concrete. *Concilium, 24*(1), 300–320. DOI: 10.53660/CLM-2680-24A04

Salman, S. A., & Okido, M. (2013). Anodization of magnesium (Mg) alloys to improve corrosion resistance. *Corrosion prevention of magnesium alloys*, 197-231.

Sames, W. J., List, F., Pannala, S., Dehoff, R. R., & Babu, S. S. (2016). The metallurgy and processing science of metal additive manufacturing. *J. International Materials Reviews, 61*(5), 315–360. DOI: 10.1080/09506608.2015.1116649

Sampath, S., & McCune, R. (2000). Thermal-spray processing of materials. *MRS Bulletin, 25*(7), 12–13. DOI: 10.1557/mrs2000.117

Sandak, A., & Butina Ogorelec, K. (2023). Bioinspired building materials—Lessons from nature. *Frontiers in Materials*, *10*, 1283163. DOI: 10.3389/fmats.2023.1283163

Sang, L., Liao, M., & Sumiya, M. (2013). A comprehensive review of semiconductor ultraviolet photodetectors: From thin film to one-dimensional nanostructures. *Sensors (Basel)*, *13*(8), 10482–10518. DOI: 10.3390/s130810482 PMID: 23945739

Sani, S. S. (2024). A single-inductor self-powered SECE interface circuit for dynamic load multi-PZTs energy harvesting. *IET Power Electronics*, *17*(15), 2613–2630. DOI: 10.1049/pel2.12815

Sankara Narayanan, T. S. N., & Seshadri, S. K. (2008). Electro-and electroless plated coatings for corrosion protection.

Sankara Narayanan, T. S. N. (2005). Surface pretretament by phosphate conversion coatings-A review. *Reviews on Advanced Materials Science*, *9*, 130–177.

Saravanan, A., Maruthapandi, M., Das, P., Ganguly, S., Margel, S., Luong, J. H. T., & Gedanken, A. (2020). Applications of N-Doped Carbon Dots as Antimicrobial Agents, Antibiotic Carriers, and Selective Fluorescent Probes for Nitro Explosives. *ACS Applied Bio Materials*, *3*(11), 8023–8031. DOI: 10.1021/acsabm.0c01104 PMID: 35019541

Scipioni, S., Dini, G., & Niccolini, F. (2023). Exploring circular shipbuilding: A systematic review on circular economy business models and supporting technologies. *Journal of Cleaner Production*, *422*, 138470. DOI: 10.1016/j.jclepro.2023.138470

Serge-Correales, Y. E., Ullah, S., Ferreira-Neto, E. P., Rojas-Mantilla, H. D., Hazra, C., & Ribeiro, S. J. L. A. (2022). UV-visible-NIR active smart photocatalytic system based on NaYbF4:Tm3+ upconverting particles and Ag_3PO_4/H_2O_2 for photocatalytic processes under light on/light off conditions. *Materials Advances*, *3*(6), 2706–2715. DOI: 10.1039/D1MA01028J

Serugga, J. (2025). Digital Twins and AI Decision Models: Advancing Cost Modelling in Off-Site Construction. *Eng*, *6*(2), 22. DOI: 10.3390/eng6020022

Shahjalal, M., Yahia, A. K. M., Morshed, A. S. M., & Tanha, N. I. (2024). Earthquake-resistant building design: Innovations and challenges. *Global Mainstream Journal of Innovation. Engineering & Emerging Technology*, *3*(04), 101–119.

Shaik, M. A., & Sneha, P. (2025). *Revolutionizing Infrastructure Resilience: AI-Driven Predictive Maintenance and Structural Health Monitoring*.

Shanley, L. A., Slaten, B. L., & Shanley, P. S. (1993). Military protective clothing: Implications for clothing and textiles curriculum and research. *Clothing & Textiles Research Journal, 11*(3), 55–59. DOI: 10.1177/0887302X9301100308

Sharma, V. K., Kumar, P., Kumar, D., Chalisgaonkar, R., & Akhai, S. (2024). Characterization of Al 6061/Al2O3/SiC composites with cerium oxide: corrosion analysis and microstructural insights. *Composites Theory and Practice, 24.*

Sharma, D., & Hiremath, S. S. (2022). Bio-inspired repeatable lattice structures for energy absorption: Experimental and finite element study. *Composite Structures, 283*, 115102. DOI: 10.1016/j.compstruct.2021.115102

Sharma, S., & Nandan, G. (2024). Casting and comprehensive analysis of nitinol-based self-healing metal matrix composite of a356 alloy prepared via clamp technique. *Engineering Research Express, 6*(2), 025568. DOI: 10.1088/2631-8695/ad52ea

Sharma, V. K., Kumar, P., Akhai, S., Kumar, V., & Joshi, R. S. (2024). Analyzing the tribological and mechanical performance of Al-6061 with rare earth oxides: An experimental analysis. *Proceedings of the Institution of Mechanical Engineers. Part E, Journal of Process Mechanical Engineering, 238*(5), 2278–2292. DOI: 10.1177/09544089231160003

Sharma, V. K., Kumar, P., Akhai, S., Kumar, V., & Joshi, R. S. (2024). Corrosion inhibition analysis on cerium induced hydrophobic surface of Al-6061/SiC/Al2O3 hybrid composites. *Proceedings of the Institution of Mechanical Engineers. Part B, Journal of Engineering Manufacture, 238*(14), 2126–2138. DOI: 10.1177/09544054231222048

Sharma, V., & Akhai, S. (2019). Mechanical Behaviour of Fly Ash Reinforced Aluminum Composite Prepared by Casting. *Journal of Advanced Research in Mechanical Engineering and Technology, 6*(1&2), 23–26. DOI: 10.24321/2454.8650.201904

Sheasby, P. G., Wernick, S., & Pinner, R. (1987). Surface treatment and finishing of aluminum and its alloys. Volumes 1 and 2.

Shehata, N., Mohamed, O. A., Sayed, E. T., Abdelkareem, M. A., & Olabi, A. G. (2022). Geopolymer concrete as green building materials: Recent applications, sustainable development and circular economy potentials. *The Science of the Total Environment, 836*, 155577. DOI: 10.1016/j.scitotenv.2022.155577 PMID: 35500705

Shen, G., Chen, X., Yan, J., Fan, L., Yang, Z., Zhang, J., & Guan, R. (2023). A review of progress in the study of Al-Mg-Zn (-Cu) wrought alloys. *Metals, 13*(2), 345. DOI: 10.3390/met13020345

Shen, H., Yin, Y., Tian, K., Baskaran, K., Duan, L., Zhao, X., & Tiwari, A. (2018). Growth and characterization of β-Ga2O3 thin films by sol-gel method for fast-response solar-blind ultraviolet photodetectors. *Journal of Alloys and Compounds*, *766*, 601–608. DOI: 10.1016/j.jallcom.2018.06.313

Shen, Y., Sahoo, P. K., & Pan, Y. (2017). A Study of Micro-Arc Oxidation Coatings on Aluminum Alloy Drill Pipe for Offshore Platform. *Marine Technology Society Journal*, *51*(3), 16–22. DOI: 10.4031/MTSJ.51.3.1

Sheoran, H., Fang, S., Liang, F., Huang, Z., Kaushik, S., Manikanthababu, N., Zhao, X., Sun, H., Singh, R., & Long, S. (2022). High Performance of Zero-Power-Consumption MOCVD-Grown β-Ga2O3-Based Solar-Blind Photodetectors with Ultralow Dark Current and High-Temperature Functionalities. *ACS Applied Materials & Interfaces*, *14*(46), 52096–52107. DOI: 10.1021/acsami.2c08511 PMID: 36346904

Shivani, K., Kaur, D., Ghosh, A., & Kumar, M. (2022). A strategic review on gallium oxide based power electronics: Recent progress and future prospects. *Materials Today. Communications*, *33*(August), 104244. DOI: 10.1016/j.mtcomm.2022.104244

Simões, S. (2024). High-performance advanced composites in multifunctional material design: State of the art, challenges, and future directions. *Materials (Basel)*, *17*(23), 5997. DOI: 10.3390/ma17235997 PMID: 39685433

Simons, G. M. (2021). *Boeing 737: The World's Most Controversial Commercial Jetliner*. Air World.

Singh Pratiyush, A., Krishnamoorthy, S., Vishnu Solanke, S., Xia, Z., Muralidharan, R., Rajan, S., & Nath, D. N. (2017). High responsivity in molecular beam epitaxy grown β-Ga2O3 metal semiconductor metal solar blind deep-UV photodetector. *Applied Physics Letters*, *110*(22), 221107. Advance online publication. DOI: 10.1063/1.4984904

Singh, S., & Akhai, S. (2024, July). Electronics in precision control for mechanical engineering: A brief overview. In *AIP Conference Proceedings* (Vol. 3121, No. 1). AIP Publishing. DOI: 10.1063/5.0221464

Singh, A., Sharma, S. K., & Batish, A. (2025). Dynamic recrystallization during solid state friction stir welding/processing/additive manufacturing: Mechanisms, microstructure evolution, characterization, modeling techniques and challenges. *Critical Reviews in Solid State and Material Sciences*, *50*(1), 77–135. DOI: 10.1080/10408436.2024.2391333

Singh, G., & Akhai, S. (2015). Experimental study and optimisation of MRR in CNC plasma arc cutting. *International Journal of Engineering Research and Applications*, *5*(6), 96–99. DOI: 10.9790/9622-07060696101

Singh, H., Chatha, S. S., Sidhu, H. S., & Sharma, K. (2011). Hot-corrosion resistance of alloy and composite coatings: A review. In *National Conference on Advancements and Futuristic Trends in Mechanical and Materials Engineering(*Oct. 7-8, 2011*).*

Singh, T. J., & Samanta, S. (2015). Characterization of Kevlar fiber and its composites: A review. *Materials Today: Proceedings*, *2*(4–5), 1381–1387. DOI: 10.1016/j.matpr.2015.07.057

Singh, T., Khaira, N. K., Repeta, M., & Mansour, R. R. (2024). Phase-change RF devices for future communications: Phase-change materials and devices for reconfigurable RF front-ends: State-of-the-art and future perspectives. *IEEE Microwave Magazine*, *25*(2), 18–38. DOI: 10.1109/MMM.2023.3332322

Singh, V., Indoria, S., Bala, M., Gardas, R. L., & Yang, Y.-L. (n.d.). Self-Healing and Self-Cleaning Nature-Inspired Surfaces. In *Materials from Natural Sources* (pp. 216–240). CRC Press.

Sinyavskii, V. S., & Kalinin, V. D. (2005). Marine Corrosion and Protection of Aluminum Alloys According to Their Composition and Structure. *Protection of Metals*, *41*(4), 317–328. DOI: 10.1007/s11124-005-0046-8

Siony, N., Vuong, L., Lundaajamts, O., & Kadkhodaei, S. (2022). Computational design of corrosion-resistant and wear-resistant titanium alloys for orthopedic implants. *Materials Today. Communications*, *33*, 104465. DOI: 10.1016/j.mtcomm.2022.104465

Sivamaran, V., Azaruddin, S., Sivaprasad, K., Ravikumar, K., Kumar, V. M., & undefined, . (2021). A Short Review on Applications of Aluminium Composites: Automotive, Aerospace and aircraft, Rail transport, and Marine transport industry. [JPIE]. *Journal of Production & Industrial Engineering*, *2*(2). Advance online publication. DOI: 10.26706/jpie.2.2.20211202

Skountzos, E. N., Ravichandran, A., & Lawson, J. W. (2024). Interfacial Characteristics of Ice-Supporting Substrates via Molecular Dynamics Simulations. *Langmuir*, *40*(51), 26864–26874. DOI: 10.1021/acs.langmuir.4c03106 PMID: 39651769

Smith, C. J. E., Higgs, M. S., & Baldwin, K. R. (2000). Advances in protective coatings and their application to ageing aircraft. In *Applied Vehicle Technology Panel (AVT) Workshop.Proceedings of RTO meeting* (pp. 15-1).

Smith, C. J. (1995). Advances in protective coatings and processes for aerospace applications. *Aircraft Engineering, 67*(5), 13–16. DOI: 10.1108/eb037598

Sobel, S. G. (2023). The Hidden Chemistry behind the Apollo 11 Mission. *After Apollo: Cultural Legacies of the Race to the Moon*, 29.

Sohn, J. M., Hirdaris, S., Romanoff, J., & Kim, S. (2022). Development of numerical modelling techniques for composite cylindrical structures under external pressure. *Journal of Marine Science and Engineering, 10*(4), 466. DOI: 10.3390/jmse10040466

Sokoluk, M., Cao, C., Pan, S., & Li, X. (2019). Nanoparticle-enabled phase control for arc welding of unweldable aluminum alloy 7075. *Nature Communications, 10*(1), 98. Advance online publication. DOI: 10.1038/s41467-018-07989-y PMID: 30626876

Song, C., Dong, B., Zhang, S., Yang, H., Liu, L., Kang, J., Meng, J., Luo, C., Wang, C., Cao, K., Qiao, J., Shu, S., Zhu, M., Qiu, F., & Jiang, Q. (2024). Recent progress of Al–Mg alloys: Forming and preparation process, microstructure manipulation and application. *Journal of Materials Research and Technology, 31*, 3255–3286. DOI: 10.1016/j.jmrt.2024.07.051

Song, G. L., Zhang, C., Chen, X., & Zheng, D. (2021). Galvanic activity of carbon fiber reinforced polymers and electrochemical behavior of carbon fiber. *Corrosion Communications, 1*, 26–39. DOI: 10.1016/j.corcom.2021.05.003

Song, G., & Feng, Z. (2020). Modification, Degradation and Evaluation of a Few Organic Coatings for Some Marine Applications. *Corrosion and Materials Degradation, 1*(3), 408–442. DOI: 10.3390/cmd1030019

Song, J., She, J., Chen, D., & Pan, F. (2020d). Latest research advances on magnesium and magnesium alloys worldwide. *Journal of Magnesium and Alloys, 8*(1), 1–41. DOI: 10.1016/j.jma.2020.02.003

Song, Z., Liao, J., Ding, X., Zhou, T., & Liu, Q. L. (2012). Stability of divalent/trivalent oxidation state of europium in some Sr-based inorganic compounds. *Journal of Luminescence, 132*(7), 1768–1773. DOI: 10.1016/j.jlumin.2012.02.013

Soni, P. K., Tiwari, A., Nayak, S., Vasnani, H., & Jha, S. K. (2024c). Ultrafine grain refinement and improved mechanical strength of compositionally modified AZ 61 alloy composite. In *Advances in chemical and materials engineering book series* (pp. 327–362). DOI: 10.4018/979-8-3373-0669-8.ch009

Sparks, E. (Ed.). (2012). *Advances in military textiles and personal equipment*. Elsevier. DOI: 10.1533/9780857095572

Srebrenkoska, S., & Dukovski, V. (2022). Defects formation in automated fiber placement technology. *Tekstilna Industrija*, *70*(4), 22–27. DOI: 10.5937/tekstind2204022S

Sridhar, M., & Kumar, M. V. (2024). Structural and Sustainability Enhancement of Composite Sandwich Slab Panels Using Novel Fibre-Reinforced Geopolymer Concrete. *Journal of Composites Science*, *8*(11), 479. DOI: 10.3390/jcs8110479

Sriniva, B. (2024). Exhaustive investigation into the corrosion resilience of aluminium-silicon carbide metal matrix composites. DOI: 10.21203/rs.3.rs-3993505/v1

Srivatsav, S., Jayakumar, V., & Sathishkumar, M. (2021). Recent developments and challenges associated with wire arc additive manufacturing of Al alloy: A review. *Materials Today: Proceedings*, *46*, 8561–8566. DOI: 10.1016/j.matpr.2021.03.542

Stanford, N. (2010). The effect of calcium on the texture, microstructure and mechanical properties of extruded Mg–Mn–Ca alloys. *Materials Science and Engineering A*, *528*(1), 314–322. DOI: 10.1016/j.msea.2010.08.097

Stankiewicz, K., Lipkowski, A., Kowalczyk, P., Giżyński, M., & Waśniewski, B. (2024). Resistance welding of thermoplastic composites, including welding to thermosets and metals: A review. *Materials (Basel)*, *17*(19), 4797. DOI: 10.3390/ma17194797 PMID: 39410368

Stepanov, S. I., Nikolaev, V. I., Bougrov, V. E., & Romanov, A. E. (2016). Gallium oxide: Properties and applications - A review. *Reviews on Advanced Materials Science*, *44*(1), 63–86.

Stier, S. P., Kreisbeck, C., Ihssen, H., Popp, M. A., Hauch, J., Malek, K., Reynaud, M., Goumans, T. P. M., Carlsson, J., Todorov, I., Gold, L., Räder, A., Wenzel, W., Bandesha, S. T., Jacques, P., Garcia-Moreno, F., Arcelus, O., Friederich, P., Clark, S., & Kozdras, M. (2024). Materials acceleration platforms (MAPs): Accelerating materials research and development to meet urgent societal challenges. *Advanced Materials*, *36*(45), 2407791. DOI: 10.1002/adma.202407791 PMID: 39239995

Studart, A. R. (2012). Biologically inspired materials that self-assemble. *Journal of Materials Chemistry*, *22*(19), 8203–8211.

Suchikova, Y., Nazarovets, S., & Popov, A. I. (2024). Ga_2O_3 Solar-Blind Photodetectors: From Civilian Applications to Missile Detection and Research Agenda. *Optical Materials*, *157*(P3), 116397. DOI: 10.1016/j.optmat.2024.116397

Suhaimi, M. A., Mahat, M. M., Yahaya, S. M., & Ramli, R. (2024). A review: Characterization of aluminium-metal matrix composite reinforced with zirconium diboride. *Journal of Advanced Research in Micro and Nano Engieering*, *18*(1), 44–51. DOI: 10.37934/armne.18.1.4451

Su, J. Y. (2024). Optimizing mechanical properties of multi-walled carbon nanotube reinforced thermoplastic polyurethane composites for advanced athletic protective gear. *Matéria (Rio de Janeiro)*, *29*(2), e20240059. Advance online publication. DOI: 10.1590/1517-7076-rmat-2024-0059

Sun, C.-Y., Liu, W., Shi, X., Rao, G.-H., & Zhao, J.-T. (2025). Stress Distributions and Luminescent Responses of Mechanoluminescent Cylinders with Various Sizes and Loading Paths. *Materials (Basel)*, *18*(2), 331. DOI: 10.3390/ma18020331 PMID: 39859802

Sun, H., Liang-cai, L., Wang, Z., Liu, B., Wang, M., & Yu, Y. (2020). Corrosion Behaviors of Microarc Oxidation Coating and Anodic Oxidation on 5083 Aluminum Alloy. *Journal of Chemistry*, *1*, 1–11. Advance online publication. DOI: 10.1155/2020/6082812

Sun, Z., Jayasinghe, S., Sidiq, A., Shahrivar, F., Mahmoodian, M., & Setunge, S. (2024). Approach Towards the Development of Digital Twin for Structural Health Monitoring of Civil Infrastructure: A Comprehensive Review. *Sensors (Basel)*, *25*(1), 59. DOI: 10.3390/s25010059 PMID: 39796850

Syrvatka, V., Rabets, A., Gromyko, O., Luzhetskyy, A., & Fedorenko, V. (2022). Scandium–microorganism interactions in new biotechnologies. *Trends in Biotechnology*, *40*(9), 1088–1101. DOI: 10.1016/j.tibtech.2022.02.006 PMID: 35346528

Tak, B. R., Kumar, S., Kapoor, A. K., Wang, D., Li, X., Sun, H., & Singh, R. (2021). Recent advances in the growth of gallium oxide thin films employing various growth techniques - A review. *Journal of Physics. D, Applied Physics*, *54*(45), 453002. Advance online publication. DOI: 10.1088/1361-6463/ac1af2

Tan, J., Xu, W., Du, S., Xu, J., Zhang, M., Song, S., Yang, B., Nie, J., & He, H. (2024). Flexible and robust aramid/octadecane phase change materials from nonaqueous emulsion template towards efficient thermal storage and camouflage. *Chemical Engineering Journal*, *500*, 157240. DOI: 10.1016/j.cej.2024.157240

Tański, T., & Snopiński, P. (2017). Effect of the Processing Conditions on the Microstructural Features and Mechanical Behavior of Aluminum Alloys. *InTech eBooks*. https://doi.org/DOI: 10.5772/intechopen.70682

Tao, Q., Xu, P., Li, M., & Lu, W. (2021). Machine learning for perovskite materials design and discovery. *npj Computational Materials*, *7*(1), 23. Advance online publication. DOI: 10.1038/s41524-021-00495-8

Taye, B., & Alemayehu, Z. (2023). Synthesis and characterization of mechanical and physical properties of false banana (ʾənäsätə) fiber reinforced composite material. *Ethiopian International Journal of Engineering and Technology*, *1*(1), 49–61. DOI: 10.59122/134CFC10

Taylor, J. R. A., Scott, N. I., & Rouse, G. W. (2019). 'Evolution of Mantis Shrimp Telson Armour and Its Role in Ritualized Fighting'. *Journal of the Royal Society, Interface*, *16*(157), 20190203. Advance online publication. DOI: 10.1098/rsif.2019.0203 PMID: 31455162

Thareja, P., & Akhai, S. (2017). Processing parameters of powder aluminium-fly ash P/M composites. *Journal of advanced research in manufacturing, material science & metallurgical engineering, 4*(3&4), 24-35.

Thareja, P., & Akhai, S. (2016). Processing Aluminum Fly Ash Composites via Parametric Analysis of Stir Casting. *Journal of Advanced Research in Manufacturing. Material Science & Metallurgical Engineering*, *3*(3&4), 21–28.

Thavasilingam, K., Sakthimurugan, D., & Giridharan, K. (2025). Aluminum alloys for aircraft structures. In *Aerospace Materials* (pp. 385–404). Elsevier.

Thilagavathi, G., Raja, A. S. M., & Kannaian, T. (2008). Nanotechnology and protective clothing for defence personnel. *Defence Science Journal*, *58*(4), 451–459. DOI: 10.14429/dsj.58.1667

Thomas, W. M., & Nicholas, E. D. (1997). Friction stir welding for the transportation industries. *Materials & Design*, *18*(4-6), 269–273. DOI: 10.1016/S0261-3069(97)00062-9

Tian, Y., Ding, R., Yoon, S. S., Zhang, S., Yu, J., & Ding, B. (2025). Recent Advances in Next-Generation Textiles. *Advanced Materials*, *37*(8), 2417022. DOI: 10.1002/adma.202417022 PMID: 39757561

Titu, A. M., Ravai-Nagy, S., & Pop, A. B. (2023). Research on the influence of Coating technologies on Adhesion Anti-corrosion Layers in the case of Al7175 Aluminum Alloy. *Coatings*, *13*(6), 1054. DOI: 10.3390/coatings13061054

Tiwari, N. A. (2024e). Prediction model for hardness and tensile strength of graphene reinforced AZ 61 alloy based composite using Metaheuristic Algorithm. *Global Journal of Engineering and Technology Advances, 20*(3), 042–052. DOI: 10.30574/gjeta.2024.20.3.0167

Tiwari, A., Kumar, N., & Banerjee, M. (2023c). Ageing characteristics of stir cast AZ 61 alloy with minor additions. *Recent Patents on Engineering*, *18*(6), e090623217838. Advance online publication. DOI: 10.2174/1872212118666230609122741

Tiwari, A., Kumar, N., & Banerjee, M. (2023i). Applications of genetic algorithm in prediction of the best achievable combination of hardness and tensile strength for graphene reinforced magnesium alloy (AZ61) matrix composite. *Results in Control and Optimization*, *14*, 100334. DOI: 10.1016/j.rico.2023.100334

Tiwari, A., Kumar, N., & Banerjee, M. K. (2023m). Effect of Hot Rolling on Microstructure and Mechanical Properties of Stir Cast AZ 61 Alloy with Minor Additions. *Indian Journal of Science and Technology*, *16*(24), 1810–1822. DOI: 10.17485/IJST/v16i24.302

Tjahjanti, P. H., Panunggal, E., Darminto, D., & Nugroho, W. H. (2013). Numerical Modeling of Ship Composite-Based on Aluminum Casting as Alternative Materials for Ship Building. *Advanced Materials Research*, *789*, 143–150. . DOI: 10.4028/www.scientific.net/AMR.789.143

Toohey, K. S., Sottos, N. R., Lewis, J. A., Moore, J. S., & White, S. R. (2007). Self-healing materials with microvascular networks. *Nature Materials*, *6*(8), 581–585. DOI: 10.1038/nmat1934 PMID: 17558429

Torres, F. G., Troncoso, O. P., Nakamatsu, J., & Grande, C. J. (2019). The challenges of bio-inspired material synthesis for aerospace applications. *Acta Biomaterialia*, *88*, 16–30.

Tripathi, D., Vyas, H. S., Kumar, S., Panda, S. S., & Hegde, R. (2023). Recent developments in Chalcogenide phase change material-based nanophotonics. *Nanotechnology*, *34*(50), 502001. DOI: 10.1088/1361-6528/acf1a7 PMID: 37595569

Tudu, S., & Ramachandran, V. (2023). Effect of graphene nanoplatelets on flexural behavior of glass fiber reinforced polymer composites subjected to different temperatures. *Materials Science Forum*, *1101*, 29–33. DOI: 10.4028/p-R09Tja

Twyman, N. M., Walsh, A., & Buonassisi, T. (2022). Environmental stability of crystals: A greedy screening. *Chemistry of Materials*, *34*(6), 2545–2552. DOI: 10.1021/acs.chemmater.1c02644 PMID: 35431438

U.S. Air Force Research Laboratory (AFRL). (2022). Advances in bio-inspired materials for military aerospace applications. Retrieved from www.afrl.mil

Uddin, N. (2013). *Developments in fiber-reinforced polymer (FRP) composites for civil engineering*. Elsevier. DOI: 10.1533/9780857098955

Udoh, I. I., Shi, H., Daniel, E. F., Li, J., Gu, S., Liu, F., & Han, E. H. (2022). Active anticorrosion and self-healing coatings: A review with focus on multi-action smart coating strategies. *Journal of Materials Science and Technology*, *116*, 224–237. DOI: 10.1016/j.jmst.2021.11.042

Ueda, N., Hosono, H., Waseda, R., & Kawazoe, H. (1997). Synthesis and control of conductivity of ultraviolet transmitting β-Ga2O3 single crystals. *Applied Physics Letters*, *70*(26), 3561–3563. DOI: 10.1063/1.119233

Ulaeto, S. B., Ravi, R. P., Udoh, I. I., Mathew, G. M., & Rajan, T. P. D. (2023). Polymer-based coating for steel protection, highlighting metal–organic framework as functional actives: A review. *Corrosion and Materials Degradation*, *4*(2), 284–316. DOI: 10.3390/cmd4020015

Urbina, M., Rinaldi, A., Cuesta-Lopez, S., Sobetkii, A., Slobozeanu, A. E., Szakalos, P., Qin, Y., Prakasam, M., Piticescu, R.-R., Ducros, C., & Largeteau, A. (2018). The methodologies and strategies for the development of novel material systems and coatings for applications in extreme environments-a critical review. *Manufacturing Review*, *5*, 5. DOI: 10.1051/mfreview/2018006

Ursache, ., Cerbu, C., & Hadăr, A. (2023). Characteristics of carbon and kevlar fibres, their composites and structural applications in civil engineering—A review. *Polymers*, *16*(1), 127. DOI: 10.3390/polym16010127 PMID: 38201792

Varatharajulu, M., Duraiselvam, M., Pradeep, G., & Jagadeesh, B. J. S. R. (2025). Exploring Epitaxial Grain Growth and Marangoni-Induced Conversion in Columnar Dendrites: A Study of Inconel 718 Additive Manufacturing via Direct Laser Energy Deposition. *Surface Review and Letters*, *32*(1), 2450102–2450115. DOI: 10.1142/S0218625X24501026

Varley, J. B., Weber, J. R., Janotti, A., & Van De Walle, C. G. (2010). Oxygen vacancies and donor impurities in β-Ga2 O 3. *Applied Physics Letters*, *97*(14), 10–13. DOI: 10.1063/1.3499306

Vasconcelos, A. R. C., de Matos, R. A., Silveira, M. V., & Mesquita, E. (2024). Applications of Smart and Self-Sensing Materials for Structural Health Monitoring in Civil Engineering: A Systematic Review. *Buildings (Basel, Switzerland)*, *14*(8), 2345. DOI: 10.3390/buildings14082345

Velardi, L., Lorusso, A., Paladini, F., Siciliano, M. V., Di Giulio, M., Raino, A., & Nassisi, V. (2010). Modification of polymer characteristics by laser and ion beam. *Radiation Effects and Defects in Solids*, *165*(6-10), 637–642. DOI: 10.1080/10420151003729516

Venkatesan, S., Jerald, J., Sundeep, D., Varadharaj, E. K., & Sastry, C. C. (2022). Evaluation of mechanical and corrosion properties of tib2-y2o3 nanocomposite fused bronze metal matrix composite. *Surface Topography : Metrology and Properties*, *10*(3), 035003. DOI: 10.1088/2051-672X/ac7faa

Verma, J., & Khanna, A. S. (2023). Digital advancements in smart materials design and multifunctional coating manufacturing. *Physics Open*, *14*, 100133. DOI: 10.1016/j.physo.2022.100133

Verma, R., Akhai, S., & Wadhwa, A. S. (2024). Use of smart materials in physiotherapy. In *Revolutionizing healthcare treatment with sensor technology* (pp. 300–319). IGI Global Scientific Publishing. DOI: 10.4018/979-8-3693-2762-3.ch019

Vetelino, J. F., & Reghu, A. (2017). Introduction to sensors. *Introduction to Sensors*, (November), 1–180. DOI: 10.1201/9781315218274

Vincent, J. F. V., Bogatyreva, O. A., Bogatyrev, N. R., Bowyer, A., & Pahl, A. K. (2006). Biomimetics: Its practice and theory. *Journal of the Royal Society, Interface*, *3*(9), 471–482. DOI: 10.1098/rsif.2006.0127 PMID: 16849244

Vinodhini, J., Sudheendra, K., & Bhowmik, S., AR, S., Trivedi, P. D., Zaveri, J. C., … & Kundu, A. (. (2022). Novel thermally insulated ultra-lightweight fireproof and high impact resistance advanced hybrid composite. Polymer Engineering &Amp. *Science*, *62*(9), 3026–3036. DOI: 10.1002/pen.26082

Vogiantzi, C., & Tserpes, K. (2025). A Comparative Environmental and Economic Analysis of Carbon Fiber-Reinforced Polymer Recycling Processes Using Life Cycle Assessment and Life Cycle Costing. *Journal of Composites Science*, *9*(1), 39. DOI: 10.3390/jcs9010039

Vollrath, F., & Knight, D. P. (2001). Liquid crystalline spinning of spider silk. *Nature*, *410*(6828), 541–548. DOI: 10.1038/35069000 PMID: 11279484

Volpini, V., & Pianesi, M. S. (2022). Characterisation of recycled quarzitic and plastic aggregates for sustainable lightweight screeds. *Key Engineering Materials*, *919*, 28–34. DOI: 10.4028/p-rf17nl

Vu, N. V., Le, V. S., Louda, P., Szczypiński, M. M., Ercoli, R., Růžek, V., & Buczkowska, K. E. (2022). Low-density geopolymer composites for the construction industry. *Polymers*, *14*(2), 304. DOI: 10.3390/polym14020304 PMID: 35054711

Wadhwa, A. S., Akhai, S., Abbass, M., Chouksey, A., Tiwari, S., & Taneja, T. (2025). Computational Intelligence-Driven Design and Optimization of Polyurethane Belt-Type Oil Skimmer for Sustainable Manufacturing Using Solidworks 3D CAD. In *Using Computational Intelligence for Sustainable Manufacturing of Advanced Materials* (pp. 445-464). IGI Global Scientific Publishing.

Wadhwa, A. S., & Akhai, S. (2014). Comparison of Surface Hardening Techniques for En 353 Steel Grade. *International Journal of Emerging Technology and Advanced Engineering*, *4*(10), 194–203.

Wagermaier, W., Razghandi, K., & Fratzl, P. (2025). A Bio-Inspired Perspective on Materials Sustainability. *Advanced Materials*, •••, 2413096. DOI: 10.1002/adma.202413096 PMID: 39757528

Wahid, M. A., Siddiquee, A. N., & Khan, Z. A. (2019). Aluminum alloys in marine construction: Characteristics, application, and problems from a fabrication viewpoint. *Marine Systems & Ocean Technology*, *15*(1), 70–80. DOI: 10.1007/s40868-019-00069-w

Walkowiak, M., Reinicke, U., & Anders, D. (2022). Numerical investigation of different core topologies in sandwich-structured composites subjected to air-blast impact. *Applied Sciences (Basel, Switzerland)*, *12*(18), 9012. DOI: 10.3390/app12189012

Wang, X., Xu, X., Sun, T., & Yin, G. (2024). A Triboelectric Nanogenerator Utilizing a Crank-Rocker Mechanism Combined with a Spring Cantilever Structure for Efficient Energy Harvesting and Self-Powered Sensing Applications. *Electronics (2079-9292)*, *13*(24).

Wang, C., Fan, W. H., Zhang, Y. C., Kang, P. C., Wu, W. Y., Wuu, D. S., Lien, S. Y., & Zhu, W. Z. (2022). Effect of oxygen flow ratio on the performance of RF magnetron sputtered Sn-doped Ga2O3 films and ultraviolet photodetector. *Ceramics International*, (August). Advance online publication. DOI: 10.1016/j.ceramint.2022.11.251

Wang, F., Chen, Y. H., Liuand, C. Y., & Ma, D. (2011). White light-emitting devices based on carbon dots' electroluminescence. *Chemical Communications (Cambridge)*, *47*(12), 3502–3504. DOI: 10.1039/c0cc05391k PMID: 21311781

Wang, F., Liu, H., Liu, Z., Guo, Z., & Sun, F. (2022). Microstructure analysis, tribological correlation properties and strengthening mechanism of graphene reinforced aluminum matrix composites. *Scientific Reports*, *12*(1), 9561. Advance online publication. DOI: 10.1038/s41598-022-13793-y PMID: 35689017

Wang, F., Sun, J., & Xu, T. (2023). OXIDE CERAMIC COATINGS ON ORTHOPAEDIC IMPLANTS: A REVIEW OF TIO2 AND AL2O3. *Ceramics-Silikáty*, *67*(4), 525–542. DOI: 10.13168/cs.2023.0052

Wang, G., Tang, Z., Gao, Y., Liu, P., Li, Y., Li, A., & Chen, X. (2023). Phase change thermal storage materials for interdisciplinary applications. *Chemical Reviews*, *123*(11), 6953–7024. DOI: 10.1021/acs.chemrev.2c00572 PMID: 36946191

Wang, H., Xie, Z., Cheng, X., Jing, K., Zhang, L., Yang, J., Liu, R., Han, L., Cao, L., Wang, X., Fang, Q., Liu, C., & Wu, X. (2023). Microstructural evolution and thermal fatigue damage mechanism of second-phase dispersion strengthened tungsten composites under repetitive thermal loads. Journal of Materials Science &Amp. *Technology*, *140*, 221–232. DOI: 10.1016/j.jmst.2022.09.007

Wang, J. H., Shi, X. M., Cui, C., Xia, X., Li, Y., Wang, Q., Zhang, L. J., Shen, Z., Li, R., & Fang, X. Z. (2024). Study on anti-penetration performance of metal matrix ceramic ball composite. *Journal of Physics: Conference Series*, *2891*(5), 052010. DOI: 10.1088/1742-6596/2891/5/052010

Wang, J. H., Shi, X. M., Wang, Q., Xia, X., Fang, X. Z., Zeng, T., Zhang, L. J., & Ye, C. J. (2023). Study on ballistic performance of metal matrix ceramic ball composite. *Journal of Physics: Conference Series*, *2478*(11), 112005. DOI: 10.1088/1742-6596/2478/11/112005

Wang, J., Jiang, P., & Qi, J. (2024). A planning method for operational test of uav swarm based on mission reliability. Computer Modeling in Engineering &Amp. *Computer Modeling in Engineering & Sciences*, *140*(2), 1889–1918. DOI: 10.32604/cmes.2024.049813

Wang, N., Yu, K.-K., Li, K., Li, M.-J., Wei, X., & Yu, X.-Q. (2020). 'Plant-Inspired Multifunctional Fluorescent Hydrogel: A Highly Stretchable and Recoverable Self-Wegst, U. G. K. Bai, H. Saiz, E. Tomsia, A. P. & Ritchie, R. O. (2015). Bioinspired structural materials. *Nature Materials*, *14*(1), 23–36.

Wang, S., Liu, L., Zhou, X., Zhu, L., & Hao, Y. (2020). The design of Ti6Al4V Primitive surface structure with symmetrical gradient of pore size in biomimetic bone scaffold. *Materials & Design*, *193*, 108830. DOI: 10.1016/j.matdes.2020.108830

Wang, T., Zhang, P., Xiao, J., Guo, Z., Xie, X., Huang, J., Zheng, J., Xu, X., & Zhao, L. (2025). Trap Assisted Dynamic Mechanoluminescence Toward Self-Referencing and Visualized Strain Sensing. *Advancement of Science*, *12*(3), 2410673. DOI: 10.1002/advs.202410673 PMID: 39604817

Wang, W., & Cheng, C. Y. (2011). Separation and purification of scandium by solvent extraction and related technologies: A review. *Journal of Chemical Technology and Biotechnology*, *86*(10), 1237–1246. DOI: 10.1002/jctb.2655

Wang, X., Wang, F., Sang, Y., & Liu, H. (2017). Full-Spectrum Solar-Light-Activated Photocatalysts for Light–Chemical Energy Conversion. *Advanced Energy Materials*, *7*(23), 1700473–1700489. DOI: 10.1002/aenm.201700473

Wang, Y. B., Xu, W. H., You, T. G., Mu, F. W., Hu, H. D., Liu, Y., Huang, H., Suga, T., Han, G. Q., Ou, X., & Hao, Y. (2020). β-Ga2O3 MOSFETs on the Si substrate fabricated by the ion-cutting process. *Science China. Physics, Mechanics & Astronomy*, *63*(7), 1–4. DOI: 10.1007/s11433-020-1533-0

Wang, Y., Liu, G., Wei, Y., & Qiao, Y. (2020). Effects of Magnesium-Calcium Alloys with Different Calcium Content on Their Mechanical Properties. *IOP Conference Series. Materials Science and Engineering*, *735*(1), 012010. DOI: 10.1088/1757-899X/735/1/012010

Wang, Z., Hou, Y., Yadav, A., Chen, T., Wu, Y., Yan, C., & Liu, M. (2025). Self-Healing Superhydrophobic Coatings with Multiphase Repellence Property. *ACS Applied Materials & Interfaces*, *17*(4), 7174–7189. DOI: 10.1021/acsami.4c19317 PMID: 39812392

Webster, D. C., Chisholm, B. J., & Stafslien, S. J. (2007). Mini-review: Combinatorial approaches for the design of novel coating systems. *Biofouling*, *23*(3), 179–192. DOI: 10.1080/08927010701250948 PMID: 17653929

Weers, M., Seggern, A. R., Vocke, H., Taffa, D. H., & Wark, M. (2024). Two Ways to more NH_2-Groups: Formation of Polymeric Carbon Nitride via Melem Tetramer Nano Sheets or Supramolecular Assembly of Melamine and Cyanuric Acid for Applications as Photocatalyst. *ACS Applied Nano Materials*, *7*(1), 1402–1415. DOI: 10.1021/acsanm.3c05516

Weiler, J. (2020). Exploring the concept of castability in magnesium die-casting alloys. *Journal of Magnesium and Alloys*, *9*(1), 102–111. DOI: 10.1016/j.jma.2020.05.008

Wei, S., Guo, Z., Niu, W., Chai, G. B., Tai, Z., & Li, Y. (2024). High-velocity impact resistance of the carbon fiber composite grid sandwich structures. *Polymer Composites*, *45*(6), 5558–5573. DOI: 10.1002/pc.28147

Williams, T. (2022). Advancing research efforts in biomimicry to develop nature-inspired materials, processes for space exploration and more efficient aircraft. In *Biomimicry for Aerospace*. Technologies and Applications. DOI: 10.1016/B978-0-12-821074-1.00014-1

Willingham, F. (2019). Boeing 707 group: A history. *Air Power History*, *66*(1), 56–57.

Win, S. L. Y., Chiang, Y.-C., Huang, T.-L., & Lai, C.-M. (2024). Thermoelectric Generator Applications in Buildings: A Review. *Sustainability (Basel)*, *16*(17), 7585. DOI: 10.3390/su16177585

Winstone, M. (2000). Technical Evaluation Report. *New Metallic Materials for the Structure of Aging Aircraft*.

Wirtz, G. P., Brown, S. D., & Kriven, W. M. (1991). Ceramic coatings by anodic spark deposition. *Materials and Manufacturing Processes*, *6*(1), 87–115. DOI: 10.1080/10426919108934737

Witucki, G. L., & Pajk, T. (2004). The evolution of silicon-based technology in coatings. *PITTURE E VERNICI-EUROPE-*, *80*, 7-17.

Wong, M. H., Bierwagen, O., Kaplar, R. J., & Umezawa, H. (2021). Ultrawide-bandgap semiconductors: An overview. *Journal of Materials Research*, *36*(23), 4601–4615. DOI: 10.1557/s43578-021-00458-1

Wu, C., Wu, F., Hu, H., Wang, S., Liu, A., & Guo, D. (2022). Review of self-powered solar-blind photodetectors based on Ga2O3. *Materials Today Physics*, *28*(July), 100883. DOI: 10.1016/j.mtphys.2022.100883

Wu, H., Meng, W., Zhu, C., Tian, Z., Ma, F., & Jiao, Y. (2024). Two-dimensional monolayer BiSnO3: A novel wide-band-gap semiconductor with high stability and strong ultraviolet absorption. *Journal of Physics Condensed Matter*, *36*(30), 305501. Advance online publication. DOI: 10.1088/1361-648X/ad3da5 PMID: 38604165

Wu, H., Shang, D., Zhang, H., Zhi, L., Sun, S., Cui, S., & Yan, C. (2024). Phase-transition materials derived photonic metamaterials for passively dynamic solar thermal and coldness harvesting. *Heliyon*, *10*(2), e23986. DOI: 10.1016/j.heliyon.2024.e23986 PMID: 38293359

Wu, Y., Zhang, C., Wang, F., & Yang, C. (2023). Theoretical and numerical study of eddy current pulsed thermography to detect damage of deep-sea manned pressure hull. *Journal of Marine Science and Engineering*, *11*(7), 1410. DOI: 10.3390/jmse11071410

Xia, Z., Wang, H., Shi, C., Sun, Z., Wang, Q., & Luo, X. (2023). Analysis and characterization of three charge thicknesses in ta1/al1060/al7075 explosive welding composite process. *Crystals*, *13*(7), 1079. DOI: 10.3390/cryst13071079

Xie, H., Zhao, H., Guo, X., Li, Y., Hu, H., Song, J., Jiang, B., & Pan, F. (2024). Recent progress on cast magnesium alloy and components. *Journal of Materials Science*, *59*(22), 9969–10002. DOI: 10.1007/s10853-024-09459-x

Xiong, Y., Han, Z., Qin, J., Dong, L., Zhang, H., Wang, Y., Chen, H., & Li, X. (2021). Effects of porosity gradient pattern on mechanical performance of additive manufactured Ti-6Al-4V functionally graded porous structure. *Materials & Design*, *208*, 109911. DOI: 10.1016/j.matdes.2021.109911

Xu, J., Zheng, W., & Huang, F. (2019). Gallium oxide solar-blind ultraviolet photodetectors: A review. *Journal of Materials Chemistry. C, Materials for Optical and Electronic Devices*, *7*(29), 8753–8770. DOI: 10.1039/C9TC02055A

Xu, J., Zhou, Y., Cheng, F., & Liu, Z. (2025). Preparation of multi-layer compound microcapsules and their application in self-healing of concrete cracks. *Sheng Wu Gong Cheng Xue Bao=. Chinese Journal of Biotechnology*, *41*(1), 448–460. PMID: 39855707

Xu, M., Zhao, Z., Wang, P., Duan, S., Lei, H., & Fang, D. (2022). Mechanical performance of bio-inspired hierarchical honeycomb metamaterials. *International Journal of Solids and Structures*, *254*, 111866. DOI: 10.1016/j.ijsolstr.2022.111866

Xu, S., Zhang, S., Shen, W., Xing huang, , Qiu, Y., & Wu, Y. (2023). Low-speed impact of glass/kevlar hybrid composites: Experimental results and numerical verification. *Journal of Composite Materials*, *57*(19), 3079–3099. DOI: 10.1177/00219983231183035

Xu, T., Yang, Y., Peng, X., Song, J., & Pan, F. (2019d). Overview of advancement and development trend on magnesium alloy. *Journal of Magnesium and Alloys*, *7*(3), 536–544. DOI: 10.1016/j.jma.2019.08.001

Xu, Y., An, Z., Zhang, L., Feng, Q., Zhang, J., Zhang, C., & Hao, Y. (2018). Solar blind deep ultraviolet β-Ga 2 O 3 photodetectors grown on sapphire by the Mist-CVD method. *Optical Materials Express*, *8*(9), 2941. DOI: 10.1364/OME.8.002941

Xu, Y., Zhang, D., Hu, S., Chen, R., Gu, Y., Kong, X., Tao, J., & Jiang, Y. (2019). Mechanical properties tailoring of topology optimized and selective laser melting fabricated Ti6Al4V lattice structure. *Journal of the Mechanical Behavior of Biomedical Materials*, *99*, 225–239. DOI: 10.1016/j.jmbbm.2019.06.021 PMID: 31400657

Yan, R., Lin, Q., You, K., Zhang, L., Chen, Y., Huang, Z., & Sheng, X. (2025). Construction strategies and recent advances of flexible EMI phase change composites. Soft Science, 5(1), N-A.

Yan, C., He, X., Yu, B., Zhou, B., Fang, S., Xu, J., Liu, B., & Wang, Z. (2024). A Mechanoluminescence-Based Stress Sensing Hydrogel for Intelligent Artificial Ligament. *Advanced Functional Materials*, •••, 2420142.

Yang, J., Zhou, Y. C., Yang, L. Y., Feng, C. P., Bai, L., Yang, M. B., & Yang, W. (2022). Exploring next-generation functional organic phase change composites. *Advanced Functional Materials*, *32*(28), 2200792. DOI: 10.1002/adfm.202200792

Yang, J., Zhu, Z., Han, S., Gu, Y., Zhu, Z., & Zhang, H. (2024). Evolution, limitations, advantages, and future challenges of magnesium alloys as materials for aerospace applications. *Journal of Alloys and Compounds, 176707*, 176707. Advance online publication. DOI: 10.1016/j.jallcom.2024.176707

Yang, N. (2022). Special Issue "Functionalities of Polymer-Based Nanocomposite Films and Coatings". *Coatings, 12*(9), 1245. DOI: 10.3390/coatings12091245

Yang, Y., Huang, W., Ma, Y.-E., Wang, S., Chen, X., & Meng, Y. (2022). Mechanical Properties and Failure Modes of Additively Manufactured Ti6Al4V Lattice Structures Under Quasi-Static Compressive Loading. *International Journal of Applied Mechanics, 14*(09), 2250081. DOI: 10.1142/S1758825122500818

Yang, Y., Xiong, X., Chen, J., Peng, X., Chen, D., & Pan, F. (2021b). Research advances in magnesium and magnesium alloys worldwide in 2020. *Journal of Magnesium and Alloys, 9*(3), 705–747. DOI: 10.1016/j.jma.2021.04.001

Yan, R., Tian, H., Gao, K., Peng, R., & Liu, B. (2022). A two-stage uav routing problem with time window considering rescheduling with random delivery reliability. *Proceedings of the Institution of Mechanical Engineers. Part O, Journal of Risk and Reliability, 237*(4), 781–797. DOI: 10.1177/1748006X221105395

Yıldırım, F., Kabakçı, E., Şaş, H. S., & Eskizeybek, V.YILDIRIM. (2022). Multi-walled carbon nanotube grafted 3d spacer multi-scale composites for electromagnetic interference shielding. *Polymer Composites, 43*(8), 5690–5703. DOI: 10.1002/pc.26885

Yin, X., Chen, Z., Bakhshi, N., Tong, O., Xiong, X., Chen, Y., Li, Y., Gao, J., Sarwar, M. S., Poursartip, A., & Madden, J. D. W. (2023). Smart roller: Soft sensor array for automated fiber placement. *Advanced Sensor Research, 2*(9), 2200074. Advance online publication. DOI: 10.1002/adsr.202200074

Yi, S., Xie, L., Wu, Z., Ning, W., Du, J., & Zhang, M. (2022). Effects of hygrothermal aging and cyclic compressive loading on the mechanical and electrical properties of conductive composites. *Polymers, 14*(23), 5089. DOI: 10.3390/polym14235089 PMID: 36501484

Yi, X., Hu, L., Li, Q., & Tang, Y. (2024). Axial crushing and energy absorption integrated design of modular filled double-hat beam composite structures. *Materials (Basel), 17*(17), 4302. DOI: 10.3390/ma17174302 PMID: 39274691

Yoshioka, S., Hayashi, H., Kuwabara, A., Oba, F., Matsunaga, K., & Tanaka, I. (2007). Structures and energetics of Ga2O3 polymorphs. *Journal of Physics Condensed Matter*, *19*(34), 346211. Advance online publication. DOI: 10.1088/0953-8984/19/34/346211

You, J., Liu, C., Li, S., Jiang, N., Feng, X., Liu, X., & Liu, X. (2024). Aromatic phase change microspheres constructed nanocomposite films for fluorine-free self-cleaning, absorption-dominated EMI shielding, and high-temperature thermal camouflage. *Composites. Part B, Engineering*, *283*, 111650. DOI: 10.1016/j.compositesb.2024.111650

Yuan, L., Dai, J., Fan, X., Song, T., Tao, Y. T., Wang, K., Xu, Z., Zhang, J., Bai, X., Lu, P., Chen, J., Zhou, J., & Wang, Z. L. (2011). Self-Cleaning Flexible Infrared Nanosensor Based on Carbon Nanoparticles. *ACS Nano*, *5*(5), 4007–4013. DOI: 10.1021/nn200571q PMID: 21466218

Yu, F.-P., Ou, S.-L., & Wuu, D.-S. (2015). Pulsed laser deposition of gallium oxide films for high performance solar-blind photodetectors. *Optical Materials Express*, *5*(5), 1240. DOI: 10.1364/OME.5.001240

Yusuf, M., Salami, O., & Amusat, W. O. (2024). *Advances in carbon-fiber reinforced polymers and composites for sustainable concrete structures.*

Zeng, Z., Bian, M., Xu, S., Davies, C., Birbilis, N., & Nie, J. (2016). Effects of dilute additions of Zn and Ca on ductility of magnesium alloy sheet. *Materials Science and Engineering A*, *674*, 459–471. DOI: 10.1016/j.msea.2016.07.049

Zhai, M., Shou, T., Yin, D., Chen, Z., Wu, Y., Liu, Y., Zhao, X., Hu, S., & Zhang, L. (2024). Bio-based polyurethane composites with adjustable fluorescence and ultraviolet shielding for anti-counterfeiting and ultraviolet protection. ACS Applied Materials &Amp. *Interfaces*, *16*(45), 62606–62616. DOI: 10.1021/acsami.4c12183 PMID: 39483089

Zhang, A., Yang, H., Liu, C., Yang, J., Yao, Y., Zhang, W., Pan, R., Zhuo, Y., Ding, J., Hu, R., Xue, M., Chen, P., & Gong, Y. (2024). Icephobic durability of molecular brush-structured PDMS soft coatings. *ACS Applied Materials & Interfaces*, *16*(16), 21001–21012. DOI: 10.1021/acsami.3c18900 PMID: 38619108

Zhang, B., Xu, W., Zhu, Q., Sun, Y., & Li, Y. (2019). Mechanically robust superhydrophobic porous anodized AA5083 for marine corrosion protection. *Corrosion Science*, *158*, 108083. DOI: 10.1016/j.corsci.2019.06.031

Zhang, C., Li, Z., Zhang, J., Tang, H., & Wang, H. (2023b). Additive manufacturing of magnesium matrix composites: Comprehensive review of recent progress and research perspectives. *Journal of Magnesium and Alloys*, *11*(2), 425–461. DOI: 10.1016/j.jma.2023.02.005

Zhang, F., Li, W., Li, X., Yang, D., Qu, W., & Li, A. (2024). Efficient 3d temperature field prediction and visualization for afp process analysis in composite structures. *Polymer Composites*, *45*(18), 16929–16944. DOI: 10.1002/pc.28941

Zhang, F., Yang, K., Liu, G., Chen, Y., Wang, M., Li, S., & Li, R. (2022). Recent advances on graphene: Synthesis, properties and applications. *Composites. Part A, Applied Science and Manufacturing*, *160*, 107051. DOI: 10.1016/j.compositesa.2022.107051

Zhang, H., Sun, Z., Zhi, P., Wang, W., & Wang, Z. (2024). Material-structure integrated design and optimization of a carbon-fiber-reinforced composite car door. *Applied Sciences (Basel, Switzerland)*, *14*(2), 930. DOI: 10.3390/app14020930

Zhang, J., Shi, J., Qi, D. C., Chen, L., & Zhang, K. H. L. (2020). Recent progress on the electronic structure, defect, and doping properties of Ga2O3. *APL Materials*, *8*(2), 020906. Advance online publication. DOI: 10.1063/1.5142999

Zhang, J., Zhang, M., Zhang, X., & Wang, Z. Z. (2022). Experimental and theoretical investigation on mechanical behaviors of tb991 weld sealant under cyclic loading. Fatigue &Amp; Fracture of Engineering Materials &Amp. *Fatigue & Fracture of Engineering Materials & Structures*, *46*(1), 125–136. DOI: 10.1111/ffe.13852

Zhang, L., Li, X., & Wang, Y. (2020). Bioinspired materials for camouflage applications. *Materials Today*, *37*, 129–141.

Zhang, L., Zhang, J., Chen, C., & Gu, Y. (2014). Advances in microarc oxidation coated AZ31 Mg alloys for biomedical applications. *Corrosion Science*, *91*, 7–28. DOI: 10.1016/j.corsci.2014.11.001

Zhang, Q., Liu, J., Tu, C., Zhai, D., He, M., & Lu, J. (2023). High-performance β-Ga2O3 Schottky barrier diodes and metal-semiconductor field-effect transistors on a high doping level epitaxial layer. *Journal of Alloys and Compounds*, *939*, 168732. DOI: 10.1016/j.jallcom.2023.168732

Zhang, S., Liu, Z., Liu, Y., Zhi, Y., Li, P., Wu, Z., & Tang, W. (2021). Electrical characterizations of planar ga2o3 schottky barrier diodes. *Micromachines*, *12*(3), 1–8. DOI: 10.3390/mi12030259 PMID: 33802423

Zhang, S., Wang, J., Ji, X., Yan, Z., Ye, L., Zheng, H., Liu, Y., Chen, X., & Li, P. (2024). Oxygen vacancies modulating performance for Ga2O3 solar-blind photodetectors via low-cost mist chemical vapor deposition. *Materials Today. Communications*, *39*(February), 108717. DOI: 10.1016/j.mtcomm.2024.108717

Zhang, X., Luo, X., Ou, W., Huang, J., & Li, Y. (2020). Effect of Er on the Characteristics of an Oxide Coating Prepared by Micro-Arc Oxidation on a 7075 Aluminum Alloy. *International Journal of Electrochemical Science*, *15*(4), 3445–3452. DOI: 10.20964/2020.04.51

Zhang, Y., Yan, J., Zhao, G., & Xie, W. (2010). First-principles study on electronic structure and optical properties of Sn-doped β-Ga2O3. *Physica B, Condensed Matter*, *405*(18), 3899–3903. DOI: 10.1016/j.physb.2010.06.024

Zhao, D., Li, K., Li, Y., Chen, X., Fan, J., & Zhu, J. (2024). Self-healing behaviour of fly ash/slag-based engineered geopolymer composites under external alkaline environments. *Case Studies in Construction Materials*, *21*, e03361. DOI: 10.1016/j.cscm.2024.e03361

Zhao, H., Li, J., Zhang, X., Xiong, B., Zhao, C., Ruan, Y., Wang, H., Zhang, J., Lan, Z., Huang, X., & Zhang, H. (2024). State-of-the-art lightweight implementation methods in electrical machines. *Machines (Basel)*, *12*(5), 339. DOI: 10.3390/machines12050339

Zhao, J., & Yang, X. (2003). Photocatalytic oxidation for indoor air purification: A literature review. *Building and Environment*, *38*(5), 645–654. DOI: 10.1016/S0360-1323(02)00212-3

Zhao, J., You, C., Chen, M., Lyu, S., Tie, D., & Liu, H. (2021). Effect of calcium oxide particle size on microstructure and properties of AZ91 Mg alloy. *Journal of Alloys and Compounds*, *886*, 160970. DOI: 10.1016/j.jallcom.2021.160970

Zhao, M., Liu, F., Zhou, H., Zhang, T., Zhang, D. Z., & Fu, G. (2023). Effect of the direction of the gradient on the mechanical properties and energy absorption of additive manufactured Ti-6Al-4 V functionally graded lattice structures. *Journal of Alloys and Compounds*, *968*, 171874. DOI: 10.1016/j.jallcom.2023.171874

Zhao, S., Chen, L., & Gao, Z. (2024). Effect of warp yarn paths on the symmetrical cyclic bending fatigue properties of 3d woven composites. *Polymer Composites*, *45*(10), 9239–9250. DOI: 10.1002/pc.28406

Zhao, Z., Qi, S., Suo, Z., Hu, T., Hu, J., Liu, T., & Gong, M. (2024). Development of a Superhydrophobic Protection Mechanism and Coating Materials for Cement Concrete Surfaces. *Materials (Basel), 17*(17), 4390. DOI: 10.3390/ma17174390 PMID: 39274780

Zhaxi, Z. (2024). Research on airframe structure, engine performance, weapon mounting capability of f-16. *Applied and Computational Engineering, 56*(1), 265–271. DOI: 10.54254/2755-2721/56/20240668

Zhi, Y. S., Jiang, W. Y., Liu, Z., Liu, Y. Y., Chu, X. L., Liu, J. H., Li, S., Yan, Z. Y., Wang, Y. H., Li, P. G., Wu, Z. P., & Tang, W. H. (2021). High-responsivity solar-blind photodetector based on MOCVD-grown Si-doped β-Ga2O3thin film. *Chinese Physics B, 30*(5), 057301. Advance online publication. DOI: 10.1088/1674-1056/abe37a

Zhou, B., Liu, B., & Zhang, S. (2021). The advancement of 7xxx series aluminum alloys for aircraft structures: A review. *Metals, 11*(5), 718. DOI: 10.3390/met11050718

Zhou, J., Zhan, J., Lv, B., Guo, Y., & Jiang, B. (2024). Deposition Contribution Rates and Simulation Model Refinement for Polysilicon Films Deposited by Large-Sized Tubular Low-Pressure Chemical Vapor Deposition Reactors. *Materials (Basel), 17*(23), 1–13. DOI: 10.3390/ma17235952 PMID: 39685388

Zhou, W., Xia, C., Sai, Q., & Zhang, H. (2017). Controlling n-type conductivity of β-Ga2O3 by Nb doping. *Applied Physics Letters, 111*(24), 1–5. DOI: 10.1063/1.4994263

Zhou, Y., Zhang, J.-H., Wang, F., Hua, J., Cheng, W., Shi, Y., & Pan, L. (2024). Recent Advances in Flexible Self-Powered Sensors in Piezoelectric, Triboelectric, and Pyroelectric Fields. *Nanoenergy Advances, 4*(3), 235–257. DOI: 10.3390/nanoenergyadv4030015

Zhu, H., Leng, M., Jin, G., & Miao, H. (2023). A Review of Research on Galvanic Corrosion of Aluminum Alloys. *Fluid Dynamics & Materials Processing, 19*(7).

Zhu, J., Xu, Z., Ha, S., Li, D., Zhang, K., Zhang, H., & Feng, J. (2022). Gallium Oxide for Gas Sensor Applications: A Comprehensive Review. *Materials (Basel), 15*(20), 7339. Advance online publication. DOI: 10.3390/ma15207339 PMID: 36295403

Zhu, T. Y., Jiang, W. J., Wu, S., Huang, Z. J., Liu, Y. L., Qi, X. D., & Wang, Y. (2024). Multifunctional MXene/PEDOT: PSS-based phase change organohydrogels for electromagnetic interference shielding and medium-low temperature infrared stealth. *ACS Applied Materials & Interfaces, 16*(12), 15372–15382. DOI: 10.1021/acsami.4c01001 PMID: 38494605

Zhu, Y., Romain, C., & Williams, C. K. (2016). Sustainable polymers from renewable resources. *Nature*, *540*(7633), 354–362. DOI: 10.1038/nature21001 PMID: 27974763

Zimmerman, G. (2018). What Aluminum Alloy Should I Use to Build a Boat, and Why? *Naimor Metal Fabrication Blog*. https://info.naimormetalfabrication.com/blog/what-aluminum-alloy-should-i-use-to-build-a-boat-and-why

Zou, J., Ma, L., Zhu, Y., Qin, L., & Yuan, Y. (2023). Gradient microstructure and superior strength–ductility synergy of AZ61 magnesium alloy bars processed by radial forging with different deformation temperatures. *Journal of Materials Science and Technology*, *170*, 65–77. DOI: 10.1016/j.jmst.2023.07.003

Zulkeply, M., Pang, M. M., Vaithilingam, C. A., & Sivasubramanian, R. (2022). The fundamental studies on the reaction kinetics of thermal decomposition of bio-composite based backsheet materials in photovoltaic (pv) panel. *Journal of Physics: Conference Series*, *2222*(1), 012002. DOI: 10.1088/1742-6596/2222/1/012002

Егоркин, В. С., Медведев, И. М., Sinebryukhov, S. L., Vyaliy, I. E., Gnedenkov, A. S., Nadaraia, K. V., Izotov, N. V., Mashtalyar, D. V., & Гнеденков, С. В. (2020). Atmospheric and Marine Corrosion of PEO and Composite Coatings Obtained on Al-Cu-Mg Aluminum Alloy. *Materials (Basel)*, *13*(12), 2739. DOI: 10.3390/ma13122739 PMID: 32560318

About the Contributors

Onur Güler received a Ph.D. in 2022 from Karadeniz Technical University, Institute of Science, in the Department of Metallurgical and Materials Engineering. Currently, he is an Assistant Professor at the same institution, where he continues his research in the field of composite materials. Dr. Güler's expertise includes enhancing the electrical and thermal conductivity, as well as the mechanical and wear resistance, of layered metallic powders produced through electroless plating methods. His postdoctoral research focuses on the characterization of coatings applied to improve wear resistance and on exploring the thermal energy storage properties of composite materials incorporating phase change materials. Dr. Güler has an h-index of 18, according to Web of Science, and has published over 60 articles in SCI-indexed journals.

Müslim Çelebi completed his Ph.D. in 2024 in Metallurgical and Materials Engineering at the Institute of Science and Technology, Karadeniz Technical University, where he now works as a Research Assistant. His research focuses on the development and analysis of composite materials, particularly investigating the mechanical, wear, and corrosion properties of metal matrix composites produced by powder metallurgy. Additionally, Dr. Çelebi has expertise in enhancing mechanical and wear resistance through advanced coatings, as well as in high entropy alloys and biocomposite materials. With over 25 SCI-indexed publications and an h-index of 12 according to Web of Science, he has established a significant presence in his field.

Abdullah Hasan Karabacak is a PhD candidate at Karadeniz Technical University, Institute of Science and Technology, Department of Metallurgical and Materials Engineering. He is currently working as a Research Assistant at the same institution and continues his research in the field of composite materials. Mr. Karabacak specializes in the study of mechanical, wear and corrosion properties of composite materials produced by powder metallurgy, as well as the machinability of composite materials. He is also working on his doctoral thesis in the field of ballistic and radiation absorption of composite armor materials. Mr. Karabacak has an h-index of 10 according to Web of Science and has published about 20 SCI indexed articles.

***

Mahapara Abbass, a distinguished academician and researcher in the Department of Civil Engineering at Maharishi Markandeshwar (Deemed to be University), Mullana-Ambala, has made significant contributions to the field of structural engineering and sustainable construction materials. Her academic and research endeavors are reflected in an impressive portfolio of publications. Dr. Abbas has authored several research papers in prestigious SCI and Scopus-indexed journals, such as Multiscale Multidisciplinary Modeling, Experimentation and Design, Journal of Building Pathology and Rehabilitation, and Lecture Notes in Civil Engineering. In addition to her journal articles, she has presented her work at renowned conferences. Dr. Abbas's expertise also extends to scholarly book contributions, where she has authored/co-authored several book chapters in critically acclaimed volumes, including Advances in Geotechnics and Structural Engineering and Lecture Notes in Civil Engineering. Further demonstrating her influence

Shalom Akhai has established himself as a prominent figure in the field of engineering, leveraging over 17 years of academic, research, and technical experience to make significant contributions. With a robust portfolio of over 50 publications in prestigious journals and conferences, Dr. Akhai demonstrates a commitment to advancing knowledge through scholarly work. Additionally, he has authored/co-authored several technical textbooks, further solidifying his expertise and dedication to education in the field. Beyond academia, Dr. Akhai's impact extends to real-world applications through the establishment of IGNIS Technical Solutions, a consultancy firm to translate his knowledge into real-world applications. His skills have resulted in the successful development of over 50 patents+, underscoring his innovative approach and influence in shaping the engineering landscape.

Jagadeesh Baskaran is currently working as a Faculty member in the Department of Mechanical Engineering, PSG Institute of Technology and Applied Research. He obtained his B. Tech in Mechanical Engineering and M.Tech in Advanced Manufacturing from SASTRA University. For his performance during the Master's program, he received the best outgoing M. Tech student award recognising his academic record and research publication from his M.Tech thesis. He has two years of work experience in SAP ERP software consulting and one year of teaching experience in Mechanical and Manufacturing Engineering. During his Ph.D. work at

NIT Trichy, he got the DORA plus scholarship to undergo research training in the Mechanical and Industrial Engineering department of Tallinn University of Technology, Estonia. His research interests include Metal 3D printing, Design for Additive Manufacturing, Laser Materials Processing, unconventional machining techniques and materials management. He has authored 4 international journal Publications in metal 3D printing in which 3 of them are the outcomes of his PhD thesis. One of the presentations from his PhD work earned a best paper award in the ICPPM conference (March 2023).

Amit Kumar Dutta is currently working as an Assistant Professor in the Department of Chemistry, Bangabasi Morning College (Affiliated to University of Calcutta). He has Completed M.Sc. from University of Calcutta in 2007 with Inorganic Chemistry specialization and obtained Ph.D. degree in the area of Nano-catalysis from Bengal Engineering and Science University, Shibpur (presently, IIEST, Shibpur) in 2012 under the supervision of Prof. Bibhutosh Adhikary. His area of research interest is the use of semiconducting nano-materials in photo-catalysis and bio-sensing application. He has published 35 research papers in different peer-reviewed journals and 3 book chapters. He has completed one UGC project and currently one research scholar is working with him. He has awarded Junior and Senior Research Fellowship (JRF/SRF) through joint CSIR–UGC (NET) and Research Associate Fellowship (CSIR-RA). Amit Kumar Dutta is currently working as an Assistant Professor in the Department of Chemistry, Bangabasi Morning College (Affiliated to University of Calcutta). He has Completed M.Sc. from University of Calcutta in 2007 with Inorganic Chemistry specialization and obtained Ph.D. degree in the area of Nano-catalysis from Bengal Engineering and Science University, Shibpur (presently, IIEST, Shibpur) in 2012 under the supervision of Prof. Bibhutosh Adhikary. His area of research interest is the use of semiconducting nano-materials in photo-catalysis and bio-sensing application. He has published 35 research papers in different peer-reviewed journals and 3 book chapters. He has completed one UGC project and currently one research scholar is working with him. He has awarded Junior and Senior Research Fellowship (JRF/SRF) through joint CSIR–UGC (NET) and Research Associate Fellowship (CSIR-RA).

Boopathy Govindarajan earned his undergraduate degree in mechanical engineering from University of Madras, his master's degree in aeronautical engineering from Madras Institute of Technology, and his doctorate from the Department of Aeronautical Engineering, St. Peter's Institute of Higher Education and Research, Chennai. At Vel Tech University, a professorship is currently held in the department of aeronautical engineering. In addition to guiding 5 Ph.D Scholars, guided 8 PG & 90 UG scholars on their significant projects, he has written in 30 international publications. He is a Life Member of the Society of Automotive Engineers INDIA, The Robotics Society and the Aeronautical Society of India.

G. Prasad is currently working as faculty in the department of Aerospace Engineering Chandigarh University, India. He has 9 years of experience in Teaching and research. He received his Bachelor of Engineering and Master of Engineering degrees with first class and distinction and Doctor of Philosophy in the field of Aeronautical Engineering from the Anna University, Chennai, India. Prof. Prasad G is the author of over 30 technical publications, 1 Patent Published and 1 Book published. His research interests include Unmanned Aerial Vehicles, Aircraft Structures, Computational Fluid Dynamics and Interdisciplinary research. He has completed three funded project sponsored by The Institution of Engineers (India) and Tamilnadu State Council for Science and Technology. Awarded Indian National Science Academy (INSA) Visiting Scientist Programme 2019 and Awarded Science Academies' Summer Research Fellowship Programme (SRPF) 2019. Further he is SPOC of NPTEL- IIT Madras, SPOC of ISRO IIRS outreach coordinator and SPOC for NASA International SpaceApps Challenge. Further key positions such as coordinator of Board of Studies, Outcome Based Education, Project Coordinator, Research Coordinator

Balaji Ganesan is a distinguished academic and researcher associated with the Hindustan Institute of Technology and Science (HITS) in Chennai, India. Known for his contributions to fields like Aerodynamics, Wind Turbiens, High Speed flow, Nose cone and Composite materials, Dr. Balaji holds a Ph.D. in Aerospace Engineering and has published numerous research papers in international journals and conferences. At HITS, Dr. Balaji has played a significant role in curriculum development and student mentoring, actively collaborating on research projects that enhance academic and practical knowledge. His work often involves interdisciplinary approaches, OBE and SDG to address real-world challenges.In addition to his research and teaching, Dr. Balaji is a sought-after speaker at national and international academic events. He is known for his passion for technology-driven solutions and is dedicated to fostering an innovative learning environment that prepares students for modern technological landscapes.

Amit Tiwari is currently working as an Assistant Professor in the Department of Mechanical Engineering at Suresh Gyan Vihar University, Jaipur Rajasthan India. He received his PhD degree from Suresh Gyan Vihar University, Jaipur, Rajasthan, India Accredited with "A+" Grade by NACC. He has 5 Indian Design Patent, 1 Australian Patent and 3 UK Design patent. He has published 50 research papers in reputed International and National Journals. He has also associated with a life member of Technical societies like The Society of Mechanical and Automotive Engineers [SMAE], LM NO – 02112, Research Scholar Hub [RSH], India, Life Membership No. - RSH/IN/RJ/LM-09/2017, International Association of Engineers [IAENG], Hongkong, Membership No.-193169, Society for Environmental Resources and Biotechnology Development, Annual Membership, Membership No. AN: 119321and Scholars Academics & Scientific Society, Membership No: SAS/RMSASS/352. Dr. Tiwari research area includes Metal Casting, Heat Treatment, Composite Material, and Optimization Techniques.

V. Srinivasan earned his undergraduate degree in mechanical engineering from S.K.P Institute of Technology affiliated to Anna University, his master's degree in CAD/CAM from Vel Tech Rangarajan Dr. Sagunthala R & D Institute of Science and Technology and currently pursuing Ph. D in Aeronautical Engineering from Vel Tech Rangarajan Dr. Sagunthala R & D Institute of Science and Technology. Currently doing research in LiDAR (Light detection and Ranging) and other remote sensing and imaging sensors, its integration with UAVs. Obstacle avoidance and autonomous navigation systems in UAV using sensor fusion are his key areas for research.

Index

Symbols

Π-Conjugated Structure 42

A

Additive Manufacturing 61, 62, 64, 67, 68, 70, 72, 73, 83, 84, 95, 97, 104, 108, 118, 119, 121, 122, 135, 136, 138, 142, 147, 158, 159, 161, 179, 202, 217, 220, 226, 234, 239, 260, 264, 274, 275, 278, 282, 287, 289, 292, 293, 294, 351
Aerospace Engineering 64, 66, 67, 69, 79, 149, 150, 267, 281
Aluminium Alloys 61, 62, 63, 64, 65, 66, 67, 68, 69, 70, 71, 72, 73, 74, 75, 76, 77, 78, 79, 80, 81, 82, 83, 84, 85, 86, 87, 88, 89, 90, 91, 92, 93, 94, 95, 96, 103, 107, 111, 112, 115, 116, 117, 118, 119, 123, 125, 130, 131, 135, 136, 138, 316
Application 2, 5, 13, 14, 15, 16, 17, 18, 25, 26, 31, 32, 39, 42, 46, 49, 51, 62, 71, 80, 81, 82, 83, 84, 85, 86, 87, 88, 89, 90, 91, 96, 98, 107, 124, 130, 138, 143, 150, 176, 177, 191, 192, 205, 208, 212, 223, 228, 232, 234, 236, 237, 238, 239, 240, 245, 246, 247, 248, 249, 250, 251, 253, 254, 255, 256, 258, 274, 275, 278, 280, 288, 298, 309, 315, 317, 318, 349, 368, 381, 386, 389

B

Bioinspiration 177
Bio-Inspired Materials 145, 146, 147, 149, 151, 154, 155, 156, 157, 158, 159, 160, 161, 162, 163, 164, 165, 168, 169, 170, 171, 172, 173, 174, 175, 178, 183, 190, 197, 198, 201, 207, 209, 210, 211, 279
Biomimicry 175, 178, 198, 206, 273, 274, 285, 292, 393

C

Catalysis-Chemical Reactions 35, 51
Chemical Vapor Deposition 2, 16, 43, 202, 360, 361, 363, 365, 366, 391, 397, 398
Corrosion Resistance 3, 4, 5, 7, 8, 10, 17, 24, 25, 29, 31, 61, 63, 64, 65, 66, 67, 68, 70, 72, 73, 75, 76, 78, 79, 80, 81, 82, 83, 84, 85, 88, 94, 95, 96, 107, 111, 112, 114, 115, 117, 122, 123, 124, 125, 126, 127, 129, 130, 131, 132, 133, 134, 135, 136, 162, 166, 168, 208, 210, 225, 226, 228, 229, 230, 231, 232, 234, 255, 309, 316, 320, 332, 341, 342, 344, 347, 349

D

Defence-Grade Materials 181, 183, 184, 210
Defense 1, 2, 3, 4, 7, 8, 9, 17, 18, 19, 20, 21, 22, 23, 26, 30, 61, 62, 93, 99, 145, 146, 147, 148, 149, 150, 151, 154, 155, 156, 157, 158, 159, 161, 163, 164, 165, 168, 172, 173, 174, 175, 176, 177, 179, 226, 227, 229, 230, 232, 233, 234, 235, 236, 237, 238, 239, 240, 241, 242, 243, 244, 245, 246, 247, 248, 249, 250, 251, 252, 253, 255, 256, 257, 258, 259, 260, 263, 273, 274, 275, 276, 279, 283, 285, 287, 290, 292, 297, 302, 303, 304, 307, 308, 309, 315, 320, 321, 332, 337, 340, 342, 345, 346, 353, 354, 355, 369, 370, 373, 374, 378, 381, 382, 385, 390
Defense Applications 3, 30, 145, 146, 147, 149, 151, 155, 157, 161, 165, 168, 174, 176, 177, 226, 230, 233, 234, 235, 236, 237, 238, 239, 240, 241, 243, 244, 246, 247, 249, 250, 251, 252, 257, 258, 259, 273, 274, 283, 292, 297, 320, 321, 332, 337, 340, 345, 353, 355, 373, 381, 382, 385

Defense Components 250, 273, 274, 275, 276, 285, 292

F

Friction Stir Welding 61, 62, 64, 72, 73, 95, 97, 104, 108, 112, 114, 118, 119, 121, 135, 139, 141, 143, 268
Full-Colour-Light-Emitting 47

H

High-Performance Composites 194
High-Strength Alloys 107, 114, 116, 118, 125, 127, 132
Honeycomb 40, 41, 42, 153, 157, 206, 209, 279, 280, 281, 282, 293, 295
Hybrid Materials 67, 68, 69, 80, 82, 96, 151, 367

L

Lattice Structure 40, 41, 293, 295
Lightweight Composites 151, 225, 226, 227, 229, 230, 231, 232, 233, 234, 235, 236, 237, 239, 240, 241, 242, 244, 245, 246, 247, 248, 249, 251, 252, 253, 254, 255, 256, 257, 258, 259

M

Marine-Grade Aluminium 107, 111, 112, 115, 118, 123, 136
Material Optimization 205, 248
Materials 1, 2, 3, 6, 7, 8, 9, 10, 11, 12, 13, 14, 16, 17, 18, 19, 21, 22, 23, 24, 25, 26, 27, 29, 30, 31, 32, 33, 35, 36, 37, 38, 39, 40, 41, 42, 44, 46, 47, 49, 51, 52, 53, 54, 55, 56, 57, 58, 59, 65, 66, 67, 68, 69, 70, 72, 74, 76, 78, 79, 80, 81, 82, 83, 85, 86, 87, 88, 89, 90, 91, 92, 93, 94, 95, 96, 99, 100, 101, 102, 103, 104, 105, 108, 111, 113, 121, 134, 135, 136, 137, 138, 139, 140, 141, 142, 143, 145, 146, 147, 148, 149, 150, 151, 152, 153, 154, 155, 156, 157, 158, 159, 160, 161, 162, 163, 164, 165, 166, 168, 169, 170, 171, 172, 173, 174, 175, 176, 177, 178, 179, 181, 182, 183, 184, 185, 186, 187, 188, 189, 190, 191, 192, 193, 194, 195, 196, 197, 198, 199, 200, 201, 202, 203, 204, 205, 206, 207, 208, 209, 210, 211, 212, 213, 214, 215, 216, 217, 218, 219, 220, 221, 222, 223, 224, 225, 226, 227, 228, 229, 230, 231, 232, 233, 234, 236, 237, 238, 239, 240, 242, 245, 246, 247, 248, 249, 250, 251, 252, 253, 254, 255, 256, 257, 258, 259, 260, 261, 262, 263, 264, 265, 266, 267, 268, 269, 270, 271, 272, 274, 275, 276, 277, 278, 279, 280, 281, 282, 285, 286, 287, 288, 289, 290, 292, 293, 294, 295, 297, 298, 299, 300, 301, 302, 303, 304, 305, 306, 307, 308, 309, 310, 311, 312, 313, 316, 317, 321, 345, 346, 347, 348, 349, 350, 351, 354, 355, 358, 359, 362, 364, 367, 368, 369, 370, 383, 385, 389, 390, 391, 392, 393, 394, 395, 396, 397, 398
Mechanical Properties 5, 43, 62, 64, 65, 66, 70, 71, 72, 73, 78, 79, 83, 84, 85, 86, 87, 88, 89, 90, 91, 92, 95, 100, 111, 115, 116, 117, 118, 119, 120, 121, 122, 127, 133, 134, 135, 151, 158, 171, 177, 185, 219, 228, 229, 231, 238, 240, 241, 242, 243, 244, 245, 247, 248, 254, 255, 259, 260, 261, 264, 266, 267, 269, 277, 278, 279, 280, 288, 293, 295, 315, 317, 318, 319, 320, 323, 324, 330, 332, 335, 337, 345, 347, 348, 349, 350
Military 2, 3, 5, 11, 12, 13, 17, 18, 19, 20, 21, 22, 23, 26, 28, 29, 30, 31, 32, 61, 65, 66, 68, 69, 72, 73, 75, 76, 86, 96, 116, 118, 126, 129, 145, 146, 147, 148, 149, 151, 152, 153, 154, 155, 156, 157, 158, 159, 161, 162, 163, 164, 165, 166, 167, 168, 169, 170, 172, 174, 175, 178, 183, 186, 187, 226, 234, 235, 236, 237, 238, 239, 240, 244, 245, 246, 247, 248, 249, 250,

253, 254, 256, 257, 258, 259, 268, 274, 275, 276, 279, 280, 281, 282, 287, 290, 297, 301, 303, 304, 305, 307, 308, 309, 310, 369, 373
Modified AZ61 Alloy 315, 323, 325, 327, 331, 336, 346

N

Nanomaterial Reinforcement 248

O

Optical Microscopy 322
Organic Semiconductor 37, 49, 52

P

Physical Vapor Deposition 2, 360
Polymer 4, 5, 6, 9, 31, 32, 33, 35, 36, 45, 46, 49, 52, 97, 151, 158, 159, 165, 169, 170, 171, 179, 186, 192, 200, 202, 208, 210, 213, 215, 216, 219, 222, 223, 226, 227, 228, 229, 232, 238, 240, 241, 244, 245, 246, 250, 251, 260, 261, 265, 266, 267, 268, 269, 270, 271, 272, 275, 278, 289, 294, 299
Power Electronics 221, 353, 355, 370, 381, 383, 395
Protective Coating 2, 3, 8, 17, 18, 23, 25, 53

S

SEM-EDX 321, 327
Sensors 12, 44, 57, 93, 94, 101, 147, 149, 151, 157, 173, 176, 187, 188, 194, 198, 205, 218, 219, 222, 224, 243, 263, 264, 303, 306, 353, 354, 355, 369, 370, 378, 379, 380, 381, 385, 389, 390, 391, 394, 395, 396
Smart Materials 33, 58, 104, 191, 203, 204, 205, 206, 217, 219, 258, 259, 266, 285
Solar-Blind Photodetectors 354, 376, 395, 396, 397
Structural Resilience 153, 183, 186
Surface Protection 78, 79, 107, 122, 123, 136
Surface Treatments 1, 2, 7, 8, 12, 14, 25, 79, 92, 111, 112, 114, 202
Sustainable Production 68, 70, 95
Sustainable Shipbuilding 129, 131, 132, 133, 134

T

Technology 1, 2, 6, 10, 15, 17, 19, 21, 26, 27, 28, 29, 32, 33, 36, 37, 38, 39, 44, 45, 51, 52, 54, 56, 58, 75, 76, 77, 81, 84, 96, 98, 99, 100, 101, 102, 103, 104, 105, 118, 119, 121, 137, 138, 141, 142, 143, 146, 148, 150, 154, 157, 158, 159, 169, 172, 174, 175, 179, 181, 203, 205, 212, 216, 217, 218, 219, 221, 225, 238, 239, 240, 242, 253, 254, 261, 262, 263, 265, 269, 270, 273, 280, 286, 293, 294, 297, 306, 307, 308, 309, 310, 312, 342, 348, 349, 350, 351, 354, 355, 363, 365, 368, 374, 378, 385, 386, 388, 389, 391
Thermal Barrier Coatings 5, 179
Thin-Film Coatings 2

U

UAV 73, 97, 153, 166, 173, 179, 254, 257, 260, 263, 270, 271, 282, 283, 306, 307, 308, 309

W

Wear Resistance 5, 6, 9, 12, 13, 123, 124, 152, 162, 187, 228, 234, 318

X

XRD Analysis 336, 346

www.ingramcontent.com/pod-product-compliance
Lightning Source LLC
LaVergne TN
LVHW080611300926
849162LV00005B/1124